Emergency Vehicle Accidents

Prevention, Reconstruction and Survey of State Law

Second Edition

Stephen S. Solomon, B.S., O.D.

Paul F. Hill, Esq.

Contributing Authors

George L. Ellis, M.D., FACEP

James G. King

Robert J. Simpson, Esq.

Steven Smyk, Esq.

Lawyers & Judges Publishing Company, Inc.

This publication is designed to provide accurate and authoritative information in regard to the subject matter covered. It is sold with the understanding that the publisher is not engaged in rendering legal, accounting, or other professional service. If legal advice or other expert assistance is required, the services of a competent professional person should be sought.

-From a *Declaration of Principles* jointly adopted by a Committee of the American Bar Association and a Committee of Publishers and Associations.

Lawyers & Judges Publishing Company, Inc.

P.O. Box 30040 • Tucson, AZ 85751-0040
(800) 209-7109 • FAX (800) 330-8795
e-mail: sales@lawyersandjudges.com

Library of Congress Cataloging-in-Publication Data

Solomon, Stephen S.
Emergency vehicle accidents : prevention, reconstruction, and survey of state law / Stephen S. Solomon, Paul F. Hill ; contributing authors, George L. Ellis ... [et al.].
p. cm.
Includes bibliographical references and index.
ISBN 1-930056-46-X
1. Emergency vehicles--Accidents--Investigation. 2. Emergency vehicles--Accidents--Prevention. 3. Traffic accident investigation. 4. Liability for traffic accidents. I. Hill, Paul F. II. Title.
HV8079.55 .S65 2002
363.12'59--dc21

2002005164

online
www.lawyersandjudges.com

ISBN 1-930056-46-X
Printed in the United States of America
10 9 8 7 6 5 4 3 2 1

Dedication

This book is dedicated to all the firefighters, law enforcement officers and EMS personnel who lost their lives on September 11, 2001.

In the aftermath of 9-11-01 firefighters have been seen in a new perspective. The image of angels surrounding a firefighter reflects the dangers of the profession and the loneliness of grief. It is an appropriate reflection for our time in history.

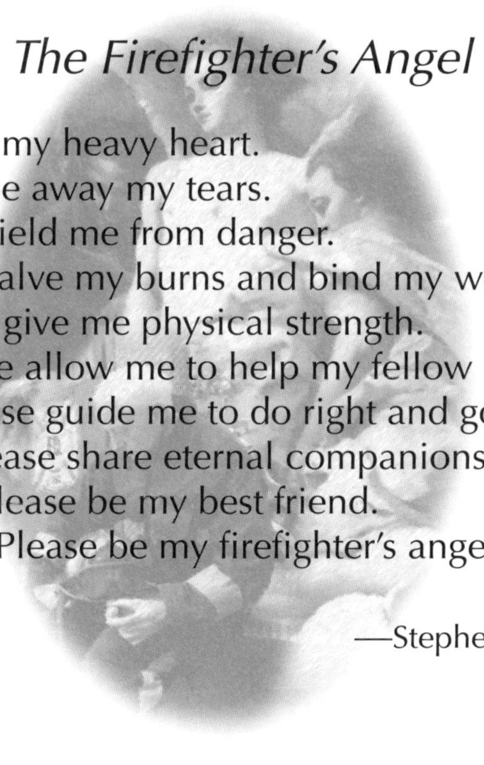

The Firefighter's Angel

Please ease my heavy heart.
Please wipe away my tears.
Please shield me from danger.
Please salve my burns and bind my wounds.
Please give me physical strength.
Please allow me to help my fellow man.
Please guide me to do right and good.
Please share eternal companionship.
Please be my best friend.
Please be my firefighter's angel.

—Stephen S. Solomon

Table of Contents

Preface

This text is dedicated to several diverse groups: to the firefighters who risk their lives for the safety of the rest of the nation, the law enforcement officers who risk their lives for the protection of the rest of the nation, and the ambulance personnel who struggle to keep us alive when we shouldn't be. It is my hope that this book will provide knowledge, awareness and training, all leading to safer operations.

To the legal profession, the book is meant to be a reference source to better understand the causes and intricacies of emergency responses, why unplanned and unsuspected accidents occur and how they could have been prevented.

To optometrists and physicians, I hope the book will provide knowledge of the interaction between vision and visibility and the hidden causes of accidents. This information should enlarge the reader's comprehension and ability to develop safer communities.

The contents include research done by the author as well as by others, and its central theme is the concept that vision is an integral component in the prevention and avoidance of accidents. When the unfortunate occurs, the well developed visual system guides the driver to minimize the outcome. The visual system leads the motor system.

In relation to the investigation of accidents, this text is designed to provide a logical format for understanding the various emergency vehicle components and how they work together to provide safe passage. Likewise, the book discusses those factors which make emergency vehicle operations more hazardous. It also points out those components which presumably improve safety, yet actually make emergency vehicle operations more hazardous.

The publication of this text is the culmination of forty-four years of experience as an active firefighter, including fifteen years as an EMT, ten years as an engine captain, and the last fourteen years as a fire commissioner. Concurrently, my profession as an optometrist spans thirty-seven years and parallels my awareness and research of visibility. Since my teens, and throughout college and optometry school, the Air Force, and my adult life, safety has been an undying interest. Consulting for industry, law, police and the fire service has added to my experiences that have shaped the contents.

Stephen S. Solomon, B.S., O.D.

Acknowledgments

I would be remiss if I did not give thanks and praise for the gifts necessary to author this text. That being said, the following were important individuals who played roles in the development of the concepts, theories, and the text.

To my wife, for her input and advice.

To the late Dr. John Carter, my unofficial mentor and repository of incredible knowledge, arcane fact, and the awareness of the contents of untold numbers of texts, posthumous thanks are extended.

Likewise, to the late Dr. Jacob Nevyas who taught and reinforced honesty, loyalty, and fairness along with the admonition that scientists must be practical, not just book-wise.

To James G. King, who coauthored some of the research and provided not only mathematical expertise, but acted as a sounding board for ideas and concepts.

To Dr. Gary J. Williams, my partner, and Dr. John Shafer, both who supplied thoughts over the years.

And to Stanley Gillmeister for word processing and computer assistance.

To the firefighter's angel, who is with me always.

Part I

Human Factors and Emergency Vehicle Operation

Chapter 1

Introduction*

Stephen S. Solomon, B.S., O.D.

Visibility cannot be separated from safety. What you see first, you react to most quickly. If it cannot be discerned from the surroundings, it is not visible.

Emergency vehicles have their origins in the evolution of transportation. It is said that the Indians used horse-drawn litters (travois) which dragged on the ground. The military employed horses hitched to wheeled carriages. Motorized transportation was developed and then modified to accommodate prone patients. Throughout this country, funeral homes often doubled as ambulance services, an approach that persisted in some locales until the government started to impose standards in the 1960s and 1970s.

With these standards, the size of ambulances steadily grew. The old "hearse style" body was quickly upgraded to a small truck chassis with a tall medical treatment body, because it was difficult to render state-of-the-art care in the lower roof construction which offered little head room common in the earlier vehicles.

Until the 1960s, an ambulance crew had some first aid skills, applied splints to broken bones, stopped bleeding, and, very occasionally, performed artificial respiration. The treatment regimen changed when knowledge of first aid was supplanted with a much more comprehensive emergency medical technician (EMT) rating. That soon expanded to prehospital cardiac care and expanded again to paramedic certification. No longer was "scoop and run" the order of the day. Stabilization of the patient, often before leaving the residence or scene, became the preferred

*The appendices of the book coordinate with this chapter. References to fire apparatus construction starting with "F-" direct the reader to the fire apparatus specifications, which are the specifications required to purchase a fire pumper. These appear in Appendix I. References to ambulance construction start with "A-" and relate to the federal KKK-A-1822D specifications. These appear in Appendix II.

treatment. More sophisticated care required larger storage space, more equipment, and more room to treat and monitor a patient while en route. The type of instrumentation used by the vehicle depends on the level of certification of the ambulance organization. It is not odd to see an ambulance equipped like an emergency treatment room. The modern ambulance is much smaller than a standard fire pumper and larger than a family automobile (A-3.1). Operator training has more similarities with the fire service than with the law enforcement agencies. Driver training varies greatly from agency to agency and may be inconsistent over time within a given department or company.

Firefighters began with bucket brigades, then employed hand carts to haul many buckets. Hose reels, ladder wagons and pumps were initially hand drawn, then horse-powered. The venerable Model T was outfitted as an early fire response vehicle. The size of the vehicle was determined by the equipment it had to carry and the number of firefighters riding on it. In the early decades of the 1900s, labor was inexpensive and large crews were the rule. Concurrently, the amount of equipment was minimal and the water tank was small. Self-contained breathing apparatus did not exist. Power tools were nonexistent, and fire pumps were puny by modern standards. Firefighters, not hardware, put out fires. Of course, the composition of the fuel of these fires was much less complex than today. While smoke was present, it was not laden with the toxic by-products of modern chemicals. A bushy, long mustache soaked with water could serve as an "air filter." Protective gear was a pair of thigh-high boots, a rubber coat, and a leather helmet. Some nozzles, hose, and hand tools, like axes and pike poles, completed the equipment list. The early fire truck needed enough running boards and tail boards to accommodate the crew. Traffic in all but the larger cities was not much of a problem and speed was limited by the vehicles' small motors and matched transmissions.

Today, fire service vehicles have specialized roles, which are reflected in the terms used to describe them. Also, the terminology applicable to fire service vehicles also differs in some respects from that used by the general public. Aside from a few exceptions, the terms "truck," "ladder," "engine," "pumper," and "tanker" have specific, well established meanings in the fire service. These distinctions are important because these vehicles vary in size, and these size differences are important when discussing visibility and maneuverability.

As used in this book, "ladder" refers to a vehicle equipped with a ladder, while "aerial device" refers to a unit equipped with a snorkel or similar equipment for reaching heights. These two types of vehicles are sometimes referred to collectively as "trucks." "Engine" or "pumper" refers to a fire service vehicle whose function is to pump water from hydrants or tankers through the fire hoses and onto the fire. "Tanker" refers to a fire service vehicle that is designed to carry water to the scene of a fire. Generally speaking, trucks (i.e., ladders and aerial devices) are the largest. Tankers are generally the next smallest and some pumpers are somewhat smaller. Tankers vary the most in size, with capacity ranging from 500 gallons to several thousand gallons of water. The term "apparatus" applies to all of the foregoing vehicles. Finally, the term "vehicle" is the most general, and includes fire apparatus along with the automobiles used by the fire chiefs and their assistants.

The federal government did not exert a regulatory influence upon the fire service until OSHA came into existence, yet many evolutionary changes occurred over the century, some easily assimilated and others hindered by tradition. The traditional and classic image of a firefighter standing on the rear step wearing his helmet, coat, and boots is a thing of the past because it was such an unsafe practice. However, there are still isolated incidents of firefighters riding on the tail board.

Today's fire truck carries an integral water tank, seating in an enclosed cab for all riders and a chassis size rivaling large municipal delivery vehicles. The changes in equipment inventories are quite impressive. Self-contained breathing apparatus for all firefighters are located on the vehicle. Personal protective garments are made of fire-resistant fabrics. In the past, the most available component was the firefighter; power equipment is now the order of the day.

A typical pumper's inventory includes the following (F-193):

- portable pump
- extension ladders
- roof ladders
- folding ladders
- straight ladders
- smoke exhaust blower, usually powered by gas or electricity

- portable electric generator
  - electric extension cord reels
  - portable flood lights
- chain saw
- deluge gun—a big bore nozzle and fittings to allow for large flows of water
- large assortment of hand tools—shovels, axes, pry bars, and crowbars
- hand lights
- first aid kit
- hose spanners (wrenches)
- hydrant wrenches
- hose adapters and connectors
- wheel chocks
- hand-held fire extinguishers
- self contained breathing apparatus
- spare air bottles (SCBA, or self-contained breathing apparatus)

In addition, a variety of hoses are needed:

- hand lines for one- or two-man operations
- hand lines for three- or four-man operations
- major water supply hoses, often 4" or 5" in diameter and in 100' lengths.
- 1000' stored in a hosebed is commonplace in rural regions. (Hose loads, including diameters and lengths, are dictated by local environments and depend on local water supply, available manpower, and the types of fires that could occur.)

The list can be increased significantly as each individual vehicle is planned by the purchaser. It should be noted that a rescue truck or an aerial device would have different inventories.

The water-carrying capacity depends upon the type of apparatus in service and the available water supply. The tank on an aerial unit may be nonexistent while a full-sized pumper in a small rural community may hold 2,500 gallons. In a major city with hydrants closely spaced on each street and little demand for coverage of rural areas, pumpers may have smaller water tanks that have a capacity of 500 gallons (F-116). The fire service also uses tanker trucks whose primary job is to supply large volumes of water to other apparatus or directly to the fire, and some of these rigs are comparable in size to fuel transport tractor-trailers.

Thus, firefighting vehicles are very large and are often operated by drivers without commercial, industrial, or transportation experience. This is especially true of volunteer departments and small paid departments. Some drivers lack sufficient experience in operating a very large, heavy vehicle under emergency response conditions. Classroom and road education cannot simulate the effect of these excitement-laden situations. Emergency vehicle operation courses (EVOC) are not consistent from department to department and may not even be similar from class to class within a department. Even today, an all too common approach to training drivers is having the trainees drive the fire truck around the fire protection area in a nonemergency mode.

Law enforcement had a slightly less complicated development in that they did not require modifications like those for ambulances or fire trucks.

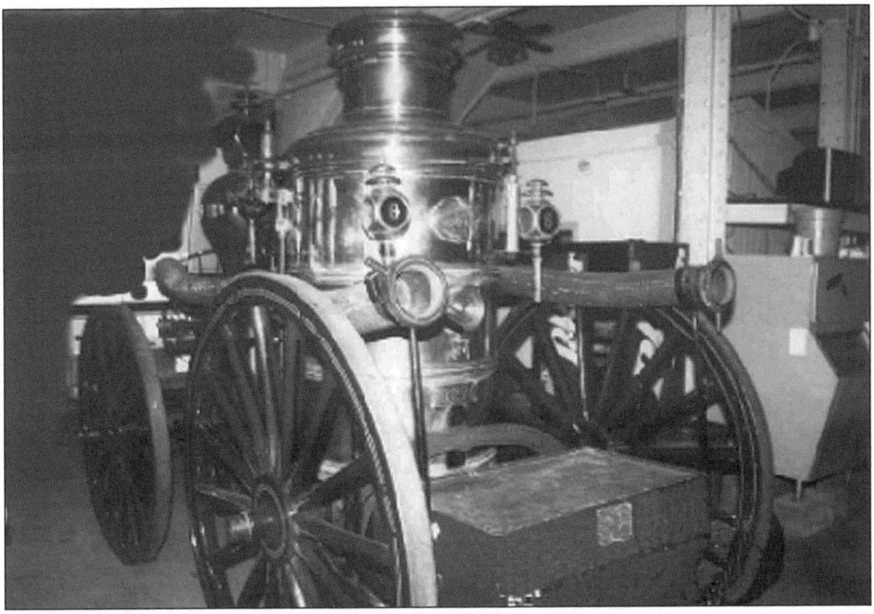

Figure 1.1 *An old coal-fired steamer with lanterns marking the vehicle. It was horse-drawn.*

While the fire service has progressed from a labor-intensive task to one that is balanced between labor and equipment, and a task that requires ever increasing vehicle sizes, the law enforcement community has evolved differently.

While two-man patrols have become less common and one-man patrols have become more frequent, the vehicle is essentially a civilian automobile. Cars have increased in size and weight, but not in the proportion that fire apparatus have grown. Equipment carried in the car easily fits in the trunk and on the seat. The weight of this equipment does not require a bigger or specially designed vehicle. Since most law enforcement personnel are employed as full-time officers there is more control over training. Many states have formal, comprehensive law enforcement schools which include an emergency vehicle operation section.

Thus, one of the greatest advantages for vehicle operation in law enforcement is that the patrol car is basically the same type of automobile which the driver privately owns and operates. In other words, he has the opportunity to drive a class of vehicle that is similar to his work vehicle.

Bells, whistles, sirens, and light were installed for all three groups, however, because of the perceived requirement of fast passage. As road-

Figure 1.2 *Lanterns mounted on an old steamer fire apparatus*

ways became more developed and traffic density increased, the emergency services became more dependent upon warning devices.

The flashing light began as a lantern, occasionally with a colored glass. A runner sometimes preceded the hand carts, alerting the populace of the approaching equipment. Hand-cranked sirens were introduced, sometimes audibly supplemented by the yelling of the operators. See Figures 1.1 and 1.2.

The practices and equipment used in the daily operation of emergency vehicles are often based upon an evolutionary transition and, in some instances, on tradition. Are the concepts that are applied today based upon research, intuition, and evaluation? Does tradition have the inertia to impede a safer approach?

Chapter 2

Construction of Emergency Vehicles*

Stephen S. Solomon, B.S., O.D.

Emergency response vehicles have specialized functions, including fire service, medical transportation and crime suppression; as a result, they have unique characteristics. Individuals not familiar with their construction might assume that there is not much difference between them and other cars and trucks on the road. In the instance of police cars, as discussed below, the relationship is close, but on the other hand, fire apparatus and ambulances are very different from their commercial counterparts.

Fire apparatus can be purchased as pre-made units, but more commonly, they are custom-made according to a user's specifications. Unlike automobiles, most large fire vehicles are individually constructed. In some cases, this means starting with two long, parallel beams which are shaped, drilled and machined. Engines, transmissions, pumps and bodies are added to this framework. Another alternative is to specify a particular brand of cab which includes a preassembled framework, engine, and transmission (F-1). In a second phase of manufacture, often at a different factory, the pump and fire body are added (F-72 *et seq.*). In any case, the voluntary standards of the National Fire Protection Association are the guidelines.[1] These standards are compiled by a consensus from an NFPA-appointed committee comprising representatives of manufacturers and users.[2] The cycle of review is approximately five years. The standard is intended to address the entire scope of the use of a fire apparatus. According to the NFPA, the current edition of the twenty-three-chapter Standard 1901 was accepted in August 1996.

---

*The appendices of the book coordinate with this chapter. References to fire apparatus construction starting with "F-" direct the reader to the fire apparatus specifications, which are the specifications required to purchase a fire pumper. These appear in Appendix I. References to ambulance construction start with "A-" and relate to the federal KKK-A-1822D specifications. These appear in Appendix II.

Ambulances are similar to fire apparatus in that they are also controlled by standards. These are known as the KKK specifications and are developed and disseminated by the U.S. Department of Transportation.[3] They deal with the function or mission of the vehicle as well as the components that are assembled to meet the purchaser's specifications. However, the individual states decide whether to accept the KKK specifications for all ambulances purchased, or to choose other alternatives. In an approach similar to fire apparatus construction, the KKK specifications for ambulances have a cab section (A-3.1.2, 3.1.3, 3.1.4) and a body or box (A-3.1.2, 3.1.3, 3.1.4). The cab is usually a commercially available small truck chassis which is delivered to a manufacturer, at which time an individually made rear portion is assembled and mounted to the frame. The rear or box is often outfitted with compartments, patient care systems, and cot provisions. These are built to the specifications of the purchaser.

In cases where the cab unit is purchased separately and then delivered to the fire apparatus body manufacturer or to the ambulance box manufacturer, a significant delay in construction and delivery of the finished product can occur. This is often due to the availability of the chassis.

Police cars have a much less restrictive format. The basic vehicle is not controlled by any national standard. Some regions may make recommendations about lights, sirens, and retroreflective material.

Most fire apparatus are individually manufactured to a fire department's specifications. A police car, on the other hand, is purchased from the same vehicle dealers that private citizens use, but with the addition of special packages for handling, cooling, and electrical requirements.

In summary, the NFPA standards for fire apparatus are developed by consensus and are considered voluntary. They are uniform throughout the United States. Ambulances are guided by the federal KKK-A-1822D (the D refers to the date of publication) specifications and if a state elects to adopt these specifications, they can apply to ambulance purchases. Police cars are not held to any standard and cities, political divisions, or geographic regions are free to establish their own requirements for performance.

The procedure for specifying (using NFPA standards) and purchasing a fire apparatus can follow several paths, but usually a small committee collects information, receives input from other interested parties, and plans for the proposed vehicle. This process commonly can take several years from conception to delivery. Alternatively, the fire chief or his representative designs and plans the desired apparatus. A starting point is the gathering of information, especially of new innovations. This can be achieved in several ways. Advertising is present in all fire service magazines. Manufacturers participate in fire service trade shows demonstrating new technology. (At these conventions apparatus used for demonstration may be offered for sale. Some departments elect to purchase these floor models without going through the steps of writing the specifications.)

Manufacturers do not usually build a fire vehicle outside the standards of the NFPA, so fire departments in the United States will usually receive a vehicle designed and built to NFPA standards regardless of the purchase path.

There were thirty voting members on the NFPA committee appointed to write these specifications of the 1996 version of the fire apparatus standard. Eleven represented manufacturers, eleven represented fire departments, and the balance came from a variety of sources, including insurance companies, fire service organizations, unions, universities and independent representatives. Some representatives have interests in more than the category they represent. For example, instructors, fire service organization representatives, manufacturing salespersons and institutional members may also be active firefighters.

The purchasing of fire fighting equipment leaves much to be desired. Purchases of fire fighting apparatus are often made by consensus, relying on information based to some degree on the salesperson's personal preferences. Fire departments receive information from a variety of sources, but often rely on the manufacturer's sales representative. Sometimes these individuals are part-time and may also be firefighters. Some experience in firefighting and an awareness of the manufacturer's capabilities help, but salespersons typically have had very little formal instruction pertaining to the capabilities of the various components and systems. Very often, the sales representative's personal experiences are the only training they have when selling a product. Unfortunately, the purchasers rely on this information and some do not independently research the problem. The introduction of scientific data into the process is difficult and is often challenged by traditional attitudes. Furthermore, once a major piece of equipment is purchased, it may remain in service for over twenty years.

By comparison, police cars are procured and traded in a relatively short period of time, sometimes remaining only a year or two in service. This provides for a rapid correction if flaws in the model or construction are encountered.

Ambulances fall into a middle ground. They usually have longer service lives than police cars, but seldom last more than ten years. A prime factor is the wear and tear of daily use. Specifications are written under the auspices of the U.S. government with the input of users and manufacturers. The healthcare community is sensitive to rapid scientific change and nontraditional applications. It is easier to introduce new ideas and have them accepted within this group. There has been a nationwide trend in which ambulance services have been incorporated into the fire service as opposed to functioning as a third, independent emergency service. This may be more prevalent in paid fire departments than in rural departments which are primarily staffed with volunteers, and the effect on traditional operations remains to be seen.

In some ways ambulance salespersons mirror their fire service counterparts, in that some are part-time, reflect personal opinion, and have not been trained formally. Again, the audience is open to his experience and often accept his advice as gospel.

Police cars are usually purchased by bid on a local level or at a state rate from certain dealers. The specification process is very limited because standard production automobiles are used. Lights and sirens are purchased as separate items, but not from the automobile dealer. As with the other emergency vehicles, the warning packages are included as part of the manufacturing process.

Endnotes

1. NFPA 1901, *Standard for Automotive Fire Apparatus*, pp. 1901–2; National Fire Protection Association, 1 Batterymarch Park Quincy, MA. 1996.

2. *Ibid.* at 1.

3. General Services Administration, KKK-A-1822 Rev D (1994).

Chapter 3

Anatomy of the Eyes and Visual Performance*

Stephen S. Solomon, B.S., O.D.

Most scientists initially describe the anatomy of the eye in the singular. That is, a single eye. However, to properly understand the function of vision, we must recognize that normal human vision is bilateral. While we will discuss individual anatomical features, this chapter will emphasize the two-eyed, binocular function as the norm. In the past, anatomists have discussed a single eye and then introduced the concept of the two eyes working together. In this text, we will begin by introducing the basic concept that vision, especially in humans as opposed to lower level vertebrates, requires the paired eyes to look in the same direction at the same time.

3.1 Basic Components and Relationships of the Visual Sensory System

Our bony skull has two frontal cavities equidistant from the midline.[1] Within these cavities are the extraocular muscles turn or direct both eyes in an orchestrated and coordinated manner. Also within the orbits are nerves, vessels, and tissues which support the globes or eyeballs.

Each eye is an assembly of intricate structures. In an overall sense, they are divided into anterior (front) and posterior (back) segments. The anterior segment includes the *cornea*, *iris* and *lens*, along with a fluid-filled space known as the *anterior chamber*. The posterior segment includes the *sclera*, *choroid*, and retinal layers along with another space known as the *posterior chamber*. In terms of relative size, approximately 20 percent of the eye is the anterior section, while 80 percent is the posterior segment (Figure 3.1).

*Drawings in this chapter are courtesy of Winchester Optical Corporation, Elmira, New York.

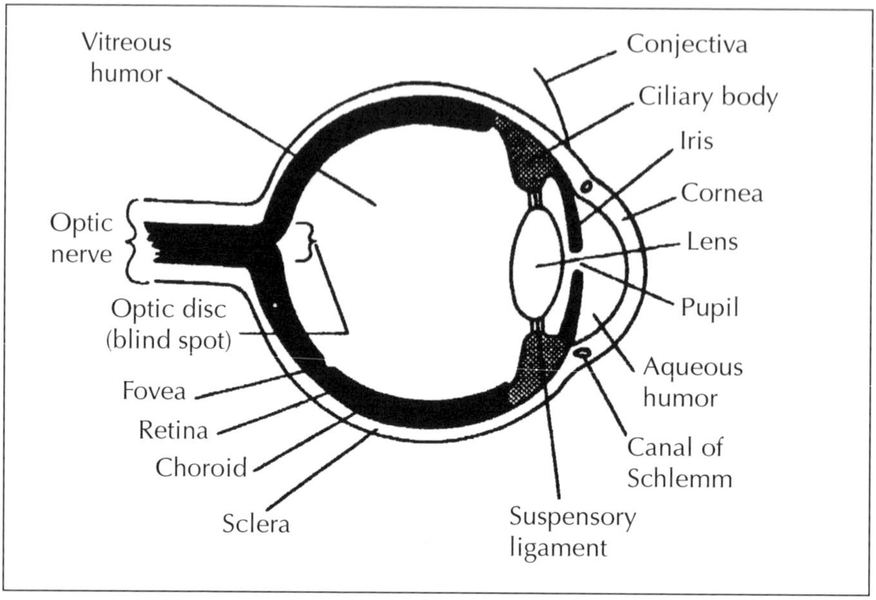

Figure 3.1 *Anatomy of the eye*

Immediately behind the eyes are the *optic nerves*, which carry the nerve signals from the eyes to the brain. The eyes are not perfectly round. Laterally, each globe is approximately one inch in diameter, while the front-to-back measurements are slightly greater than the lateral dimensions. The smallest dimensions are the vertical measurements. The key anatomical structures from front to back are:

Anterior segment:
- Cornea
- Iris
- Anterior chamber
- Lens

Posterior segment:
- Sclera
- Choroid
- Posterior chamber
- Retina

The cornea is a clear dome at the front of each eye.[2] A contact lens floats on the tears just in front of the cornea. The back of the cornea forms the front of the anterior chamber. This chamber is kept in shape by a fluid known as the *aqueous humor*. (The tears flowing over the outer or anterior cornea are not the same composition as the aqueous humor, and the fluids are manufactured in different ocular locations.)

The conjunctiva is a thin, transparent membrane laced with small-caliber blood vessels. This sheet covers the white sclera in each eye where they are very loosely attached. With certain allergy conditions or trauma, fluid can fill the area between the conjunctiva and the sclera, giving the appearance of a fluid-filled sac or blister.

The iris (plural: irides) is the pigmented or colored structure that is behind each cornea.[3] When someone is referred to by eye color, the reference is truly to the irides. When viewed "face-on," the irides are shaped like disks with holes in the center. These holes, or *pupils*, allow light to enter the eyes. The amount of light entering is gauged or metered in a complex neurological manner which ultimately causes the irides to dilate, thus enlarging the pupils (a reaction to low levels of light). Conversely, the contraction, or miotic response, of the irides causes the pupil diameters to decrease (a reaction to high light levels).

The sclera is the fibrous protective coat of the eye.[4] It is white in appearance and is located under the conjunctiva and small conjunctival blood vessels. The sclera helps define the dimensions of the eye and originates at its junction with the cornea's edge.

The *choroid layer* lies beneath the sclera and is highly vascular.[5] Nutrition is carried to the inner eyes or retinas to some degree by the choroids. At the anterior origin of the choroid is the *ciliary muscle*, which shaped like a life saver and provides the movement to help focus the eyes. Radiating from the ciliary muscle are spoke-like fibers that attach to and support the *crystalline lens*. This transparent lens tissue changes shape as a result of tension on the fibers. This tension is derived from the *ciliary body*. As we age, the ability of the lens to flex is reduced. This compromised function leads older individuals to require spectacles for close work such as reading. In addition, as we age, the lens loses its natural transparency and becomes more translucent. This is known as *cataract* formation. (Figure 3.2)

The retina is the innermost layer within each eye.[6] It attaches to the choroid. While retinas receive nutrition from the choroidal blood supply, they also have their own blood vessels, many emanating from the region of the optic nerve head. The retina is the layer on which light is focused and converted from light energy to electrical impulses. The center zone of the retina, or the fovea, is a small area on which images are focused. Retinas are composed of two types of light receptors—*rods* and *cones*. If we

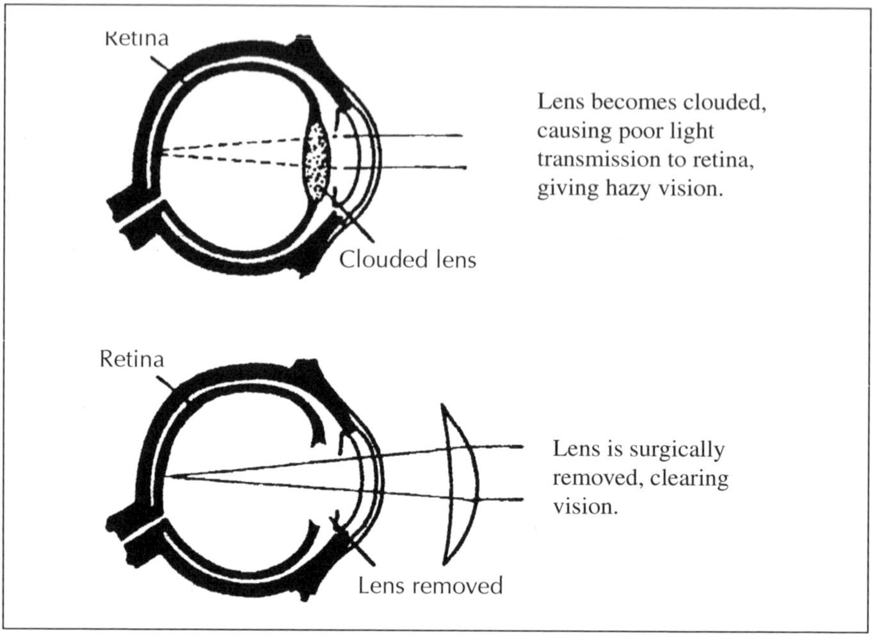

Figure 3.2 *Cataracts*

visualize a retina as a target with concentric rings, the fovea is the bull's-eye and the peripheral area is all of the other concentric rings.

Cones are clustered predominately in the fovea. As the distance from the fovea increases, the cone population decreases,[7] while concurrently, the rod population increases. Cones are responsible for daytime vision and color vision. Therefore, by having a very high cone population, the fovea is the center of resolution or sharpness of vision (visual acuity), size, day vision, and color vision. On the other hand, the rods, while increasing in number as the distance from the fovea increases, never reach the most dense area of receptors in the fovea. Rods have greater sensitivity in dim light and in shape, brightness, and awareness of motion. One reason that rods have high sensitivity is that groups of rods share a single nerve fiber which connects to the brain. If any rod in a cluster is stimulated, the nerve is stimulated and the signal is transmitted. By comparison, cones do not share nerve fibers, so that each cone has its own nerve fiber and must be individually stimulated. Sensitivity is related to the minimum energy input necessary to trigger a neural response. High resolution is a function of a high number of receptors with unshared fibers. These nerve

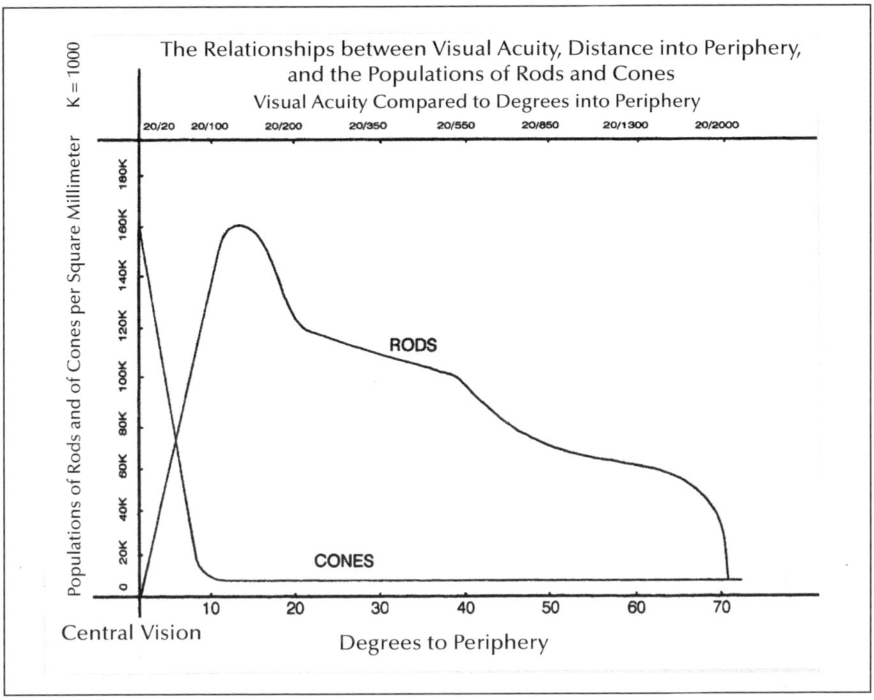

Figure 3.3 *Rods and Cones*

fibers all converge to one area at the back of the eye, called the *optic disc*. Here, the millions of fibers coalesce into a bundle forming the origin of the optic nerve (Figure 3.3).

While our central vision is related to acuity or clarity of image,[8] the detection of objects occurs most often in the periphery. When an object is detected, the foveae are directed by the brain to focus on the object. There is constant competition among all of the things that fall within the field of view. We will notice what stands out. To be detected most efficiently, that object must be contrasted to its environment.

The eyes would be poor sensory organs if they were not relatively constant in shape. An elastic or soft globe would have irregular, misshapen corneas and warped or wrinkled retinas. An eye maintains its shape because it (a) is constructed of corneas and scleras that do not readily stretch, (b) has a fluid in the anterior chamber and (c) has a gel in the posterior chamber. The rigidity of the eye is derived from the pressure within the eye.[9] When a basketball is filled with air, it does not markedly expand, but gets firmer. The eye's firmness is a result of the presence of

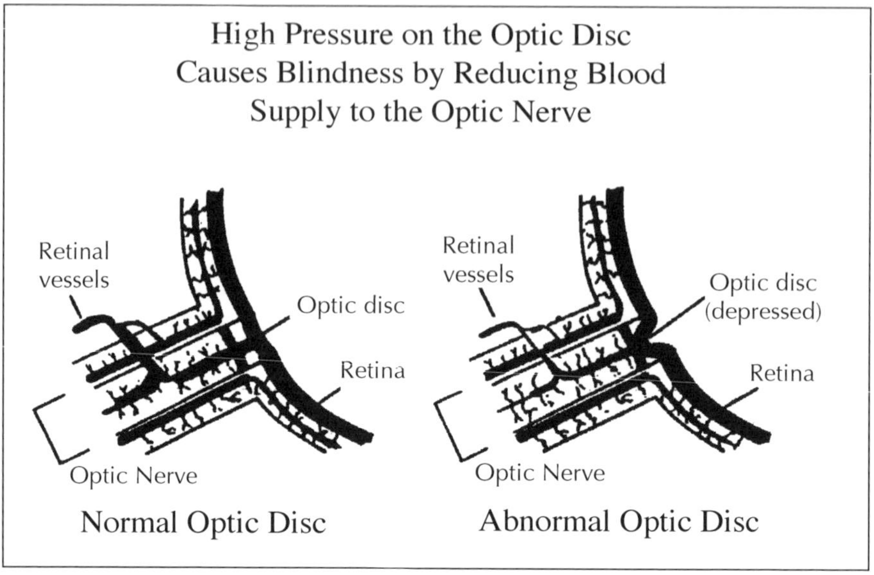

Figure 3.4 *Glaucoma*

fluid (aqueous) in the anterior chamber. The eyes are closed containers with drains, and the fluid manufactured by secretion within the anterior chamber creates a pressure that is transmitted equally throughout the entire inner eye. This pressure helps establish and maintain the rigidity of the globes which assures the eye's ability to focus (refract) light. If however, the pressure is too great for too long, then the retinal nerve fibers, because they form the optic disk edge, can become damaged. This is known as glaucoma and usually results first in peripheral nerve loss (Figure 3.4). The posterior chamber gel is called the *vitreous humor* and is transparent. It contributes to the shape of the eyes just as the filling of a cushion contributes to its shape.

Seeing occurs in a field of view. We have central, peripheral, and binocular fields. Under normal circumstances, we see binocularly with depth or stereopsis within the 120 degrees of central vision. Peripheral vision occurs between 120 degreeas and 180 degrees. While we see peripherally with each eye, we do not see binocularly to the sides until the head and eyes are turned in that direction. The range of the visual fields is discussed here in the lateral dimensions, but it is important to remember that these fields actually extend in a 360-degree cone of view (Figure 3.5).

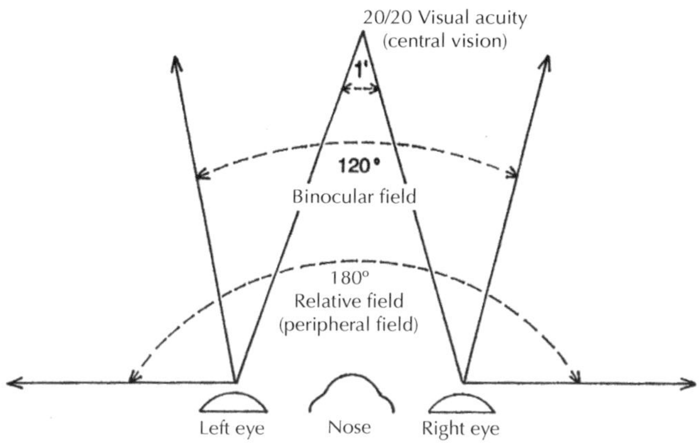

The Central Relationship between
the Limits of Peripheral Vision, Binocular Vision and
Central Visual Acuity

Figure 3.5 *Peripheral vision*

The sharpest vision occurs in the very center of the field.[10] Sharpness of vision is measured as acuity. Normal vision is defined as "20/20," which means that what the individual sees at twenty feet is the same as what the average person in the population sees at twenty feet. A person with 20/100 vision sees at twenty feet what the average person in the population can see at 100 feet. Visual acuity is measured for each eye and then for both eyes together.

While visual acuity has been commonly measured using the Snellen Scale (e.g. 20/20),[11] it may also be evaluated using contrast sensitivity. This is done by having the patient view alternating black and white line patterns. The finer the lines that can be discriminated, the better the contrast sensitivity. It is possible to have good acuity but poor contrast resolution. Beyond acuity and contrast, any disruption in the relationship of the two eyes distorts not only vision but binocularity.

When someone cannot see efficiently, he may need spectacles or contact lenses. These devices compensate for the person's limited ability to focus. *Myopia* (nearsightedness), *hyperopia* (farsightedness), *astigmatism* (distorted sight), and *presbyopia* (aging sight) are often oversimplified and do not accurately explain how people really see.

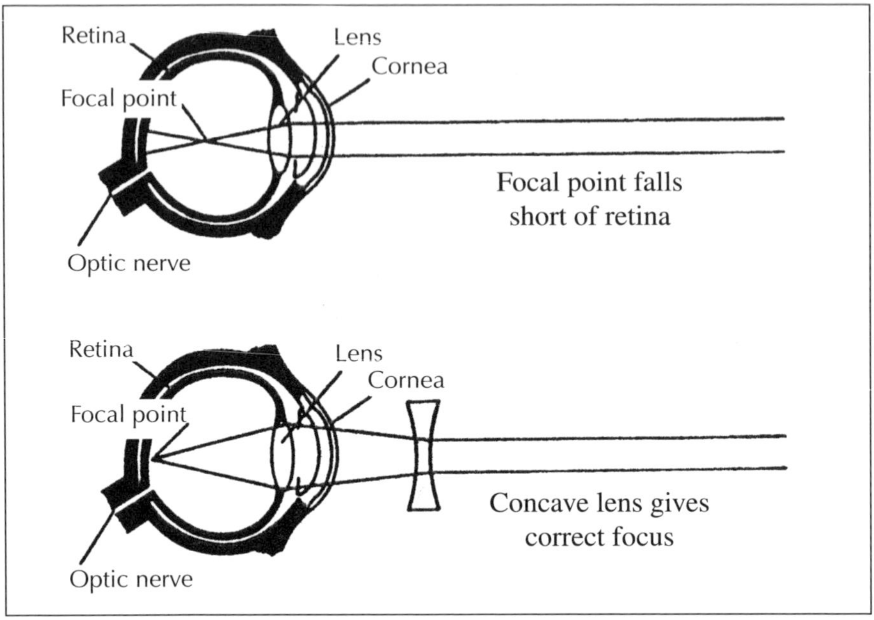

Figure 3.6 *Nearsightedness*

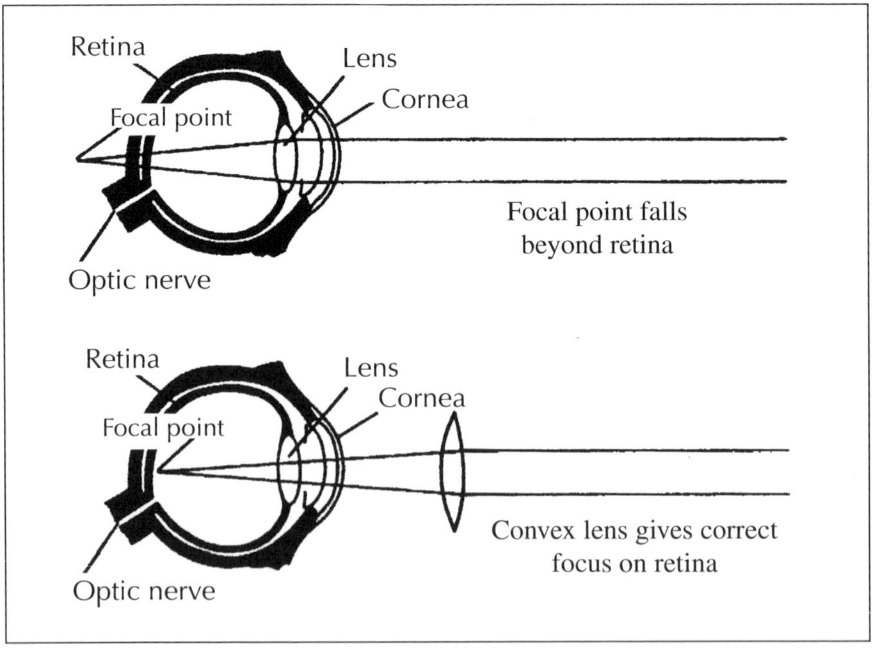

Figure 3.7 *Farsightedness*

Myopia is an optical condition where light comes to a focused point before or in front of the fovea. Some research suggests that this results from an elongated globe where the retina is abnormally far from the cornea.[12] An out-of-focus stimulus is cast upon the fovea and blurry distance vision is the outcome (Figure 3.6).

Hyperopia is an optical condition that is not truly the opposite of myopia. Light is cast upon the fovea but it is also out of focus. In this case some researchers feel that the globe is shortened so the light does not come into focus by the time it reaches the fovea.[13] With many individuals, the natural focusing system within the eye compensates for small to moderate amounts of hyperopia. Headaches, eye aches, double vision, excessive tearing, and intermittent blurred vision can result. Many of these individuals can force their eyes to see clearly, but possibly not at all times. As fatigue, stress, and hunger intensify, focusing requires still more effort. While many of the symptoms can be resolved with lenses, a segment of the population may reject this solution, especially if they can force adequate acuity. Because of the well-knit neurological relationship between

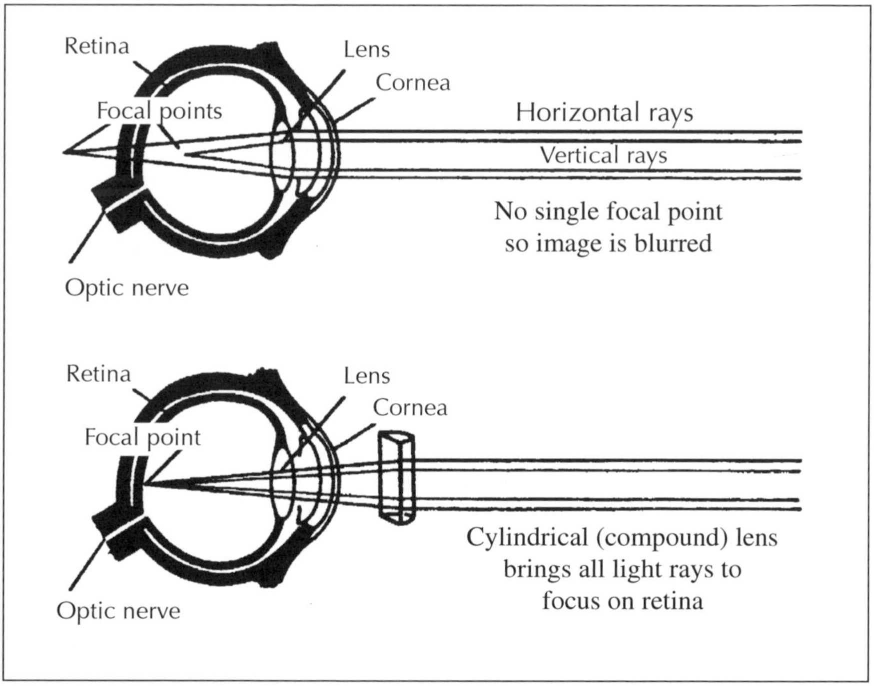

Figure 3.8 *Astigmatism*

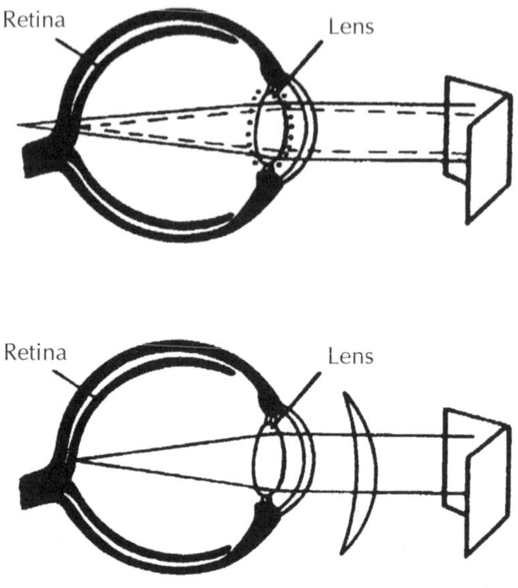

Figure 3.9 *Presbyopia*

the two eyes, as the individual forces clear vision, an overreaction often occurs between the two eyes and results in double vision or shifted depth perception. This then distorts true distance and time judgments (Figure 3.7).

Astigmatism occurs when light from different meridians is focused at different distances from the fovea. One common cause is a change in the shape of the cornea.[14] If the cornea is shaped like the side of a softball then light would be focused at a single point, but if that cornea is shaped more like the side of a football, light rays would be focused at many different points. Some astigmatic patients can force focus, just as hyperopic patients do. It is important to note that, in most patients, myopia and astigmatism or hyperopia and astigmatism occur concurrently (Figure 3.8).

Presbyopia, or aging sight, refers to a different type of focusing phenomenon. The lens within the eye is relatively flexible in youth. But as we age, this flexibility declines.[15] This decreases the ability to see clearly at the distance measured from extended fingertips to the face. The onset and severity of presbyopia vary from person to person. In the early stages, patients may still be able to see well at close distances, but with increased effort. Often these individuals exerience blurred vision whent the object shifts. After looking at far distances for a period of time, it may take longer to focus on near targets (dashboard, text, instruments). Conversely, after looking at a near target, the transfer to distance can have a time delay before the onset of crisp, binocular seeing (Figure 3.9).

In considering the conditions which affect sight, it is also necessary to be aware of ocular diseases and how they may affect how an individual sees. *Cataracts*, which involve a loss of transparency in the lens of the eye, reduce vision by limiting the amount and color of light entering the eye or eyes. Some cataracts act primarily in daylight while others are most debilitating in the dark. *Glaucoma* can affect vision by damaging nerve fibers, characteristically those first in the periphery and finally all fibers if left untreated. However, some of the medications traditionally prescribed can constrict the pupil, which leads to very poor nighttime vision. Therefore, in any review of an accident, the vision of the individuals involved must be very carefully analyzed. An in-depth examination must be performed and should include as a minimum the following.

- History including systemic and ocular diseases
- Medications in use or prescribed
- Visual acuity at distance, with and without glasses
- Peripheral-field sensitivity
- Central field study
- Refractive error
- Color vision
- Depth perception on near and distant targets
- Ocular health including retinal integrity and eye pressure
- Perceptual deficiencies
- Pupil diameters in dark and illuminated conditions
- Accommodative flexibility
- ANSI-rated sunglasses (if used)

Vision is more than just seeing clearly. It is also a relaxed, easy focusing or shifting of ranges of seeing and processing, and the health of the eyes must be factored into this visual equation. Pedestrians, cyclists, and motorists need not only see clearly, but they must also be able to sustain focus for long periods of time with comfort and efficiency. They must be able to quickly shift gaze from near to far, far to near, and at different far or near points. All this must be done effortlessly and quickly. As effort, strain, and stress increase, more deterioration can occur in effective vision.

In this section we have briefly discussed an introduction to vision. However, for vision to be complete, the brain must be able to receive the impulses from the retina (nerve transmission) and then properly and efficiently interpret them (visual perception).

3.2 Visual Perception—The Second Part of the Vision Equation

Visual perception is the process through which the brain interprets the retinal impulses from the eyes. Those signals from the eyes vary slightly from one eye to the other. Not only are the signals interpreted, but previous experiences shape the interpretive response. For example, if an individual sees a rose, and then closes his eyes, a visual image of that picture can be retained. Upon a suggestion, either internally or externally initiated, the characteristics of our visual image can be shifted. If, on a subsequent exposure, that individual is asked to visualize a flower, then they might draw upon their previous experience with a rose, and call back that image. Yet perceptions of the color, size, and even the fragrance may all vary. Another individual asked to visualize a flower, may interpret that as "flour," and see a product used in baking. Visual perception depends in part on what we have learned to bring to the retinal image.

Perception also encompasses how senses intermingle or integrate. If we hear a sound, such as a bell, we can see in our mind the shape and even the size of the bell. Visualization (the mind's eye) may include where the bell is located—a tower, an ice cream truck, or a school room. Taste, smell, touch, image, kinesthesia, and sound all play roles in visual perception. The images on the retinas are transferred to the brain by electrical signals and can be altered by perception. Depth perception, as a visual function, is commonly accepted as a binocular function. Because of the separation between the right and left eyes, the brain is capable of comparing the slightly different images from the right and left eyes even though they are focused on the same target. These slightly different images provide the brain with clues that allow for a judgment of depth.

It should be evident that what we bring to the image or scene is shaped by prior experience. A hunter in the woods may see "ears" dimly illuminated and assume "rabbit." In daylight the ears may well have been two leaves of a branch. The experience and imagination of the hunter provided the bulk of the information to complete the image. If part of a car is seen driving away, the observer can complete the shape. This is visual completion and can be accurate or misleading depending upon prior learning.

Contrast between an object and its surroundings is necessary for the object to be seen. This contrast is established by five factors—size, shape, color, brightness and motion.[16] An example of each is:

Size—a toothpick in a display of popsicle sticks
Shape—a star shaped cookie on a plate of round cookies
Color— a yellow ball among a box of red balls
Brightness—a bright flashlight shining across a field amongst the lights of dimmer flashlights.
Motion—a crawling ant among other ants which are stationary

In each of these examples, the object is compared with other objects in the field of view. However, this only draws contrast between the objects. The background must also be considered.

Size—the toothpick with the popsicle sticks would be less visible if viewed against a background of a wooden table
Shape—the shape of the cookies could be camouflaged by placing them on a plate of similar color
Color—the detection of the yellow ball would be more difficult if the backdrop were a yellow table cloth
Brightness—a bright beam becomes less effective as daylight increases
Motion—the moving ant is more difficult to detect if viewed against a dark anthill

We have described two categories of contrast. The first is the object versus other objects, and the second accounts for the scene in which the objects were viewed. Contrast increases visibility and, conversely, a lack of contrast produces camouflage. Camouflage is a very misleading concept in that the object of regard may be apparent when viewed as an independent entity, yet not apparent when viewed against the field of view. An example is the modern fire truck which is painted red, or red and white. This vehicle is equipped with sirens, flashing lights, and retroreflective tape, yet research consistently shows that this type of vehicle is effectively camouflaged in the common environment. Research shows that these vehicles have higher accident rates than similarly equipped apparatus painted lime-yellow.[17, 18, 19]

Size, shape, colors, brightness and motion can all exist simultaneously and the interrelationship can constantly change. In addition, the rest of the field of view is composed of a three-dimensional background and foreground, which also can be in a constant flux. The end result is a highly complex interaction that is often underappreciated. To illustrate, the very complex task of driving presents a constantly changing scene. In that scene, vehicles move at different speeds. Their colors are different as well as their shapes and sizes. Some may have headlights on and brake lights may be flashing as individual drivers control their position on the roadway.

Pedestrians, cyclists, and animals can appear in view at any time and often from an unanticipate direction. At night, oncoming headlights can overwhelm the visual systems, and brake and turn signals can intermix. A visual stimulus, in order to be seen, must be more attention-getting than all other competing stimuli and cause the nerve fibers of the peripheral retina to fire and signal the brain to direct central vision in a new direction. A split second later, new signals are detected and the shift of vision is induced again. This occurs every time we drive.

Nighttime driving presents some obvious changes. As darkness increases, the peripheral or rod-receptor vision is more sensitive. The neurological response is a switching to more rod (peripheral) receptors and fewer cone (central) receptors. The time delay for dark adaptation to be reasonably complete can take upwards of 15 minutes in a healthy, nourished, rested, alert person. In a practical sense, it is very rare that any driver is 100 percent adapted to the dark, because of the influence of dashboard illumination and the back-scatter of light from the vehicle operator's headlights. However, even the less-than-total dark adaptation is critical for safe nighttime driving. In rural areas, with fewer oncoming cars, fewer street lights, and sparser advertising displays, dark adaptation is greater and it takes longer to readapt after oncoming headlight beams impinge upon the driver. The time to shift out of dark adaptation is very fast, requiring only a few seconds.

A simple example of dark adaptation and deadaptation occurs when one enters a movie theater. Immediately upon arrival, it is difficult to see available seating in the darkened auditorium, yet within five to ten minutes it is much easier. Soon after, it may be possible to identify the features of people in the area around you. When exiting, especially into daylight, there will be several moments of discomfort, but cone vision and central acuity rapidly recover. Deadaptation is not a prolonged response, but occurs in a very brief period of time.

Especially in rural driving, with no street lights or advertising displays, the world beyond and outside the headlight beams is cloaked in darkness and therefore camouflaged. Objects approaching the operator's vehicle from the side may not be detected at all or only after they enter the headlight beam. Objects along the path of travel may be illuminated by the headlights. However, being illuminated does not necessarily mean being seen. If a small, dark stationary object is on the road, it may not be detected until it is struck. The first stimulus could well be an auditory input (thump) or a tactile input (the vehicle shakes or shudders). If oncoming traffic is present then the driver has to deal with bright headlights striking his rods and cones. Even if one looks away from the headlights of an approaching vehicle, light still enters the eyes and produces dark deadaptation. It bears repeating that, for a period of time, perhaps as long as fifteen minutes, the driver is not effectively adapted to night vision and is less sensitive to visual stimuli.[20] Within realistic reaction times (see Chapters 7 and 13), it may be impossible to avoid hitting an object illuminated by the headlights. This can be true even at low speeds.

Street lighting can sometimes play a cruel and unusual trick on observers because some bulbs emit light in a relatively narrow band of wavelengths. In order to see, we must receive reflected light. If an object has a particular color, for example red, and the nighttime illumination source contains no red wavelengths, then that object does not reflect and is dark or camouflaged. In particular, sodium and mercury lights do not emit a wide band of wavelengths and colors can be distorted, dimmed, or not seen.

Adaptation to light or dark is a normal visual function; however, excess or extraneous light may overwhelm the visual sensory system. Excess light beyond what the eyes can handle is called glare and can have external and internal origins. An example of external glare would be the intense dazzling of sunlight shining on snow in front of the observer. Internal glare occurs when a normal amount of light enters the eyes, but it is not focused sharply on the retina; the light become extraneous, and is distributed on many, diffuse nerve receptors. This can be caused by imperfections in the eye, such as cataracts, corneal scars, or floaters.

The cardinal fact of glare is that it reduces contrast and thereby decreases acuity. When glare reduces the ability to see objects, it is known as *disability glare*. External glare is commonly present and is often countered with sunglasses. In vehicle accidents it is important to determine the time of day and direction of travel of the involved drivers. From this, sun position and the potential for glare at the accident scene can be determined. While filters such as dark glasses and windshield tints reduce some of the effects of glare, they may not be sufficient to offset the reduced ability to see clearly. Many drivers actually block the sunlight by pulling down the visor. If a visor is used, then the body size of the operator should be considered: a very short driver may not benefit as much as a taller person, and conversely, if the driver is quite tall the visor could obstruct too much of the view.

External glare can be blinding. Decreased contrast has been mentioned and that is quite common. An example would be illustrated by a driver heading west into a late afternoon sunset. Approaching an intersection controlled by a traffic light, the driver does not realize he has a red signal and proceeds, only to collide with another vehicle entering the common area at a right angle to the first car. Because of the sunlight's intensity, not only would the red and green signals not be visible to the westbound driver, but the driver may not even see the entire traffic signal structure if he or she was not familiar with the area. Similarly, an individual would also have great difficulty in seeing flashing lights on emergency vehicles as well as individuals standing in the roadway, either as pedestrians or for traffic control. External glare is blinding at times, but what caused the events in any given situation may not be evident by the time investigators arrive on the scene. The distinction between the time of the accident, the time the accident was reported, and the time of arrival of the police should be understood. This will help to determine what the sunlight conditions were at the various times observations were recorded.

One of the time-honored methods of handling daytime external glare is to use sunglasses of one of two categories, prescription or over-the-counter. Those prescribed by an eye doctor are made in a laboratory using materials that pass industry standards for transmitting light of varying wavelengths. Over-the-counter materials are labeled for beach or driving. Only those labeled for driving are guaranteed to transmit traffic signal colors, while beach glasses may not. Most purchasers are not aware of the differences and usually shop for style and look, not realizing that some traffic lights will appear to emit no light when viewed through unlabeled beach types. When investigating an accident it is prudent to evaluate all types of eyewear for appropriate function.

Glare can be increased by dirty or scratched windshields. The imperfections on the surface of the glass disperse incident light, causing it to be spread out and be less focused in the eye. While washing windshields and side windows helps reduce glare, only replacement of the panes remedies scratches. Since the glass areas of cars are often washed with rough cloths, dirty rags, or coarse paper towels, the surfaces become rough or scratched over time. Road debris, pebbles and sand regularly strike the windshield and cause further scratches and pitting. Old wipers often become coated or impregnated with debris and abrade the glass when used. Any older vehicle can be assumed to have a less than optically clear windows and windshields.

Night driving is not immune from glare problems. Bright headlights often dazzle oncoming drivers. In recognition of this, vehicles are equipped with dimmer switches. The same imperfections of glass and eyes cause similar problems at night, but with one extension. Dark adaptation and deadaptation, as discussed above, becomes operational.

Internal glare is related more to the health of the eye than the amount of incident light. Just as the dirty and scratched windshield disperses rays and causes glare, cataracts and ocular imperfections such as corneal scars, vitreous opacities, iris imperfections, and retinal degenerations produce internal glare. Imperfections are not only caused by disease, but also by surgery. Some persons attempt to free themselves from constant spectacle or contact-lens use by submitting to corneal reshaping, either by laser, surgical blades, or a combination of both. The end result may lead to corneas that are less-than-excellent transmitters of light energy. As a result, glare is sometimes a by-product and can have varying effects on the patient. At night the pupils dilate and allow more light to enter the eyes. With wider pupils, more peripheral light is allowed to enter the eyes and more corneal area comes into play. Therefore, corneal imperfections and irregularities may have a greater role at night than during daytime.

The cornea is the target tissue for refractive surgery. The first generation procedure was a mechanical process where blades penetrated the cornea in a radial like pattern; this was called radial keratotomy. The incisions healed and formed deep corneal scars that reshaped the eye, which then caused a change in the way the eye focused. The scars were faint but permanent, and diffused light. The end result in some patients was varying degrees of glare. As the pupil dilated at night, the scars were more in line with the larger opening and allowed for greater sensory discomfort and increased visual interference. (Some articles discuss the fragility of the eye after radial keratotomy, and some doctors feel even the impact of a deploying air bag is vigorous enough to cause the corneal scars to rupture and result in great damage to the eyes and vision.) Some patients have had this type of surgery.

The second generation of corneal surgery, and the one currently in vogue, employs a laser to reshape the surface of the eye, and the result is anatomically unlike the radial incisions which were very deep. The glare related to scaring was diminished, but if the cornea does not heal smoothly, vision may be affected.

The third generation of refractive surgery requires a horizontal flap like a manhole cover on the center of the cornea to be lifted, allowing a laser to reshape the underlying tissue. The flap is then placed back over the treated area and allowed to heal. One advantage of this procedure (called LASIK) is in hepling highly nearsighted individuals. Glare caused by improper corneal healing can be a side effect. In all three types of corneal reshaping the amount of glare can be affected by the skills of the surgeon, the disruption of the smoothness and clarity of the cornea, and the time of healing. It is normal for a successful patient to have some temporary glare problems that may vary in intensity. Unfortunately, there is a growing body of patients with permanent glare interference and it can vary from day to night and even from week to week. Sometimes it can be permanent.

Corneal surgery changes the focusing, or refractive, power of the eyes by altering the cornea. When a patient has a cataract removed because it is reducing visual acuity, the cloudy, natural lens is removed and replaced with a transparent, plastic disk inserted in the eye behind the iris. This is called an *intraocular lens implant*, or IOL. This alters the eye's ability to focus. Unwanted distortions and imperfections can cause glare and because most, but not all, IOL patients are elderly, reaction times as well as vision may be further compromised.

For whatever reason, glare can be a significant factor in vehicular accidents; it is directly related to visual function in that with increased brightness or ocular imperfections, contrast is reduced, signals are overridden, and vehicle operation becomes more dangerous.

Drivers have the responsibility to operate their vehicles in a safe and conventional manner. They have the right to expect predictable behavior from other drivers. Stopping at a red light is responsible behavior even if no other traffic is present. As drivers approach a green light, intersecting operators are expected to stop at the red signal. When the responsibility of one driver and the predictability of the other driver sharing the road are not present, the chances for accidents increase.

Predictability and responsibility also extend to other highway users, including pedestrians, motorcyclists, and bicyclists. The potential for injury is much higher for these groups because their visibility is markedly reduced. Most notably, they are much smaller and have less apparent color. They can change direction more sharply and stop more quickly. Therefore, along with lower visibility, they are less predictable.

When pedestrians and cyclists see approaching vehicles, they assume that they in turn are seen. This is not true. A car or truck is large and presents a relatively greater visual stimulus. The pedestrians often do not mentally compute that while they see this large stimuli, they present a much smaller stimuli. Furthermore, although individuals may have the same visual acuity, they are not presented with equal stimuli. Nighttime driving compounds this problem. A pedestrian or cyclist may see the headlights of an approaching vehicle long before the headlight beam reaches him or her. The pedestrian or cyclist may think the situation is safe when in reality extreme danger may exist.

The headlights themselves offer some clues to the relative closeness of the approaching vehicle. The smaller the apparent separation between the right and left headlights, the farther away the vehicle appears. One pitfall arises from different vehicle designs, in which small cars have less separation between the lights than do larger vehicles. When other clues are lessened at night, headlight patterns gain importance. A small car, with a narrower span between the headlights, could be interpreted as a larger, more distant vehicle.

All aspects of judgment are mediated by fatigue, hunger, medication, alcohol, street drugs, prior experience, and mental state. Reaction times are increased, predictability is decreased, responsible behavior is distorted, and peripheral vision contracts.

Peripheral vision, or the ability to detect stimuli from the corners of the eyes, is important for normal functioning in daily life. Signals that are detected peripherally are weighed by the mental process and given a higher priority than what is currently being focused centrally, thereby shifting one's gaze to the new target. The time taken to shift the eyes from one stimulus within the overall field to another is about 0.25 second. A vehicle moving at 60 mph will travel twenty-two feet during the 0.25-second fixation-shift time.

A pedestrian approaching a highway at 90 degrees to the traffic flow increases his or her stimuli to the driver's periphery and subsequent visibility by:

1. proceeding further into the central cone of view of the oncoming driver;
2. having exaggerated motion such as a swinging arm movement or a leg-stepping stride;
3. wearing bright colors, especially yellows and whites;
4. adding fluorescent and retroreflective material to garments and footwear.

A driver's peripheral detection system is extremely important. Yet, vehicle operators often do not notice a reduction of the peripheral field of view caused by fatigue in the short-term and aging in the long-term. Most persons underestimate its importance. Furthermore, many states do not even have a licensing requirement for it.

Adding to the problems of vision are certain components of vehicle construction. The windshield-mounted rearview mirror, standard in most cars, effectively blocks part of the peripheral vision that is just to the right and above eye level. A vehicle's corner posts, known as "A" posts, are the supports that connect the anterior roof corners to the car body. These relatively vertical supports block side vision. Because the left post is closer to the driver it blocks a relatively larger angle of view than does the right-side post. Therefore, the driver's left-side view has a significantly larger blind spot. This increases the possibility of an accident because oncoming traffic is more prevalent on the left side. The obvious conclusion is that accidents that occur in the left front zone should be investigated with this blind spot in mind. The smaller the object in any blind spot, the easier it is to obscure, and the greater the risk of an accident. A small object, such as a pedestrian or cyclist, is much less detectable in the area obscured by the "A" post.

In the foregoing discussion of driving tasks, we have assumed that the driver has binocular vision and good depth perception. (We should note that the presence of binocularity is not a foolproof guarantee that good depth perception simultaneously exists.) However, one-eyed drivers are certainly present on the road, and no state denies them licenses for private vehicle operation. These drivers do not possess true binocular vision, but they have adapted to certain visual clues that allow them to make certain depth judgments.

Monocular clues to depth perception include overlay, size, color intensity, and background. Overlay is in play when part of a distant object is partially obscured by a nearer object. The breakup of the distant outline provides information about the relative closeness of one object to another. Because common objects are of a predictable size, a distance judgment can be made based upon how small that object appears. However, an inherent flaw is present with automobiles. Very small cars can be interpreted as bigger but farther away. In this case, depth judgment is skewed because the small vehicle is actually closer than anticipated. From the perspective of color, the farther away an object is seen, the duller or grayer it appears. Background is the appearance of the terrain. Mountains at a great distance do not appear crisp, but hazy and their colors are muted.

Earlier in this chapter it was mentioned that the senses function in an integrated manner. An example of this is the interreliance of hearing and vision while walking. Most persons easily identify the visual detection components, but few consider the auditory-visual relationship. A pedestrian approaching an intersection often scans quickly in each direction, but also listens for oncoming traffic that is close by. Children with hearing deficiencies of recent onset, such as those occurring with a head cold, lack the awareness that their senses are dulled. In that situation it is more easily understood why some children step into traffic without looking. If the sensory loss is longstanding or chronic, the individual goes through a learning curve and relies on other senses to function safely.

Endnotes

1. Goss, C.M., *Gray's Anatomy of the Human Body*, (Philadelphia: Lea and Febiger, 1959) 168, Fig. 186.

2. Wolff, E., *Anatomy of the Eye and Orbit*, (Saunders: Philadelphia, 1961) 300, Fig. 21.

3. *Ibid.*, 30, Fig. 21.

4. *Ibid.,* 31.

5. *Ibid.,* 54

6. *Ibid.,* 88

7. Hart, W.M. Jr., *Adler's Physiology of the Eye*, (Mosby: St. Louis, 9th ed., 1992) 595.

8. Borish, I.M., *Clinical Refraction*, (Professional Press: Chicago, 3rd ed., 1970) 401.

9. Hart, *op. cit.*, 228, 248.

10. *Ibid.,* 941.

11. *Ibid.,* 538.

12. Borish, *op. cit.*, 70.

13. *Ibid.*, 118.

14. *Ibid.*, 124.

15. *Ibid.*, 162.

16. Solomon, S.S., "Visibility of firefighters," *Journal of the American Optometric Assoc.*, 1974; 45:140.

17. Solomon, S.S., "The safety color," *Firehouse*, 1984; 9 (June), 106.

18. Solomon, S.S., "Lime-yellow color as related to reduction of serious fire apparatus accident—The case for visibility in emergency vehicle accident avoidance, *Journal of the American Optometric Assoc.*, 1990; 61: 827–831.

19. Solomon, S.S. and King, J.G., "Influence of color on fire vehicle accidents," *National Safety Council-Journal of Safety Research*, 1995; 26: 41–48.

20. Hart, *op. cit.*, 511.

Chapter 4

Color Vision

Stephen S. Solomon, B.S., O.D.

Humans have an elaborately developed system of nerves in order to see color. While we have many differences from lower primates and other vertebrates, the ability to see colors is not limited to man alone.[1] From an evolutionary perspective, those sensory improvements that permitted a more advanced detection and warning system probably helped in the succession and survival of that species. Any stimulus that causes an impulse to register in our brain more quickly improves reaction time. Color is a band of energy within the electromagnetic spectrum that is detected by our visual apparatus and imparts more comprehensive information than a monochromatic view. It is necessary to understand some anatomy of the eyes in order to comprehend the role of color vision and how it affects our perceptions, interpretations, and reactions.

To review our discussion in Chapter 3, the eye is composed, in part, of a sensory layer of nerve endings and attached nerves known as the retina. The sensory endings of the nerves have been named rods and cones. They are the basic receptors of light, and in the process of converting light energy into electrical impulses, they participate in a reversible chemical reaction. After the impulse is discharged, a reconversion of the end-receptor chemistry occurs, setting up that particular rod or cone to be able to respond again, after a time lapse, to new stimuli. Some individuals perceive a different color called an afterimage during this recharging period.

The cones are most populous at the very center of the retina, called the fovea, and they rapidly diminish in number as one traverses the retina toward the periphery. Rods increase in a population pattern as measured from the central fovea, where practically none exist, toward the middle of the periphery where they are most dense. Then they taper in number as the far periphery is approached.[2] The cones are responsible for central vision, acuity, size and color vision. By contrast, the rods are responsible for peripheral vision and detection of gross presence, such as shape and motion. They are most sensitive in dim to dark illumination.

Recognizing that rods do not react to color during the daytime does not mean that they do not respond in daylight. If two colors are compared for peripheral retinal sensitivity, the color of more reflectance will be detected farther from the center of vision,[3, 4, 5] and therefore will register more quickly. This is considered to contribute to a faster reaction time. Remember that most detection of what needs to be seen, and requires a reaction, occurs in the periphery first. After peripheral detection occurs, the brain decides whether the stimulus has strength enough to cause a shift from what had previously occupied central vision to what has now been detected by the peripheral detectors. This can be likened to an early warning system.

Therefore, when a visual stimulus needs to convince the brain that it needs scrutiny, it must present sufficient attraction. Contrast is the ability for an object to sufficiently stand out from its background and force central vision to center on the new stimulus. Because different colors possess different levels of brightness, it is possible to rank them in order of strength. Obviously, the brighter colors are detected more quickly, but there is a complication which can cause confusion: colors detected in the periphery are seen first as levels of brightness, but not with an associated color. Therefore, a color is first detected as a brightness, and afterward, when the most peripheral cones are triggered, as a color. Those colors that stimulate the periphery first are those which receive the first reaction. Among the colors that are detected farthest into the periphery are the narrow band of wavelengths from the greenish-yellows to the yellowish-greens. The reds are detected much later. It does not matter what type of vehicle is stimulating the periphery, only that the vehicle is detected. For the survival of the approaching driver there is no difference between fire trucks, mail trucks, cement delivery trucks, or gas tankers. One does not identify the type of vehicle, and decide to miss it, but rather detects any hazard and then avoids contact.[6] As we know from other material presented in this book, sirens, air horns, bells, two-tone paint, flashing lights, strobes, and retroreflective material may not provide an exclusive advantage for earlier detection.

Color provides an additional stimulus and facilitates early detection. Because light does not suffer projection distortion as does sound, and because some special colors are particularly difficult to camouflage, they provide added safety—that is, the greenish-yellows to the yellowish-greens.

At night, when the rods are the primary light receptors, colors are again detected most quickly by brightness. When the central vision is finally directed toward a target and reaction decisions must be made, then the warning or safety color plays a role in that those that can be seen first, noting the relative sensitivity, provide for a quicker response. It has been stated repeatedly in this text that what is detected first has the best opportunity to trigger a reaction first. It is not necessary that the object sharing the road be recognized as a particular type of vehicle, but rather, the driver should recognize that a hazard exists, detects it, and avoids it. The eye is a very poor detector of red color in daylight, darkness, or poorly illuminated conditions. As a result, red cannot induce a quick reaction.

Two-tone paint schemes create further problems. Red and white two-tone patterns create double camouflage in that red is very poorly seen and white easily blends into many backgrounds. Further, white lacks attention-getting capability. The combination of the two colors breaks the silhouette or outline of the vehicle.[7] By breaking the silhouette into two parts, each with a different color, this further distorts and slows the judgment process. Some colors are perceived as closer than others and in using a two-toned pattern the observed object may be perceived as two separate segments, and not equidistant from the observer. This is called *chromostereopsis*, and is caused by chromatic aberration.[8]

During an accident investigation, one of the first questions asked by the police will be whether the flashing lights and siren were operating, not realizing that the civilian motorist may not have seen nor heard them. As discussed elsewhere in the book, if the driver and crew of the emergency vehicle hear the sirens and horns, and see the flashing lights, then the assumption is made that all others on the shared roadway will similarly perceive these signals. The law enforcement official will often have the same impression.

Color has a unique significance as a warning stimulus. It can provide added protection day or night, not only visually, but psychologically. Among those arguing for red as the symbolic color of danger, the following should be considered. Red is one color that can produce multiple, inconsistent reactions, in that it may signify "stay away" or "avoid me," but can also imply "come closer." Red is used in the animal world as a warning, and this seems to have transferred to human signals. However, red is also a significant attractive signal. Red meat is considered a safe choice, and red is also a provocative, enticing attraction in a sexual connotation. These mixed signals detract from the safety arguments for red and provide another perspective for choosing a color based on its implied meaning. At best, red does not provide a single, accepted, universal message but rather, offers a mixed message and is incapable of performing an unambiguous safety-oriented role.[9]

Color vision is a normal component of vision for most people, but not all. Almost five percent of the population (mostly males) are color-defective. Phrased differently, these individuals have more difficulty in seeing selected colors, and red is among the colors most affected. It has been reported in the literature that color-defective individuals as a group have more vehicle accidents compared to a similar group without color impairment.

Among the explanations is the fact that reaction time is hindered because the message or the detection of a signal is delayed. It simply takes longer for the color-defective individual to respond to a given situation where color is included in the message. More information is provided for decision making using color and coupled with that is brighter colors are reacted to more quickly.

Color-defective vision can be studied in two major categories, genetic and acquired. Genetic defectives are also known as *classical* in etiology, and of which there are three subdivisions, each possessing three types of photoreceptors (red, green, and blue) that must be considered. They are *protans*, which deal with the color red, *deuterans*, which deal with the color green, and *tritans*, which deal with the color blue. They can be further subdivided into weaknesses of color detection known as anomalies, or as complete color losses using the suffix *opia*. Combining the type of color involved with the degree of color loss gives a term that describes each condition. Therefore, an individual who sees red, but with decreased sensitivity is known as a *protanomal*, and a red-blind individual is known as a *protanope*. Likewise, the person with weak green vision is a *deuteranomal*, while green blindness is *deuteranopia*. The rarest situation of the three pairs, a weakness in detecting blue, is known as *tritanomaly*, while true blue blindness is *tritanopia*.[10] Almost all classical common color deficiencies, be they partial or complete, are sex-linked to an "x" chromosome and are transmitted from the grandfather to the grandson, through the mother. Very often, the patient's history will reveal the possibility of a genetic color defect. In the same vein, the second major division of acquired color losses may often be detected in the history, but from another point of view.

Illicit drugs and prescribed medications can alter normal color sensitivity. It has been established that cocaine abuse can lead to diminished green and blue interpretation. Learning this from the subjective responses of the patient is unlikely, but must be kept in mind. Since blue losses or weaknesses are genetically rare, it should cause the practitioner to consider chemically induced weakness or loss as a possibility when a patient presents with this sign. We must also remember that some prescribed pharmaceuticals can change color vision sensitivity.[11] Appropriate testing can assist in determining whether a vehicle operator was under the influence of illicit drugs or taking medication by prescription. In our aging society, we can expect increasing numbers of people to have altered color vision associated with medications. Unlike genetic defects, which are lifelong, many of the chemically induced deficiencies are reversible. After a reasonable, drug-free period that varies with the dosage, duration, and the individual, the ability to detect color becomes relatively normal. While we accept that slightly more than 8 percent of the male population, and less than one percent of the female population, have a genetically associated color vision problem, there is no accurate data yet available predicting the percentage of male and female drivers who have chemically induced color vision deficiencies. It is reasonable to assume that it is far more common than previously considered. Color vision testing using the Pseudoisochromatic Plate Series can assist in the diagnosis.

Endnotes

1. Hart, W.M. Jr., *Adler's Physiology of the Eye*, (St. Louis: Mosby Year Book, 9th ed, 1992) 594.

2. *Ibid.* at 595.

3. Traquair, H.M., *An Introduction to Clinical Perimetry*, (St. Louis: C.V. Mosby, 1949) 10–11.

4. Allen, M.J., *Vision and Highway Safety* (Philadelphia: Chilton Book Company, 1970), 138–139. An expanded version of this book has been published as Allen, M.J., et al., *Forensic Aspects of Vision and Highway Safety* (Tucson: Lawyers & Judges Publishing Co., 1996).

5. Leonard F.M. and R.B. Sleight, *Conspicuous colors for police vehicles*, Century Research Corporation. Paper delivered at the Highway Research Board, 46th annual meeting, Washington, DC, January 1967.

6. Olson, P.L., *Forensic Aspects of Driver Perception and Response* (Tucson: Lawyers & Judges Publishing Co., 1996), 173.

7. Allen, M.J., et al., *Forensic Aspects of Vision and Highway Safety* (Tucson: Lawyers & Judges Publishing Co., 1996), 181.

8. Borish, I.M., *Clinical Refraction* (Chicago: Professional Press, Inc, 1970), 736.

9. Wise, B.K. and J.A. Wise, *The Human Factors of Color in Environment Design: A Critical Review*. NASA grant number NCC 2-404. March, 1988, 10.

10. Hart, *op. cit.*, at 718.

11. Onofrey, B.E., et al., *Ocular Therapeutics Handbook.* (Philadelphia: Lippincott-Raven, 1998), 105.

Chapter 5

Flashing Lights*

Stephen S. Solomon, B.S., O.D.

Since before biblical times, light has come from the sun and occasionally, the moon. These sources emit or reflect a steady light. When fire was seen, the light did not radiate steadily like sunlight, but flickered. The sun, moon, and fire were relatively steady when compared to lightning. In nature, the random and infrequent lightning was easily differentiated from the more common sources. Man probably noticed how these flashes grabbed a person's attention. What accented or enhanced the flash intensity was the fact that it was viewed against an empty visual field, such as clouds and sky.

Humans invented flashing lights to attract a person's attention from a distance. Among the first to use these were mariners, who saw these lights against another empty visual field, the ocean. They wanted to determine what flash intensity and duration was visible or detectable at a great distance. Not only was this important for observers spotting a vessel, but for the vessel's navigators to see, identify, and locate lighthouses.

With the advent of airplanes, it was found that flashing lights aided pilots in detecting fellow flyers. The aviation industry became a prime investigator of the functioning and perception of flashing lights.

There are some basic similarities between the flashing lights used by mariners and aviators. They both are viewed against a primarily empty field. The direction of travel is of importance to both groups, and lights were added to delineate the right and left sides. Aircraft have a red wing light and a green wing light. Two different types of information were now

* The appendices of the book coordinate with this chapter. References to fire apparatus will start with "F-" and are printed near the left margin of Appendix I. They direct the reader to the fire apparatus specifications. References to ambulances will start with "A-" and are the paragraph and section number in the left margin of Appendix II. They relate to the federal KKK-A-1822D specifications.

presented, with the first being detection and the second being direction. The observer had to receive different information from groups of flashing lights.

The aircraft beacons manufactured to aeronautical requirements found their way, with modifications, to ground-based emergency vehicles. Obvious differences in purpose affect flashing light performance. The flashing lights of ships and planes are meant for detection. They are point light sources and have little competing background. They are intended to be detected at great distances and in absence of many other nearby flashes. Combined with them are the lights intended to indicate direction of travel. Although these lights may be seen from the ground or land by casual observers, the individuals having to observe and respond to these light signals are specially trained. Further, the amount of time available to make a decision and change the course of an aircraft or vessel is usually quite long compared to the time available in street and highway traffic. This decreases the importance of reaction time in marine and aviation situations as compared to vehicular traffic.

Emergency vehicle flashing lights are primarily meant to convey an emergency message, such as requesting right of way (F-42, F-43 and F-160—F-164) (A-3.8.2). These lights are used in an environment of other flashing lights, headlights, taillights, turn signals, and possibly highway flares. Often, a very quick response is required to avoid an accident, and therefore compound reaction time must be as short as possible. According to a National Bureau of Standards (N.I.S.T.) report on emergency lights,

> particularly because people become flustered in emergencies, it is important to keep the number of different warning light configurations that must be remembered to a minimum. Severely limiting the number of possible signals will tend to reduce the time required for a target driver to recognize any one of the signals and to recall the appropriate reaction.[2]

The philosophy of all the emergency services is to facilitate detection and establish right of way by using a number of flashing lights, blinking in an uncoordinated manner, and using a mix of colors. The appearance of motion is caused by lights oscillating, rotating, or having moving reflectors.

A flashing light is detected more quickly if all the other lights in the background are steady. But, the advantage is lost if even one background light is flashing (such as brake lights or other flashing emergency lights).[3] Furthermore, in dim light with good brightness contrast, the color of the signal light has little effect on reaction time.[4] Police agencies sometimes favor a blue light and ambulance volunteers often employ green. There seems to be little understanding that attempting to identify the type of service the responding vehicle represents by the use of specially colored lights only adds to the confusion of what the emergency message is and how the civilian driver should react. In reality, it doesn't matter which agency is approaching or operating at an accident scene, as long as the civilians provide a safe path and can drive safely around the incident.

In response to increasing accident rates, it was assumed that more flashing lights would help. The fire service, responding to the NFPA standard, increased the number of flashing lights on fire vehicles.[5] Yet we have

Figure 5.1 *This head-on shot shows what a driver might see in the rear view mirror of his automobile. The siren is mounted in the middle of the bumper and the air horns are mounted on the fenders. The red emergency lights are installed just inboard and slightly above the headlights. There is no center line light.*

found no research that defines the optimal number or type of lights for fire vehicles. Opinion articles appear from time to time and offer justifications for the increased number and placement of flashing lights on emergency vehicles. Also, flashing light manufacturers are represented on the voluntary standards committee for fire apparatus construction.

To see a flashing emergency light best, it needs to be viewed against a background without glare. Light bars mounted on roofs are often viewed against sky or sunlight and become washed out or invisible because of the high light levels of ambient illumination. No flashing light can rival the sun in energy output and the sun's glare markedly reduces their efficiency. In the same vein, flashing lights mounted near the headlights are similarly affected and at night the flashing red lights cannot easily be seen because of the overpowering glare. See Figures 5.1, 5.2 and 5.3.

A single flashing emergency light on the midline in front of the grill is removed from the headlights and therefore has maximum visibility. It is also more easily seen in the rear view mirror of the civilian automobile directly ahead.

Figure 5.2 *The illustration shows red lights just above the headlights. There is no center line emergency light.*

Figure 5.3 *In the center of the grill is a center line mounted emergency light. It is not near the headlights and is installed at a height to reflect in the windshield-mounted rearview mirror of a passenger car.*

Figure 5.4 *The intersection light in this example is mounted on the front bumper and is facing sideways. The light energy is produced by a strobe discharge.*

In an attempt to further improve detectability of emergency apparatus (without admitting that detectability is a problem), bright strobe lights have been added to the emergency vehicle lighting package. However,

> flashing signal lights are detected more quickly than steady signals when all other (background) lights are steady, but the advantage is lost if even one background light is flashing. If more than half of the background lights are flashing, the reaction time for a flashing signal is > 250 msec longer than the reaction time for a steady signal.[6]

An emergency vehicle presents many emergency lights which are used to impart different messages. It is not reasonable to assume that the unsuspecting public will know how and what to decode. By having several emergency vehicles in the same general location, with operating warning signals, the intended message will be further delayed in reception, understanding, and finally in reaction. (See Figure 5.4.)

The three types of ambulance designs showing the location of emergency lights are illustrated in Figures 5.5 through 5.7. It should be noted

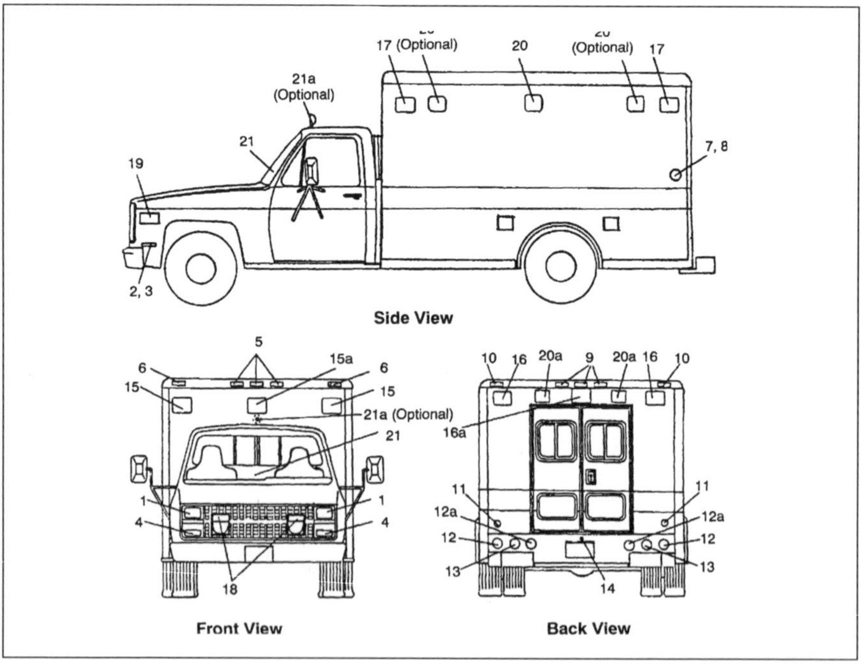

Figure 5.5 *Type I Ambulance (See the specifications in Appendix II.)*

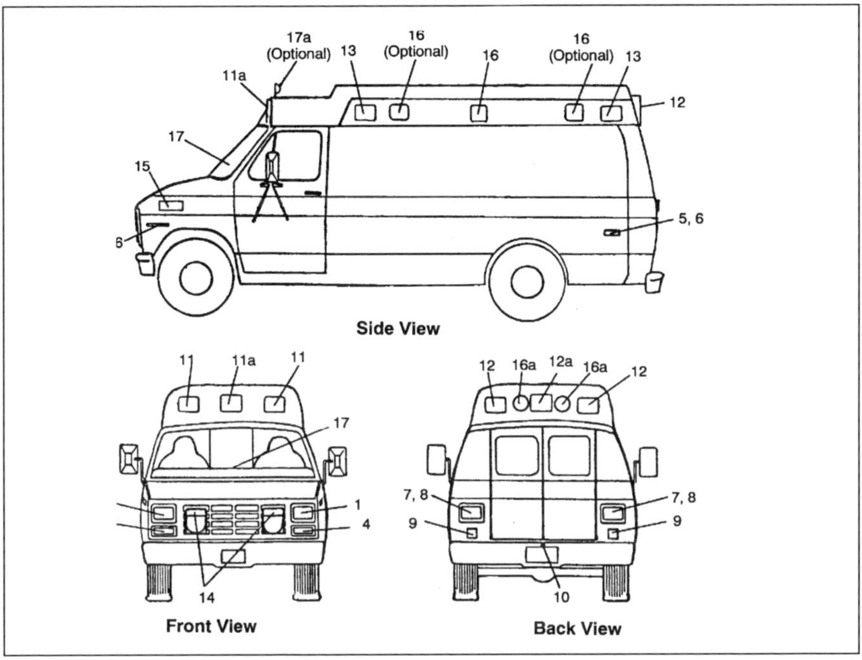

Figure 5.6 Type II Ambulance (See the specifications in Appendix II.)

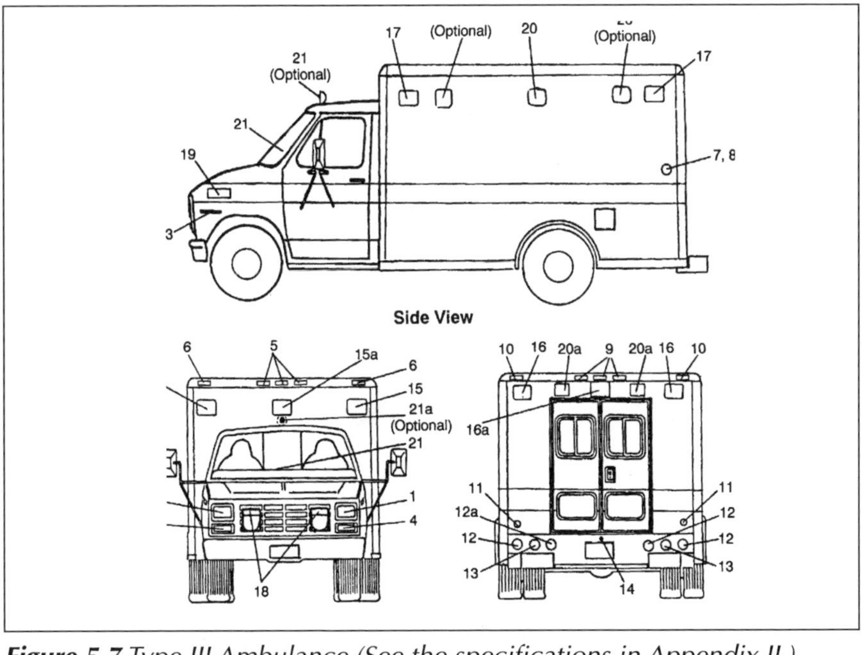

Figure 5.7 Type III Ambulance (See the specifications in Appendix II.)

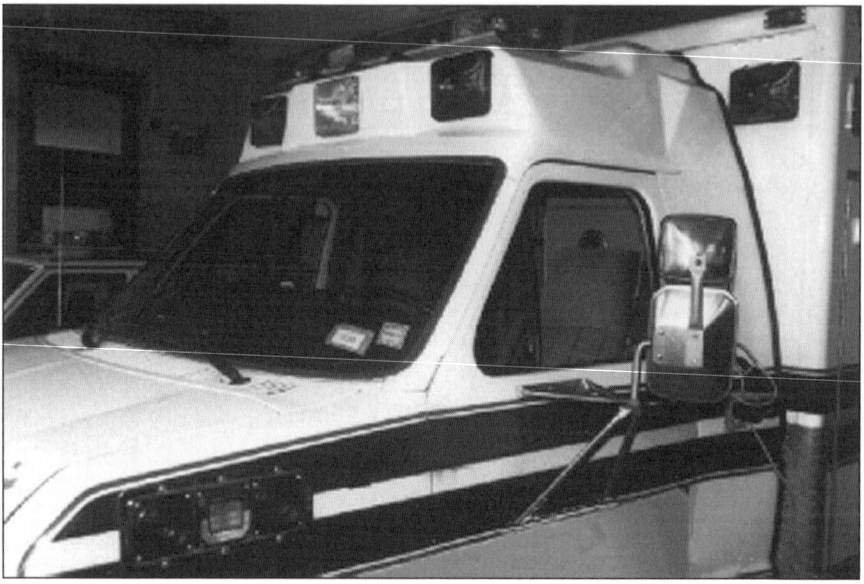

Figure 5.8 An ambulance with a roof-mounted light bar and rectangular flashers immediately below. Refer to the text for the discussion on an overabundance of emergency lights and the resulting decrease in warning signal potential.

Figure 5.9 A roof-mounted light bar illustrating the relationship between rotating bulbs/reflectors and mounted mirrors

Figure 5.10 *The only red lights below the roof are adjacent to the headlights. Roof lights and those high on the body of the vehicle are not visible to the vehicles directly in front of the emergency vehicle.*

Figure 5.11 *A centerline emergency light which is visible to traffic directly in front of the emergency vehicle*

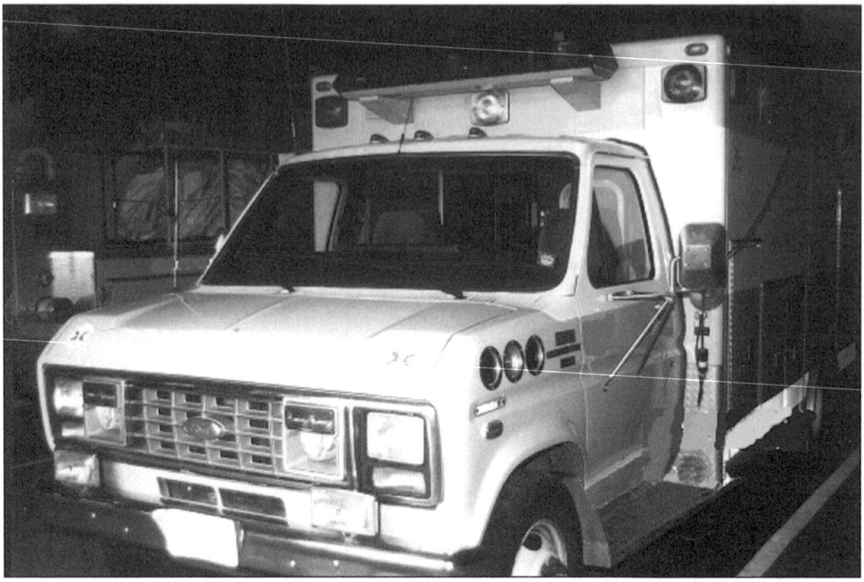

Figure 5.12 *An ambulance with a raised roof-bar, rectangular flashers, intersection lights in the fender, and grill lights mounted along with siren speakers just above the bumper*

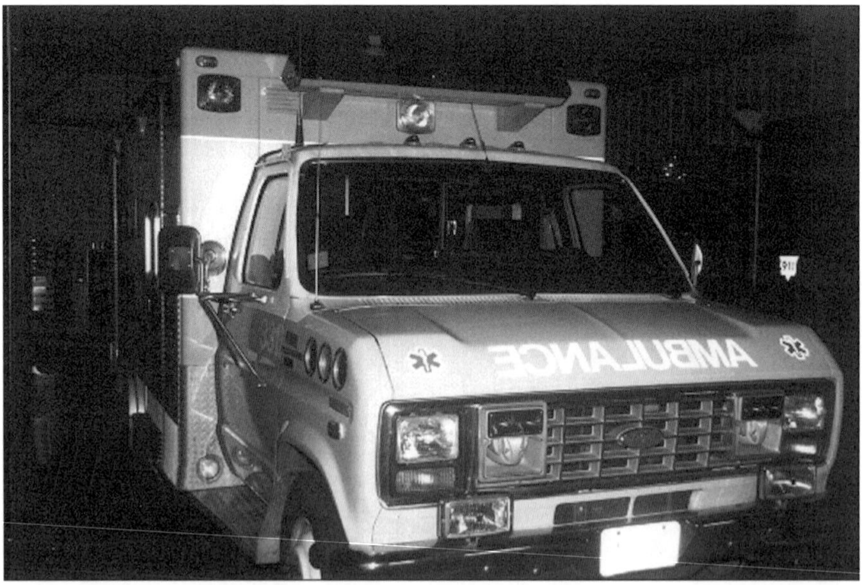

Figure 5.13 *An ambulance without a centerline emergency light*

that these drawings and the associated descriptions are from the KKK-A-1822D federal specifications. Also, see Figures 5.8 through 5.13.

While discussing emergency lights as a group, it may be helpful to review the differences in the construction of the bulbs themselves. Incandescent fixtures consist of a filament or wire in a partial internal vacuum containing an inert atmospheric gas. The filament is heated until it glows. There is a time lag starting with the application of electricity until maximum light output is reached and a similar time lag when the electrical circuit is interrupted and the filament both cools and dims. For that reason, a filament bulb cannot recycle with rapidity by itself. To try to make these lights attract more attention, the bulbs are often coupled with rotating mirrors or rotate themselves, all to impart more flash-like impulses.

Halogen bulbs are similar to incandescent but have an internal environment in which the natural mixture of atmospheric gases has been replaced by one of the halogen family of gases (fluorine, chlorine, bromine, or iodine). The halogen bulb, therefore, is pressurized. In an incandescent bulb, the high temperature which causes the tungsten filament to glow also causes the tungsten to evaporate and eventually coat the inner surface of the transparent glass container. After a period of time this coating decreases the transparency of the bulb and reduces its output. It can be seen as a blackened material on the inside surface of the bulb. The halogen bulb protects the inner surface of the glass by promoting the redepositing of the evaporated filament back onto the wire itself. For this to occur, the temperature must be much higher than the traditional incandescent light. The end result is a light source that provides a more consistent level of brightness during its life span.

"Strobe" is the shortened and modified nickname for stroboscope, which was initially an electronic device used for timing. It found some entertainment applications in "freeze action" views seen in discos of the seventies and eighties. As a light source, the device is also known as a gaseous-discharge flash lamp. It is not filamental, but rather employs electrodes separated by a space. The electrical discharge occurs in a specific gas-filled environment, and is not controlled by heating and cooling to establish a flash rate or frequency.

The signature flash of a strobe can be described as of very high intensity and extremely short duration. This pattern has some undesirable side effect effects.[7]

> The electrical activity of the human brain has certain characteristic frequency components. Most prominent among these components is the alpha rhythm, which has a frequency in the neighborhood of 10 Hz. It is found that when a human subject looks at a light flickering at a frequency of about 9–12 Hz, the alpha component of his brain waves is increased; this effect is known as photic driving. The subjective experience that accompanies photic driving is usually unpleasant, and may include dizziness, nausea, nervousness, and strong urges to escape the situation. The essential nature of the disease known as epilepsy is an instability of the brain's electrical activity. When epileptics are exposed to lights flickering near the frequency range of 9–12 Hz, it is common for the resultant photic driving to trigger an attack, including convulsions in the more serious cases.[8]

Neurologists commonly use strobe lights to precipitate seizures in patients to help diagnose the presence of light-sensitive epilepsy. *Given this, why use such lights on emergency vehicles?* It should be noted that a single strobe may have a flash rate below the danger range, yet when more than one strobe is mounted on a vehicle (which is very common), or when two or more strobe-equipped vehicles are near, then the combined effect of the two lights may precipitate photic (alpha) driving, seizures, or both. (Note: The flash intensity does not burn or damage the nerves at the back of the eye.)

Early conventional wisdom centered on the fact that if a flash rate was slow enough, then a seizure would not be precipitated. However, it has been found that a single, bright flash or even sudden exposure to sunlight may initiate a seizure in susceptible individuals.[9] The safest policy would be to assume any bright light that moves from a sudden, steady exposure to a rapidly flashing display has the potential of inducing a seizure in susceptible individuals. The brightest flashes have the strongest capability of causing this, and strobes are the instruments most commonly associated with the brightest outputs.

Some individuals feel there is a social stigma associated with seizure-related disorders and fail to submit to diagnosis and testing. Others do not seek treatment because it may mean the suspension of their driver's license. Therefore, some persons remain prone to seizures but are undiagnosed. Any person with a history of brief "memory lapses" or "fainting spells" should be suspected of having more serious problems, and an investigator of any accident in which such a person is a participant should consider this segment of that person's medical history. A recent onset of seizures in a previously unaffected individual is a sign of potentially more serious medical problems and should be evaluated by a neurologist.

A strobe's flash duration is so brief that often an unsuspecting observer may not at first detect any signal, and then, because that signal is in the peripheral and not the central field of view, he cannot locate the source of the flash. In other words, many flashes have to occur in order for an unsuspecting operator to detect, locate, identify, and finally, react. The end result is an increase in reaction time.

One significant goal of flashing lights is to garner attention. Once that is done, the driver must be given enough information from which he or she can locate the emergency vehicle. This includes its distance from the driver, where on the roadway the vehicle is operating, or how far off the roadway on the shoulder it is located. All of these factors aid in the judgment process of braking and maneuvering.

During the daytime, judgments of depth and distance are less complicated because the flashing lights are relatively diminished in brightness and more peripheral clues are present. At night, especially in locations where there are no street lights, many of the scene clues are absent, making depth and location judgments less reliable. In this environment, if a flashing light is introduced, it will draw the approaching motorist's vision toward it, first by stimulating the periphery and then the central field. Strobe flashes are of such a short duration that fixation on them quickly is very difficult. Aside from decreased reaction times already discussed, perception of depth and localization is markedly hindered. It is difficult to find where the emergency vehicle is. Adding other flashers or strobes further confounds the driver's judgment

A commonsense approach to using warning lights, especially at night, would be in part to use a slowly rotating beacon. Since the beacon does not blink or flash, it provides a constant visual signal. The rotation imparts motion and separates it from the lights of normal traffic, while the beams of horizontal light emanating from the fixture help the observer locate the source. Realize that like a light house, even when the beam is not directed toward the observer, it can still be seen and located with a minimum of time and perception. As mentioned earlier, flashing lights mounted near the headlights on the front of emergency vehicles are often diminished in effect, because the output of the headlights is much greater than that of the warning lights. The glare of the white lights masks the red lights from being quickly seen. The result, especially at night, is an inability to attract attention.

Unlike red lights, an amber display does not stimulate extreme interest, given that the public has been conditioned to associate red with an accident or emergency. Therefore, although red emergency lights may be used to request a clear path to an emergency, after the emergency vehicle is parked at the scene, amber should be displayed to provide warning of a hazard without attracting undue attention. The other lights should be switched off. Emergency services using this approach operate in a less dangerous mode.

There are other side effects which must also be considered. When a driver approaches the scene at night and his or her eyes are adapted to nighttime driving, the brightness of the flash causes the retina to de-adapt to night vision. Since readaptation is much slower than de-adaptation, the driver is then proceeding from the scene with a reduced visual function. This can contribute to nighttime accidents beyond the immediate scene. An example of dark adaptation is what occurs when you enter a darkened theater. It may take up to fifteen minutes to identify the number of fellow theater-goers surrounding you. De-adaptation is similar to going outside into daylight from that theater. After a brief moment of discomfort, the average individual can quickly see with accuracy. Any very bright light can cause this, and the strobe has an extremely bright flash.

The emergency services offer the defense that they use strobes attract attention and the other lights for imparting a message. However, coupling messages together causes an increase in reaction time and often the message or flashing pattern is lost when mixed with other flashing lights.[11] Anything that lengthens reaction time increases the chances of an unwanted event. Flashing lights are thought by emergency services providers to attract attention or convey information, neither of which may be entirely true. Their operation can have other unwanted effects. Because bright flashes act as a visual stimulus, some drivers have a tendency to continue to look at them after the lights attract their attention. As a consequence, they steer where they are looking. This attraction has been referred to as hypnotic, like a moth being drawn to a flame.[12] This appears to be a good explanation as to a number of the accidents involving parked emergency vehicles.

The underlying message about flashing emergency lights is that there is a wide range of opinion pertaining to how they function and the assumed aura of safety they convey. However, there is precious little *bona fide* research to support the sale and use of the ever expanding numbers and intensities of these lights on emergency vehicles. While some emergency warning lights are necessary, there should be a limit to the number placed on a vehicle. The available data suggests that emergency vehicles be equipped with fewer lights which flash less rapidly and less brightly, and which convey a minimum number of messages.

It is wise to consider emergency responses within the context of the research by Hunt, et al., who have concluded that ambulance responses using red lights and sirens saved very little time when compared to similar runs without red lights and sirens. "The 43.5 second mean time savings does {*sic*} not warrant the use of lights and siren during ambulance transport, except in rare situations or clinical circumstances."[13]

Endnotes

1. Vos, J.J. and A. Van Meeteren, "Visual processes involved in seeing flashes." Presented at International Symposium, Imperial College, London, *The perception and application of flashing lights*, (Toronto: University of Toronto Press, 1971), 3.

2. Howett, G.L., et al., *Emergency vehicle warning lights: State of the art*, (Washington D.C.: National Bureau of Standards, publication 480-16, 1978), 5.

3. Boff, K.R.and J.E. Lincoln, "Visual warning signals: Effect of flashing." *Engineering data compendium: Human perception and performance*, (AAMRL, Wright-Patterson AFB 1988), 2408.

4. *Ibid.* at 2398.

5. National Fire Protection Association, Standard 1901, 22 (Quincy, Ma., 1996).

6. Boff, *op. cit.*, at 2408.

7. Brewer, F.C., "Aircraft flashing lights—A summary of Canadian forces investigations," Presented at Imperial College, London, *The perception and application of flashing lights*, (Toronto: University of Toronto Press, 1971), 284.

8. Howett, G.L., et al., *op. cit.* at 57.

9. Livingston, S., M.D., *Living with Epileptic Seizures*, (Charles C. Thomas 1963), 53.

10. Boff, *op. cit.*, at 2408.

11. Howett, *op. cit.*

12. Younger, J.D., "The moth and how to beat it," *Car and Travel*, (Garden City, New York American Automobile Association), Oct. 1997, 16.

13. Hunt, R.C. et al., "Is ambulance transport time with lights and sirens faster than that without?" *Annals of Emergency Medicine*, (American College of Emergency Physicians), April 1995, 507.

Chapter 6

Sirens and Audible Devices*

Stephen S. Solomon, B.S., O.D.

Sirens and other audible devices generate sounds of various frequencies, intensities, and patterns. The general assumption is that audible warnings alert others sharing the roadway and pedestrians in proximity to it. Other drivers should be able to hear the warnings soon enough to react and give way to the emergency vehicle. However, for the siren to be heard inside the automobile, it must project into the vehicle at a sound level greater than the existing noise.

All devices that generate sound share common shortcomings. Chief among them is the directional nature of sounds. In other words, the sounds bounce or echo from external surfaces. While the full spectrum of light is reflected from a polished surface, it does not bounce from reflector to reflector unless they are carefully aligned. A color is reflected from a surface hue directly to the eyes and is perceived, but these color displays do not ricochet from panel to panel. By contrast, sound reflects from wall to wall with ease. The result is that the location of a sound source is often identified as the position of the last reflector encountered by the sound before entering the ears, which may have no direct correlation to the location of the originating source. If the individual searches for the sound's origin, valuable reaction time is lost and the probability of an accident increases. It is very difficult and time consuming to accurately identify the location of a siren, bell, or air horn. Emergency vehicle operators often develop an attitude, based on their experiences, that others are just as aware as the operator of the vehicle's sirens and horns. However, while most individu-

* The appendices of the book coordinate with this chapter. References to fire apparatus will start with "F-", and are printed near the left margin of Appendix I. they direct the reader to the fire apparatus specifications. References to ambulances start with "A-" and are the paragraph and section number in the left margin of Appendix II. They relate to the federal KKK-A-1822D specifications.

als do eventually hear and recognize them, the question is how soon they are detected.

Modern vehicles are replete with systems that either filter out noises, provide entertaining distractions, or devices that require auditory attention. They compete with sounds from outside the vehicle, some unwanted and some alerting in nature. Unfortunately, there is little distinction between distracting, entertaining, warning, and external noise. Automobile manufacturers regularly advertise the insulating qualities of their products and demonstrate how effectively they provide a quiet ride. Air conditioning adds an internal background sound that further masks outside sound. Within the environment of the modern vehicle are radios, stereos, tape players, and disc players. Recently, cellular telephones have become commonplace, and with them come another level of distraction. Not only is sound involved, but two-way conversation is required. This higher level of attention requires a shift from driving skills to conversation and decision skills. All this subtracts from attentiveness to driving and increases the compound reaction time.

As mentioned earlier, a siren will not be heard inside an automobile unless its intensity, measured in decibels, exceeds that of the other competing sounds in the passenger compartment. If an emergency vehicle siren is activated, it can produce about 115 decibels of sound, measured 50 feet in front of the siren. Inside an automobile with the windows closed, that intensity is reduced to 82 decibels. With the siren at 200 feet from that automobile, the inside sound intensity was 63 decibels. By comparison, a sample car, traveling at 35 miles per hour, without the air conditioner operating and the radio off, had an internal noise level of 56 decibels. At the same speed, but with the air conditioner on high, the noise level was 63 decibels. Operating the radio and air conditioner simultaneously could generate sound levels as high as 90 decibels. In this example, the siren would not be heard even in extremely close proximity, for in order to be detected, the siren would have to produce a higher decibel intensity in the cab than that generated by road sounds, radio, and air conditioning.[1] (See Tables 6.1–6.3.) Also, if a driver has hearing loss, then that must be factored into the situation.

We are an aging population. Our senses dull with time, and hearing is not excluded. Most of us recognize the effect of aging on vision by either wearing bifocals or reading glasses or knowing individuals who do. These

Table 6.1
Sound Intensity of Wailing Siren as Measured inside an Automobile with the Windows Closed and the Engine Off*

Distance of siren to auto	Sound intensity (db)	Siren-to-auto angle
50 feet	82.2	0 degrees
100 feet	79.2	0 degrees
200 feet	63.2	0 degrees
300 feet	53.7	0 degrees

Table 6.2
Same as Table 6.1 but with the Siren at 45 Degrees (Intersection) to the Automobile

Distance of siren to auto	Sound intensity (db)	Siren-to-auto angle
100 feet	57.2	45 degrees
200 feet	55	45 degrees
100 feet	61.7	315 degrees

The "yelp" sound is slightly less effective than is the "wail" function at all angles (0–315 degrees).

Table 6.3
Sound Levels inside 1979 Automobile (Oldsmobile) at 35 mph

Air conditioner	Radio	Internal sound level
Not in use	Not in use	56 db
Maximum	Not in use	63 db
Maximum	Volume low	72 db
Maximum	Volume higher (not max)	90 db or more

The information in Tables 6.1 through 6.3 is adapted from the *Student Workbook* of the New York State Department of Health, Emergency Medical Services Program—Ambulance Accident Prevention Seminar 12–88.

sensory-assisting devices are readily accepted by almost all segments of modern society. Yet hearing loss is another story. Hearing aids have not yet reached a comparable level of acceptance, even though auditory impairment can also be associated with aging. Many drivers in our communities are hearing impaired and sirens or other audible devices are more difficult for them to detect. Drivers with many years of experience often unknowingly become hearing-impaired and are not aware of it. When sensory ability, either vision or hearing, is compromised, reaction times increase.

As reported in fire service literature, by the time a siren can be detected by an automobile operator, it is often too late to stop at an intersection to avoid an accident.[2,3] There is now some recognition that sirens and other acoustic warning devices do not provide the early warning once assumed. However, emergency vehicle operators still heavily rely on them. As has been discussed elsewhere in this text, expectation can lead to a presumed reaction or performance. If emergency vehicle operators assume civilian drivers will respond by providing right of way or safe passage as they normally do, then they will eventually be surprised by the driver who does not operate according to previously reinforced expectations. Because the entire crew on the emergency vehicle can always hear the audible devices very clearly, it is difficult for them to imagine that other users of the road can not detect the sound at great distances. Also, because the source of the sound is on the emergency vehicle, it is difficult for the crew to remember that the sounds are nondirectional. Emergency vehicle drivers are trained to exercise the awareness and caution needed when in the proximity of other drivers who occasionally seem to react suddenly and in an unpredictable manner. Accidents still occur as a result of unexpected responses, whether drivers are surprised or only responding to the best of their abilities.

Pseudoscientific information available to the fire service often prevents these problems from being solved. Recently, in one fire journal, a fire maintenance spokesperson stated in an article that sirens of the older style (please read "traditional") were easier to locate than the newer, transistorized models.[4] This probably struck a harmonious chord with a segment of his audience. Because the fire service has no scientific body to evaluate such things, there was no published rebuttal. Consequently, committees planning equipment purchases will be exposed to this false information and, because it was reported in a print medium, it will be accepted as truth. Sirens are available in traditional mechanical and electronic versions. With the latter, a control panel in the driver's reach allows for a selection of assorted siren sounds so an auditory choice is available. There does not appear to be any solid, scientific evidence of which sound is best and some states may not legally recognize the use of some electronically generated siren sounds (Figures 6.1–6.4).

It can now be understood how different audiences exposed to siren sounds have distinctly different interpretations of siren and audible device function and their efficiency.

Hearing losses in the fire service have been discussed and studied. The conclusion was perhaps siren and horn sounds contributed to these hearing losses. In response to this problem, the National Fire Protection Association, through its committee on fire apparatus, has recommended that sirens and air horns be removed from roofs of vehicles and placed at the front of the vehicles.[5] (F-166, F-167, F-169, F-6, A-3.14.6) (Figure 6.4)

Figure 6.1 *The siren speakers are mounted on the front bumper. In this case, the emergency lights are mounted as part of the shell of the speakers. The headlights are in close proximity to the emergency lights and there is no centerline emergency light.*

Figure 6.2 *The siren is mounted in the bumper and a midline flashing light has been special ordered and installed.*

Figure 6.3 *This presents a combination of auditory devices. There is an electronic siren mounted in the bumper, an electromechanical siren mounted on the bumper, and a large bell. (The bell is for traditional reasons and is no longer an emergency warning device).*

Figure 6.4 *An electromechanical siren is mounted on the bumper and two air horns are built into the bumper.*

By following this approach, the siren's effectiveness may be further diminished by having the sound waves deflected by objects ahead, such as other vehicles. Police cars often have sirens of similar size also mounted on the roofs, yet to date the police agencies have not reported officers' hearing loss. A side issue may be discovering whether firefighters have hearing losses from other job-related influences.

Sirens have detrimental side effects, including increased muscle tension, (which can cause the emergency vehicle driver to further depress the accelerator), decreased peripheral vision, and increased adrenaline output.[6] Experience, perceived seriousness of the call, length of service, fatigue, and other personal factors can affect an emergency vehicle driver's response to his own audible devices. It has been suggested that, psychologically, the presence of the siren and horn sounds in the cab may actually cause emergency vehicle drivers to go faster. While emergency vehicle operator training is assumed to help solve this problem, no research has yet identified whether that is true. The urgency conveyed by the auditory devices operating can affect the emergency vehicle driver, who slightly accelerates more and more. A cascade affect can occur when a motorist immediately ahead pulls to the right to allow passage, but does not sufficiently slow down. To get past, the emergency vehicle must speed up. Sometimes, an emergency vehicle does not slow down after getting around the last car and the cycle has the possibility of repeating.

Pedestrians observing emergency vehicles approach them often estimate the speed of that vehicle to be much faster than if it were approaching without audible devices. The testimony of untrained observers is suspect in this situation. Pedestrians may also be affected by the presence of auditory warning sounds in that determining the direction of approach can be questionable. Pedestrians have been known to step off the curb into the path of emergency vehicles. Alcohol consumption, drug use, hearing impairment, medications, and the possible combination of several of these factors need to be considered.

Patients being transported in ambulances can be adversely affected by audible device sounds penetrating the treatment compartment.[7] Many are already stressed or anxious. Those with heart-related problems are among the most susceptible in this environment. By stabilizing the patient before transporting him, a nonemergency run can be made with greater comfort and less anxiety for the patient.

Dispatching is the function of the agency receiving the report of a problem and involves alerting the appropriate emergency service. While most individuals may have some understanding of how this occurs (usually from watching television or movies), what is not evident is the type of training the dispatcher possesses. Before notifying the emergency service or services required, the dispatchers gather the basic information. Some of this may be automatically obtained from an enhanced 911 system, while more specific issues must be queried. If a dispatcher is trained to determine the level of urgency, then an appropriately guided response can be initiated. A child with a fever who needs to be transported to a medical facility for an evaluation may not require a response with red lights and sirens. However, if someone has uncontrolled bleeding or respiratory arrest, a more aggressive approach may be appropriate.[8] Therefore, the sorting or grading of the calls by the dispatching agency can be extremely important. The more an emergency vehicle can operate in the regular flow of traffic in nonemergency mode, the safer it will be, both for the crew and the others on the shared roadway.

Endnotes

1. Mechanical Engineering, staff report, Feb 1986, pp. 38–39.

2. Hatton, L., "What fire truck?" *Fire Engineering*, (Nov. 1986) 139:58.

3. New York State Department of Health, Student Workbook, Ambulance accident prevention seminar, circa 1989, p. 75.

4. O'Connell, K. "Electronic sirens, a failed experiment from the 1960s," *American Fire Journal*, Feb. 1996: 28–29.

5. National Fire Protection Association, *Standard for automotive fire apparatus*, (Quincy, Ma., 1901, 1996 ed.) p. 25, para 9-9.

6. New York State department of Health, *op. cit.*, p. 76.

7. *Ibid.*

8. Operational update 88-20, New York State Department of Health, Sample standard., Oct 1988, p. 1.

Chapter 7

Why Fire Vehicle Color Is a Safety Issue*

Stephen S. Solomon, B.S., O.D., and James G. King

7.1 Vehicle Color and Safety

Safety is a dominant issue within the fire service. It receives constant attention from federal, state, and local officials as well as from the national fire service organizations. Safety recommendations that bear on all aspect of apparatus safety have been discussed, composed and published. Driver training and emergency vehicle operator courses are in active use. Driver candidates often undergo physical examinations and reviews of driving records. The NFPA periodically publishes updated standards for fire apparatus.

However, a disproportionate number of fire vehicle accidents continue to occur. An estimated 14,670 fire vehicle accidents happened in 1995, with 950 firefighter injuries.[1] Fire service records do not report the number of civilians injured or killed in these accidents, but a conservative estimate is that 1000 civilians were injured or killed in 1995. The *Journal of Safety Research*, a National Safety Council publication, estimates that accidents involving red or red-and-white fire vehicles occur at a rate of 62.1 accidents per million miles.[2] By comparison, the rate of municipal and commercial truck accidents is 21.6 accidents per million miles.[3]

The fire service has long relied on audible devices, flashing lights, and retroreflective material to make a vehicle's presence noticed. Yet, fire service research has pointed out that audible devices are nondirectional and are diminished by sound-proofing materials and internal noise in the passenger compartments of automobiles.[4]

* The appendices of the book coordinate with this chapter. References to fire apparatus start with "F-", and are printed near the left margin of Appendix I. They direct the reader to the fire apparatus specifications. References to ambulances will start with "A-" and are the paragraph and section number in the left margin of Appendix II. They relate to the federal KKK-A-1822D specifications.

Flashing lights lose effectiveness when viewed against a scene filled with other flashing lights. Examples include a roadway crowded with brake lights, flashing turn signals, headlights,[5] roadside commercial displays, and seasonal decorations. "The advantage of a flashing signal is lost if even one irrelevant light is flashing."[6] In bright daylight emergency lights fade and the proper signal is lost.[7]

What goes unsaid is that what the firefighter hears and sees is not directly transferred to what a civilian motorist hears and sees. All too often the firefighter hears his sirens and air horns and cannot imagine anyone else not hearing them. Often he views a flashing or blinking light at great distances and can only reason that flashing lights are highly visible at any distance. Photo and TV images often show retroreflective stripes appearing to glow, but they don't. The safety decisions made about lights, sirens, and retroreflective tape are based to some extent upon what firefighters presume these things do, not on how they really function. Because the fire service is very focused on their equipment and their role in a community, they have a tendency to overrate the interest and awareness the general population has toward the fire service.

Some emergency lights are mounted relatively low on the corners of emergency vehicles and are presumed to facilitate detection at intersections; but, they are often completely blocked by other traffic parallel to the fire vehicle. By the time a motorist sees these lights, he may be unable to avoid an accident.

Retroreflective stripes have been added (F-188 through F-190)(A-3.16.4), yet they do not add any visibility during daytime when 80 percent of the fire-response accidents occur. Much like the "intersection" lights, they are screened by parallel traffic. We are strong advocates of the use of retroreflective material, but we hasten to add that it may be more effective on a stationary vehicle and primarily at night.

For years, the response to higher accident rates was to buy louder sirens and brighter lights. We have succeeded in three things: (1) a greater risk of hearing loss in firefighters (the NFPA has stated in its standard that sirens and air horns be moved off the vehicle's roof to help protect against hearing loss; however there is some debate as to what is responsible for the hearing loss); (2) making vehicles less rather than more visible because the ability to discern the warning message decreases as the number of lights on an emergency vehicle increases; (3) driving the cost of the vehicle upward without a proven decrease in accidents. After all, if the number of fire vehicle accidents were stable or declining, we would not be increasing the number of warning devices on the vehicles.

7.2 How to Improve Visibility

Color's role in accident prevention is keyed to vision as our main sense: whatever is most visible receives our first response. The more quickly we detect a hazard, the more time we have to avoid it. Thus, the ability to quickly detect the presence of a fire vehicle decreases the chances of an accident. For years, flashing lights, audible devices, and retroreflective stripes have been employed to enhance the visibility of fire vehicles. However, most recent fire apparatus safety recommendations do not mention vehicle color. Yet, in terms of published scientific and fire service research, vehicle color has received more attention than lights, sirens, and air horns as a major factor in accident prevention.

Under normal circumstances, all people have difficulty seeing red. During daytime, it is one of the most difficult colors to see.[8] At night, in the dark-adapted state, everyone is practically "red blind."[9] However, the range of yellow colors from greenish-yellow to yellowish-green are the most easily detected, day and night.[10]

More than 8 percent of the male population is color blind.[11] While most of these individuals have trouble seeing red and green, they detect yellows more readily.[12] Color-blind drivers have more accidents than similar driving groups with normal color vision.[13]

Peripheral or off-center vision is most often responsible for early detection. Lime-yellow is seen significantly faster than red in the peripheral view. (Figure 7.1)

As mentioned in a previous chapter, we see by contrasts against background. Color and shape are features of contrast. According to a NASA study, these bright colors, which are rarely present in nature, attract the most attention: they stand out.[14] The average background in both rural and municipal settings has been judged to be dark,[15] and against these dark backgrounds, a light object is more easily seen than a dark one.[16] The dark colors include red and black; the light colors, yellow and white. When colors and outlines are not contrasting, they are camouflaging. Silhouette (shape) also is a consideration in detection. The straight lines of architecture and highway designs act to decrease detection of vehicles which are

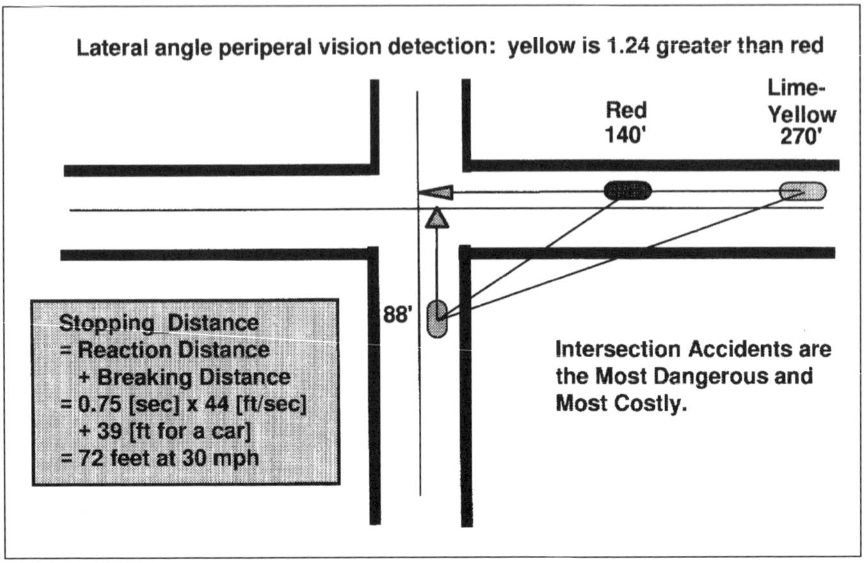

Figure 7.1 *Lateral angle peripheral view*

also manufactured and marked with straight lines. The fire service has attempted to make vehicles more visible by painting vehicles in two-tones, using straight line patterns—most notably red and white. The thinking is that because solid red is not visible, both tradition and visibility can be served by using two popular colors. Two-tone red and white does *not* make a vehicle more visible but rather acts to camouflage that vehicle by breaking up its silhouette. The National Safety Council has published a ranking of the relative visibility of two-tone colors. Red-and-white placed tenth out of twelve pairs.[17]

A visibility accident is one in which at least one of the drivers does not receive sufficient visual information, and causes an accident. For instance, if a civilian auto driver does not see a fire vehicle and the fire vehicle operator performs a defensive reaction which leads to a fire vehicle rollover, it is a visibility accident. The actual accident involved only one vehicle, yet there was an interaction between at least two.

When differently colored groups of fire vehicles, both equally equipped, with drivers similarly trained, are analyzed for accidents, the lime-yellow vehicles are involved in fewer and less severe accidents than are red or red/white two-toned ones. In those accidents that did occur, the damages were less and injuries were not as serious with lime-yellow. (Figure 7.2)

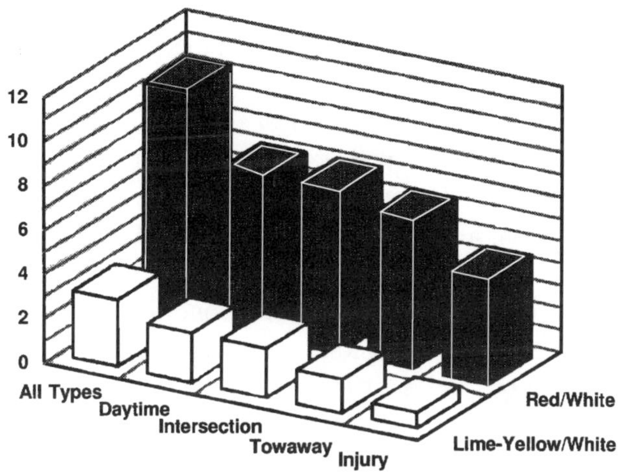

Lime-yellow/white fire engines are safer than red/white in all accident categories

Figure 7.2 *Accidents per 100,000*

Intersection accidents are usually visibility accidents. Data indicate that intersection accidents comprise about seven out of ten fire apparatus accidents.[18] It is reported that intersection accidents are usually the most costly, both in damage and injury. However, this means three out of ten fire apparatus accidents occur away from intersections. Neither of those categories can be ignored in accident prevention.

A study published in 1984 compared intersection accidents in nine major cities. In that report, red vehicles were involved in twice as many accidents as lime-yellow ones.[19] Why does vehicle color play a prominent role in fire vehicle accident reduction? It can, in part, be understood by the steps (reaction time) a driver follows to avoid an accident:[20]

1. **Detection**—something is in or approaching the path of travel.
2. **Identification**—is it a hazard or merely sharing the roadway (note: identification does not mean deciding whether it is a fire truck but rather deciding whether a hazard exists to the driver).
3. **Decision**—what should be done?
4. **Action**—introduction of an appropriate driving maneuver.

The faster the first step, detection, takes place, the more quickly the subsequent steps can occur. Detection is a function of visibility.

These four steps comprise the major components of reaction time calculations. In the past, reaction time was assumed to be 0.75 seconds and that value was often used to determine braking and stopping distances for vehicles. That value was not a practical or accurate number, yet many safety judgments are based upon it. A more realistic estimate takes into account a wide variety of factors, including the driver's age, vision, hearing, health, nutrition, alertness, time of day, speed, highway design and construction, weather conditions, and internal vehicle distractions (for example: cell phone use, radio, stereo, air conditioning, noisy children.) Therefore, the use of the 0.75 second value is suspect and it should not be used for computations unless it is very carefully documented. It should never be used routinely. (See Chapter 13.)

The mere knowledge that an event will occur is enough of a clue to improve reaction time: if a driver is warned and expects to encounter a hazard, he or she will react more quickly. Computations using the 0.75 second reaction time will significantly underestimate the total stopping distance in real life situations. Fire service literature contains apparatus safety advice that uses the ideal rather than a realistic reaction time.

It bears repeating: *people will react first to that which is most visible and color plays a central role in allowing us to detect objects more quickly, thus allowing more time to react to and avoid a hazard.* With that in mind, DuPont recently measured fire department red (DuPont Imron 674) and lime-yellow (DuPont Imron 7744-1) paints. This was done under two conditions, daylight (Figure 7.3) and nighttime (Figure 7.4), because our eyes function differently at night.

This new DuPont information indicates that lime-yellow has a peak eye response (grabs the attention) approximately 4.9 times greater than fire department red during daylight and approximately 93 times greater at night.[21] Red is practically invisible. Within a paint type, one fire truck color is no more expensive than another (F-68, F-189 through F-191) (A-3.16.2).

Firefighters often argue that if a fire truck is not red, people won't recognize it. As alluded to above, recognition is not the issue. Research at the University of Michigan Traffic Research Institute indicates that drivers do not identify a fire truck and then decide to avoid it, but rather our survival sense demands that we avoid any hazard. It does not have to be identified as a fire truck or any other specific hazard, but only that it needs to be

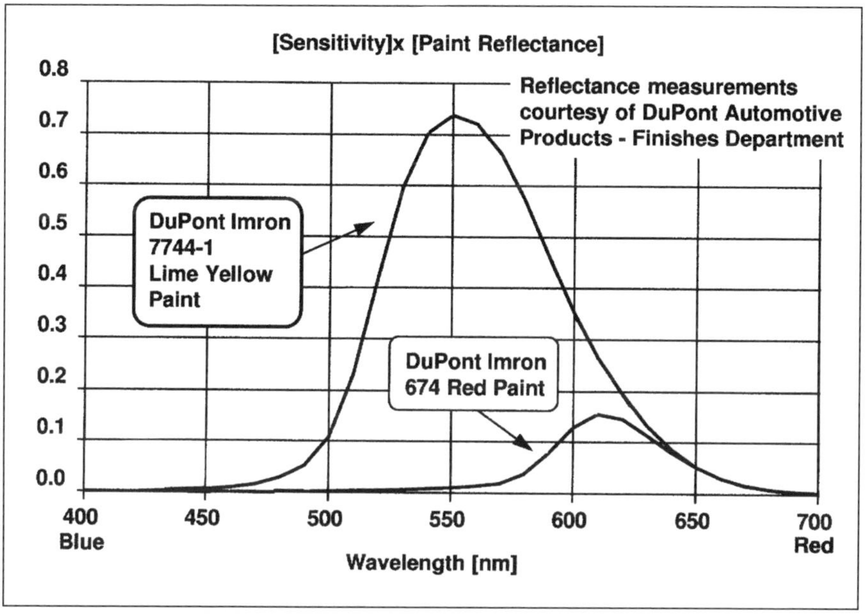

Figure 7.3 Daytime eye response

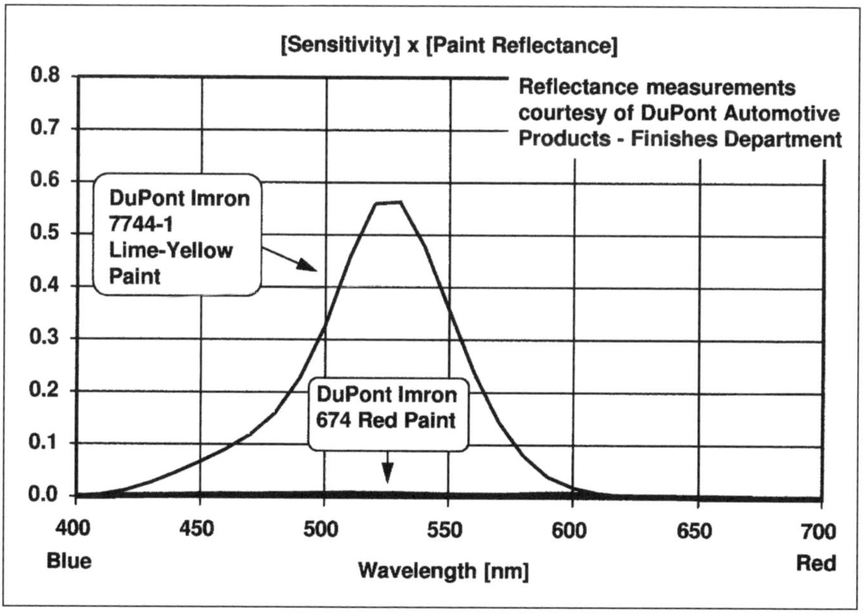

Figure 7.4 Nighttime eye response

avoided. Many cities have never had red fire apparatus, and they do not suffer the expense of delayed response or increased accidents.

Red may be emotionally important to some firefighters but it is not a safe color. The fire service preaches safety and prevention, yet falls back to an unsafe and outdated tradition as a symbol of the profession.

The message to the fire service should be quite clear. Even if there is a strong sense of tradition, it should be placed into the proper perspective. Not only do firefighters risk their own lives but also those of civilians in the path of response. Supporting the use of red is a popular position for leadership to adopt. Simultaneously, it is a position not supported by common sense and research.

Firefighters seem to select apparatus color based on personal tastes rather than what is best for the community. Red is a color that is not visible, is poorly detected, and has an artificially assigned safety image. The fire service cannot continue to turn a blind eye to a safer, cost-neutral alternative, especially since they have a mandate to protect others as well as themselves. The prevention of even one accident is reason enough to use lime-yellow (F-187).

Endnotes

1. Karter, M.J., Jr. and LeBlanc, P.R., "U.S. firefighter injuries—1995," *National Fire Protection Journal*, (November/December, 1996) 90.

2. Solomon, S.S. and King J.G., "Influence of Color on Fire Vehicle Accidents," *Journal of Safety Research*, (Spring 1995), 41–48.

3. National Safety Council, Accident Facts, (1991 edition), 84.

4. Hatton, L.F. "What fire truck?" *Fire Engineering* (Nov. 1986) 139:58.

5. Hargroves, R.A. "A survey of the use of flashing lights on roads and road vehicles," presented at the international Symposium, Imperial College London, *The perception and application of flashing lights*, (Toronto: University of Toronto Press, 1971) 183.

6. Boff R. and Lincoln, J.E., *Engineering Data Compendium: Human Perception and Performance*, (AAMRL, Wright-Patterson Air Force Base, OH, 1988), 2408–09.

7. *Ibid.,* 2398.

8. Southall, J.P., *Introduction to Physiological Optics*, (NY: Dover Publications, 1961), 273.

9. *Ibid.,* 273.

10. *Ibid.*

11. Allen, M.J., *Vision and Highway Safety*, (Philadelphia: Chilton Book Company, 1970), 88.

12. Lahr, L.E. and Heinsen, A.C., "A field study of the relative visibility of various colors," *California Fish and Game Quarterly* (1959) 45:208–209.

13. Cox, J.J., "Viewing of railway flashing light signals," presented at the International Symposium, Imperial College, London, *The perception and application of flashing lights*, (Toronto:University of Toronto Press, 1971), 196.

14. Beach, L.R. et al., *The human factors of color in environmental design: A critical review,* (NASA Grant No. 2-404, 1988), 10.

15. Allen, M.J., *Vision and Highway Safety*, (Philadelphia: Chilton Book Company, 1970), 139, 174, 179. An expanded version of this book has been published as Allen, M.J., et al., *Forensic Aspects of Vision and Highway Safety* (Tucson: Lawyers & Judges Publishing Co., 1996).

16. *Ibid.*

17. Nathan, R.A., "What's the safest color for a motor vehicle?" *Traffic Safety*, (Sept. 1969) 69:13.

18. New York State Department of Motor Vehicles, *Summary of motor vehicle accidents involving fire vehicles (in) 1988*, (MV-144A, 1990), 2.

19. Solomon, S.S., "The safety color," *Firehouse* (June 1984) 9:106.

20. Olson, P.L., *Forensic Aspects of Driver Perception and Response*. (Tucson: Lawyers & Judges Publishing Co., 1996), 198.

21. Private Correspondence. Reflectance Measurements. Courtesy of DuPont Automotive Products—Finishes Department, David H. Alman, Ph.D. Private correspondence of 20 October 1995. Used with permission.

Chapter 8

Flourescent, Retroreflective and Phosphorescent Materials*

Stephen S. Solomon, B.S., O.D.

8.1 Fluorescence

Emergency vehicles use visual devices in an attempt to improve conspicuousness, which in turn contributes to safer operations. While flashing lights come quickly to mind, there are other choices also available to assist in promoting visibility. Fluorescent material converts light energy, often ultraviolet, into light energy of a different color or wavelength. Fluorescent products are usually brighter than their comparable colors in nonfluorescent products.[2] In a recent study it was found that against a multicolored, spring foliage background, fluorescent colors were more quickly detected than their nonfluorescent namesakes.[3] And when compared as a group, fluorescent lime yellow (yellow-green) was the most detectable.

Since fluorescent material needs natural light energy, it is only active during daylight hours.[5] During the periods of the day in early morning and late afternoon when visibility is compromised by low visible light levels, fluorescent brightness is an asset. The same holds for hazy conditions. However, no matter how bright a fluorescent color is during daytime, it does not radiate or light up at night. Education is a key factor of safety and users of these materials must be taught about the loss of protective qualities after sunset. Unfortunately, many users of fluorescent materials fail to recognize the limitations imposed by nighttime. Not only are they less visible, but assume they are visible to others. The increased risk here is with assumed visibility that is not, in fact, present. The wearer may take

*The appendices of the book coordinate with this chapter. References to fire apparatus will start with "F- 9.1" and are printed near the left margin of Appendix I. They direct the reader to the fire apparatus specifications. References to ambulances start with "A" and are the paragraph and section number in the left margin of Appendix II. They relate to the federal KKK-A1822D specifications.

chances on a highway that he normally would not take if he knew the oncoming driver was oblivious of him. There is a false sense of security imparted by the misunderstanding of fluorescent material.

While fluorescent material is highly effective when fresh, it deteriorates under regular exposure to sunlight. Therefore, it must be renewed, either by repainting or reapplying tape strips. The life-span of these materials must be constantly reviewed to maintain optimum efficiency.

8.2 Retroreflectors

Retroreflectors are nighttime functioning materials. These use a combination of microscopic beads, lenses, and prisms imbedded in a matrix or added to a paint-like base. They all have the ability to reflect light back along the path of the light's origin.[6] (See Figures 8.1 and 8.2.) While it is somewhat similar to the tail reflector on a bike fender, it is more like the nighttime views of highway speed and direction signs. As the viewer moves off axis from the path of the incident beam, the apparent brightness decreases. A common misconception is that retroreflective material glows. This usually is assumed after viewing a photograph taken with a flash attachment. Since the flash is in line with the camera lens, the resultant photo illustrates a very bright retroreflective surface. However, if the flash would be decentered off axis in relation to the camera lens, the reflex would be markedly diminished.[7] Therefore, the photograph of retroreflective material may act to impart a substantial bias toward the degree of observer brightness and presumed conspicuousness. (See Figures 8.3 and 8.4.)

Because of the widespread use of news photographs illustrating the presumed brightness of garments worn by public safety personnel, the question of whether this makes the officer a more noticeable target. However, by definition, the retroreflective function is operational when the illuminating beam is in line or on axis with the viewer. Enhancing a target is not a problem because any weapon sighting that would occur would be made evident or disclosed by the light beam's path.

Retroreflective material is often applied to vehicle surfaces. (F-9.1) (A-9.1) Originally it was displayed as a single horizontal band, but recently the applications have taken a more diverse route. (See Figures 8.5 and 8.6.) While single horizontal bands are still popular, diagonals and zigzags have come into use. Camouflage can be a difficult concept to un-

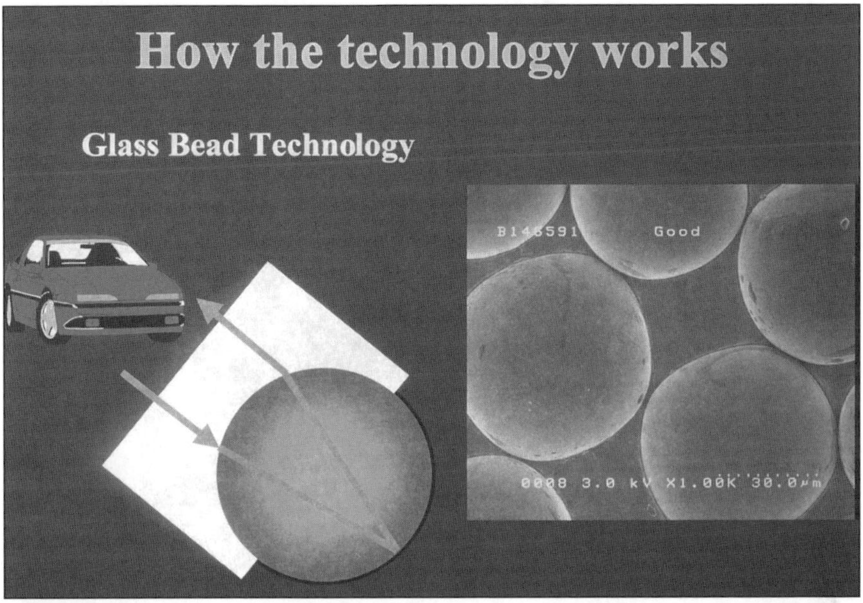

Figure 8.1

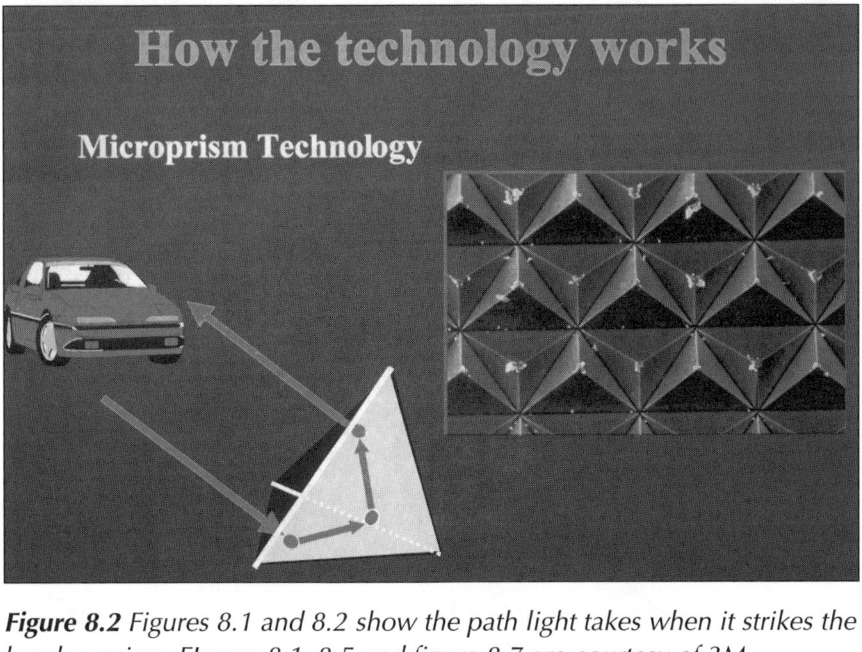

Figure 8.2 *Figures 8.1 and 8.2 show the path light takes when it strikes the bead or prism. FIgures 8.1–8.5 and figure 8.7 are courtesy of 3M Corporation.*

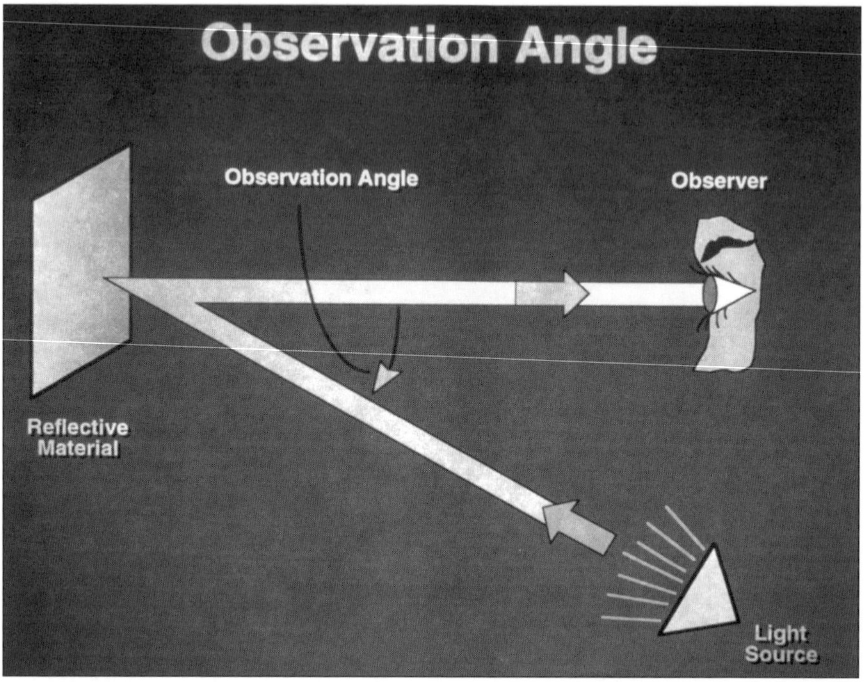

Figure 8.3

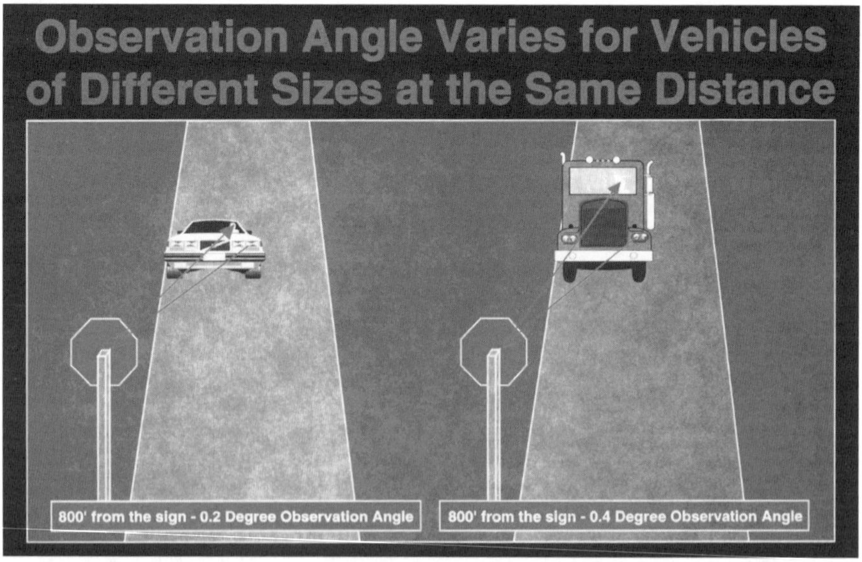

Figure 8.4 *Figures 8.3 and 8.4 illustrate that as the observation angle increases between the light source and the eye, the brightness of the retroreflector decreases.*

Figure 8.5 *A single horizontal band does not provide adequate outline information when viewed from the side, as in an intersection. The two right angle patterns do provide more visual information when approaching this vehicle from the rear.*

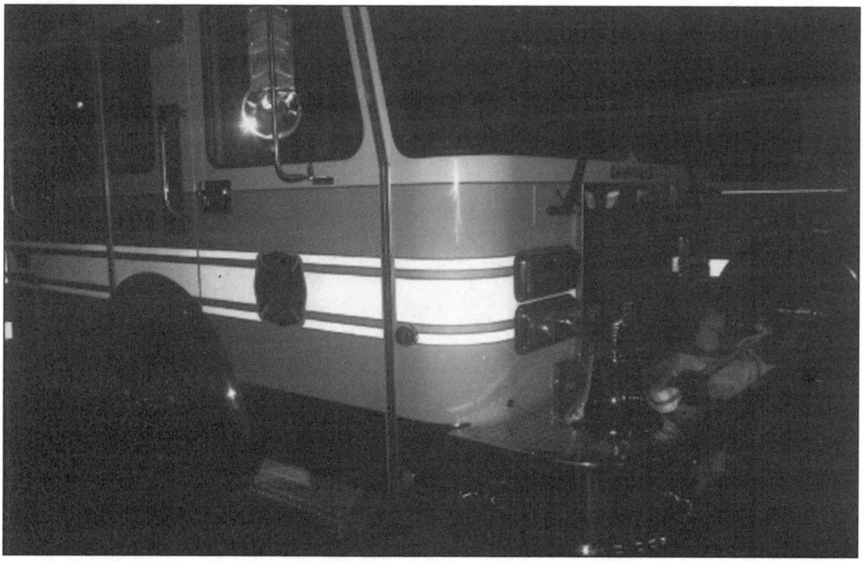

Figure 8.6 *Retroreflective material mounted horizontally*

derstand, especially when using high visibility materials, but in this case the daytime viewing of these patterns act just that way. A component of identification is by seeing a contrasting silhouette or outline. Using patterns that do not follow the general lines of vehicle shape add to the camouflage effect. When a fraction of a second can be the difference between an accident or safe passage, any advantage is desired. Contrast for visibility (the opposite of camouflage) is between that vehicle and the background against which it is observed. While the vehicle is in motion, that background is constantly changing. The vehicle must be seen as an overall shape, and this is enhanced by preservation of the natural outline. Extraneous decorations, two-toned colors, and dark colors all contribute to camouflage. Many public safety officials see their vehicle and the color schemes used as contrasting to other motor vehicles and assume they are very visible. This is not entirely true, for that vehicle must be seen against the general background of the environment and geographical location. Civilian drivers do not see a fire truck, police car, or ambulance, identify it as an emergency vehicle, and then drive defensively to avoid it, but rather see a vehicle, determine if it is a hazard to them, and decide to avoid it. This is the survival reaction that is a normal component of driving. To a civilian, an emergency vehicle is no different than a cement truck or a dump truck when he or she must avoid a hazard.

This is not to say that appropriate use of retroreflective material is dangerous or ineffective. When a vehicle is parked on or along the roadway, these markings help oncoming drivers in dark conditions to detect and localize the hazard, and is most effective when outlining the vehicle. The light reflected back is steady, not flashing or blinking. It is important to remember that flashing lights often confuse, rather than inform, the unsuspecting operator. By offering a steady light pattern, there is little distraction to the oncoming driver, and when the retroreflective tape pattern outlines the vehicle, the visual information is highly informative.[8] The steady nature of the signal allows the driver to more rapidly localize the object, make a judgment as to where the vehicle is parked, and how big it is. The faster this is done, the more rapidly a defensive action can begin. A flashing light forces the approaching observer to concentrate on it to figure out the warning. When looking intently at a target, it is not uncommon for a driver to actually steer toward it. This is not the desired effect. As stated in the chapter on flashing lights, the more blinking lights contained

Figure 8.7 *The large amount of visual information can confuse a driver. If a poorly marked vehicle were included as part of the roadway scene, the confusion would soar.*

in a scene, the more difficult it is to sort out the message. The consequence is a longer reaction time and an increased chance of an accident. Retroreflective material used with a scientific rationale and an appropriate pattern helps present a more safe nighttime scene. Today's approach using a single, horizontal band presents a night pattern of a single stripe, not unlike a traffic barricade constructed of wood. A driver may not realize until too late that what appeared as a minor obstruction is in reality a large vehicle, which possibly has attracted pedestrians to the immediate area. Compounding the horizontal application with zigzags or diagonals only presents a more complex pattern that increases confusion and reaction time. Reiterating, the pattern of tape that highlights the silhouette is one the most effective nighttime retroreflective applications. (See Figure 8.7.)

The human visual system has a built-in microscan movement that is composed of a horizontal component and a much smaller vertical component. The resultant motion is basically of horizontal oscillations that are slightly slanted toward the vertical. It is thought this exists so the rods and cones are not singularly fixed on the focused image. Allowing a single group of neuro-receptors to steadily fixate on an image causes a loss of sensitivity, that is to say, if the eyes did not have these small oscillations, the image would eventually fade into darkness. Therefore these motions, also known as *micronystagmus*, provide for an opportunity for stimulated rods and cones to be refreshed.

Recognizing that as the eye's rods and cones are constantly moving in a predictable pattern, certain images are better capable of attracting attention. Since the sweeps are basically horizontal, vertical lines have the opportunity of presenting a stronger stimuli. Objects that display vertical stripes as part of a visibility package have an enhanced safety role. However, this should not be extrapolated to mean that diagonal lines should replace groups of horizontal and vertical lines. The basic concept that the silhouette or outline of the object be visible is still key. This aids in early recognition which, in turn, allows for a faster reaction time.

Worthy of mention is the obvious fact that retroreflective material and high visibility paint applied to vehicles do not need a power supply from the vehicle. Loss of electrical power is not unknown on emergency apparatus and when this does occur, all lighting fails. Often the only operational visual warnings available during these events are the striping patterns and the paint color. These are passive warning methods that attract attention quickly, economically, and function whenever the vehicle is in use, without needing the operator to switch them on, and regardless of power availability.

8.3 Combination Fluorescent/Retroreflectors

The manufacturers of retroreflective material have very successfully combined them with fluorescent qualities. These materials have traditionally been used for firefighting garments, so they bend and conform to the underlying fabric. This two-in-one product is commonly available and in widespread use. The fluorescent portion, as previously mentioned, must be monitored for fading and the material must also be inspected for cleanliness and degradation from heat exposure. As the material ages and losses function with use, the visual characteristics deteriorate. The obvious next step is to replace the striping, or more likely, retire the garment and replace it with new protective clothing. Because clothing has a shorter functional life than vehicles, the tape was effective for the former, yet not practical for the latter.

Circa 1999, The 3M company released a new fluorescent and retroreflective product for vehicle application. This material was initially designed for the federal government as a new, federally approved traffic signage color. It is beginning to be seen throughout the United States on signs warning of nonmotorized activity on or near roadways. This includes school zones, pedestrian crossings, and bicycle use. The daytime display is fluorescent lime-yellow and the nighttime component is retroreflective lime yellow. The brightness of the night portion rivals or surpasses the previously manufactured silver or white retroreflectors. While this within itself is significant, equally impressive is the enhanced life span of the fluorescence. While fading had been a problem in the past, the new material functions well for many years.

Of emerging importance to the emergency services is the application of this highway sign material to vehicles. Fluorescent material that faded was a problem with past products and so did not lend itself to vehicle application as the maintenance and replacement was too expensive and time consuming. Now the conspicuousness provided is markedly improved and the striping has a much longer operational life. As such, vehicle applications are more practical. Therefore, for safer emergency vehicle operations, the new 3M material or the equivalent product of other manufacturers should soon become the industry standard. The application should follow the guidelines discussed earlier and outline of the vehicle, not be placed as a single horizontal band.

8.4 Phosphorescence

Phosphorescence is a cold light emission from a material after exposure to a radiation source. The light output continues after the incident radiation is removed.[9] The period of time it operates varies upon the material, but none has been found that is bright enough or radiates long enough to be practical in an environment of emergency vehicle operations—either as warning devices or vehicle labels such as department information. The time lapse for a 50 percent decrease in brightness is known as half life and defines the decay or dimming rate. The process is renewable by re-exposure to a radiation source. A common example of a phosphorescent application is the numbering on some watch dials. There is ongoing research to find ways to enhance both the energy output and lengthen the rate of decay. These materials may have greater application in the future.

Fluorescent material is functional only with daylight, retroreflective material is operational only in a dark environment, and phosphorescence is not yet a practical application. Concepts used to highlight emergency vehicles can unknowingly contribute to their camouflage. Sufficient tape use, comprehensive pattern application, and appropriate paint color selection all need to be understood.

Individuals that have an incomplete knowledge of the scientific rationale of these aids or make decision based primarily on tradition can worsen the situation with inaccurate assumptions and purchases.

Endnotes

1. Solomon S.S. Visibility of firefighters, *Journal American Optometric Association*, 1974; 45:142.

2. *Ibid* 1: 142.

3. Zwahlen, H.T., Schnell, T. *Visual detection and recognition of fluorescent color targets versus nonfluorescent color targets as a function of peripheral viewing angle and target size*, presented at Transportation Research Board 76th annual meeting, Washington, DC, Jan. 1997.

4. *Ibid* 3.

5. *Ibid* 1: 142.

6. *Ibid* 1: 142.

7. *Ibid* 1: 142–143.

8. Allen, M.J., Abrams, B.S., Ginsburg, A.P. and Weintraub, L. *Forensic Aspects of Vision and Highway Safety*. Tucson: Lawyers & Judges Publishing Co. 1996, 188.

9. *American Heritage Dictionary*, New College Edition. Houghton Mifflin, 1980: 986.

Chapter 9

Emergency Response and Vehicle Operation

Stephen S. Solomon, B.S., O.D.

The attitude of a law enforcement officer toward an emergency vehicle operator requires special understanding. Civilian operators encounter this relationship in a negative sense, when they find that the traffic officer has shown deference or partiality to a fellow officer or EMS vehicle operator with whom the civilian has collided.

A review of fire apparatus accidents shows that civilian drivers are often cited for failure to yield right of way and driving inattentively.[1] It appears as if the investigative logic follows the line that if an emergency vehicle is proceeding with lights and sirens and is involved in an accident with a civilian vehicle, then the civilian *must* have failed to yield *and* could not have been paying attention.

Given the dangers faced by fire, EMS and police personnel in protecting the safety and security of a community, and the potential of a strong bond between these providers, the unfortunate civilian involved in an accident with an emergency vehicle may have significant inequalities to overcome. Often, a bias or prejudice can come into play because the investigating officer is employed by the same municipality as the involved officer, firefighter, or EMS operator. These services commonly function together, which can distort the sense of impartiality. In volunteer services, it is common for some members to be involved with two if not all three agencies. They may volunteer as fire and EMS providers while being employed by a police agency.

The independent investigation of these types of accidents must be conducted with knowledge of, and sensitivity to, these interrelationships. Most accident reports contain information pertaining to the response mode of the emergency vehicle. This is usually stated as responding with lights and sirens or responding in an emergency mode. The unstated thinking is that the civilian operator must have seen or heard the ambulance, police car, or fire truck. This assumption is not always supported by the facts in these cases. As has been covered elsewhere in this text, vehicle colors can act as camouflage, sirens are nondirectional, and red, blue, white, or yellow lights may not be perceived as emergency warnings.

Another implied conclusion is contained in the statement of "responding with lights and sirens," and that relates to the activity of the emergency vehicle. The emergency vehicle may have entered an intersection against the light or may have been driven in lanes not commonly used in routine driving. Excessive speed and unsafe operation are certainly factors to be analyzed. Emergency vehicles are allowed to proceed safely through an intersection when facing a red traffic light, but if an accident occurs, was this a "safe" operation of that vehicle? Motorists should operate their vehicles in a prudent and predictable manner, and in fact that is the way almost all drivers behave. In other words, they obey the laws. In return, they expect other drivers to do the same. If this pattern is broken, then accidents occur. It is expected in our society that red traffic lights prevent drivers from entering the intersection until a green signal is received. Emergency vehicles are the exception to the normal traffic pattern and flow. In recent years, emergency vehicle operator courses have emphasized the dangers of intersections.[2] To the credit of municipal and private providers, they are aggressively focusing on this issue, but teaching these dangers is not uniformly practiced in all agencies.

The contradiction now becomes apparent. On one hand, the emergency vehicle is proceeding in an emergency mode and its operator expects that others on the roadway will give clearance. Having been given clearance on almost all occasions, the operator has learned to expect that civilians will always yield. The more one sees this response, the more one expects it to happen. From the second perspective, the civilian operator expects that all traffic will obey the rules of the road, and that includes stopping at red traffic lights and staying in predictable lanes of traffic. The more they experience this, the more they expect it. When an accident occurs, the investigating officer is confronted with judgments such as who had right of way, was vehicle operation safe, was vision unimpeded, and who was at fault.

A moving police car often has obstructions to vision that are integral to police operations. On the dash board can be mounted a radar transpon-

Figure 9.1 *A radar gun and control box mounted on the dash board of a patrol vehicle. Note that it is on the midline of the vehicle.*

Figure 9.2 *A radar gun which is dash board mounted near the driver. Note the zone of usable vision between the top of the radar and the centerline rear view mirror.*

Figure 9.3 *An "A" post mounted flood light in a patrol car. The external reflector head is movable and controlled by twisting the handle or shaft from inside. The position of the head determines the size of the blind spot.*

der, control unit, and a video camera. In some cases the barrel of a shot gun stored vertically and a flood light may protrude into the field of view. These items are not seen in private vehicles and they can all obscure vision. In an accident involving police vehicles, consideration should be given to the reduced visual space available. Blind spots can be an easily overlooked contributor to accidents.

A single emergency vehicle responding in an emergency mode is commonplace in most communities. When two such vehicles are proceeding in the same direction at the same time, it is important that adequate spacing be maintained. Civilian drivers can not distinguish quickly enough the presence of both and may reenter traffic after the first vehicle passes but before the second vehicle. This can complicate an accident analysis. One the one hand, if the driver yielded for the first emergency vehicle, he should have been aware of the second, or have carefully checked before entering traffic. On the other hand, insufficient spacing between two emergency vehicles may cause all the sirens and flashing lights to appear to blend together. This may mislead other drivers into believing that there is a single vehicle, with the result that detection of the second vehicle may be inhibited.

All accidents have the potential of being tragic, but when two emergency vehicles collide with each other, not only is there an increased opportunity for more serious injuries, but the community may lose protection in the sense of damaged apparatus that may not be able to continue to the call.[3] If two emergency vehicles are approaching an intersection at ninety degrees to each other, the approaching sirens can not be heard. If each operator has the expectation that civilian traffic will provide the right of way and yield to an emergency vehicle, then an accident is possible. With civilian traffic in each direction yielding, the reinforced perception could be that a clear path through the intersection will exist. Emergency personnel are trained for such events, but accidents still occur.

Ambulance and fire personnel, especially volunteers, often respond to an alarm using private vehicles, and go either directly to the scene or to the station. Officers may employ emergency lights and sirens, while the active membership often uses blue or green lights. While the laws vary among the states, the general concept is that the use of colored lights by volunteers are more for courtesy to help identify the vehicle as responding to an emergency, rather than signalling a legal requirement for civilians to yield. Accidents involving these individuals are common. Unless otherwise allowed by law, they should have been driving in a predictable manner and obeying the traffic laws in place.

Endnotes

1. State of New York, Department of Motor Vehicles, MV-144A, Jan.–Dec. 1995.

2. New York State Department of Health, *Ambulance Accident Prevention Seminar, Student Workbook*, (circa 1990), p.18.

3. Solomon, S. and J. King, "Influence of color on fire vehicle accidents," *Journal of Safety Research*, National Safety Council, Vol. 26, No 1 (Spring, 1995) p. 41.

Chapter 10

Medical Control

George L. Ellis, M.D., FACEP

In most areas of the country, including Pennsylvania and New York, which is the area with which I am most familiar, regions are established for control of EMS activities. These regions are usually formed along county lines, often with several counties being combined into one EMS region. Each region has a council composed primarily of representatives of the various EMS agencies within the region and often with one or more paid employees providing support for the council. The council usually has a medical advisory committee primarily composed of physicians. In most circumstances, only the physicians may vote in this committee, but others may be welcome to attend as nonvoting members. The council usually deals with operational issues; the medical advisory committee deals with medical treatment issues and the development of regional medical protocols. Quality-assurance issues involving medical care are often considered by the medical advisory committee, or by a subcommittee of this group, although ultimate responsibility for the quality of care rendered by an advanced life support (ALS) service unit rests with the agency's medical director. Basic life support (BLS) service units are, in many states, not required to have an agency medical director (although those using automated external defibrillators may be required to have physician oversight of automatic electronic defibrillation (AED) use).

10.1 Medical Command

"Medical command" refers to instructions provided to the prehospital personnel regarding medical treatment of patients. It can be "on-line" or "off-line." "Off-line" medical command refers to written protocols, usually divided according to the symptom with which the patient presents (for example, shortness of breath, chest pain, or seizure). The protocol will detail what the responding EMS personnel should do, such as applying oxygen, starting an IV, giving bronchodilator therapy, and so forth. The written protocols usually include an instruction to contact medical command at some point in the protocol, usually at the point when an increased level of medical judgment is needed. When the EMS personnel contact medical command, they then begin to receive "on-line" medical command. These are instructions given directly to the EMS providers by the physician, who is usually at the receiving hospital's emergency department. This is usually accomplished by radio, but may also be done by cellular phone or even from the patient's home phone before leaving the scene.

The protocols used are usually developed by the region's medical advisory committee. They are written protocols and are distributed to the prehospital providers. The physicians involved usually include representatives from the emergency departments of hospitals in the region, medical directors for advanced life support agencies, and other physicians, such as trauma surgeons, who are interested in the care of patients in the prehospital setting. The degree to which this physician group interacts with the dispatch center varies, an issue that will be discussed later.

10.2 Time-Critical Conditions

Emergency medical services can be divided into first-responder services, basic life support services, and advanced life support services. These can be further subdivided in some states, usually to define the various levels of advanced life support that can be administered. The differing capabilities of the various levels significantly affect what services can be provided in the field and, therefore, the urgency with which the patient must be transported to the hospital. For example, if a patient with partial upper airway obstruction due to tissue swelling, and who is being cared for by a basic life support unit, the patient would need transport to the hospital as quickly as possible. The basic life support unit could apply oxygen and could use a bag-valve mask to assist ventilation, but in the event of further obstruction to the airway, the unit would have little to offer the patient in the field and, therefore, arrival at the hospital in as short a time as possible is essential. On the other hand, if the patient were being treated by an advanced life support unit, intubation could be undertaken in the field before transport and before complete obstruction could occur. Once the airway is secured, the transport could be undertaken without emergency lights and siren.

As another example, a cardiac patient being treated by a basic life support service could have oxygen applied, but in the event of ongoing chest pain, would have little else available until arrival at the hospital. On the other hand, an advanced life support service could start an intravenous line in case emergency drugs become needed, administer nitroglycerin to relieve the chest pain, monitor the patient's cardiac rhythm so as to detect and treat dangerous arrhythmias, and apply oxygen. In the event of complications of myocardial ischemia, the ALS crew could treat congestive heart failure should it occur, and could follow the American Heart Association's advanced cardiac life support protocols, or regional protocols should cardiac arrhythmias or cardiac arrest occur. These capabilities, while not negating the need for prompt transport to the hospital, may mitigate against the use of lights and siren in cases where the saved time would not be substantial, especially since the lights and siren may further frighten the patient and actually worsen his cardiac condition.

With this as a background, we will consider various emergency medical conditions and the extent to which the travel time to the hospital may be critical. Looking first at traumatic conditions, the American College of Surgeons have long discussed the "golden hour." In a significant trauma, the differences between what BLS and ALS can provide in the field are, in most cases, less important than in medical conditions. Both BLS and ALS must appropriately extricate and immobilize the patient before he is transported, must secure his airway, assure appropriate ventilation, control bleeding, and expeditiously transport him to the hospital. ALS is critical when securing the airway includes the need for intubation. ALS units are also able to start intravenous lines and start fluid resuscitation, and are usually able to decompress the chest in the event of tension pneumothorax. Except in extraordinary circumstances, ALS interventions are performed en route to the hospital so as to minimize the time to surgical intervention, should it be needed. In such a "scoop and run" scenario, lights and siren are recommended if the injury is significant, especially if the injury involves chest, neck, head, or abdomen and if vital signs are abnormal. In cases in which the patient is awake and alert, has essentially normal vital signs and is not complaining of respiratory distress, severe abdominal pain, or uncontrolled bleeding, the amount of time that would ordinarily be saved by use of lights and siren would not be expected to affect the patient's outcome. Patients having major burns, particularly when there is an inhalation component to their injuries, and unconscious patients with head injuries, may warrant lights and siren in settings in which the saved time is more than a few minutes. Likewise, patients having a pulseless limb or requiring re-implantation of a severed body part may warrant lights and siren if the setting is such that significant time would be saved.

The next group of patients for whom time is crucial comprises those having myocardial ischemia. For these patients, the saying is "time is muscle," referring to the increasing destruction of heart muscle the longer it is deprived of sufficient oxygen. With the advent of thrombolytic therapy, the so-called "clot-busting drugs," and the availability in some institutions of emergency angioplasty (in which a balloon is threaded into the coronary artery and inflated at the site of obstruction to allow return of blood flow) time becomes critical, whereas it was less so when treatment was only supportive. Before the advent of these therapies, an ALS unit capable of monitoring and treating arrhythmias and heart failure in the field, while en route, could provide the patient with appropriate therapy en route to the hospital, and there was no urgency to save a few minutes because this was basically what would be done for the patient at the hospital anyway. With the advent of these new therapies, however, there is the attempt to reduce the time of therapy wherever possible. There are, however, other ways of decreasing the time to therapy other than racing through the streets with lights and siren. Some EMS services are performing 12-lead EKGs in the field and transmitting them to the receiving emergency department. This allows the emergency physician at the receiving hospital to determine whether the patient is a likely candidate for thrombolytic therapy or angioplasty, and to mobilize the hospital resources that are going to be needed upon the patient's arrival. This could save substantial time if the patient is going to be transported a significant distance, if it takes more than a few minutes to get an EKG once the patient arrives, if personnel have to be summoned, if the drugs need to be mixed, or if the catheterization lab must be readied. If the patient does meet the EKG criteria for one of these therapies, the prehospital providers can be asked to review other criteria with the patient before his arrival, such as his medical history and duration of symptoms to determine whether the patient is an appropriate candidate for these new therapies. Assuming that these functions are performed en route to the hospital so as not to delay the transportation, the time savings that can be achieved through these means are often substantially greater than the time that would be saved by lights and siren in many clinical settings. As stated previously, the effects of lights and siren on patients having a heart attack may be detrimental.

Recently, new therapy for the stroke patient has emerged as well, again using thrombolytic therapy, the so-called "clot-busting drugs." This has led to the public education strategy of classifying a stroke as a "brain attack." It is essential that patients get to the hospital quickly following the onset of a stroke if they are to receive the potential benefit of thrombolytics. Initial studies suggest that, to be beneficial, these drugs must be given within three hours of the onset of stroke symptoms. Upon arrival at the hospital, the patient must have a radiographic study called computed tomography (the so-called CT or CAT scan) to make sure that there is no bleeding into or around the brain and that there is no significant swelling (which, in most cases, suggests that the stroke began more than three hours earlier.) This, of course, does take some time and not all institutions offer computed tomography or thrombolytic therapy. If the patient is transported to an institution not providing these services, then the patient would need to be transferred to another institution still within the three-hour window of opportunity if he or she is to receive this therapy. It is, therefore, essential that the patient be transported to the hospital immediately. This does not necessarily mean that the patient needs to be transported with lights and siren if there is not a substantial time savings, but does mean that the patient should be promptly transported without delay from the scene. In the past, patients would often delay calling for the ambulance and the prehospital personnel would have no sense of urgency and might even wait for the patient's family to pack a suitcase for the patient, feeling that there was nothing urgent that was going to occur even after the patient arrived in the emergency department. This is no longer the case in patients having an acute stroke and, while in most cases lights and siren will not substantially decrease the time to arrival, in those cases in which there are significant savings and the stroke is less than three hours old, this should be considered a time-critical emergency.

A cardiac arrest has always been the condition that most lay persons considered the ultimate time-critical emergency. In fact, it still is, with time to defibrillation probably being the most important factor in determining successful resuscitation. Response time to the scene, bringing the defibrillator to the patient, is the paramount concern. Certainly, lights and siren on the way to such a scene is perfectly appropriate. A BLS service arriving at the scene of a cardiac arrest will assure an open airway, assure ventilations and cardiac compressions, and transport as quickly as possible to the hospital. In this scenario, lights and siren are appropriate. Unfortunately, in the patient with full cardiac arrest brought in with CPR being performed, the survival rate is dismal. A BLS crew arriving who has an automated external defibrillator will apply the device, again assure an open airway and ventilations, and will follow the instructions of the machine regarding defibrillation. If ALS is not available as backup and the BLS crew will be transporting the patient, transportation should occur with lights and siren although the results will again be dismal in most cases if they are unable to regain a pulse with use of the defibrillator.

An ALS crew arriving at the scene of a cardiac arrest, on the other hand, has with them almost all of the devices and drugs available in the emergency department. These crews have been well trained in resuscitating a the cardiac patient, can intubate the patient to improve ventilation, can interpret the cardiac rhythm and defibrillate or treat with medications as appropriate, and can start intravenous lines and give fluid as needed. In short, they can perform resuscitation in the field. In most settings these days, it is unusual for a patient who has not been able to be resuscitated in the field to be successfully resuscitated from full cardiac arrest upon arrival in the emergency department. While these patients will undoubtedly be transported with lights and siren, if for no other reason than for the public's and family's perception that everything possible is being done, the fact of the matter is that undue risk should not be taken for patients with prolonged downtime or in whom ALS field resuscitative efforts have been fruitless.

Few prehospital providers have significant experience with childbirth in the field. Certainly, patients with high risk pregnancies or those presenting with complications of childbirth such as a prolapsed cord, premature labor, or arrested delivery, should have their transit to the hospital expedited as much as possible. In the case of an ALS crew, trained in delivery, transporting a low-risk patient in labor in whom delivery is not imminent, the risk associated with lights and siren would not be advised. On the other hand, a BLS crew without such training or an ALS crew transporting a patient in whom it is anticipated that delivery could be performed in the emergency department as opposed to in the ambulance if lights and siren were used, may appropriately choose to use lights and siren.

Patients in apparent respiratory distress should be transported with lights and siren if this would decrease their time out of hospital and they are being transported by BLS. If the patient is being transported by an advanced life support crew, to whom a number of interventions are available to relieve respiratory distress, lights and siren often can be avoided. Such interventions would include intubation if the patient is in extremis, administration of bronchodilators for the patient with bronchospasm, administration of nitroglycerin, Lasix, morphine, and other cardiovascular-active agents in the setting of congestive heart failure. This is not to imply that transport can be undertaken without concern for expediency, but rather to consider the risk/benefit ratio of the use of lights and siren and the potential for motor vehicle accidents which would further add to the morbidity, mortality, and delay in reaching the hospital.

The patient with uncontrolled hemorrhaging of significant quantity requires prompt transportation to the hospital. In the case of BLS this may well require lights and siren. While ALS can start intravenous fluids for replacement and support of circulation, this is only a temporizing measure and does not replace the blood that is being lost. Most sources of external hemorrhage can be controlled with pressure, and bleeding at fracture sites can often be controlled with traction splints. Sources of internal hemorrhage such as severe gastrointestinal bleeding are not amenable to field treatment. If the bleeding is substantial and the time saved by use of lights and siren is substantial, then they should be used. Again, an analysis by the prehospital provider of the risks versus benefit is essential.

Patients who are having ongoing seizures or who have repeated seizures without regaining consciousness are said to be in "status epilepticus," which is a medical emergency. These patients should be transported to the hospital as quickly as can be safely done. While ALS units often can give Valium which may help control seizures, status epilepticus is often difficult to control and may require multiple drugs or large doses of drugs.

A pulseless extremity is a medical emergency. The patient should have blood flow reestablished as quickly as possible. However, it would be an unusual setting in which the time saved by use of lights and siren would have affect the morbidity and salvage of the limb, because most settings would save only a few minutes by such means.

Table 10.1
Medical Criteria for Use of Lights and Siren

1. Vital signs (patients > 8 years old)
   a. Systolic BP < 90 mm Hg with possibly related disease or trauma
   b. Respiratory rate > 36/min with patient as relaxed as possible
   c. Respiratory rate < 10/min
2. Airway
   a. Inability to establish or maintain a patient's airway
   b. Upper airway stridor
3. Respiratory
   a. Severe respiratory distress unresponsive to BLS/ALS treatment
4. Cardiovascular
   a. Cardiac arrest
   b. Severe, uncontrolled hemorrhage of any cause
5. Trauma
   a. Penetrating wound to head, chest or abdomen except for obviously superficial wounds
   b. Two or more suspected proximal long-bone fractures
   c. Major amputations including two fingers, three toes, or above the wrist or ankle
   d. Penetrating or blunt neck trauma except obviously mild or superficial injury
   e. Neurovascular compromise of an extremity
6. Neurologic
   a. Glasgow Coma Scale score < 13, only if acute change of any cause
   b. Seizure activity not controlled by BLS/ALS treatment
7. Obstetric
   a. Intrapartum emergencies including, but not limited to, cord prolapse, premature labor, and arrested delivery
8. Pediatrics
   a. Upper airway stridor
   b. All patients ≤ 8 year of age individually based on the mechanism of injury, degree of distress, and the EMS personnel's experience with patients of this age; when in doubt, seek advice from medical command, transport as an emergency, or both.
9. Other
   a. Emergency transport should be used in any situation in which the most highly trained EMS provider believes that the patient's condition could be worsened by a delay equivalent to the time that could be gained by emergency transport. In all cases using this option, documentation of the reason for this on the trip must be recorded.

KEY:	ALS = advanced life support	BLS = basic life support
	BP = blood pressure	EMS = emergency medical services

Source: Kupas, D.F., D.J. Dula, and B.J. Pino, "Patient outcome using medical protocol to limit 'lights and siren' transport," *Prehosp Disaster Med* 1994;9(4):227, Table 1.

How then should the use of lights and siren be directed? There is no substitute for good medical judgment of the prehospital provider. The risks to the ambulance crew, patient, and community of the use of lights and siren must be weighed against the benefit to the patient of arriving at the hospital faster. The particular setting in terms of distance from the hospital, traffic patterns, number of traffic lights, time of day, weather conditions, and so forth, will determine what "faster" means. This judgment can be aided by the use of protocols. Likewise, the paramedics' judgment does not come into play until they have evaluated the patient, and, therefore, it is pertinent to the decision to use lights and siren in transporting the patient *from* the scene to the hospital. There remains the question of whether to use lights and siren in responding *to* the scene. This clearly relies upon protocols.

The use of lights and siren in transporting from the scene to the hospital was addressed in a study by Dr. Kupas et al., in 1994.[7] In this study, medical criteria that very closely parallel those discussed above were used to determine the use of lights and siren transport from the scene to the hospital (see Table 10.1). It was noted that an analysis of Pennsylvania EMS trip sheets indicated that 58 percent of all emergency calls in 1991 culminated in emergency transport, although only 19 percent were believe to be life-threatening. With the use of the medical criteria as shown in Table 10.1, 92 percent of the patients transported during this study were transported in accordance with nonemergency procedures. There were no adverse outcomes related to a non-lights-and-siren transport.

The issue of response to the scene clearly depends upon dispatch protocols. The use of protocols in determining the use of lights and siren was addressed by the National Association of EMS Physicians' Emergency Medical Response Task Force in 1994. In a position paper entitled "Use of Warning Lights and Siren in Emergency Medical Vehicle Response and Patient Transport," the National Association of Emergency Medical Services Physicians and the National Association of State EMS Directors supported the following position statements:

1. Emergency medical services' medical directors should participate directly in the development of policies governing emergency medical vehicle (EMV) response, patient transport, and the use of warning lights and siren.

2. The use of warning lights and siren during an emergency response to the scene and during patient transport should be based on standardized protocols that take into account situational and patient problem assessments.
3. EMS dispatch agencies should utilize an emergency medical dispatch priority reference system that has been developed in conjunction with and approved by the EMS medical director to determine which requests for prehospital medical care require the use of warning lights and siren.
4. Except for suspected life-threatening, time-critical cases, or cases involving multiple patients, lights and siren response by more than one emergency medical vehicle usually is unnecessary.
5. Use of emergency warning lights and siren should be limited to emergency response and emergency transport situations only.
6. All agencies that operate EMV or are responsible for emergency medical responders should institute and maintain emergency vehicle operation education programs for the EMV operators.
7. Emergency-medical-vehicle-related collisions occurring during an emergency response or transport should be evaluated by the EMS system managers and medical directors.
8. A national reporting system for EMV collisions should be established.
9. Scientific studies evaluating the effectiveness of warning lights and siren under specific situations should be conducted and validated.
10. Laws and statutes should take into account prudent safety practices by both EMS providers and the monitoring public.
11. National standards for safe EMV operation should be developed.

Obviously the thrust of this position paper is that the need for lights and siren is severely limited, and they should be used only in time-critical emergencies. EMS medical directors (often working together through regional or state medical advisory committees) should be intimately involved in the development of protocols regarding the use of lights and siren, and further study regarding the risks and benefits of emergency lights and siren should be undertaken.

While protocols based upon criteria such as those shown in Table 10.1 can help to determine the means of transport after the patient has been evaluated by prehospital providers, the more difficult question involves the means by which the ambulance responds to the scene—that is, before benefit of hands-on evaluation.

10.3 Medical Dispatch

As recommended in the position paper by the National Association of Emergency Medical Services Physicians and the National Association of State EMS Directors, protocols should be developed that would direct the appropriate use of lights and siren *to* the scene. It was further suggested that EMS dispatch agencies use an emergency medical dispatch priority reference capable of determining which requests require the use of warning lights and siren. There are several medical dispatch priority references in use today, of which the Clawson system is the most widely known. The Medical Priority Dispatch System (MPDS) concept was initially developed from 1976 to 1979. It was designed with the recognition that the vast majority of ambulance requests are not for life-threatening situations, do not require on-scene treatment of a critical nature, and do not require a lights and siren response. Its goal is to "send the right thing to the right person in the right way at the right time."[3] Jeff Clawson, M.D., the originator of the Medical Priority Dispatch System points out the need to distinguish between "saving significant time" with "RLS (red lights and siren) response" since no published data has proven that RLS response does save significant time. He reported that in an unpublished study conducted in Salt Lake City, it was demonstrated that "fire pumpers and paramedic-staffed engines responding within their initial response districts in urban or suburban areas experience a 9 percent to 28 percent reduction in response times. Reductions depended on the time of day in concentration of semaphores, or visual signalling apparatuses, encountered. It was found that the maximum saving occurred during rush hour, when emergency vehicles traveled in the direction of main flow and when there was a traffic light at every block. It is interesting to note that the average time saved in these in-district responses was approximately 30 seconds."[3] He points out that the use of red lights and siren can incur significant risk. As many as 12,000 emergency medical vehicle accidents have been estimated to occur each year in the United States and Canada as a direct result of using red lights and sirens. The so-called "wake effect" of emergency vehicles disrupting, confusing, and startling other drivers may cause up to five times as many accidents as involve the emergency medical vehicles themselves.[11] This could bring the total number of accidents to 70,000 or 75,000 per year, resulting from the use of red lights and siren. In 1983, after adopting the MPDS, Salt Lake City reported the emergency medical vehicle accident rate had dropped 78 percent.[3]

Emergency medical dispatchers are trained to extract pertinent information from untrained, often excited, observers at the scene. Using the MPDS and based upon the obtained information, they can assign a response level appropriate for the situation. The response level assigned is associated with a level of capability, that is, BLS versus ALS, and the response time appropriate for the call, that is, respond "hot" or "cold." In some cases, the closest BLS unit or paramedic "fly car" responds "hot" with the transporting ambulance responding "cold" so as to bring a trained provider to the scene as quickly as possible while the transporting vehicle arrives shortly thereafter. In determining whether the response should be "hot" or "cold," the system must ascertain whether time will make a difference in the outcome, how crucial time is in responding to this type of problem, how much time can be saved using lights and siren in the particular locale and circumstances of the call, and the significance of the time saved by use of lights and siren when compared with time spent waiting for care, performing tests, and so forth, once the patient reaches the hospital.

When an emergency medical dispatcher receives a call for help, he must first determine the location and callback number. He then asks the caller to identify the problem. The age or approximate age of the patient is determined and then the two most important questions are asked: Is he or she conscious? Is he or she breathing? If the patient is unconscious or not breathing, or if it cannot be confirmed that he or she is conscious and breathing, a maximal response is sent immediately before proceeding to a specific dispatch priority card. If the patient is conscious and breathing, the dispatcher will move on to the priority dispatch card appropriate for the chief complaint. The dispatch cards have further questions aimed at identifying significant problems. They also have information that can be relayed to the caller to provide treatment until the EMS providers arrive. There is information listed for the dispatcher that may guide him to other,

19 HEART PROBLEMS — SHUNT — AMPDS v103 NAE — Issued 1-SEPT-98

CLASSIFICATIONS-TYPES-CAUSES | DEFINITIONS-AXIOMS-RULES-LAWS

Instructions for Taking a Pulse

1. "Find the Adam's apple on her/his neck.
2. Feel on either side of it for a pulse.
3. Be careful not to push too hard.
4. Count the pulses for 15 seconds (I'll time you).
5. How many did you count?" (multiply x 4 = rate)
   - ≤ 25 x 4 = Rate < 100
   - 26-32 x 4 = Rate 100-129
   - ≥ 33 x 4 = Rate > 130

Heart Attack, Information Validation

If there is a question of inaccurate information due to patient denial or caller confusion, the EMD may read the following to the person on the phone to help clarify the signs and symptoms of a possible heart attack:

"Sometimes the symptoms of a heart attack are difficult to recognize. Many times it feels like severe indigestion. Other feelings may be tightness, heavy pressure, constricting band and crushing discomfort in the chest with the spread of these feelings to arms, jaw, neck or back. They may also be nauseated or sweating a lot. Are any of these feelings present?"

Rules

1. The caller should be directed to take a pulse whenever it is physically possible (age, location, comprehension) using the appropriate "Instructions for Taking a Pulse" script.
2. If a patient has a very rapid heart rate (≥ 130) paramedics (ALS) should be sent.
3. Automatic Implanted (internal) Defibrillators (A.I.D.) are becoming more common. If reported as firing or causing problems, a maximum response is sent, as this may mean the patient is having significant arrhythmias or ventricular fibrillation (cardiac arrest will occur if unconverted by this device).

Axioms

1. Heart Problems are considered a specific "diagnosis." Heart problem situations range from old rheumatic fever through benign forms of congestive heart failure to acute angina or serious heart attack (myocardial infarction). It is occasionally even reported as the chief complaint in cardiac arrest.
2. Obtain symptoms and medical history (through complete interrogation) which more accurately indicate the urgency of the problem and therefore can be prioritized.

19 HEART PROBLEMS — SHUNT — For use through Exclusive License only — NAE — Patent Pending ©98

19 HEART PROBLEMS — SHUNT — AMPDS v102 NAE — Issued 1-SEPT-98

KEY QUESTIONS | POST-DISPATCH INSTRUCTIONS

KEY QUESTIONS	GOTO
1. Does s/he have chest pain? — Y —	10
2. Is s/he breathing normally? — N —	6
3. Is s/he alert (able to talk)?	
4. Has s/he had heart problems before?	
a. (If yes) Does s/he have a pacemaker or implanted defibrillator?	
i. (If A.I.D.) Did it fire (go off) in the last 30 minutes?	
5. I want you to take her/his pulse (heart rate). See Additional Information for instructions on taking a pulse.	
6. (If suspected AND heart rate ≥ 100) Did s/he take any drugs (medications) in the past 12 hours?	
a. Cocaine (or derivative)	

POST-DISPATCH INSTRUCTIONS	GOTO
a. Ensure ABCs.	
1. Assure airway.	G1
2. Verify breathing.	G4
3. Treat for shock.	G8
b. Allow her/him to assume the most comfortable position for her/him.	
c. If her/his condition worsens in any way, call me back immediately for further instructions.	
d. If s/he is on any medications, please get them out for the paramedics. Write down the name and phone number of her/his doctor. Put any family dogs away. Can someone meet the paramedics please?	

The Advanced Medical Priority Dispatch System™

DETERMINANTS | RESPONSES

		Determinant	Response
A	1	Heart rate < 130 (without priority symptoms)	1
B	1	Unknown symptoms (3rd party situation)	1
C	1	Not alert	1
	2	Cardiac history	2
	3	Heart rate ≥ 130 (without priority symptoms)	3
	4	Cocaine	4
D	1	Firing of implanted defibrillator	1

19 HEART PROBLEMS — SHUNT — For use through Exclusive License only — NAE — Patent Pending ©98

Figure 10.1 *Example of an MPDS card. (Reprinted with permission of the National Academy of Emergency Medical Dispatch.)*

less apparent causes for this chief complaint. Finally, there is a reminder for practical instructions such as having someone meet the paramedics, putting any family dogs away, and providing the paramedics with any medications the patient may be on as well as the name and phone number of his or her doctor. An example of a MPDS card is shown in Figure 10.1.

Studies are underway to determine the accuracy with which dispatchers can determine, by interrogation of callers who in most cases have no medical training, the severity and time urgency of patients' conditions. Use of such dispatch protocols to assess the time urgency of calls should result in the use of lights and siren in responding to the scene being avoided in many, probably in most, cases. Once the EMS personnel have arrived on scene, the most highly trained provider can determine need for lights and siren in transporting to the hospital based upon patient condition, treatment options available, local driving conditions, and discussion with "on-line" medical command. Again, in most cases lights and siren can be avoided.

Endnotes

1. Auerbach, P.S., J.A. Morris, J.B. Phillips, et al., "An analysis of ambulance accidents in Tennessee," *JAMA* 1987; 258(11):1487-1490.

2. Clawson, J.J., "The maximal response disease," *JEMS* 1987;10:28–31.

3. Clawson, J.J., "Running 'hot,'" *JEMS* 1991;16:11–13.

4. DeLorenzo, R.A. and M.A. Eilers, "Lights and siren: A review of emergency vehicle warning systems," *Ann Emerg Med* 1991;20(12):1331–1334.

5. DeLorenzo, R.A., "Bright lights, big noise," *JEMS* 1992;17:57–63.

6. Hunt, R.C., L.H. Brown, E.S. Cabinum, et al., "Is ambulance transport time with lights and siren faster than that without?" *Ann Emerg Med* 1995;25(4): 507–511.

7. Kupas, D.F., D.J. Dula, and B.J. Pino, "Patient outcome using medical protocol to limit 'lights and siren' transport," *Prehosp Disaster Med* 1994;9(4):226–229.

8. Lacher, M.E. and J.C. Bausher, "Lights and siren in pediatric 911 ambulance transports: are they being misused?" *Ann Emerg Med* 1977;29(2):223–227.

9. National Association of Emergency Medical Services Physicians and the National Association of State EMS Directors, "Use of warning lights and siren in emergency medical vehicle response and patient transport," *Prehosp Disaster Medicine* 1994;9(2):133–136.

10. Pirrallo, R.G. and R.A. Swor, "Characteristics of fatal ambulance crashes during emergency and non-emergency operation," *Prehosp Disaster Med* 1994;9(2):125–132.

11. Clawson, J.J. et al., "The Wake Effect—Emergency vehicle-related collisions," *Prehospital and Disaster Medicine*, 1997;12(4):274–277.

Part II

Accident Reconstruction and Litigation

Chapter 11

Emergency Vehicle Accident Profiles

Stephen S. Solomon, B.S., O.D.

Accidents involving emergency vehicles are particularly tragic in that these vehicle operators have spent a considerable portion of their adult life in preparing for the job they do. The dedication and commitment they bring to their employment is beyond what civilians realize. They have all taken the position of helping others, especially under trying and dangerous circumstances in which people of lesser resolve would pass by. The daily risks exceed what most workers experience in the workplace during a lifetime. Volunteers can be found doing all the same tasks that their paid counterparts do. These people, volunteer or paid, are the savers of life and property. We call them by dialing 911, which indicates we have an emergency and want them "right now."

In an instant they become victims rather than caregivers. Because they are operating with lights and sirens, their jeopardy is increased. It is common for an emergency vehicle operator in an accident to suffer injury, cause injury or death to fellow crew members, or injure or kill others sharing the roadway. As a consequence, they may be arrested on criminal charges, face civil actions or both. In their view, they were just trying to help someone in need or protect their property, yet they are "transformed in the blink of an eye" from responsible citizens to individuals facing possible financial ruin and criminal records. This consequence must be added to all the other costs of emergency vehicle accidents.

Therefore, it is critically important for all agencies that offer emergency services to provide initial and ongoing emergency vehicle driver training, and it should be at a higher skill level than for other classes of drivers. Matching documentation must be complete and accessible. This provides protection to the vehicle drivers and at the same time establishes a criteria for litigators.

While there may be no such thing as a routine or commonplace accident, accidents can be categorized or grouped, and by reviewing groups of similar accidents investigators can compare a specific incident to the group. Officials of emergency service agencies can apply the data to their accident reduction and prevention programs. The law enforcement community can use this data to plan future enforcement efforts and needed manpower. Highway officials can spot deficiencies in engineering and construction and appropriate funds to upgrade or for repair.

Being able to predict accidents aids in the planning and implementation of preventive strategies. In the previous chapters of this book, factors associated with emergency vehicle operation are discussed. This chapter brings together those various topics into a discussion of accidents involving ambulance, police, and fire vehicles. The larger category of all motor vehicles is included to provide comparisons.

In 1995, the most recent year for which data is available, the National Highway Traffic Safety Administration's Fatal Accident Reporting System (FARS) reported that about 20,000 accidents involved police cars, 3,300 accidents involved fire trucks, and 3,800 accidents involved ambulances nationwide.[5] A typical fire department that offers ambulance service as part of its program may have seven to ten ambulance calls for each fire call. (The rate of fire vehicle accidents, extrapolated from the FARS data, is greater than the rate of ambulance accidents.) A different perspective was published by the National Fire Protection Association. For 1995, they reported that 14,670 accidents nationwide involved fire department vehicles and an additional 1,690 accidents involved firefighters' personal vehicles. Additionally, 950 firefighters on department vehicles and 190 firefighters using personal vehicles were injured in these accidents.[6] The NFPA does not collect data on the number of civilians injured in these accidents. They do report that the number of fire department accidents is about 0.1 percent of the total number of responses. The NFPA data is collected by a voluntary survey and from that sampling projections are made for the nation as a whole. It is possible that even more accidents occur than are estimated, and it is certain that the numbers of injuries and fatalities are higher than the numbers reported, because no civilian data are recorded in this particular category.

The State of New York uses a uniform method of reporting accidents by field agents. The accident investigation Form 104A is coded for con-

Table 11.1
Comparison of Causal Factors in Vehicular Accidents

Causal factors	Conditions	Motor vehicle accidents[1]	Police vehicle accidents[2]	Ambulance accidents[3]	Fire apparatus accidents[4]
Weather	clear	55%	59.2%	60.6%	49.0%
"	coudy	19.5%	21%	17.4%	25.5%
Road surface	dry	65.1%	68.7%	68.9%	67.1%
"	wet	20.6%	22%	21.1%	23.5%
Road character	straight and level	69.3%	77.9%	79.6%	76.5%
"	straight and grade	10.2%	11.1%	8.1%	10.7%
Location	intersection	52.5%	57.9%	69.8%	72.2%
"	non-intersection	37.3%	35.6%	25.9%	24.8%
Light conditions	daylight	67.5%	55.3%	61.6%	70.7%
"	dark, road lighted	20.7%	35%	28.9%	21.1%
Operator error	alcohol	1.9%	3.1%	0.9%	1.6%
"	driver inattention	8.5%	9.8%	7.0%	9.4%
"	failed to yield right of way	9.5%	9.4%	11.1%	15.5%
Vehicle defects		3.1%	2.4%	3.6%	3.7%

cise, defined responses. That data is entered into a centralized computer system which includes virtually all reported accidents statewide. The state outputs the data on Form 144A and provides a vast database across a broad band of categories. For instance, it is possible to focus attention on tabulations for a single city or the data can be expanded to a statewide view. Information can be provided based on geography, vehicle type, weather, road conditions, times, illumination, operator error, locations, and other categories. Presented below are a few pertinent categories for different vehicle types as reported for statewide accidents. The first category of motor vehicle accidents takes into account all reported accidents for all vehicles in the state.

Some interesting generalities can be extrapolated from these accident data. The typical accident occurs in the following conditions: clear weather, dry roads that are straight and level, at an intersection, and during daylight hours. Alcohol involvement is very slight, as are vehicle defects.

Apparent driver contributing factors are most heavily weighted toward inattention to driving and failure to yield the right of way.

The category that demonstrates a great degree of variability is accidents in intersections, where ambulances and fire vehicles have accidents in much higher numbers. While the police car accidents in intersections are greater than the general motor vehicle category, they are less than ambulances and fire vehicles. A possible explanation could be that ambulances and fire trucks are larger than automobiles, which translates into less maneuverability and greater stopping distances.

While New York State's data has been used in this chapter to provide an example of accident profiles, many other states have similar information. To those not familiar with these types of forms, it should be brought out that hundreds of entries are shown for each report. The illustration used in this chapter just introduces a small sample of the available information.

In New York, special printouts are available for each of the three emergency services, and it should be kept in mind that almost every accident report (MV104A) is completed by a police officer. As mentioned in an earlier chapter, in an accident involving an emergency vehicle and a civilian one, a bias may exist in favor of the emergency vehicle operator. This may be most evident in MV144A Table 13 (apparent accident contributing factors) and MV144A Table 22 (apparent driver contributing factor). Failure to yield the right of way, human factors and inattention to driving are frequently cited.

Emergency responses are not as safe as nonemergency travel. According to data from the National Safety Council, emergency vehicles have

Table 11.2
Reported Vehicle Accidents per Million Miles Traveled

Municipal police cars:[7] 36.34
Passenger cars (national):[7] 3.85
City trucks:[7] 21.60
Red or red-and-white fire trucks:[8] 62.1
Lime-yellow fire trucks:[8] 28.2
Ambulances *

* No data is available because these accidents have not been studied.

higher accident rates than nonemergency vehicles of similar size. Reported accidents per million miles traveled are detailed in Table 11.2.Because of a spiraling accident rate, the fire chief in one major city ordered his department to use emergency lights or sirens only on the most urgent calls.[9]

It should be noted that many fire, police and ambulance organizations keep internal records of accidents and runs. When studying this information, it becomes apparent that there is no standard method of data analysis or presentation. Some fire departments record activity by runs, which is the number of responses per vehicle. Other departments chronicle activity in terms of alarms. An alarm may require a single vehicle or a multiple vehicle response. Therefore, an alarm can account for a single run or many runs. Neither runs nor alarms define the actual distance traveled. (In the United States the National Safety Council records accidents per million miles traveled.)

A run in a major city may be only a block or two in distance, while rural departments may have runs of several miles. As a rule of thumb, fire stations are usually at about the center of populations or areas of higher value, and more alarms occur nearer the station. The common run can be very short and each run has two segments, an emergency response and a nonemergency return to quarters.

Police vehicles may be thought of as in service or in circulation. They move about an area or territory and only a small segment of their miles traveled are emergency responses.

Ambulances are similar to fire trucks in that they are in quarters until a call is received. Often, the ambulance call comprises three segments: (1) from the station to the scene, (2) from the scene to a medical facility (hospital), (3) a nonemergency return to quarters.

In order to be able to compare accident rates between groups, a common denominator is required, which has already been established and is in use. In one of its statistical presentations, the National Safety Council tabulates accidents and fatalities based upon miles of vehicle travel. Other studies have examined response times.

One study of ambulance calls found only a minimal difference in the time of response (that was felt to be insignificant), when no lights and sirens were in use. Two lines of thinking emerge from this study:[10]

1. The use of lights and sirens do not significantly affect response time. Patient sickness does not degrade because of differences in time of response.
2. The use of lights and sirens may actually make transit more dangerous.

Also, the ambulance siren sounds may easily penetrate the patient compartment. Some patients' condition may worsen when exposed to this type of noise. Anxiety levels can increase in an already worried patient.

Typical accidents for all vehicles, including emergency ones, occur near or at intersections. Based upon streams of traffic flow, this is logical. However, some incidents are peculiar to these special classes. Fire and ambulance responses often start from a station where these vehicles suddenly enter a street or road in an emergency mode. However, a civilian has little time to see or hear what is happening off to the side of his car, where a person would not expect to have an emergency vehicle appear and require the right of way. This is very different from an emergency vehicle coming down a street or entering an intersection.

Police vehicles have their share of hazards or misadventures in travel. High speed pursuits can finish with a successful stop or a collision, often with intense media coverage and negative public exposure. High speed pursuits can have very high profile results. Once initiated, it is difficult to break off and equally difficult to observe normal driving regulations. Among the issues are:

1. Is the pursued driver likely to cause further injury or death if not immediately apprehended?
2. Was the driver identified before the pursuit?
3. Does the nature of the crime warrant the high risk of a high-speed pursuit?
4. If a community bans high speed pursuits, does this encourage more crime by signalling to the criminal element that they should feel less threatened?

If a high speed chase can be avoided, it should be. There is danger to the suspect, the police officer, and the unsuspecting users along the route. The agency should define the parameters within which a high speed pursuit should occur.

A less publicized but very dangerous maneuver occurs when a patrol vehicle is screening traffic while parked on the shoulder of the road or in the median. When a violator is detected, the officer starts to pull out into traffic and this presents some problems. By crossing into oncoming traffic they increase the chances of an accident. If the officer is focused on the civilian vehicle, he or she may lose peripheral sensitivity. As the excitement or intensity level increases, the chances of peripheral detection decrease. Judgment is often affected by the level of intensity and excitement of a situation. As the adrenaline level rises, the width of scope decreases. A T-type or broadside impact can result and the chances for serious injuries in this scenario are high.

Emergency vehicle operation can in part be determined by intelligent dispatching. By properly screening requests for help, the dispatcher can often determine whether flashing lights or sirens are warranted. When the EMS personnel arrive and assess the patient, a revised decision can be made about the urgency of the trip to the hospital. Only life-threatening injuries or illnesses which *cannot* be stabilized by the ambulance personnel warrant an emergency trip.

A typical ambulance call could unfold in the following way. The local high school has a football game on a weekend afternoon and an ambulance with a trained crew is present for the game. Their medical responsibilities include the players and staff as well as the spectators. During the game one player seriously injures his lower arm. The game is stopped, he is assisted from the field and after a brief evaluation the trainer asks the ambulance crew to examine the injury. It becomes apparent the injury site is swelling and is quite tender to touch, but there are no life- or limb-threatening circumstances. After a medical history form is completed, the certified emergency medical technician recommends a visit to the hospital emergency department for x-rays and the school officials and parents agree. The time elapsed is approximately ten minutes. The patient is placed on a cot and loaded into the ambulance for transport. The vehicle slowly exits the stadium and enters traffic, where the siren and red lights are activated for the run to the hospital.

This type of scenario occurs every day in communities throughout the nation. The type of injury and the location of the incident may vary, but this can be a common and routine call. Except that nothing is ever routine! The question is: How vital to the patient's health is a trip to the medical facility using flashing lights and siren? If the trip is made in an emergency mode, will it be more hazardous than a nonemergency trip? Research shows that ambulance runs in the emergency mode do not save any great amount of time.[11] Previous chapters have discussed the risks present when motorists are surprised, startled, or confused by an approaching emergency vehicle. Accident data show that emergency vehicle drivers have more intersection accidents than do civilian drivers. Does the condition of the patient warrant the possibility of increased trauma or illness that can result in an emergency run? Training ambulance crews to appropriately use lights and sirens is very important. In a recent study it was found that lights and sirens were used inappropriately in 39.4 percent of the runs.[12]

All branches of emergency service experience common or routine calls that may not have to be answered with the fastest, most dangerous responses. Dispatchers are trained to understand and analyze the incoming request for assistance. A patient who has fallen from bed and needs help to get up does not need a flashing-light-and-siren approach. The fire call that is transmitted from a box with a high number of malicious false alarms should be responded to with utmost caution. If the first unit on the scene ascertains that a true emergency exists, then a radio request can upgrade the response. Often, a dispatcher will get several reports of a serious fire from different citizens. They may use the street box, home phone, or even a cellular phone. Were this to occur in the vicinity of an address of prior false alarms, the fire department should be advised that the chances of a real alarm has increased.

Police departments are constantly responding to burglar alarms. Many occur when a merchant opens his establishment for the day and forgets to deactivate his system, and even more occur when homeowners try to enter or leave their homes without first disabling their alarms.

Intelligent dispatching is a key to safe emergency vehicle response. The more that emergency vehicles can operate in traffic in a nonemergency mode, the safer everyone will be. Experienced dispatchers often recognize patterns of incoming calls that help them grade or classify the requests. Civilians do not have the necessary experience to allow them to comprehend or understand which calls rate a high priority and which do not.

As discussed previously in the text, accidents between vehicles occur more frequently when driving behavior is unexpected. Emergency vehicles operating with lights and sirens, entering intersections against the signal, changing lanes, going the wrong way on a one-way street, traveling against opposing traffic, or moving around the traffic flow instead of staying in the flow are all unpredicted and unexpected behaviors and as such increase the risk of accidents.

Volunteers often have to respond to a call by using a private vehicle, which opens up a new perspective. The operation of private vehicles while en route to emergencies or duty stations is the source of many accidents. The control officials have over volunteers is tenuous during the transition from civilian status to volunteer emergency providers. Some go directly to the station and some go to the call, depending on the relative locations of the responder, the station and the accident scene. These private vehicles are not emergency vehicles, even if they display courtesy lights to identify a responding volunteer. Regardless of the seriousness of the potential situation, these individuals must drive according to all the traffic laws governing private vehicles. As emotion-laden as a call can be, it is often difficult to control the adrenaline rush and excitement that a responder feels, and the operation of his or her vehicle will reflect those feelings. However, the dispatching systems currently employed are not foolproof and often what sounds like a major event can rapidly de-escalate to a nonemergency.

When all the factors are tallied, the public safety requires that emergency vehicle operators be trained well, and that equipment be properly selected so that it functions in an expected manner. The emergency response agencies must avail themselves of the latest scientific discoveries, but not be blinded by unsafe or misplaced tradition. The ultimate tradition is to save lives and property, and lesser traditions must be set aside to achieve this end.

Endnotes

1. New York State Department of Motor Vehicles, *Summary of motor vehicle accidents,* MV 144A. New York Statewide (1995).

2. New York State Department of Motor Vehicles, *Summary of motor vehicle accidents, MV 144A—Accidents with police vehicles.* New York Statewide (1995).

3. New York State Department of Motor Vehicles, *Summary of motor vehicle accidents, MV 144A—Accidents with ambulances*. New York Statewide (1995).

4. New York State Department of Motor Vehicles, *Summary of motor vehicle accidents, MV 144A—Accidents with fire vehicles*. New York Statewide (1995).

5. National Safety Council, *Accident Facts, Chicago*, 102 (1997).

6. Karter, M.J., LeBlanc, P.R., "U.S. firefighter injuries 1995," *NFPA Journal*, Vol. 90, No. 6, 105 (1996).

7. National Safety Council, *Accident Facts 1991,* 84. Chicago. (1991).

8. Solomon, S.S., King, J.G., "Influence of color on fire vehicle accidents," *Journal of Safety Research*, Vol. 26, No. 1, 47 (1995).

9. Schaper, F.C., "'On the quiet': St. Louis moves to reduce fire apparatus accidents," *Firehouse Magazine*, 1995:5, 88.

10. Hunt, R.C. et al., "Is ambulance transport time with lights and siren faster than that without?" *Annals of Emergency Medicine*, April; 25(4): 507–511 (1995).

11. *Ibid.*

12. Lacher, M.E., Bausher, J.C., "Lights and siren in pediatric 911 ambulance transports: Are they being misused?" *Annals of Emergency Medicine*, 29(2):223–227 (Feb. 1997).

Chapter 12

Accidents and Litigation—Case Studies

Stephen S. Solomon, B.S., O.D.

This chapter describes seven emergency vehicle accidents. Questions are listed at the end of each summary to assist in analyzing the accident. They are intended to stimulate thought, to assist the readers in finding potential flaws in the reports presented, and to develop an awareness of the reported causes. Also, the questions listed for one case can be used for the other cases as well.

12.1 Accident #1

On a sunny, clear, dry April day in 1975 the Springfield Fire Department in Massachusetts was alerted to a vehicle fire and dispatched a fire truck to the scene. While en route, the fire truck made a left hand turn and so crossed a lane of oncoming traffic. Approaching the intersection from the opposite direction was Raymond Stoddard on a motorcycle. A collision occurred when Stoddard did not see the fire truck in time and could not stop. As a result of the accident, Stoddard was seriously injured and permanently disabled.

Stoddard sued the truck's manufacturer in part because the truck was red and not as highly visible as it could have been. At the time of the accident sovereign immunity in Massachusetts was in effect and the municipality was not included in the lawsuit

One of the major contentions of the motorcyclist's claim against the manufacturer was that the fire apparatus should have used all methods available to assure it would be maximally visible. As a component of visibility, the truck should have been painted lime yellow.

According to the attorney who reviewed the case in an article,

> documents filed with the court showed that approximately 55 percent of accidents with emergency vehicles occur in intersections, that approximately 80 percent of fire apparatus accidents occur during clear weather conditions, that approximately 60 percent of fire apparatus accidents occur during daylight hours and that a substantial contributing factor to these accidents is the poor visibility characteristics of red-colored emergency vehicles.

It was then reported that "further testimony was that lime-yellow is a more visible color and is more likely to prevent and/or significantly reduce the number and severity of accidents involving emergency vehicles. Thus the truck that struck Stoddard was not properly visible and, therefore, defective." According to Krenzer, "Stoddard was represented in his action by the firm of Cargill, Masterman, and Culbert of Boston, Mass., with the case being settled out of court with Stoddard receiving a substantial settlement." Mr. Kenzer went on to state that

> it is apparent that operating a red fire truck can substantially increase your chances of a lawsuit. The use of red and white apparatus will not provide a valid visibility argument—you may still be held liable in court. The number of intersection accidents involving fire apparatus has increased from 1970, which means that more multiple vehicle accidents are occurring. This means more lawsuits. . . . Statistically, lime-yellow will stand up better in court than the traditional red. Can you imagine telling a judge and jury that your vehicles are red because it's "tradition"?[1]

From an accident analysis perspective, some questions that could be asked are the following.

1. Was the motorcycle operator as visible as other motor vehicles?
2. Was the fire truck maximally visible?
3. Did sun angle or time of day play a role?
4. Were the flashing lights mounted on the fire truck placed to maximize its ability to attract attention?

5. Could a different color of fire truck have played a larger role than what the attorney suggested?
6. If the case were not settled out of court could the award have been substantially more? Or, perhaps, less?
7. Since the accident did not occur at a signal-controlled intersection, was the fire truck less or more exposed to an accident?
8. Was the nature of the emergency significant enough to warrant a response with red lights and sirens?

12.2 Accident #2

The second example[2] is representative of an accident involving two fire vehicles. On May 3, 1994 at 0845 hours the Tulsa Fire Department received notification of a fire in an apartment. Among the equipment dispatched were a reserve pumper, Engine 29, and an aerial unit, Ladder 29. Both units were dispatched simultaneously from the same station. Based upon the location of the fire, two separate routes of travel were available, and the units took separate paths. There were three firefighters on board each vehicle. While en route, a message was received that this was a working fire and instructions for engaging the incident were transmitted.

About a block from the scene, the response routes crossed at 90 degrees and both rigs entered the intersection at the same time. An impact occurred when the front of Ladder 29, the aerial unit, struck the rear quarter of the pumper on the passenger side. Engine 29 struck a median curb and was propelled into the air where it flipped and rolled. It landed on its wheels but facing opposite the original direction of travel. It sustained large scale damage and equipment from the truck was strewn on the ground. Two of the three crew members were wearing seat belts. The unbelted member, the officer in charge of the vehicle, had unsnapped his, so that he could finish putting on firefighting gear. He was reported to have been thrown about the closed cab, but he was not seriously injured. The driver was also not seriously injured, and the rider in the jump seat suffered broken ribs.

The aerial unit came to a stop in the upright position without becoming airborne. The three crew members suffered no serious injuries. Both vehicles were immediately out of service and not capable of providing any support at the fire.

Another engine company shortly thereafter came to the intersection, viewed the accident, and then proceeded to the fire. The first question that immediately comes to mind is why the two vehicles took different routes to the same call? With experienced operators familiar with the streets, why didn't they see each other before they entered the intersection? Both trucks were red, had flashing lights and sirens, and daytime visibility was unencumbered by weather.

When interviewed, the pumper operator said he would be more cautious but would not do anything differently. This raises a serious question: this individual survived a very serious and violent vehicle accident which could have caused serious injury to at least six firefighters and any number of civilians, and the driver states he would do nothing differently other than be more cautious.

Even more disturbing is the fact that the operator of the aerial unit said "I still like to get there fast." From an accident analysis perspective the following issues should explored.

1. Is "fast" necessary?
2. Should gear be put on before leaving the station?
3. Should both vehicles respond using the same route?
4. If separate routes are better, should there be special safety policies in effect?
5. Were the sirens capable of being heard?
6. Were the apparatus maximally visible?
7. Could the flashing lights be seen?
8. Were the officers in the passenger seats observing intersection traffic or were they distracted?
9. Was the peripheral vision of either apparatus operator restricted by the urgency of the call?
10. What potential exposure did the fire department incur by an interrupted response?
11. Since the engine was a reserve unit, did it fulfill all the requirements made of the front line unit it was replacing?

A number of issues in the two incidents discussed above and are also common to those that follow.

12.3 Accident #3

The Mt. Laurel Fire Department[2] in New Jersey received a fire alarm and dispatched ladder truck number 3615. The date was October 5, 1994, and the time was 1706 hours (slightly after 5 P.M.). The truck was a "quint," which is a combined aerial unit and pumper. Such vehicles are among the heaviest rigs in municipal fire service. The truck was red and white and had a horizontal stripe, assumed to be retroreflective.

As the unit was responding it approached an intersection with the traffic light turning red. The driver proceeded into the intersection when an automobile that had the green light also entered and interfered with the fire truck. The apparatus driver took defensive action and swerved the large quint, lightly struck the moving car and then overturned on top of two stationary automobiles. One was empty, but the other had one occupant, a female, who was trapped in her vehicle, under the midsection of the quint. The five firefighters riding in the truck were not seriously injured and immediately evaluated the scene and initiated rescue efforts for the trapped civilian.

Command-level individuals immediately took charge, requested heavy rescue trucks and a commercial crane to lift the quint. Engine 3612 members assisted at the scene. Air bags and cribbing were employed as a second crane, which also had been summoned, working in concert with the first crane, lifted the wreckage. At that point, it was judged for safety reasons to pull the auto out from under the mass and then extricate the driver. She was removed from the car using special rescue tools, placed on a long spine board, and airlifted to a hospital. During this time, the original five-man crew from the wrecked fire truck were taken to a hospital to check for undiscovered injuries. A critical stress debriefing team was made available for the involved firefighters to help them handle the problems of the event. The volunteer driver was placed on administrative suspension. The chief, when interviewed, stressed the importance of a fire department's records for driver training and maintenance. He felt his department was fulfilling fire service standards.

The following questions should be asked.

1. Was the fire truck safely entering the intersection?
2. With siren and red lights in operation, did that absolve the driver of responsibility when entering the intersection against the light?

3. Was the fire truck maximally visible?
4. Was the response in the emergency mode necessary?
6. What was the written department policy for entering intersections?
5. Why was the driver suspended?
6. What was the front seat officer doing at the time of entering the intersection?
7. Did siren operation cause the driver to loose periphery or assume clear passage?
8. What were the fire chief's concerns?

12.4 Accident #4

On the night of October 10,1987, in Minot, North Dakota, an ambulance responded to an automobile accident, began treatment on a patient and started to transport the patient to a medical facility. At about 9:30 P.M., while this ambulance was proceeding using red lights and siren, it entered an intersection and struck a passenger vehicle which was traveling in a path perpendicular to the ambulance. After the collision, the automobile, which contained two people, spun around and came to a stop. The ambulance, with four attendants and one driver, was knocked on its side and slid into permanent obstructions. As a result, live electrical lines fell to the ground and traffic lights were rendered nonoperational. In addition to the usual hazards that are produced when an accident occurs, the live wires and inactive traffic controls increased the dangers at the scene.

The local police began an investigation. The ambulance driver said he did not see the passenger car until the accident happened, but also said he had gone through a red light. A witness who was travelling parallel to the involved auto stated she saw red lights and assumed an emergency vehicle would enter the intersection and began slowing. The investigating officers believed that the ambulance driver had seen the auto in question earlier than he had stated. However, even though North Dakota law allowed for an emergency vehicle operator displaying emergency lights and sounding his audible device to have the right-of-way, the operator must drive with a due regard for the safety of all persons using the highway. Based upon the information available, the ambulance driver received a summons for failure to exercise the "care required."

Some questions to be asked:

1. What was the speed of the ambulance when it entered the intersection?
2. If one civilian driver saw the ambulance, why didn't the other?
3. How did the officers calculate the spot at which the driver saw the automobile? What reaction time did they use?
4. What sounds were present inside each auto as they approached the intersection?
5. Did the ambulance driver attend an emergency vehicle operators' course and how much emergency vehicle driving experience did he possess? What was his driving record?
6. Did any of the drivers have impaired vision or hearing?
7. Were all drivers in good health and "fit to be driving" at the time of the accident?

From the author's perspective, emergency vehicles enter intersections countless times each day without incident. However, when the overall intersection accident rates are reviewed and a comparison is made with nonemergency vehicles, it is apparent that these emergency vehicles are prone to a greater accident rates.

12.5 Accident #5

In Rochester, New York, the day of Halloween in 1992, a school bus, with twenty-nine children ages five to twelve, was about two blocks from its destination, a local school. The children were reported to be excited because of the planned parties that afternoon, but they were not reported as being excessively noisy. About one-fourth of the children did not speak English as their primary language.

Simultaneously, the fire department had received a report of a fire and dispatched appropriate vehicles. One was an aerial ladder unit that weighed in excess of 20 tons. As reported by witnesses, the fire truck approached the intersection with its siren in operation and its red lights flashing. As it entered the intersection, the driver saw the school bus in his path and crossing at 90 degrees in front of him. Short skidmarks were left on the road surface. The estimated speed was reported to be at about 35 miles per hour and slowing. An instant later the fire truck hit the bus broadside, almost exactly in the center. The bus slid to a sidewalk and toppled onto its side. This blocked the normal exit door at the front of the bus. The rear door was opened, as was the roof hatch, and the children and accompanying adults were removed through these exits. Thirty-five persons were injured, including firefighters, children and adults on the bus. Several individuals required hospitalization, but there were no life threatening injuries. (The construction of school buses is well regulated and although they may appear flimsy, they are extremely durable. Some attribute the rugged construction as the reason why injuries are not generally more numerous or severe.)

The emergency services available in Rochester had completed a mass casualty drill about a month before this incident, and officials attributed the smoothness of the response to this type of training. Firefighters, EMS staff, and bystanders all cooperated in the rescue effort. Ample ambulances were available and the treatment and removal of the patients was accomplished professionally.

The fire chief was quoted in the newspaper as saying that the driver of the fire truck "went through a red light." At another time the chief was quoted as saying that "in cities, people don't respect sirens, air horns and flashers anymore." The fifty-eight-year-old school bus driver was reported to have said that she did not hear the siren of the fire truck, but looked up to see that vehicle a moment before impact. She did not see the fire truck before that. She was proceeding through the intersection with a green light. A window near the driver was sometimes left opened, but was closed that day. Her hearing was reported to be normal.

After an investigation by the police, their report concluded that both drivers "share blame for the accident but neither should be charged with a crime. . . . The city fire truck failed to use due regard for safety as the truck approached a red light." The school bus driver "failed to observe and yield to an emergency vehicle."

The district attorney said it was likely he would follow police recommendations and not file criminal charges. Traffic citations were not ruled out.

The fire chief stated that the accident would be reviewed by a department board of review and said that "more than likely" the board would recommend no disciplinary against the apparatus operator.

Some questions to be asked:

1. Was the fire truck maximally visible?
2. Could the school bus driver hear the siren? (consider inner bus noise, driver hearing capability, and sound localization)
3. What was the school bus driver's peripheral vision capability?
4. What influences of favoritism, if any, are present?
5. Was the fire truck operating safely?
6. Was the school bus driver actually "not yielding"?
7. What observations of witnesses would be prejudiced against the bus driver?
8. What observation of witnesses would be prejudiced against the fire truck operator?
9. Did speed play a role in the accident?
10. Was the fire truck operated safely? If so, what did the bus driver do wrong?
11. If the fire truck was not operating safely, what was the driver doing wrong?
12. Was the witnesses' testimony accurate? Did it affect police decisions on cause and effect?
13. What is the fire department's role in a mass casualty situation?
14. Did the time of day (sunlight, traffic, etc.) affect visibility?
15. Would retroreflective tape have improved visibility in this accident?
16. How many flashing lights were visible from the front of the fire truck?
17. What number of flashing lights were mounted in a forward-illuminating manner?
18. Could the fire truck's color have played a role in the accident?

12.6 Accident #6

On an October night in 1993, a seventeen-year-old male and his seventeen-year-old girlfriend drove to a friend's house, arriving between 8 P.M. and 9 P.M. While there they consumed an unknown number of beers. At about 11 P.M. they left to sit in a car belonging to one of the girl's relatives, and in what seemed an unplanned choice the male started the car and proceeded to back out of the driveway. His companion requested that he stop and he replied by saying, "No, I'm okay." He entered the state highway and drove east toward his home about 1.5 miles away. In a subsequent interview, the female stated that she asked him to pull over and let her drive. She added that she was aware that the right headlight was out.

The county sheriff provides road patrols and law enforcement services in addition to being responsible for the operation of the county jail. At 11 P.M., two deputies in a single patrol car left to begin their shift and proceeded west on the state highway. At 11:08 P.M., the deputies spotted an eastbound automobile without an illuminated headlight on the passenger side. Radar showed the vehicle was traveling at 50 mph in a 30-mph zone.

The deputies pulled their patrol vehicle to the right shoulder and waited for the automobile to pass them. In their report they stated that the area was dark and there was a sprinkle of rain.

As the car passed the waiting patrol car, the driving deputy stated he saw a male driving as he pulled out, crossed to the eastbound lane and turned on his flashing lights and siren. At this time the male driver saw the deputy's vehicle and increased his speed. A pursuit was now in effect. After several blocks the speeding teenager went through several red lights without incident. The sheriff's car pursued and also went unscathed through the intersections.

When initiating pursuit, the deputies radioed their dispatcher and reported the unfolding events. This department has a protocol in place for conducting pursuits. The teenager's escape route passed the front door of the sheriff's department's headquarters and another deputy was exiting that facility after hearing the details of the direction of travel. He later reported he estimated that the car was traveling about 70 mph and the patrol car was three blocks behind and going about 40 mph This corresponds with the original deputy's report which stated that he backed off so as not to push the other vehicle, but instead kept the vehicle's taillights in view. At no time did he lose sight of the pursued vehicle. He further stated he did not exceed 40 mph during the chase.

About one-half mile east of the headquarters is a railroad crossing which is graded above the level of the roadway, so that there is a sudden elevation of several feet up to the tracks while on the other side there is a more pronounced downgrade. Ahead of the speeding teenagers was a couple in their car also east bound and on the other side of the tracks. They were planning to make a left-hand or northward turn onto another street. The teenage male was now only a few blocks from his residence and approaching the tracks (the topography of which he was familiar). Given the substantial speed at which he was moving, there was enough velocity to launch his car well into the air and, as he descended on the far side of the crossing, his vehicle landed on the turning car ahead of him.

From the time he left his friend's house and proceeded east, until the accident, the elapsed time was about ninety seconds. The crash pinned the older couple in their car. They were extricated by the fire department and transported to the hospital, one by ambulance and the other by air evacuation helicopter. The teenage male was fortunate and suffered no long term serious injuries. His girlfriend was uninjured.

The speeding youth was tested for blood alcohol and found to have a level of .14 which was considered DWI. He was arrested, ticketed, and given appearance citations for court. The older couple survived their injuries.

The sheriff's department involved in this pursuit had basic policies in affect that addressed emergency driving and pursuits. Specifically, the department order for emergency driving refers to section 1104 of the New York State Vehicle and Traffic Law which requires the operator of an emergency vehicle to drive with "due regard for the safety of all persons." The operator of an emergency vehicle is not protected "from the consequences of their reckless disregard for the safety of others." Maximum speed is determined by traffic, weather, road conditions and other hazards. Further, the emergency vehicle must come to a complete stop at red lights, flashing red lights, and stop signs. The operator can proceed when he decides it is safe to do so. The decision to respond using red lights and sirens is based upon the operator's judgment of:

1. The nature of the call.
2. The urgency of police presence at the scene.
3. The danger of collision with other emergency vehicles.

When contemplating a pursuit response, the deputy should consider the following.

1. The seriousness of the offense.

2. The road conditions.
3. Traffic flow and conditions.
4. Time of day.
5. Type of vehicle being pursued.
6. Knowledge of the area of the pursuit.

The pursuit policy cautions that because of the level of emotions, personal control is important. The sheriff goes on to caution that the opportunity for pursuit is often tempered by the seriousness of injury to the public. It is also stated that a decision to stop a pursuit if the conditions deteriorate is a proper choice.

As a result of the accident, the older couple filed a lawsuit against the county, the sheriff's department, the pursuing deputies, and the teenage operator along with his passenger, whose family owned the car. After lengthy litigation, the case was settled with the county paying an amount less than $100,000.

Some questions to be asked include:

1. Did the deputies do the right thing by pursuing the automobile?
2. Could the deputies see the driver when he passed in front of them?
3. Were the rules of pursuit obeyed?
4. What extrication methods were used to remove the older couple from their car?
5. What was the elapsed time from reported entrapment to entry into the hospital?
6. Were sirens and emergency lights in use?
7. How accurate was the speed judgment of the deputy exiting the building and observing the pursuit?
8. Did the deputies observe the pursuit rules and not endanger others using the roadway? For questions 9 and 10, assume nonlife-threatening injuries:
9. Did the ambulance transporting the patient need to have lights and sirens in use?
10. Was a helicopter transport required?

12.7 Accident #7

One afternoon in the spring of 1984, a burglary was committed at a residence. After it was reported to the village police, an officer dispatched to the scene discovered that the thief exited the home through the back door. At the time of the crime, a new snowfall had blanketed the ground, providing the officer with fresh tracks to follow. These tracks led from the rear exit to a neighborhood home.

The officer approached the home, attempted to have someone inside admit him, but was unsuccessful and concluded no one was at home. He returned to the police station and earned that a juvenile who resided there had a criminal record.

Before moving into the county, this suspect has been arrested, tried, and convicted of armed robbery during which a clerk was shot in the back and killed. It was conjectured that the current suspect shot and killed the clerk. He was sent to a youth home and served a short sentence. After moving to his current residence this individual was thought to have become involved with a loosely knit group of local teenagers, all with some common interests of questionable social and community value.

With the suspect's background disclosed, the officer contacted the district attorney to advise him of the desire for a search warrant which was subsequently issued. A group of officers returned to the suspect's residence and were met at the door by the parents. When explaining the reason for the visit, the mother requested permission to call an attorney and was instructed to do so. The step-father was very cooperative and directed the police to first search the son's room which was on the lower level.

After going to the son's room and knocking on the door, they entered and discovered a neat and orderly room. During their search, they discovered evidence of the stolen materials which had been concealed from direct view. The mother joined the group and asked if she could speak with her son for a moment. This was permitted in a corner of the nearby family room. As this was in progress, the juvenile suddenly bolted from the room and up the stairs with his mother slowly following him. It gave the appearance that she may have been trying to impede the police from apprehending her son. He passed a table or counter on the way out of the front door, climbed into his family's car, rapidly accelerated and fled the scene. It is not known whether his mother had given him the keys or instead had grabbed them as he left the house.

The officer who started the investigation earlier that day ran upstairs, struggled past the mother, and pursued the juvenile suspect in his marked police vehicle. Once en route, he notified by radio other law enforcement agencies of the situation. The path of travel was east on a two-lane state highway. Near the six-mile point the officer noted that he no longer saw taillights ahead and assumed the fleeing suspect may have turned his lights out and was trying to provoke an accident to possibly disable the patrol car. He immediately braked. The chase lasted about six miles at high speed, ending when the fleeing driver struck the rear of another eastbound car and then collided head-on with a westbound vehicle. The third vehicle may not have had headlights on and, after testing, that driver was found to be DWI. In the aftermath of the crime, search, escape, pursuit and multiple vehicle crash, two people died. The suspect expired several days later of his injuries and the driver of the westbound car died at the scene.

The ensuing lawsuits were settled several years later with few details made public.

Some questions to be asked:

1. Was the pursuit appropriate?
2. Did the pursuing officer use appropriate measures? (assume 1984 knowledge and rules)
3. Would it be possible to examine the headlight filaments and determine if they were illuminated?
4. What role did alcohol play in this accident?
5. Were the mother's actions a contributing factor?
6. Was the suspect capable of serious harm to others?
7. Did weather and time of day have any relation to the accident?
8. What emergency measures did the ambulance personnel initiate to sustain life?
9. Which police jurisdiction investigated the incident?

Endnotes

1. Krenzer, C.A., "Seeing red over fire engines," *Emergency Magazine*, 19:11 (Feb. 1987).

2. "American Heat," (Carrollton, Tex.: Fire Education Television Network, Feb. 1995).

Chapter 13

Accident Analysis Considerations

Stephen S. Solomon, B.S., O.D.

The investigation of an accident is never "cut and dried." As soon as an investigator concludes that a collision is no different from any previous one, the investigator's effectiveness in finding facts is blunted. This chapter will address the many factors that must be considered in analyzing an accident. But first, to show how the unique and unusual characteristics of accidents require individualized analysis, we begin with a brief discussion of two different accidents, not because they were serious, but because they were unusual.

13.1 A Review of Two Accidents

The first accident was between a fire fighting vehicle and a civilian automobile in an intersection. The emergency vehicle was responding to a call and was using lights and sirens, while the civilian was stopped at an intersection to provide right of way. As the fire truck approached the intersection, its right front tire and wheel assembly separated as a single piece from the truck, rolled across the intersection and struck the civilian's car with substantial force (hence the classification of an intersection accident). Equipment failures account for very few fire apparatus accidents; in Chapter 12 it is reported as less than 4 percent.

The second fire apparatus accident, reported as one causing personal injury, involved an unusual chain of events. As the unit in question pulled up in front of the presumed address of the emergency, the crew stepped off the truck and began performing their assigned tasks. One individual, as common practice, placed a chock against the wheels to prevent rolling. (When a pumper is set up to pump water, the engine continues to run but the power is transferred from the rear wheels to the pump. There have been instances in which a pumper has inadvertently shifted into road gear from pump gear and the result is a driverless vehicle in motion. NFPA standards address safety considerations of fire apparatus operation while in pump mode.) At that moment, the driver was notified that he was parked at the wrong address and called his crew to remount. The individual who had placed the chock started to remove it and the truck simultaneously moved forward, rolling over the firefighter's finger.

13.2 Accident Analysis: The Role of Vehicle Maneuvers and Size

Accident analysis should take into account the concepts and examples presented in this chapter. Points are included to help planners and trainers establish safe vehicle operation programs. This is not meant to be an exhaustive outline for a driver training program, but it should serve as a useful supplement for such programs. For investigators, it should provide additional details specific to emergency vehicles.

The largest group of emergency vehicle accidents comprises those classified as minor and, as such, they do not appear in state motor vehicle accident reports. Backing a rig into quarters is a maneuver responsible for damage on many occasions. Accident-reduction programs suggest that a second crewman, standing to the rear of the vehicle and in view of the driver, help direct the vehicle safely into the garage or bay. The large mirrors mounted on the sides of the vehicle doors often strike objects or people, or are themselves bent or broken while reversing. Rear bumpers often strike obstructions such as the rear walls of the station. Since backing the vehicle into the bay may be a transition from daylight into darkness, visibility is markedly reduced by poor illumination and the difference of contrast. Stripes are often painted or glued to the floor to aid the driver, but need to be refreshed on a regular basis.

At any time that the vehicle is driving in reverse, a spotter or assistant to the driver should be deployed at the rear of the vehicle and positioned so he is visible to the driver in the side view mirrors. Standardized hand signals should be used to facilitate communications between the observer and the driver. Yelled commands are often not effective in that vehicle noise and background sounds can obliterate voice communications. The driver must be aware of the spotter's exact location. If the spotter is not visible, the driver must stop the vehicle.

While relatively minor incidents are often dismissed, the expense of medical treatment or vehicle repair is still significant. Accident investigations are difficult in these situations because record-keeping requirements for collisions may not be as stringent as for other types of accidents.

The larger vehicles present problems of a greater magnitude and among them is driving in reverse. One of the key reasons is the size of the blind spot to the rear. In the case of a police car with a rear window and a windshield-mounted rearview mirror, the areas around the vehicle are much more visible. Given how our automobiles are designed, the corner posts are the visual obstructions, with the windows providing reasonably wide views in all directions. Conversely, in an ambulance or fire truck, the cab does not have a functional rearview mirror mounted on the windshield and there is no way to view directly to the rear of the vehicle. (Closed circuit television with a small camera mounted on the rear of the vehicle and a monitor located in the cab is available, but not in widespread use.) A compounding factor arises from the driver's greater height above the road surface, which causes a shift in depth perception in some individuals. The police officer, as stated previously in the text, drives the same type vehicle at work as he does during his time away from work. He is much more familiar with the handling characteristics of his vehicle, he has more experience driving it, and it is smaller than an ambulance or a fire truck.

However, the police car does have one significant structural problem that does not exist in private automobiles. Radar equipment is often mounted on the dashboard of police vehicles and it often obstructs the officer's vision. The radar gun, often mounted near the midline of the vehicle is large enough to block important fields of view. So, although the operation of the police vehicle is very similar to a private automobile, the vision obstruction is not. Safe operation can be compromised by this equipment.

Generally speaking, the larger the class of vehicle, the greater the number of accidents per million miles traveled. Special driver's licenses, known as commercial driver's licenses (CDL) are required for operators of large trucks and other large vehicles. These include vehicles having a gross vehicle weight over 13 tons, vehicles towing units of more than 5 tons, vehicles carrying 15 or more people, vehicles transporting passengers under Article 19-A, Section 509(a), and vehicles of any size carrying hazardous materials. An exception has arisen in some states because of delays in implementation or exemptions in the motor vehicle regulations for the drivers of fire trucks. The CDL requirement is tied to federal regulation (Federal Commercial Motor Vehicle Safety Act of 1986-CMVSA/86) and states may be limited in the duration of this exemption. At this time, it is very possible that some fire trucks are being operated by drivers who do not have CDL certification

13.3 Reaction Times

Accidents which are far more serious to life and property require some special knowledge in accident analysis. The investigation of an emergency vehicle accident involves many of the techniques used to investigate other vehicle accidents, and there are some statistics about civilian automobile collisions that assist the analysis of these events. Reaction times need to be understood and carefully considered.[1] The use of a 0.75-second reaction time for these calculations is extremely suspect and any investigation in which it is used must be viewed skeptically. (See Chapter 7 for a discussion of reaction times.) Actual reaction time, rather than some assumed time, should be considered before setting up calculations for accident reconstruction. Driver perception and reaction is a very complex process that is affected by many variables. These include inputs from the driver, the roadway, the environment and the vehicle. A profile of each of these sources of input will help us to analyze any accident.

The driver profile should take into account the following factors:

- age
- sex
- health
- nutrition
- state of rest or fatigue
- observed alertness
- medications
- drug or alcohol consumption
- years of driving experience
- hearing
- vision (components of vision testing are listed in Chapter 3)
- prior driving record

The roadway profile should include:

- limited access or intersections
- intersections, including traffic controls or signs
- posted signs or traffic regulations
- parking available or no parking
- one-way or two-way use
- straight or curved
- inclined or level
- surface material and condition of the material (establishes contrast and conspicuousness)
- debris on road surface (can cause distractions)
- traffic volume and speed
- condition of paint that marks lanes, center line, shoulder
- highway illumination
- highway lighting including type of bulb emission (incandescent, sodium, mercury, halogen)
- obstructions to line of sight
- unusual activity in the immediate environment (construction, accident, fire)

The environment profile should contain:

- weather
- accumulation on roadway
- precipitation by type and intensity
- temperature
- atmospheric transmissivity (clear, hazy, fog, smoke)
- sun position (time of day and time of year)
- lunar quality (full moon, obscured, partial moon)
- profile of off-road surroundings (rural, forested, commercial, advertising displays, residential)

The vehicle profile should include:

- size of vehicle (establishes eye height and subsequent field of view)
- age of vehicle
- windshield condition (glass surfaces rapidly degrade and scratch)

- windshield wiper condition including windshield washer solution volume
- location and mounting of mirrors
- angle from driver's position to the front corner posts, ornamentation, hanging obstructions, storage of packages obscuring vision
- seat settings compared to driver's stature
- windows opened or closed
- stereo, radio, tape or disk player, air conditioning, cell phone (any or what number in use and at what setting or volume)
- headphone use
- antilock braking system

After considering the reaction time of the operator, we can calculate the estimated stopping distances, which are described by the basic formula: SD = RD + BD where SD is stopping distance, RD is reaction distance, and BD is braking distance. Further, the RD is a function of reaction time and the speed of the vehicle. (See Tables 13.1 and 13.2.) The compound reaction time increases as the complexity of the driving task increases. A lane change on a high speed expressway will encompass more judgment time than braking for a ball rolling into a rural roadway. Approaching an emergency scene in a private vehicle may be a very complex driving task. (Reaction distances are added to the other components of total stopping distances.)

The reaction time estimates have been discussed by Olson[1] and he points out that two main categories exist, standard and nonstandard. Within the standard category, relatively straightforward and simple situations with an alert and healthy driver may yield a reaction time between 0.75 to 1.50 seconds. However, as discussed previously in this book, the 0.75 second time is suspect when considering the variety of factors that can come into play. It would be difficult to support the use of a 0.75-second reaction time unless very careful documentation exists, showing that the following criteria are met: young, alert driver; ample driving experience; healthy; rested; no drugs or alcohol; a simple scene with a large, visible hazard; and good visibility. In short, the use of a 0.75-second reaction time is justified when the driver is *acutely* aware or is informed that on a specified signal he or she is to commence braking. However, this scenario does not describe a large number of vehicle operations.

Table 13.1
Distance Traversed as Related to Reaction Time and Speed (expressed in feet)

Speed (mph)	1.5 sec.	2.0 sec.	2.5 sec.	3.0 sec.	3.5 sec.	4.0 sec.
10	22	29	37	44	51	59
15	33	44	55	66	77	88
20	44	59	73	88	102	117
25	55	73	92	110	128	147
30	66	88	110	132	154	176
35	77	103	128	154	180	205
40	88	117	147	176	205	235
45	99	132	165	198	231	264
50	110	147	183	220	256	293
55	121	161	202	242	282	323
60	132	176	220	264	308	352
65	143	191	238	286	333	381
70	154	205	256	308	359	410
75	165	220	275	330	385	440
80	176	235	293	352	410	469

The four phases of the perception and reaction process are detection (when awareness first occurs), identification (when the a hazard's existence is ascertained), decision (when an appropriate response is chosen), and response (when the central nervous system signals the body to respond).[2]

Departure from such nonstandard reaction times occurs in a more cluttered or complex situation. As a general rule, the perception-reaction times increase as the number of competing stimuli increase. A second rule is that as the situation declines in familiarity, perception-reaction time increases. The faster the nervous system can detect a stimulus and interpret its message, the shorter the time lapse and the quicker the response. We mentioned in our earlier discussions of red-and-white two-toned color schemes, flashing lights and audible devices, that investigators assume

Table 13.2
Speed Conversion Chart

Miles per hour	Feet per sec.	Feet per 1/2 sec.	Feet per 1/4 sec.
10	14.7	7.3	3.7
15	22	11	5.5
20	29.3	14.7	7.3
25	36.7	18.3	9.2
30	44	22	11
35	51.3	25.7	12.8
40	58.7	29.3	14.7
45	66	33	16.5
50	73.3	36.7	18.3
55	80.7	40.3	20.2
60	88	44	22
65	95.3	47.7	23.8
70	102.6	51.3	25.7
75	110	55	27.5
80	117.3	58.7	29.3

that the message received by the civilian operators may not have the neurological effect the emergency vehicle operator expects. *Using those methods and devices previously discussed in this book to make emergency vehicles more conspicuous may be counterproductive—they may actually present less-than-informative stimuli which in turn may increase reaction time.*

The experience of driving near an emergency scene or sharing the roadway with an emergency vehicle is not a common occurrence for many drivers, most of whom are more than willing to react in a law abiding manner, when they ascertain what the warning message means. In the panicked attempt to do the appropriate thing, they often make an unexpected move, placing them in harm's way. Reviewing the incident in a calmer moment, it is not surprising to hear them state something like they didn't know what they were thinking of when they suddenly saw the responding emergency vehicle near them. The introduction of something not routinely experienced increases the reaction time.

The stopping distance's second component is braking distance, which is the measurement that starts when the brakes are first applied and ends when the vehicle comes to rest. At the moment the brakes are applied, the perception-reaction phase has been completed. This is a critical juncture because there is a shift from a primarily human function to a mechanical activity. At this juncture between the two components of the equation, namely reaction distance and braking distance, the time required from the moment the appropriate response is decided until the actual initiation of the physical response occurs. In other words, it is the time it took for the driver's arms to start to turn the steering wheel or the foot to move toward the brake pedal. Some investigators include this in the total perception response time and this seems appropriate.

Braking distance is equally as complicated as the reaction distance. There are a number of tables that provide some of these calculations and break them down into vehicle sizes. Some refer to vehicle construction and delineate vehicle groups by the number of axles in use. These tables can differ. Further, there is not a universal requirement among the fire, ambulance, police, and commercial trucking organizations which specifies the required braking distance for emergency vehicles.

The National Fire Protection Association requires that a fully laden fire fighting vehicle have a stopping distance of 35 feet when the vehicle is traveling at 20 mph For overall stopping distances, the reaction time distance must be added. For a commercial tractor-semitrailer with air brakes, the Office of Motor Carriers requires a stopping distance of 40 feet when the vehicle is traveling at 20 mph The same reaction time comments apply.

The American Trucking Associations (ATA) has provided information on typical air-braked tractor-semitrailer applications. The basic criteria they established was an initial brake lag of 0.5 second, brakes in reasonable adjustment, smooth, dry and clean pavement, typical tires, full pressure on brakes, appropriate load, and the front brakes operational. At

30 mph a vehicle fulfilling these requirements and operating on the described surface would have a braking distance of 82 feet. At 60 miles per hour, with the same criteria, the total braking distance would be 284 feet. (A loaded truck will keep its tires more firmly on the road with less locking of the brakes.) Federal Department of Transportation tests (60 mph) reported by the ATA, with a loaded vehicle yielded results of 230–280 feet and with an unloaded truck, 300–410 feet. It should be remembered that these findings do not include perception-reaction distances. The stopping distance for an automobile at 60 mph is about 180 feet (perception-reaction distance must be added). Table 13.3 presents the ATA data combined with the reaction distance. These measurements are generalities and should not be viewed as absolutes.

This chart gives estimates for stopping distances of a heavy truck similar to a large fire truck such as a tanker, pumper, or aerial ladder unit. The driver reaction time is 1.5 seconds. With very heavy vehicles there can be a lag in initiating braking of approximately 0.5 second, and this is included within the calculations. Brake fade, the lost effectiveness of brakes when they heat, and which is greatest at higher speeds, has not been added to the data. There are many variables and this table is meant as

Table 13.3
Heavy Truck Stopping Distances: Air Braked, Tractor-Semitrailer

Speed (mph)	Reaction dist.	Mechanical	Total stopping
20	44	40	84
25	55	60	115
30	66	82	148
35	77	107	184
40	88	135	223
45	99	168	267
50	110	202	312
55	121	242	363
60	132	284	416
65	143	330	473

a general guide only. The brake performance data has been adapted from material supplied by the American Trucking Association, Alexandria, Va., 1996. (Variables that affect total stopping distance can include brake adjustment, brake pedal pressure, pavement character, tire condition, vehicle load, and fade.)

A second table with similar vehicles and speeds is presented to illustrate another version of total stopping distances. A modified version was published for the emergency services and was updated in this text to apply appropriate reaction times. The concept being emphasized is that no single chart or table will provide the ultimate, accurate data. Estimates are readily available and must be adjusted to the individual situation.

13.4 Skidmark Patterns

At the time of an accident, skidmarks are often evident on the road surface or shoulders. Careful measurement and documentation by a reconstructionist is important to future outcomes. This section is not meant to be a complete "manual" on skidmark analysis but will point to several critical factors. While specific facts are being discussed it is important to keep in mind that speed calculations from these measurements are inexact. ABS (antilock braking systems) equipped vehicles do not leave skidmarks in panic stops. Also, a wet highway may not have any visible skidmark, even though vigorous braking occurred.

Table 13.4
Stopping Distances

Speed	Cars	Ambulances	Fire trucks
20 mph	69 feet	74 feet	84 feet
30 mph	121 feet	133 feet	158 feet
40 mph	193 feet	213 feet	253 feet
50 mph	298 feet	335 feet	365 feet
60 mph	432 feet	492 feet	532 feet

These stopping distances are computed by combining the braking distances and the reaction distances based on a reaction time of 1.5 seconds. These figures are approximations of stopping distances on flat, dry surfaces. (Recognizing that authorities differ on stopping distances, this table reflects estimates only.)

Some observations that help distinguish these patterns from others on the highway include the appearance of the skid, which occurs when the wheels lock, causing friction and heat to increase which forces the tires leave a pattern. The investigator should look for multiple, matched stripes from several wheels that are usually straight in appearance. These stripes should show for each side of the vehicle and often there are definite marks from all wheels of a passenger vehicle. The tread grooves should be evident as part of the imprint. The pattern usually has a definite starting point and a fairly distinct endpoint. They do not gradually fade away until they are no longer visible. If the skidding car stops because of a collision or if the wheels lost contact with the road surface (i.e., the vehicle rolled over or became airborne), the vehicle's skidmarks do not reflect the accurate distance that the vehicle would have skidded to reach a complete and uninterrupted stop. In this situation the estimation of speed is markedly more complicated. Some investigators believe that using a test vehicle to make similar skidmarks helps determine the speed of the vehicle in the accident, but there is significant difficulty in matching all the variables. Among them would be tire wear, tire pressure, brake condition, road surface, temperature, braking pressure, and initial speed. Some of these variables are thought to be of minor significance, but if accurate speed determinations are required then these items can not be totally discounted. (Performing skid tests at the scene of the previous accident can be an extremely dangerous exercise.)[3]

One purpose of making test skidmarks is to provide data for determining the coefficient of friction. Knowing this value helps us to compute stopping distances. However, if the accident occurred on a slight grade, then that value is affected by the influence of gravity.

Nomographs are available which predict speed based on the length of the uninterrupted skidmarks and the drag factor, which is derived from the coefficient of friction.

13.5 Auxiliary Braking and Traction Devices

Climate and difficult terrain affect the construction of emergency vehicles. The severe requirements put forth by snow, mud, sand, off-highway terrain, and seasonal roads find their way into the purchasing processes. All-wheel drive similar to that found on sports utility vehicles can be ordered for emergency vehicles. With power being delivered in all quadrants instead of just the rear, significant assistance is available when traction is needed. However, if ice is encountered, the extra traction of all-wheel drive may not be enough to overcome the problem.

Unlike automobiles, heavy trucks often have an auxiliary braking device known as a retarder, which acts like a braking system in that it interacts with the engine, transmission, drive line, or exhaust system. The purpose of these components is to slow down a heavy vehicle merely by taking pressure off the accelerator. It is believed that the use of retarders extends brake life and wear is reduced because the engine and transmission participate in the slowing process. Furthermore, brakes generate heat which is usually dissipated by air currents beneath the vehicle. As brakes heat up their effectiveness tends to decrease; this is termed "brake fade." Retarders reduce this overheating by absorbing some of the kinetic energy of the moving vehicle and thus preserving the brakes' capability. During periods of prolonged braking, such as descending a long grade, or repetitive stopping or slowing in traffic, the retarders are a significant aid in extending brake life and reducing brake temperatures, which results in safer vehicle operation. These devices often have the capability of being inactivated or activated by the operator. See Figures 13.1 and 13.2.

Figure 13.1 *The dashboard mounted toggle switch on the far left is to activate the retarder system.*

Figure 13.2 *A retarder switch mounted on the dashboard*

Antilock braking systems (ABS) are available on newer models of vehicles and may be found on police cars, ambulances, and fire trucks. It is thought that these systems allow more control when stopping during a skid or on slippery surfaces. Braking habits for cars without ABS, including pumping the brakes, are not recommended with ABS brakes. Traditionally, when braking on a slippery surface or when the wheels begin to skid, the normal response is to rapidly apply and then release the brake pedal; however, with ABS constant brake pedal pressure is applied. The cycling of the braking-releasing activity is performed automatically by the system rather than the driver.

An additional option is provided in trucks with dual rear axles. The driver can interconnect both axles for added traction or braking and can control this from the driver's seat by a operating a switch or lever. According to one major manufacturer the interaxle setting should be in the "off" setting for normal highway use and engaged when slippery or wet conditions are encountered. It can be engaged for long durations and at normal operating speeds. (See Figure 13.3.)

Single- and dual-rear-axle vehicles may be purchased with a differential lock. The differential is the pumpkin-shaped device at the rear end of the drive shaft; it converts the torque of the engine and transmission so

Figure 13.3 *The dashboard switch illustrated in this picture is the axle interlock activation lever. It connects both axles in a tandem rear setup to improve traction.*

that relatively equal amounts of energy are distributed to the rear wheels. When the differential lock is engaged, it reduces the tendency of the rear wheel to spin or slip when traction is reduced. It is manually activated by the driver while the vehicle is moving at low speeds and should not be used for speeds in excess of 25 miles per hour. For larger vehicles with a differential lock, turning is affected and the radius of the maximum turn is increased.

When accidents are investigated, it is important to ascertain the presence of these devices. If the appliance is not present on that vehicle, it is important to determine whether the vehicle was specified to omit this, and why. The age of the vehicle requires special attention in this inquiry. Some of these items may not have been available at the time of purchase and some older devices may not be in use on newer vehicles. Some older vehicles have a special brake control switch which is labeled for dry or slippery road surfaces. When the slippery road setting is used, the braking capability of the front brakes is reduced. It is also important to review driver training procedures to assure that appropriate education was provided in the use of these specialized attachments.

In the event of an accident, the switch position or settings for these devices should be recorded. The training procedures and documents should be reviewed to determine when these devices are to be activated or deactivated, and the specifications of the vehicles should be studied to determine whether they were purchased with any of these devices. Further, was specific training and documentation for these devices provided to the driver, and what was the normal and customary training administered by the involved agency?

13.6 Vehicle Transmissions

The transmissions of vehicles have origins in early automotive history. The use of a manual transmission was prevalent in emergency vehicles until fairly recently. Most lighter units, such as cars and ambulances, have been equipped with automatic transmissions for many years, but the heavier fire vehicles have not been so equipped until recently. There are still fire vehicles in service with the older, standard transmission. Depending upon the age, experience, and training of the driver, the transmission can the driver's ability to control the vehicle and safely brake. (While older drivers in our society may have learned to drive when standard transmissions were commonplace, the younger generations have decidedly less exposure and experience.)

If total stopping distances are relevant in an accident analysis, the topics discussed in this chapter should be considered. The use of a simple chart without qualifying comments does not do justice to the complexity of issues that affect these measurements.

13.7 External Traction Aids

For years trucks and busses had built-in sanders that discharged material upon demand. These units were usually mounted in front of the main drive wheels and had delivery tubes that extended downward. The operator had a control device that released the sand or grit. There may still be some sanders in use today, but they don't seem to be a special order item in new vehicle construction. Built-in chains offer another means of improving roadway traction.

Historically, tire chains were common years ago and when snow fell in sufficient amounts, automobile operators mounted chains on the rear wheels of their cars. With the advent of more aggressive tire tread designs, better highway surfaces and more efficient municipal snow removal, the use of tire chains on autos has markedly declined. They may still be found in more rural or isolated areas. Emergency vehicles do not have the luxury of declining travel in extremely bad weather or of waiting until the plows open the road. (Volunteers still have the problem of getting to their stations in case of a call and some fire and EMS departments may even provide space for them to bunk when bad weather threatens.) Tire chains for emergency vehicles are currently available and are in use by many departments. Some chains are applied only to the outer tires on dual wheeled vehicles, others are applied over both sets of the dual wheels, and a third type has a tension band on the chains that fits between the inner and outer wheels of the dual chain sets. Obviously, it can be a time-consuming task to apply these devices and, during challenging snow conditions, they should be in place as a matter of routine, based upon local procedures. Further, it is important to use a chain that matches the tire size.

As important as it is for chains to be mounted in anticipation of snow-covered roads, it is equally important to promptly remove them when the roads are clear. Driving on chains makes for less safe vehicle operation when the roads are bare.

A new device for vehicles has made a significant penetration into the traction market—chains permanently mounted under the truck. However, unlike those that were mounted on the tires, these new chains are mounted on a shaft and twirled by a small rotor. The rotor's rate of spin is established by a device that contacts the tire, so the spin varies directly with the vehicle speed. When they are not needed, they are retracted like the landing gear of an airplane. When deployed, they spin and fan out under the rolling tires, providing a succession of chain strands to drive on and thus enhancing traction. (See Figures 13.4, 13.5.) There is some question, however, about the performance of the automated chain strands in deep snow. The manually-affixed tire chains perform well in snow of varying depths but obviously are more difficult to apply and remove.

When traction is an issue in an accident investigation, the construction of the emergency vehicle should be considered, with special consideration of how local conditions affected the vehicle's original design. The written procedures for the use of traction devices, other responses to bad weather, and drivers' training records should be reviewed.

Figure 13.4 *Permanently mounted traction chains for use in snow are shown in the deployed position.*

Figure 13.5 *The chains in this view are the same ones in the previous shot, but in the inactive or retracted position.*

13.8 Bulb Analysis

It is common to read in an accident report that one of the vehicles involved in the accident did not have certain critical lights in operation. Often cited are inoperable headlights, directional signals, and brake lights. Gathering data with regard to bulb operation can be very informative and should be collected as soon as possible. The gathering and protection of this evidence provides more complete information to make a quality analysis. A standard procedure for many fire departments responding to vehicle accidents is to disconnect the battery cables of damaged vehicles to decrease the risk of a subsequent fire. This may aid in preserving some bulbs if the switches are still in the on position. Firefighters often turn off the ignition and lights as part of stabilizing the vehicle, and police reports do not often mention these actions. (See Figure 13.6.)

Therefore, a record should be made of the positions of activation switches inside the vehicle and the condition of all the bulbs on that vehicle as soon as possible. This will be helpful when attempting to prove

Figure 13.6 *The large round mechanism is the master switch which inactivates almost all of the on-board circuits. When an emergency vehicle is shut off, that switch protects the electrical system from excess discharge and subsequently drained batteries.*

whether certain lights were on or off. Also, the fire officer in charge may have to be asked about what stabilization steps were taken. If a question arises about certain lights being illuminated, and those lights are completely destroyed or not available, a test on its sister bulb on the other side or corner of the vehicle may provide information. When gathering data, make every effort to find and identify bulb bases that often remain in sockets after the rest of the bulb is destroyed. For bulb remnants and bases that are still in the sockets, do not attempt to separate them but rather cut the supply wires and remove the whole piece. This also applies to bulbs that appear whole but may be loosened or cracked by the impact.

Recall that tungsten filament bulbs emit light by heating the coil of wire contained in a partial vacuum. The hot filament is more ductile and flexible than a cold filament and will react differently when exposed to atmospheric gases. The relative age of a bulb can be estimated by the discoloration of the wire filament, the condition of the bulb's base, and the blackening of the inner surface of the glass shell. By careful inspection of the device, some determinations can be made of the state of operation of the bulb at the moment of impact. If the moment of impact was severe and tungsten filament bulbs were illuminated, the hot filaments may deform in the direction of the impact, forming a bowed or curved appearance, with the center of the deformation pointing in the direction of the impact. If the bulb was illuminated at the time its glass shell fractured, the filament may have a black coating, while a cold bulb filament may appear shiny. If the filament has a range of colors, it was hot from the heat of the second filament within the same bulb, but was not illuminated itself. Occasionally, small particles may appear on the filament. If they adhere to or are fused on the wire, the bulb was incandescent at the time the shell fractured. Debris that easily falls off is not fused to the wire.

The halogen bulb also contains a tungsten filament but of much smaller dimensions. The bulb is pressurized and if it is cracked or has a poor seal, it will explode or rapidly degrade if supplied with electrical current. White coating on bulb remnants is characteristic of halogen bulbs that structurally failed during operation.

Fresher bulbs will have a characteristic appearance. The glass shell will be quite clear and without a hint of blackening. (Recall from Chapter 5 that heat causes the tungsten to evaporate from the filament and coat the inner glass.) The filament will not have a gravity-induced sag caused by long use of the bulb, and will not appear roughened or pockmarked. The base will have a bright, clean appearance.

If a bulb burned out sometime before the accident it might be possible to detect a separation along the filament, accompanied by a related darkening of the glass shell. The remaining filament will not be blackened, as in the case of a bulb burning out because air was introduced into the vacuum. Careful examination of the separated ends of the filament may show a thinning or even a small bead-like accumulation of tungsten where the arc that occurred during the moment of failure caused an instance of higher temperature. This appearance is different from other types of filament failure. It helps determine whether the lights were illuminated appropriately before the impact.

In a careful evaluation of an accident, not only broken bulbs should be examined, but also the undamaged ones. However, to determine whether a bulb in a vehicle is operational, do not turn on the circuit. If the bulb is cracked or loosened from its base, it will rapidly burn out and destroy evidence. Subsequent testing in a laboratory using very low, adjustable voltages can establish bulb integrity.

Bulbs, bases, filaments, and glass pieces can be removed at the appropriate time using evidence-gathering procedures. High-power magnification, including the use of microscopes, significantly assists the examination of a filament's condition and connection points, as well as fused glass particles and the glass shell.[4]

13.9 Traffic Control Devices

Traffic control devices may play role in an accident. When an accident involves an emergency vehicle the odds are greater that one of these signals may be involved. One question that should be raised is the ability of each driver to see color in an appropriate manner. As previously discussed, drivers with color defective vision, as a group, have more accidents than a similar group of drivers with normal color vision. However, beyond the physiological aspects, other factors can alter color vision, even in normal drivers.

Tinted windshields alter colored light entering the vehicle. Flashing emergency lights can sometimes actually appear to change color or go through a metamorphosis because of some tints used in glasses as well as tints present in windshields. Some drivers have claimed that the windshield tint actually neutralized the red in traffic signals, rendering them nonilluminated or blank. Manufacturers state that automobile windshield tints pass traffic signal colors without problems. After an accident, the windshield fragments should be recovered as evidence and tested.

In the past, traffic control lights have emitted light over a broad range of the spectrum, which usually allowed some colors to transmit through filters. When the newer devices switched from bulbs to LEDs (light emitting diodes), the light given off was of a narrow band and not easily passed through some sunglasses. ANSI standards were established to specify the transmission values of spectacle lenses so traffic signals would be seen. However, some inexpensive, popular sunglasses which are nonprescription and sold over the counter may not pass these ANSI standards. Use of such glasses may render a driver blind to narrow-band traffic signals. Of particular concern are over-the-counter sunglasses sold for sports activities such as skiing or beach use. They may not pass the standards because they were not manufactured for driving. The potential problems are great, given that many more pairs of the OTC sunglasses are sold and in use than prescription, professionally fitted models. What remains unstudied, but potentially dangerous, is the color-defective driver who wears sunglasses.

Persons wearing these nonprescription, over-the-counter sunglasses may also have trouble perceiving flashing emergency vehicle lights. Therefore, flashing lights that were logically thought to be reasonably perceivable may actually be markedly dimmer or darkened with the wearing of some sunglasses.

When analyzing an accident, the investigator should not only notice whether the driver was wearing sunglasses and check the sunglasses' ability to pass recommended standards for the transmission of traffic control light wavelengths, but also to the person's need and use of prescription lenses or contact lenses. All individuals involved in an accident should be examined for eyesight problems requiring correction by these devices. If a driver is assisted by a second individual in the front, right seat then the passenger should also be included in the investigation.

When determining whether an individual was required to wear some form of corrective lenses, not only should the driver's license be checked, but the optometrist's records as well. If the patient wears contact lenses according to his optometrist, then the ability of investigators to accurately detect their presence on the eyes is suspect and should be reviewed care-

Figure 13.7 *Initially, the steering column was manually adjustable to shift the height of the steering wheel for different drivers. It has malfunctioned and was repaired by using two horizontal metal constricting bands. It is now in a fixed position.*

Figure 13.8 *The driver's seat has adjustable settings for operators of varying sizes and is controlled in this case by an air switch.*

fully. As difficult as it is for an untrained examiner to see the rigid contacts when worn, the soft contact lenses present a significantly greater challenge. If vision is an issue, then vigilance in determining whether a corrective prescription was necessary and in use is paramount. Simply asking the individuals involved is not enough, and a physical confirmation that contact lenses were in use, is essential.

13.10 Adjustable Seats and Steering Wheels

Just as with civilian automobiles, emergency vehicles have adjustments available so drivers of different sizes can operate the vehicle safely and comfortably. Two of the most common are the height of the seat and the corresponding height of the steering wheel. It is appropriate that each driver be aware of how to adjust each. (See Figures 13.7 and 13.8.)

Endnotes

1. Olson, P.L., *Forensic Aspects of Driver Perception and Response*. Tucson: Lawyers & Judges Publishing Co., 1996.

2. *Ibid.*

3. Baker J.S. and L.B. Fricke, *The Traffic Accident Investigation Manual.* Evanston, Ill.: The Northwestern University Traffic Institute, 9th ed.

4. *Ibid.*

Chapter 14

The Police Chase

Robert J. Simpson, Esq.

The police chase has long entertained the movie-going public. From the earliest crude efforts of the bungling Keystone Kops to the technical wizardry of today's cinematography, the increasingly sophisticated and dangerously scripted chase scenes have continued to thrill audiences. Who is able to resist the spectacle of police vehicles with lights flashing and sirens wailing pursuing the fleeing felon? Cameras whirl as highly trained stunt drivers participate in elaborately choreographed mayhem. We watch with fascination as vehicles careen through congested urban streets. The drivers barely avoid catastrophe as they speed through red lights and stop signs, race the wrong way on one way streets, sideswipe the unwary, and plow down sidewalks throwing vendors' carts into the air and demolishing fire hydrants, all climaxing when the culprit is finally nabbed an in often fiery and always spectacular conclusion. In this make-believe world, property damage is maximized and personal injury nonexistent.

The recent development of television's ability to instantaneously transmit pictures into our homes of live ongoing police action has added a new viewing dimension. Commencing with O. J. Simpson's well-documented Bronco drive down a Los Angeles freeway, the simulated dangers of motion picture action have had to make room for the eye-in-the-sky helicopter continuously monitoring the streets below, ready to immediately transmit images of police intercepts. As a supplement to live coverage, the Fox Television Network now attempts to satisfy the public's insatiable appetite for actual violence by broadcasting a series entitled *The World's Scariest Police Chases*.

In the often mundane world of law enforcement, a chase can be defined as an officer's active attempt to stop a motor vehicle, after the driver becomes aware of this attempt and resists apprehension. Although the police chases depicted in movies and on television normally occur in major metropolitan areas, they are, in fact, events which happen with alarming frequency in every county, town, city and village in the United States. The death, serious physical injury and heartbreak visited upon innocent bystanders, their families and loved ones echo throughout these communities. The Association of Trial Lawyers of America in 1996 filed a brief in the United States Supreme Court case of *County of Sacramento v. Lewis*, 523 U.S. 833, (1998), asserting that there are some 100,000 police chases a year in this country, and federal officials had documented 5,306 deaths in the preceding sixteen years as a direct result of police pursuits. These fatalities and the accompanying serious injuries are the consequences ignored by the entertainment media.

Tioga County, New York, is not unlike any other community in experiencing the destructive results of police chases. Part of an area commonly known as the Southern Tier, Tioga County is rural in character, abutting Pennsylvania midway between New York City and Buffalo. It has approximately 50,000 inhabitants with its largest population centers being the Villages of Owego and Waverly, each with perhaps 4,000 residents. Each village has its own police force and the county is patrolled by the New York State Police and the Tioga County Sheriff's Department. Each law enforcement agency is independent of the others and to the extent not governed by state law, rule or regulation, operates within its own procedures and guidelines or, in some instances, in the absence thereof.

Tioga County has an average of one or two homicides a year with a similarly unremarkable number of other violent crimes. History does not record the last time a police officer in the line of duty shot a suspect. Unfortunately, the county has experienced its share of fatalities and serous injury resulting from high speed pursuits. What follows are three representative occurrences.

On January 19, 1982, the Village of Owego Police Department, acting in conjunction with the Tioga County District Attorney's Office, secured a search warrant for the residence of Charles Winston. The police had probable cause to believe that this nineteen-year-old suspect had in his home the proceeds of a recent burglary. Winston was well known to local authorities. At approximately 9:30 in the evening, several uniformed officers arrived at his dwelling. One of the policemen knew Winston and greeted him by his first name. After announcing their purpose and displaying the warrant, the officers proceeded to the suspect's bedroom. Although informed of his constitutional rights, Winston was neither under arrest nor restrained in any fashion.

As the search progressed, Winston left his home and fled in his automobile heading in an easterly direction on New York State Route 434, a two-lane rural roadway having a maximum speed limit of 55 miles per hour. The road was clear and dry and visibility was good. As Winston accelerated, he was pursued by a member of the Owego Police Department. The officer operated his vehicle with lights and siren activated. During the entire pursuit covering a distance of seven miles, both vehicles maintained speeds in excess of 90 miles per hour. The police officer was able to keep Winston's vehicle in constant visual contact. He observed that Winston was driving in a "reckless and dangerous manner" as he passed vehicles on curves. Winston was aware that he was being chased and on a handful of occasions came perilously close to colliding with the cars that he passed. Finally, Winston's vehicle did make contact with an automobile he was overtaking, causing him to lose control as he entered the oncoming lane. He smashed head on into an approaching vehicle being operated by nineteen-year-old Timothy Shiner. Shiner died instantaneously. Winston was thrown from his car and died three days later in the hospital.

The accident and resulting deaths were investigated the Tioga County Sheriff's Department. After reciting the factual circumstances, the sheriff's department report attributed the direct cause of the accident to Winston's:

1. speed
2. initial behavior—unsafe passing
3. delayed perception
4. faulty evasive action

The report went on to conclude that: "The other two vehicles involved in the accident had no chance to react to conditions created by the Winston vehicle, due to the fact of the high speed reckless manner it was being operated at."

Although the culpability of Winston was beyond question, the report was silent as to any contribution that the tactics of the police officer may have had in precipitating the collision.

The death of Timothy Shiner immediately elevated the awareness of the public to the unintended consequences of police chases and provoked a healthy but contentious debate. After reviewing the police files, the district attorney, while accepting that the police officer "had a snap decision to make and he made it," advocated a thorough review of the free-wheeling nature of such chases. Observing that there was a general absence within most law enforcement agencies of written procedures governing pursuits, the prosecutor called for a New York State study leading to the establishment of guidelines to be followed when the police are confronted with a pursuit decision. An uproar ensued.

The editorial page of the area's largest regional newspaper, the Binghamton, New York, *Sun-Bulletin*, supported the proposal, stating that a "uniform set of guidelines would protect both police and public from such senseless accidents." *The Sayre Evening Times*, which circulates in the western part of Tioga County, offered a different perspective. While acknowledging that most law enforcement agencies had no standing criteria to govern when a pursuit should be instituted, it unwittingly recognized the obvious by stating that

> a police officer is only human; faced with a split-second decision under the stress of the moment, he often reacts by pure reflex, and his decision may not always be the proper one.

Placing blame squarely on Winston, it notes that once the suspect flees at a high speed in order to avoid questioning, "he constitutes a danger to other drivers encountered along the escape route." Since the suspect's primary desire is to flee, the newspaper concluded that the operator may speed whether or not "he sees flashing lights in the rear view mirror." The newspaper postulated that under such circumstances a tragedy may result even in the absence of a police pursuit.

In addition to the editorial conflict, the readers weighed in pro and con with their letters. The views were generally well reasoned and expressed very valid arguments both in favor of and against such chases. Lost in the finger-pointing was any discussion of guidelines and training.

As with the public, local and regional law enforcement agencies were oblivious to the prosecutor's underlying suggestion. In defense of his officers, Owego's Chief of Police simply iterated that the pursuit was "justified and within the law." He emphasized the prevailing view within the law enforcement community that each situation was different and the final decision to chase must be left to the sound discretion of the individual line officer. Like many police agencies in the Southern Tier, the Owego Police Department had no written policy relating to pursuits and, as the chief so emphatically stated, had "no intention of writing one." A spokesman for the neighboring Broome County Sheriff's Department concurred that the decision to give chase had to be left to each officer's discretion. A representative of the Tioga County Sheriff's Department, which also had no written policy, offered as justification "cops are human . . . . [T]he adrenaline gets pumping and people jump right on it."

The second incident occurred on June 17, 1988. At approximately 2:20 A.M., the Owego Police Department received a report of a stolen automobile. Within minutes, one of its officers observed the car heading out of the village at a high rate of speed. The village officer together with members of the Tioga County Sheriff's Department engaged Sean Cornett, the driver of the stolen car, in a high speed pursuit as he headed north towards Ithaca, New York. Over a distance of approximately twenty miles, the vehicles maintained speeds exceeding 100 miles per hour. Shortly after entering Tompkins County, Cornett lost control and was involved in a head-on collision. The driver of the oncoming vehicle was permanently and seriously injured. Sean Cornett, who was legally intoxicated with a blood alcohol content of .15, was uninjured.

The final example occurred shortly after 11 P.M. on October 17, 1993. Deputies of the Tioga County Sheriff's Department were proceeding west in a 30-mph zone in the village of Owego. They observed a vehicle being operated by Jeremy Young approaching at 50 miles per hour with one headlight out. The deputies turned, activated their red lights and siren and attempted a stop. Young's vehicle accelerated through a red light in a predominantly residential area and continued at a high rate of speed through two red lights in the downtown business district. Aware of the danger to pedestrian and vehicular traffic, the police vehicle slowed. Young, at speeds exceeding 70 miles per hour, proceeded through another residential area forcing an oncoming car off the road. As he crossed a series of railroad tracks, his automobile became airborne, striking an oncoming car. A passenger in that vehicle suffered severe and permanent head injuries.

No reasonable person can argue that all police chases should be prohibited. Certainly the police must pursue a fleeing murderer or other violent criminal. It is equally well accepted that a policy of no pursuit encourages even minor violators to seek to elude the police. In a newspaper article entitled "Using Adrenaline to Fight Car Theft," written by David M. Herszenhorn and appearing in the May 7, 1998, edition of *The New York Times*, the reporter addresses in the following language the positive effect of a policy implemented by Newark, New Jersey, authorizing the pursuit of stolen vehicles:

> Founded in December 1991 to combat Newark's standing as the nation's auto theft capital, the task force has helped the Newark area drop to 16th from first in the nation in car thefts, according to statistics released last month by the National Insurance Crime Bureau. In just over six years, the number of cars stolen in Newark alone has dropped from more than 15,000 to 5,745 last year. The task force had made about 2,200 arrests and recovered nearly 2,400 stolen cars.

Once the focus of the argument is shifted from whether or not there should be police chases, policy makers can begin to address the more pressing issue of what criteria the road officer should consider before actually engaging in this dangerous activity. The knee-jerk response that considers only "discretion," "adrenaline" or "testosterone" virtually ignores the serious underlying question of defining in which situations is it justified to risk the life and limb of the innocent bystander. In the absence of a sound written policy, each office is left to his own devices and intuition to determine whether or not a high speed chase is appropriate under the circumstances. Training and guidelines would provide the police officer with assistance in making this often fateful determination.

Police officers and other employees of this state and its political subdivisions are generally required to observe the same rules of the road as are private citizens. Section 1103(a) of the Vehicle and Traffic Law states:

> The provisions of this title applicable to the drivers of vehicles upon the highway shall apply to drivers of all vehicles owned or operated by the United States, this state, or any county, city, town, district, or any other political subdivision of the state, except as provided in this section and

> subject to such specific exceptions as are set forth in this title with reference to authorized emergency vehicles."

Operators of authorized emergency vehicles do not have an unfettered right to violate the laws of this state based merely upon whim or fancy. Section 1104 of the Vehicle and Traffic Law establishes the parameters of the exception in the following terms:

> (a) driver of an authorized emergency vehicle, when involved in an emergency operation, may exercise the privileges set forth in this section, but subject to the conditions herein stated.
>
> (b) The driver of an authorized emergency vehicle may:
>
> (c) stop, stand, or park irrespective of the provisions of this title;
>
> (d) proceed past a steady red signal, a flashing red signal or a stop sign, but only after slowing down as may be necessary for safe operation.
>
> (e) exceed the maximum speed limit so long as he does not endanger life or property;
>
> (f) disregard regulations governing directions of movement or turning in specified directions.
>
> (g) The foregoing provision shall not relieve the driver of an authorized emergency vehicle from the duty to drive with due for the safety of all persons, nor shall such provisions protect driver from the consequences of his *reckless disregard* [emphasis added] for the safety of others.

A police vehicle is without question an authorized emergency vehicle. Emergency operation is defined in Vehicle and Traffic Law Section 114-b as follows: "The operation . . . of an emergency vehicle, when such vehicle is engaged . . . pursuing an actual or suspected violator of the law . . . ."

In summary, a police vehicle engaged in a pursuit has authority to disregard the rules of the road so long as the officer operates the vehicle in such a manner as does not recklessly disregard the safety of others. In theory, a police officer who operates his vehicle in violation of this standard could be arrested. The reality is of course quite different.

A review of the case law in New York finds scant evidence of the prosecution of any operators of emergency vehicles, much less police officers. The reason for this is strikingly apparent. When men and women are risking their lives to enforce the laws, neither fellow police officers, prosecutors not the courts are ever going to interfere with an exercise of discretion. There are several reported prosecutions in New York but they offer little assistance in determining the boundaries of emergency operation.

In *People v. Bisig*, 46 Misc.2d 299, 259 N.Y.S.2d 452, a driver was issued a citation while operating his ambulance at a speed of 75 miles per hour in a 50-mph zone. In his defense, he asserted the privilege set forth in Section 1104, alleging that he was responding to an emergency call and was not endangering life or property. In its decision, the court agreed that the operator could be convicted of speeding if he exceeded the limit in the absence of an emergency. The judge opined, however, that it was not for the courts to determine whether an actual emergency existed. It is the operator's reasonable belief that counts. Even if an emergency situation did exist, this did not foreclose a conviction since the driver was prohibited from exceeding the speed limit if he endangered life or property. The court determined that the exemption applied and found the defendant not guilty because the operation of the ambulance, though at an excessive rate of speed, occurred at dusk on a four-lane highway in fairly heavy traffic, caused no accident.

A different result was reached in the case of a game protector employed by the Bureau of Law Enforcement of the Conservation Department who was ticketed for speeding (*People v. McIntosh*, 231 N.Y.S.2d 714). In that proceeding, the officer also asserted his exemption from prosecution under the constraints of Section 1104. The arrest arose after the defendant was informed of an anonymous telephone complaint that someone was unlawfully spearing bass within the officer's jurisdiction. As McIntosh headed to that location in his state-owned vehicle, he passed though the Village of Sacket's Harbor. He was stopped by the Chief of Police and issued a citation for traveling 60 miles per hour in a 30 mile per hour zone (so much for professional courtesy). Following trial, the defendant was convicted. On appeal, the County Court affirmed, stating that "whether the defendant's trip . . . in response to an anonymous message was an emergency call in connection with the performance of his duties or the pursuit of an actual or suspected violator of the law, was a question of fact for the police justice. It is implicit in the finding of guilt that it was not so considered." Recognizing the limits on the definition of pursuit, the court stated that

> on the entire record, it does not appear that there was any pressing necessity or urgency for the game protector to travel at almost 60 miles per hour in a 30 mile per hour zone. Neither was there actually any suspected violator of the law being pursued.

People v. Emmi, 146 Misc.2d 399 550 N.Y.S.2d 787, involved the actual prosecution of a police officer for allegedly committing a traffic violation in the line of duty. The officer appealed his conviction arising in the city of Auburn for the infraction of "following too close." On the evening in question, Emmi was operating his police vehicle in downtown Auburn. He observed an automobile heading in the opposite direction with improper inspection and registration stickers. He turned and pursued with intent to ticket the violator. When an intervening vehicle stopped abruptly while entering an intersection, the office was late in braking and a rear-end collision resulted. In reversing the underlying conviction, the county court stated in its decision:

> The record clearly states that appellant was operating a police vehicle and following a possible violator who he Intended to stop. These facts were conceded in a letter from the district attorney to the Auburn City Court. While it is true that appellant made a distinction between "following" and "pursuit" in his testimony . . . such distinction is without legal justification. . . .

It was, at the very least, the duty of the prosecutor to disprove the emergency vehicle theory. The record is devoid of such proof; nor did the lower court apparently consider the emergency vehicle doctrine in its decision."

Obviously, the criminal statutes do not provide fertile ground for regulating police conduct in emergency operation situations. The real battles over death and injuries sustained as a consequence of police chases are fought in the area of civil litigation. When the innocent bystander is left to seek monetary damages as the only method of compensating for the loss, the police officer and his municipal employer make inviting targets because of their "deep pockets." Unfortunately for the aggrieved party, the obstacles to a financial recovery are substantial and as a consequence, the results generally unsatisfying.

The courts of New York and its legislature realize that operators of emergency vehicles have a primary obligation to respond quickly to preserve life and property and to enforce the criminal laws. In recognition of this need, the qualified exemption was granted to police officers. The state's highest judicial authority, the court of appeals, has determined that Section 1104 severely constricts the imposition of civil liability on either the officer or his employer unless the officer's conduct rises to the level of recklessness. *Saarinen v. Kerr*, 84 N.Y.2d 494, 620 N.Y.S.2d 297.

In *Saarinen*, a police officer employed by the Village of Massena observed a dark-colored van, operated by Kerr, fishtail and squeal its tires. The officer followed the van, watched it run a stop sign at a speed of approximately 30 miles per hour and then proceed into a shopping center parking lot. After the officer activated his emergency lights, the van pulled away. During the course of the pursuit the van entered a public road, drove into the oncoming lane of traffic and passed a red light. Kerr collided with Saarinen's vehicle causing serious physical injuries. The plaintiff sued Kerr and the Village of Massena.

The trial court dismissed the lawsuit noting that "recklessness" was the appropriate standard for assessing the police officer's conduct. The lower court held that the plaintiff had failed to raise any inference of "reckless disregard for the safety of others" by the police officer. The court also rejected the claim that deficiencies in the village's training of its officers was a proximate cause of the collision. It determined that there was no proof that it was the patrolman's conduct in pursuing Kerr rather than Kerr's own erratic and illegal conduct that was the direct cause of the collision.

On appeal the intermediate appellate court reinstated the claim on the basis that Kerr's vehicle had been traveling in a "highly populated area" and there was a "possibility" that there would be other traffic on the roadway. The court also held that there was a justiciable issue that the village's policies regarding police pursuits were inadequate. Finally, the court believed that allowing officers to engage in vehicular pursuits without proper training was "inherently dangerous and unsafe."

A further appeal by the Village of Messina resulting in the court of appeals reversing and dismissing the complaint. In its analysis, the court stated:

> We hold that a police officer's conduct in pursuing a suspected law breaker may not form the basis of civil liability to an injured bystander

> unless the officer acted in reckless disregard for the safety of others. The standard demands more than a showing of lack of "due care under the circumstances"—the showing typically associated with ordinary negligent claims. It requires evidence that the actor has intentionally done an act of an unreasonable character in disregard of a known or obvious risk that was so great as to make it highly probable that harm would follow and has done so with conscientious indifference to the outcome . . . .

The court was cognizant that emergency personnel must routinely make conscious choices that will "necessarily escalate the overall risk to the public at large in the service of an immediate, specified law enforcement or public safety goal." In rejecting the ordinary negligence test in determining liability and imposing a much higher standard, the decision held that it would:

> Lead to judicial "second guessing" of the many split second decisions that are made in the field under highly pressured conditions. Further, the possibility of incurring civil liability for what amounts to a mere failure of judgment could deter emergency personnel from acting decisively and taking calculated risks in order to save life or property to apprehend miscreants. The "reckless disregard" test, which requires a showing of more than a momentary judgment lapse, is better suited to the legislative goal of encouraging emergency personnel to act swiftly and resolutely while at the same time protecting the public's safety to the extent practicable.

In measuring the officer's conduct against this standard, the court in *Saarinen* determined that the officer had not exceeded the limits of the statutory-qualified privilege. Simply exceeding the speed limit on a wet highway while other vehicular traffic was present did not elevate the officer's conduct to recklessness. Likewise, in dismissing the plaintiff's claim that the officer was ill-trained in the art of pursuit, the court observed that the plaintiff had not met the heavy burden of demonstrating how any changes in the training program might have altered the course of events that evening. Finally, the court rejected the argument that the Village's pursuit policy gave the officers too much latitude and discretion, holding that these are procedures established by governmental policy and may not be reviewed in the absence of irrationality.

As the holding in *Saarinen* would clearly suggest, the judicial landscape is littered with decisions rejecting monetary claims by innocent bystanders killed or maimed in police chases. In *Simmen v. State*, 81 A.D.2d 398, 442 N.Y.S.2d 216, a state police officer observed a vehicle pass a red light at speeds estimated between 65 and 70 miles per hour. A head-on collision occurred following a pursuit on a two-lane road at speeds ranging from 70 to 110 miles per hour. In rejecting a claim of liability against the State of New York, the court acknowledged the legal principal that recklessness must be considered in the context of the time and under the circumstances in which the chase occurred, not in the light of subsequent facts or in retrospect.

In *Palella v. State*, 141 A.D.2d 999, 530 N.Y.S.2d 650, a fourteen-year-old passenger was injured while joyriding in a vehicle being operated by his fourteen-year-old friend. The state police officer pursued the stolen vehicle for approximately fifteen minutes at speeds reaching 75 miles per hour. The court held that the officer was not required to abandon the pursuit simply because the youths slowed down whenever the officer did. In the final analysis, the court adhered to the well-accepted doctrine that the proximate cause of the accident was the fourteen-year-old driver's erratic operation as opposed to the police officer's pursuit.

In *Bellows v. Amsterdam*, 157 A.D.2d 1006, 550 N.Y.S.2d 486, the subject of the police pursuit attempted to recover damages for injuries he sustained after he, himself, initiated the flight. After committing a traffic violation, Bellows led the police through the City of Amsterdam at speeds approaching 85 miles per hour. He ultimately struck a tree. In his lawsuit seeking damages for personal injury, he contended that the police were negligent in undertaking and conducting the chase. He argued that there was no need for the pursuit since he was known to the officers and his conduct did not rise to a level justifying a stop in the first instance. He also claimed that the pursuit violated agency guidelines. With an understandable lack of sympathy, the court rejected the argument stating:

> Finally, the record demonstrates that the actions of defendants in their attempt to stop or intercept plaintiff's automobile were not a proximate cause of the accident and subsequent injuries sustained by plaintiff [citation omitted], nor is there any showing that the procedures followed by defendants were not in accordance with established police procedures.

In *Mullane v. City of Amsterdam,* 212 A.D.2d 848, 622 N.Y.S.2d 346, a bystander was seriously injured when struck by a vehicle traveling approximately 95 miles per hour while being pursued for a traffic violation. The night was clear and the road conditions were dry. The court dismissed the cause of action holding that the officers were duty-bound to investigate this recklessly operated vehicle. Although acknowledging the speed of the vehicle created a risk of possible harm to others, the court found the officers were entitled to take that risk in the interest of stopping a vehicle whose presence on the road posed a clear and present threat to the public safety. Once again, the court concluded that it was erratic driving of the suspect that was the proximate cause of the accident and not the manner in which the police conducted the pursuit.

The victim was equally unsuccessful in *Szczerbiak v. Pilat,* 90 N.Y.2d 553, 664 N.Y.S.2d 252. There, the officer was responding to a report of a fight in progress when he struck a bicyclist crossing in front of him. The collision occurred as the officer accelerated over a distance of more than 800 feet before attempting to engage his emergency lights and sirens. As he glanced down from the roadway to activate his emergency equipment, there were no pedestrians in sight and he was several blocks from any intersection. In denying culpability, the court held that even if the officer were negligent in glancing down, this "momentary judgment lapse" did not alone arise to the level of recklessness required for liability to attach.

In *Naue v. Higgins*, 662 N.Y.S.2d 527, the complainant was injured during a pursuit that lasted approximately seven seconds at 2:35 a.m. in a quiet neighborhood with no traffic. The officer failed to activate his siren or lights during the pursuit while traveling approximately 10 mph over the speed limit. This did not constitute reckless disregard for the safety of others.

Although a reading of the case law could easily lead to the conclusion that departmental policy and training is irrelevant in assessing police officer and municipal liability, this is not true. In *Kerwin v. County of Broome,* 134 A.D.2d 812, 521 N.Y.S.2d 871, the appellate court ruled that juries are entitled to review the particular departmental training which the officer has received as an aid in determining the issue of recklessness. Likewise, expert testimony is admissible to clarify proper police practice expected in a given police emergency, despite the fact that no specific departmental rules or guidelines prevail and despite the jury's common understanding of driving standards generally. *Selkowitz v. County of Nassau,* 45 NY.2d 97, 408 N.Y.S.2d 10.

Although not a police pursuit incident, one case that helps elucidate the standard of recklessness is *Hudson v. Boutin,* 656 N.Y.S.2d 781. There the state police officer parked his vehicle partly on an interstate highway behind a disabled tractor trailer. The court held that an issue of recklessness existed for a determination by a jury because the trooper's emergency lights were activated but there were no flares, it was night time and the weather conditions included freezing rain with snow.

On May 26, 1998, the United States Supreme Court weighed in with its decision in *County of Sacramento v. Lewis*, 523 U.S. 833, (1998). The case arose in the federal courts following a 1990 chase. An eighteen-year-old motorcycle driver ignored police orders to stop. The pursuit lasted about 75 seconds at speeds of 100 mph. After the motorcycle tipped over it was struck by the deputy sheriff's patrol car, resulting in the death of the motorcycle passenger. Stymied by state law, the parents sued in federal court alleging deprivation of a substantive right protected by the due process clause of the Fourteenth Amendment to the United State Constitution, which provides that no state shall deprive any person of "life, liberty, or property without due process of law." In its decision, the Supreme Court ruled that there was no constitutional violation unless the police acted with "a purpose to cause harm." In dismissing the lawsuit altogether, the Court noted that under any analysis of the evidence, the officer's behavior "does not shock the conscience." In a line of reasoning paralleling the New York standards, the Court termed the motorcycle operator's actions "outrageous" and observed "the officer's instinct was to do his job as a law enforcement officer, not to induce Willard's lawlessness, or to terrorize, cause harm, or kill."

In the absence of a reported case defining reckless conduct by a police officer in a chase situation, it is difficult to present a factual circumstance that exposes clear liability. For the sake of argument, one would presume that the courts would find recklessness if, for instance, the police undertook a vehicular pursuit through a crowded school yard. In spite of this generally gloomy outlook for plaintiffs as chronicled in the foregoing cases, all is not necessarily lost. For fear of greater financial exposure, many claims are settled by municipalities before trial for substantial sums of money even though liability is very suspect. By way of illustration, the injured parties in the *Winston*, *Cornett*, and *Young* pursuits each entered into a financial settlement with the respective municipal employers.

Since the death of Timothy Shiner in 1982, there have been some very positive steps taken by Tioga County police agencies with respect to chase procedures. At that time, the New York State Police Department was the only law enforcement organization having jurisdiction in Tioga County that had written guidelines for their officers. Today, all the other agencies have adopted and implemented their own written procedures governing these activities. This advancement has been stimulated by the ongoing efforts of each department to put better trained and more qualified officers on the road. Considerable impetus has also been supplied by self-insured municipalities and their insurance carriers in a continuous effort to sidetrack potentially enormous liability claims.

A sampling of the police policies in Tioga County demonstrates that although the goal of guidelines is always to protect the safety of the police officer and the public, the methods of implementation often differ. By way of illustration, it is helpful to review the existing policies and guidelines of two arbitrarily selected agencies, the New York State Police and the Owego Police Department.

The *New York State Police Field Manual* requires that its members consider a number of factors before the initiation of a vehicle pursuit. Some of these are:

(1) That pursuit can result in accidents or personal injury to members, the persons pursued, or an innocent bystander.
(2) The chances of an incident occurring is increased as the speed of the vehicles increase.
(3) In determining whether or not to pursue, a member should consider the following factors:
    (a) the nature of the offense
    (b) the safety of the officer and the safety of other passengers and pedestrians
    (c) the driving capabilities and tactical preparations of the officer
    (d) the type of vehicle pursued
    (e) road and weather conditions
    (f) traffic conditions
    (g) time of day or night

(h) officer's knowledge of the area
(i) population density of the area
(j) knowledge of the violator

State police officers are required to activate all emergency warning lights and audible devices when participating in a pursuit and they are instructed to terminate the pursuit if they consider the danger to be unacceptable. The rules specifically provide that if the suspect is known and there is no immediate danger to the public, termination of the pursuit is required and an arrest warrant should be obtained.

Officers are strictly prohibited from duplicating reckless and hazardous driving maneuvers taken by the pursued driver. Members must at all times drive in a manner exercising reasonable care for their own safety and that of all other persons and property within the pursuit area. No more than two vehicles are allowed to become actively involved in a pursuit unless directed by a supervisor. Pursuits by unmarked patrol vehicles cannot be undertaken unless no marked patrol vehicles are available. Termination of the pursuit is the responsibility of the primary pursuit vehicle unless otherwise directed by the supervisor.

Termination shall occur if:

1. the immediacy of apprehension is clearly outweighed by the unreasonable danger to the public
2. an accident occurs in the pursuit with an uninvolved citizen and there are no other police or medical personnel readily available
3. adequate information is available to obtain a warrant of arrest and there is no immediate danger to the public

Although the Owego Police Department's guidelines are not as detailed as those of the state police, they allow for many of the same contingencies. It is mandated that the pursuing officer immediately notify the dispatcher and provide as much information as possible. The supervising shift officer may assert control over the pursuit and order immediate termination. The pursuing officer is instructed to constantly consider the risks and weigh:

1. the time of day
2. volume of vehicular and pedestrian traffic
3. location of the pursuit
4. weather conditions
5. road conditions
6. speeds involved

The Owego Police Department rules allow pursuit by an unmarked vehicle only when the fleeing vehicle represents an immediate and direct threat to life and property. Officers may not pursue suspects the wrong way on controlled-access highways. Extreme care must be used whenever disobeying traffic control devices even when a statute specifically permits such conduct.

Each officer is directed to familiarize himself with Section 1104 of the Vehicle and Traffic Law. All considerations regarding justification or risks assumed during a chase must be geared to the public safety. A pursuit will be terminated when

(a) the level of danger created by the pursuit outweighs the necessity for immediate apprehension
(b) the suspect's identity has been established and there is no longer any need for immediate apprehension

There can be no doubt that appropriate and well-conceived guidelines can assist an officer in making a rational decision on whether to initiate a police chase or continue it. The importance of departmental policies cannot be overstated. Given the seemingly insurmountable burden of proof imposed by the "reckless disregard" standard, adherence to well-defined chase guidelines offers potential victims some limited protection. In retrospect, were regulations in existence in 1982, one could argue that the pursuit of Winston, who was well known to the police, could have been avoided and an arrest warrant secured. The necessity of a chase becomes more problematical with respect to Cornett. Furthermore, by backing off, the deputies appeared to have acted correctly the pursuit of Young.

In any event, guidelines are only useful if the individual officers are aware of their existence and possess adequate training in their proper interpretation and application. An informal survey in Tioga County disclosed that very few officers had more than a passing knowledge that their respective departments had any rules and regulations whatsoever governing chases. There was very little indication that any of the departments provided specific training that would aid their officers in making a knowledgeable decision based upon awareness of departmental standards.

Based upon the method in which it is used, an automobile is unquestionably a deadly weapon. Given its enormous potential to cause devastating injury, a great paradox exists in the absence of training in its appropriate use. Almost every police agency requires that its officers be recertified in handgun and other weapons usage at least once if not twice a year. The public would find it extremely disconcerting if an officer was allowed to discharge his firearm in the line of duty without necessary training in the use of deadly physical force. Even with training, it is inconceivable that a policeman would fire his weapon while wildly pursuing a suspect under circumstances that might expose an innocent bystander to death or serious physical injury. Although the potential risk is far greater, either through oversight or neglect, we readily condone a complete absence of training in the skills required for police chases. Examining this issue in a historical perspective leads to the conclusion that this deficiency should be addressed.

Without advocating the banning of police pursuits altogether, minimum efforts can be taken to reduce the degree of danger posed to the innocent bystander. A productive first step would be to require each police agency to devote adequate time in informing each officer of the rules and regulations governing police interdictions. Additionally, each officer should be thoroughly instructed in a variety of factual situations and the choices available in response. They should be taught that in certain circumstances it is appropriate not to pursue because the danger outweighs the potential benefit. Finally, recognizing that a motor vehicle is a deadly weapon, officers should be given at least as much training in high speed pursuit driving as they receive in weapons instruction. This would include an ongoing system of qualification on aggressive and defensive driving techniques.

Death and serious physical injury cannot be eliminated as a by-product of police chases. However, there is little question that police departments can do more to protect the safety of innocent users of the highways. In setting the threshold for monetary recovery so high by requiring proof of "reckless disregard," neither the courts not the legislature have done anything to encourage training. Until this deficiency is corrected, the only practical remedy for lowering the incidents of death and injury is being ignored. Since most police agencies are neither inclined to nor capable of providing the rudimentary training that logic dictates, the legislative adoption of statewide standards should be considered.

Chapter 15

Emergency Vehicle Accidents: Representing the Innocent Bystander

Stephen Smyk, Esq.

15.1 Introduction

The steady progression of modern-day society has produced fundamental and monumental changes in tort law and civil litigation. In what now seems light years ago, municipalities often enjoyed complete immunity from lawsuits arising out of municipal or governmental activities. In many jurisdictions negligence suits against a municipal government, school district, or even not-for-profit charitable organizations were prohibited. This traditional concept arose from one of sovereign immunity; namely, that the King (read government) could do no wrong. In many jurisdictions sovereign immunity has been nullified by legislation, and in others, by court decisions. The basic rationale prevalent today is that any party injured should have his or her rights adjudicated by the appropriate judicial forum.

The necessity for legislative or judicial intervention in abrogating sovereign immunity was based on the notion that a person injured by the harmful conduct of another should not be barred from recovery but should be afforded the right to be recompensed if the claim withstands the scrutiny of a judge or jury.

The issues one encounters in analyzing and assessing a potential claim are complicated by myriad factors—the most fundamental of which is the law of the jurisdiction which will be applied. Each state and territory has its own rules to be applied. The first step is to understand the specific rules for your jurisdiction. Unfortunately, the specific rules can change even as the subject accident may be occurring; courts hand down decisions, legislatures pass laws, and executives and departments promulgate regulations daily. What may have been yesterday's rule may not be today's. Even a general understanding of the rules may not be enough as there is an exception for every general rule. However, in assessing and analyzing the facts to determine whether liability exists, one must accept the general propositions and one must factor in the anomalies which may exist.

Primarily, a recovery for an injury (or death) arising out of an emergency vehicle accident requires the application of tort or negligence law initially. As in all such cases the claimant must establish negligence,[1] proximate cause, and damages.

The most common legal definition of *negligence* is the failure to exercise reasonable care. A failure to exercise reasonable care is usually defined as doing that which an ordinarily prudent person would not do under the circumstances or not doing that which a reasonably prudent person would do under the same circumstances. The legal definition is fairly consistent throughout all jurisdictions.

Proximate cause essentially requires that the damages must arise from the negligent conduct; it is often delineated as the "but for" test—the negligence, even if proven, must be the competent producing cause of the injuries claimed. One can lack reasonable care (be negligent) but if that lack of reasonable care is not necessarily the competent producing cause of the damages which ensue, the failure to prove the causal nexus will bar recovery.

If negligence and proximate cause exist, one must have damages. Damages which may be awarded vary considerably throughout the various jurisdictions. In some jurisdictions medical bills, even if paid by an insurance provider, may be proved and recovered; in some, the judge will deduct them from the final award; and in others, the jury is informed of the payment but may make no award for the bills so incurred. Similarly, lost wages, sundry expenses, and other damages are subject to each jurisdiction's particular rules. Damages may relate to pain and suffering (past, present, and future), permanency, loss of enjoyment of life, emotional and mental distress, wages or loss of earnings (past, present, and future), loss of consortium, companionship or affection, and so forth. Because the trial of a lawsuit may require the production of evidence which can easily cost thousands of dollars, it is incumbent upon the plaintiff's attorney to assess a case from a number of different vantage points. There are numerous considerations, the chief of which are the law, the facts, and the practicality.

15.2 Legal Principles

As stated above, the first legal issues which may arise are related to statutory or governmental immunity from suit. In addition, care must be taken because immunity may not necessarily be conferred upon the driver of an emergency vehicle, prompting the scenario that one may not be able to sue the municipality, but yet still sue the operator of the vehicle.

Most jurisdictions' statutes give right-of-way privileges to emergency vehicles. In some jurisdictions that may include a right of way at a controlled intersection or within certain speeds. The statutes of many jurisdictions provide that ordinary motorists must yield the right of way to an emergency vehicle being operated while engaging its signals, lights, sirens, or other indicators of emergency use. Many exemptions for emergency vehicles from traffic regulations may arise from the situation presented and some statutes may actually demarcate rules granting one type of emergency vehicle a right of way over another. Generally speaking, drivers of emergency vehicles must obey traffic signals in the absence of a specific statutory exemption. Many states have determined that having the right of way does not authorize the emergency vehicle to enter an intersection against a red light without stopping and ascertaining the conditions of the intersection. In New York, the Vehicle and Traffic Law, Section 1104, allows the driver of an authorized emergency vehicle to proceed past the steady red signal, a flashing red signal, or a stop sign after slowing down "as may be necessary for safe operation." The statute also provides that the driver of an authorized emergency vehicle is not relieved of the duty to drive with "due regard for the safety of all persons" and is not otherwise protected "from the consequences of his reckless disregard for the safety of others." Obviously, "due regard" and "reckless disregard" have been determined to impose liability only for reckless disregard, not a general lack of due care under the circumstances.

Due regard is generally defined as being that degree of care which a reasonably prudent person would have used in the discharge of official duties of like nature and underlying circumstances. Further protection for the driver of the emergency vehicle has generally been afforded by the courts in that a "momentary judgment lapse" does not satisfy the reckless disregard test and the injured party must prove the driver of an authorized emergency vehicle intentionally disregarded a known or obvious risk so great as to make it highly probable that harm would ensue and with conscious indifference to the outcome.[2]

Various jurisdictions have carved out other requirements for due regard. In some, a suitable warning of approach to the intersection is enough; in others, audible or visible signals are not enough but also require a duty to maintain a reasonable and proper lookout in a reasonable belief of safety before proceeding.

Thus, in many jurisdictions the actions of an emergency vehicle driver may be measured by the very same rules as that of other drivers. In essence, even with an audible or warning signal, an emergency vehicle operator must exercise reasonable precaution against the dangers of the situation which the very performance of their duties creates. In many jurisdictions the type of the emergency vehicle may dictate the standard of proof to be applied. Fire department vehicles, police vehicles, and ambulances may all have separate standards to define due regard. The type of a call on which an emergency vehicle is engaged may give rise to a determination as to whether it should be accorded a separate legal status in that jurisdiction. Generally speaking, fire department vehicles are usually accorded the highest degree of latitude by way of exemption from the rules of the road and being granted a right of way over other vehicles; however, care should be taken when a volunteer fire department is involved, because not all are accorded the same rights as a municipal fire department under statute. Likewise, a fire fought beyond the general municipal limits or volunteer fire department limits may provoke different legal rights and issues, and in some jurisdictions the presence in the vehicle of a superior (captain, lieutenant, etc.) with the right to control the manner of the operation of the vehicle may invoke a different standard of care than that of an operator following (or leading) other vehicles to the scene.

The type of trip and whether it is truly an emergency may give rise to statutory exemptions. As a general rule, although what may prove later to be nonemergency (false alarm, etc.) will not cause a waiver of any statutory exemption from traffic regulations, a return trip may require the vehicle obeying the vehicular rules of an ordinary motorist.

The type of exemption or excuse from statute has to be determined on a situation-by-situation basis. Other factors to be examined include the adequacy of the warning, the actual speeds involved, the type of traffic signal which was disregarded, the negligence of the other vehicles' operators, the failure of other drivers to pull their cars over, injuries to pedestrians, the physical infirmities of a pedestrian which may have caused the pedestrian's inability to detect sirens or lights, the careening of vehicles striking the pedestrian, and a pedestrian stepping out to visualize the commotion caused by an approaching warning: all of these may give rise to legal precepts affecting recovery.

Of significance may be the manner in which the forum applies contributory negligence (i.e., the negligence of the plaintiff-claimant). In some jurisdictions contributory negligence may be a complete bar to any recovery. Some jurisdictions use comparative negligence with the negligence of all the parties adding up to 100 percent and award damages according to the percentage of each wrongdoer. Some jurisdictions prevent any recovery for a plaintiff who is 50 percent or more at fault, regardless of the remaining allocation of negligence.

Police vehicles are generally governed by the same rules of the road as fire vehicles, but issues may arise in many courts as to departmental procedures, the training received by the operator, and guidelines adopted by the municipality. Suffice it to say, the innocent occupant of a moving vehicle injured as a result of a police chase is more likely, these days, to be afforded his or her day in court today than in times past. On the other hand, in New York recently, a bill was introduced which would bar the operator of a vehicle being pursued by an official vehicle from making any claim for damages.

The liability of a policeman operating a vehicle may also hinge on whether the accident occurred at a location other than an intersection, such as when making a turn from a highway into a private drive, or passing or cutting off another vehicle. Such factors as excessive speed, loss of control at high speed, skidding (or fishtailing), swinging into oncoming traffic, and the like will create potential claims which need to be assessed in accordance with the prevailing case law in that jurisdiction. In such cases the negligent operation of yet another vehicle may bar or substantially reduce the likelihood of a verdict. In most jurisdictions, the police are granted a paramount right of way at intersections, so that recovery (as in most cases) is based on the particular facts and circumstances of the accident.

The expanding use of hands-off or cellular phones, car faxes, sound systems as good as any recording studio, windows which effectively reduce outside interference or noise, may all create issues which must be properly assessed as soon as possible. Suffice it to say, inattention caused by smoking in a vehicle, changing radio stations, popping in a new tape, and so forth. All should tend to put the attorney on guard that a jury may be less sympathetic to a claimant, absent some "utter disregard" exercised by the emergency vehicle operator.

As a corollary, police sometimes respond in silence because the use of any type of warning might tip-off a violator to police presence. Such action must also be analyzed in accordance with the facts and circumstances. For example, a pedestrian's actions will be measured by whether he was crossing at an intersection with the right-of-way or jaywalking.

Ambulances generally do not have right-of-way privileges as do other emergency vehicles. The type of trip on which the ambulance was engaged may determine whether the ambulance will be afforded privileged status or exemption from statute. The same general rules with regard to the warning signals, speed, failure to obey traffic signals, negligence of other motorists or pedestrians, or the failure to yield right of way all must be considered.

In summary, the type of vehicle, its mission, and the situation presented must be examined to establish whether statutory provisions will bar a claim.

15.3 Facts and Circumstances

As a general rule, the elements of proof of negligent operation of an emergency vehicle depend on facts and circumstances tending to establish lack of reasonable care. As indicated above, the precise facts and circumstances necessary to prove negligence may depend on the type of emergency vehicle and will be predicated upon the particular facts and circumstances of each case. For a claimant to succeed, certain general elements must be examined. First, was the injured party one who can maintain an action? If a car being pursued crashes, the statutes may bar the operator of that vehicle, but not the passenger. Did a special right of way arise? Was the emergency vehicle exempted from traffic regulations? Since an emergency mission is usually claimed, was there any reasonable presumption or knowledge that such an emergency did not exist or was there an unreasonable belief that an emergency did exist? Was the type of trip, the vehicle, or its driver legally privileged? Not every fire is consequential. Not every trip is in the exercise of an emergency. In some volunteer fire departments, the volunteers have equipped personal vehicles with flashing lights to enable them to get to the fire (or firehouse); those are not always accorded emergency status. The vehicle operator may have failed to recognize or appreciate the danger of injury to others by disregarding traffic regulations by excessive speed, driving on the wrong side of the road, not signaling turns or changes of lanes, passing on the right, or failing to signal its claim to a right of way. Certainly, failure to recognize existing conditions, whether they be weather, other traffic, pedestrians, or impaired visibility, are significant issues. Also, was adequate warning given at all? If so, was it intermittent or continuous? Was it too late to avoid the consequences? Was speed appropriate under the existing conditions? Was caution taken at an intersection or a signal? Was speed decreased where warranted or required? Was the vehicle under control? Did the operator make reasonable assumptions as to the conduct of other motorists, pedestrians, the audibility or visibility of his own warning devices or the ability of other vehicles to timely appreciate and be able to control their vehicles? Other factors which may appear in specific situations may arise from vehicles pulling out of side streets or driveways, blind spots for motorists when being overtaken, sudden directional changes, cutting off or overtaking other vehicles, or collisions with vehicles making a proper turn whose sight or auditory recognition may be reduced.

The specific facts and circumstances which may give rise to a claim of liability necessitate an appreciation not only of the law, but of what witnesses to the operation of the emergency vehicle say. Eyewitnesses to any collision, the circumstances of the collision, speed, warning, or approach, must all be identified. In most cases, conflicting testimony from independent and unbiased eyewitnesses is common. Some witnesses will recall a warning (even if one was not given) others will remember none (even if eardrums were pierced by a warning signal); such is the nature of the beast, and contradictions arising from other persons' perceptions and recollections are routine. Certainly, an appreciation of the weather, road, lighting conditions, general topography, and geography in which the accident occurred is essential. Estimating and evaluating the presence of other traffic, and so forth, is also necessary.

Assuming the suit has been commenced, the testimony of the driver of the emergency vehicle and other occupants of that vehicle is essential to determine their awareness of guidelines or regulations, safety rules, emergency status, operation of the vehicle itself, and use of warnings (visual and audible), and speed at the various benchmarks during the operation of the vehicle. Logs, reports, and recordings must be reviewed to establish times and distances covered so that a recreation of the actual speeds, and events can be mathematically determined.

Lastly, an issue arises as to the use of expert witnesses in any such case. Primarily, consideration must be given to whether to employ expensive experts on accident reconstruction, police protocol, regulations, guidelines, the appropriateness of those guidelines, and prior similar situations.

It should be noted that even when a municipality's rules or regulations are violated, in many jurisdictions, such a violation is not necessarily negligent conduct unless those rules and regulations state that it is.

Certainly, the lack of any guidelines at all regarding the type of activity may demonstrate a lack of reasonable care. In other words, the absence of guidelines will not necessarily excuse a municipality from a claim of negligence where it can be shown that a standard of care exists, either in the jurisdiction or at some other level, which normally would be promulgated to those officials in the municipality expected to be aware of such standards. Persons likely to be required to be aware of such standards include officers in a fire or police department, whether it be municipally organized or volunteer. Likewise, independent ambulance services must and are expected to be aware of the rules of the road and prevailing standards and customs of emergency ambulances nationwide with respect to the operation of vehicles.

15.4 Practical Considerations

Even with a full appreciation of the legal principles expected to govern at trial and with full appreciation of the facts and circumstances surrounding the accident, a responsible plaintiff's or claimant's attorney must understand the practicalities of the case. Generally speaking, at trial a jury usually decides the facts as it wishes and can usually accept or reject them as it sees fit. Obviously, in jury selection the plaintiff's attorney would be leery of jurors who have had emergency responses to their own personal, family problems, neighbors, or close friends' problems. Certainly jurors who have been the victim of criminal behavior would be expected to be lenient toward a policeman who is justifiably claiming that he was engaged in an emergency situation. One who has called a fire department for what proved to be a nonemergency call may be a very poor juror from a plaintiff's standpoint. One who has been rushed to the hospital in an emergency ambulance and who underwent treatment should be avoided as a juror in most circumstances.

The practicalities to be encountered at trial may run the gamut of jurors' (and judges') prior contacts and one should anticipate jurors to have substantial misconceptions about the rights and obligations of an emergency vehicle. While care has to be taken in eliciting from jurors their preconceptions and understandings, one must remember that most jurors tend to be accepting of the municipality engaged in responding to an emergency. This does not mean that the plaintiff will never recover damages, because that certainly is not so. However, one must understand that approximately half of all cases which are tried result in a defendant's verdict, regardless of the type of accident. The expense of litigation also may be a factor to be calculated with respect to the likelihood of a verdict, and the subsequent deduction from that verdict of the expenses associated with obtaining it. Likewise, the prospects of appeal, difficulty in collection, and so forth, all have to be appreciated. Contrary to the perceptions of many people, trials involving an injured party are rarely ever about principles. The client will likely not regain his or her health because of a trial. In fact, the opposite will most likely occur, in that the stress of the trial may prove to be temporarily disabling or debilitating. Also, the expense of going to trial is not a particularly attractive proposition when there is a likelihood of failure because the attorney usually ends up bearing the expense.

Lastly, it is not unusual for a victorious defendant to contemplate suing a plaintiff or the plaintiff's attorney for a frivolous action in our present day. Thus, the utmost care must be taken with respect to analyzing all of the factors before a suit is contemplated, much less started.

Of certainty is the fact that, as time progresses, concepts and evidence will likely change or modify. The recent spate of the antitobacco verdicts has been promoted not by sympathy, but by the discovery of the existence of materials from the tobacco industry establishing the dangers and exposures of people to the effects of smoking. The rules which govern legal principles are highly fluid and tend to be subject to the same political and practical pressures as most endeavors surrounding governmental (or quasi-governmental) activities. Taxpayers may not want to return verdicts which might affect their taxes. Those who have been stranded by governmental agencies may feel disgruntled. It is likely that common sense application of reasonable standards is usually predicated upon subjective, not objective, determinations. As with most situations, liability will swing from time to time for or against injured persons vis-à-vis emergency vehicles. Before trial, it is essential to have an awareness of the general prevailing attitude of the community in which the case is going to be tried.

15.5 Conclusion

The specific issues to be addressed are of the "gray" variety as in almost all potential lawsuits; however, the "black" and "white" areas can be readily identified by an understanding of the facts and circumstances surrounding the initial response by the emergency vehicle, the occurrence of the accident, and the reasons for it. Taking those facts and applying the existing law should determine the lengths to which a suit should be considered. One final word on the subject. An attorney should not be surprised to find his case to be the "ideal" case to test the prior law if the municipality (as often) hopes to extend its immunity for engaging in a governmental function and not a proprietary one—a legal distinction that in most jurisdictions makes the difference between getting to a jury or having the case dismissed on motion before any discussion of settlement can be had.

The issues which arise run the gamut from unclear law and inconsistent facts to a practical slant which may economically determine whether it is worthwhile even taking on the case in the first instance. Despite those drawbacks, the field should expand because one injured in a situation involving an emergency vehicle is likely to be seriously injured (or dead) as a result of the forces, speeds, and masses involved. Accordingly, most cases involving such a situation are entitled to a long hard look by the attorney.

Endnotes

1. While other theories of liability may be adapted (perhaps successfully), we have limited this exposition to the traditional negligence concept. The California courts had sustained a jury verdict for a sixteen-year-old passenger on a pursued motorcycle who was struck and killed by a police vehicle, basing the verdict on a "deliberate indifference" or "reckless disregard" legal test (discussed hereinafter); however, a unanimous but extremely divided (five separate concurring opinions supporting the ultimate reversal) United States Supreme Court found that the due process requirements required that the test to be applied was whether the police officer's "[actions] shock(ed) the conscience of the Court so as to find that the actions of the municipality through its police officers had deprived the youth of his life without "due process of law." *Estate of Lewis v. Smith*.

2. The review and decision in the *Lewis* case by the Supreme Court puts into question the extent to which individual jurisdictions will be affected. In the absence of specific legislation codifying the municipality's duties to be performed, will the states adopt the "shocks the conscience" test or continue to follow previously decided law? Second (and more important), whose conscience need be shocked—the jury, the trial judge, or the appellate court? As in most things legal, further definition will arise eventually, but one must plunge on—usually hoping not to represent the test case.

Chapter 16

Parked Emergency Vehicles: Highway Hazards

Stephen S. Solomon, B.S., O.D.

Emergency vehicles are often called upon to operate in dangerous locations, namely on or near highways. The proper conspicuousness, placement, and traffic controls enhance safety while in those vulnerable locations. It is accepted that anytime a vehicle is parked or placed in or near any approaching traffic, the danger level rises greatly. It is common for emergency vehicles to be operating in or near traffic based upon the very duties they perform. However, their common use does not make them any less liable for being a participant in a collision or promoting a collision between other vehicles.

Under normal circumstances civilian drivers negotiate the daily tasks of motor vehicle operation in a relatively safe manner. However, when the scene is compounded by an unusual event or an activity that attracts attention, the routine driving function deteriorates. Flashing emergency lights and pedestrian activity on the part of emergency workers draw a driver's attention. Vision is often directed away from the driving activity to one of scanning an event for interesting information. This raises reaction times and increases the possibility of an untoward or unwanted situation.

From firefighting, EMS, and law enforcement perspectives, apparatus are *placed* at an emergency. *Placement* refers to the strategic and tactical positioning of the motorized equipment to allow for them to operate efficiently and in support of other units. At some incidents, the apparatus are *parked* pointing to the anticipation of pending use. At a highway or road incident the emergency vehicles may be *placed* and examples are an ambulance in close proximity to the patients or a rescue unit located near a heavily damaged civilian vehicle with trapped occupants. Other equipment may be *parked* waiting assignment. Law enforcement vehicles may be placed to warn oncoming motorists of danger ahead, performing enforcement activity or parked while investigating incidents.

Placement and parking should follow some rules. At any situation with fire, EMS, and police working on or near a street or highway, if possible the civilian traffic should be detoured or completely shut down. Unfortunately, this often can not be easily accomplished. The classic example would be a major vehicle accident on a multi-lane expressway. The severity of the accident may require a number of emergency vehicles to be in attendance, which should dictate road closure, while the designation of the route as a major expressway points to keeping lanes open for through traffic.

Therefore the cardinal plan would be to shut down and stop or detour all traffic away from the scene. However, when this is not practical a series of rules should come into play. Among them should be warning devices placed well in advance of the scene to warn motorists to steer safely into an adjacent lane. A second rule is all emergency vehicles should be placed on the same side of the highway. This eliminates distractions on both sides of the road and eases the confusion of civilian drivers. Third, all flashing emergency signal lights should be turned off, with the exception of amber, nonstrobe flashers. The multitude of blue, red, and red/white lights are a very confusing visual picture, moreso if more than one emergency vehicle is on the scene. Four, at night, headlights should be extinguished because the light intensity can dazzle or temporarily blind oncoming drivers. Traffic control officers in the roadway and signal warning lights can thus be obscured. The brightness from the headlights will wash out the nearby red flashers so they will be much diminished in effectiveness. Five, if possible all fire and EMS vehicles should be placed or parked off the road surface on the shoulder.

If there is no room for off-road parking, then the entire lane should be blocked. The most dangerous parking or placement is with a partial lane obstructed. The reason centers on the decision process of the approaching driver. Valuable reaction time is lost because some drivers will evaluate if they can squeeze by while not changing lanes. Several scenarios evolve where the driver can decide too late that they can not get by and must veer to one side or the other, or come to a complete stop. Regardless, there is a secondary affect on traffic in the adjacent lanes as a driver drifts over the dividing line into another driver's space. There is also the possibility that the approaching driver will crash into the emergency vehicle. By blocking the entire lane, the decision process is shortened where any doubt about getting past is removed. Evasive action can take place more quickly and safely.

At no time should emergency vehicles be placed or parked where personnel become pedestrians and made to cross active high speed lanes. When personnel exit an emergency vehicle, they should do so from the door that opens onto the median whenever possible. Climbing out in the face of approaching traffic is extremely risky, and just because the emergency services individual can see the oncoming vehicle, there is no guarantee that the driver in turn sees the potential pedestrian. Because of the high attention level of the activity on the scene, and the attendant warning lights, the driver's attention could easily be distracted.

When an ambulance is placed at a highway scene and the rear doors are opened to off-load equipment to treat the patient, the doors should be closed as soon as possible. There are several reasons for this, one being that the opened doors can block from view some of the warning lights. Another reason is when the doors are left opened, the space comprising the exposed interior is visible and actually is cave-like and appearing very dark. This enhances camouflage by reducing the brightness of the painted and more visible rear surface area of the ambulance.

For fire and EMS units that are to some degree staffed by volunteers, the issue of private vehicles driven to the emergency scene needs to be addressed. Obviously, the fewer number of emergency and personal vehicles at the call, the safer the environment. If private vehicles must be driven to the site, and this often is the case, they must follow the rules for all thc emergency vehicles. If possible, it is better to park beyond the incident and on the same side, and then walk back to the scene. As with the apparatus, do not leave headlights on and extinguish any warning lights such as red, blue, or green flashers. (States often have regulations pertaining to the use of a specific color warning light for volunteer responders, but there is no nationally accepted convention.)

16.1 Placement of Warning Devices

Warning devices, including signal arrows, signs, cones, and flares should be deployed as soon as possible after the arrival of the emergency vehicles to the scene. All equipment and manpower may not arrive at the same time, and the first actions may usually to be devoted to the on-scene evaluation, life threatening interventions, and fire suppression. And in some instances, there may be a delay until traffic control can be established. This underscores a span of time where increased danger exists. As part of any preplanning, traffic control must receive priority attention starting with education of the personnel. As soon as possible, aggressive traffic warnings and controls must be instituted. If possible, as part of the initial evaluation that is transmitted to the dispatchers, a description of the traffic conditions should be included. Command decisions can be made to implement traffic control in a timely manner with that extra information. It may be prudent to establish early traffic control with the first arriving personnel.

Placement of traffic warning devices requires thought and planning. A 2,000-foot distance from the first symbol until the accident scene may be excessive on a 30 mph street in a village, while that same distance many be minimally effective on a superhighway. At the same time, sight distance plays a role in that warnings should be placed in advance of blind turns or curves. With a range between 3 to 6 seconds of reaction time being possible, then some approximations of the first placement of warning devices can be developed based on the speed of the cars operating on that route. (See Tables 12.1–12.3) The placement of warning devices should be calculated based on the stopping distance for the largest vehicle using that type of road and traveling at the highest possible speed. Do not use a total stopping distance for cars and do not assume the posted speed limit will be heeded. On a highway posted for 55 mph, the stopping distance for a tractor-trailer operating at 65 mph could be over 600 feet. (With a reaction time of 3 seconds at 65 mph, 286 would be traversed. The mechanical stopping distance would add on another 330 feet.) On highways with longer reaction times and higher speeds, it may be necessary to place the first device at least 1,200 feet from the start of the accident scene. If traffic is heavy and fast, the start of the placement should be even further back and one mile may be a reasonable distance. If more than one lane is blocked on a highway of three or more lanes, then first warnings may have to be placed several miles before the accident site. However, based on the lack of sufficient staff there may be a delay for the optimal placement, but it should start as soon as manpower becomes available and they should work from the accident back into approaching traffic. After a safe distance is determined based upon highway characteristics, vehicle flow density, speed, scene activity, emergency vehicle quantity, weather, and sight visibility, the devices should be placed to form a diagonal pattern gradually shunting traffic into an adjacent, safe lane. The devices should continue along the safe lane past the emergency scene, and then taper back to allow traffic to resume flow in all lanes. The spacing of the devices should be such so that a vehicle can not cut back into the protected lane prematurely. The higher the speed of the traffic the further the spacing. All warning devices do not have to be the same type and a line of cones may be preceded by flares or sign boards.

Emergency crews should also be aware of the arrival of additional apparatus as they may need clear travel past slowed traffic, in order to have access to the scene. On an undivided, two-way road, traffic control must be established in each direction. The law enforcement and fire police agencies should take over traffic control as soon as they have sufficient manpower available.

16.2 Traffic Cones

Traffic cones do not usually have lights and their function is limited to daylight hours unless they are banded with retroreflective material. Sufficient cones should be stored on the apparatus. Fluorescent material affixed to the cones will not aid in nighttime visibility as has been discussed previously. An advantage of the cones is if they are struck by the civilian driver inadvertently, the impact's sound may warn and alert the driver to what lies ahead.

16.3 Road Flares

Highway flares are often used in proximity to emergency scenes. The flare has some negative characteristics: (1) the ability to ignite flammable materials (2) generation of copious amounts of smoke and (3) emission of a red- orange flickering flame.

The flammability issue is well known to emergency service providers, but in the excitement of the moment, the hazard may be overlooked. Some explosive or flammable gasses may not dissipate in the atmosphere but pool or collect in low spots. If an ignition source is presented then the results will compound an already troubling situation.

Public safety officials can sometimes overlook the effects of flare smoke. If there is not a breeze, the haze produced can obscure vision to some degree. It is often the case that many flares will be burning simultaneously. Instead of providing an early warning system, they can generate a copious degree of low lying, thick, chemically produced fog, resulting in very poor visibility. This obstruction to vision need not blanket the whole scene to cause a hazard, but merely obscure temporarily any other signal to warn the approaching vehicles.

The flickering flame can act as a direct detractor to the overall attention demanded by emergency lights. As discussed previously, flashing lights in the background may reduce or negate the effectiveness of other flashing lights. Road flares may cause more problems than suspected, because the smoke contributes to a visually cluttered background.

However, because flares are inexpensive and readily available they will continue to be used. Several considerations must be taken into account, including the assurance there is no exposure to flammable or explosive products. Because of the smoke they generate, there should be sufficient air currents or wind in the area to dissipate the collection of the by-products of combustion. Flares are rated by their burn time and a traffic control person should be assigned to monitor their placement and replenishment. Obviously, a thirty-minute flare will not provide coverage for a two-hour incident. The locations for flare placement should be based on the same distance considerations for cones with the exception of not being used in the proximity of dangerous materials.

There are positive aspects of flare use. They illuminate the immediate area in which they are placed and the glow they generate in the atmosphere, especially at night makes them highly visible if the conditions are right. Carrying flares is easier and lighter than carrying cones so the flares can be deployed more quickly with less personnel. If they are kept from the immediate emergency zone, and placed well in advance of the incident, then they can be used with the understanding of their limitations.

16.4 Traffic Sign Boards

A separate class of warning signs exist as sign boards. They are available in various colors and can use fluorescent or retroreflective applications. These panels are often manufactured in an a-frame design. They fold flat for storage and are easily deployed. The printed message is applied to the sign at the factory. They are larger than traffic cones and more difficult to quickly set up along a highway. However, with their much larger size they provide a more powerful visual stimuli and offer some degree of increased safety. Combining traffic sign board display with traffic cones and other alerting devices is an acceptable strategy.

16.5 Arrow Displays

In situations involving dense traffic, it may be useful to use a vehicle that displays a blinking amber arrow to signify a lane closure. The entire signal should be illuminated at one time. The arrow should be displayed for several seconds, followed by a shorter period of darkness. The lights, which should only be amber, must not be so bright as to blind or dazzle the oncoming nighttime driver and the lights should not be sequenced across the arrow width. The horizontal dimension of the signal may exceed 4 feet. The higher the arrow is mounted above the road surface the more effective it is. The vehicle on which it is mounted may be placed on the shoulder or in the lane to be blocked depending on traffic density and speed. This message is easily detected and understood. A horizontal string or bar of lights blinking in sequence is not an acceptable substitute for the arrow design. If flashing lights on a civilian's automobile or sport utility vehicle are mounted on the rear shelf facing traffic or even above the rear deck but still inside the passenger compartment, the warnings are too low and will be obscured from view of all but the civilian driver immediately behind the emergency vehicle.

16.6 Special Considerations

As a general rule, the longer the time frame for the emergency personnel to be working at the scene, the more durable the traffic controls should be. If it is an uncomplicated scene and the activity may encompass an hour or less, then flares or cones may be sufficient. As the event time escalates then consideration should be given to cones in sufficient numbers. In prolonged activities, saw horses or traffic barrels from the highway departments may need to be placed and can often be brought to the scene upon the request of fire, police or sheriff departments.

If long-term and complicated activities are contemplated, then the roadway must be closed. While the closure-detour scenario is not usually considered in routine emergency operations, it must be implemented when there is a great risk to the lives of the personnel at the site. This occurs when there are a large number of vehicles involved in the accident and there are many rescue and support individuals present. It is also the only alternative when hazardous material involved, such as a chemical spill, nonbreathable gas release, or a potential explosion as well as extreme environmental or weather conditions like thick smoke, fog, or roadway ice.

A special category of traffic control must be implemented when a medical helicopter is being employed for patient evacuation and there may regional guidelines in place to help with these cases. Not only is scene safety of consideration, but safe landing and take-off zones must be established. On a multilane interstate it would not be unusual for traffic in both directions to be stopped, with the helicopter choosing to land on the paved surface of the uninvolved side. The pilot has the final say of where he will land and the ground personnel have the responsibility of securing the landing area. The distractions to drivers are greatly magnified with the operation of a helicopter at the site. The attention civilians pay to their safe transition through the emergency area is still more challenged when debris is blown into the air by the rotor blades causing poor visibility as well as hazards from airborne particles. Depending on circumstances, the aircraft may land hundreds of yards from the immediate patient location and this consideration plays into establishing traffic controls.

16.7 Nighttime Lighting

Working at night has an obvious limitation on visibility. Previously, the discussion had focused on using devices and applications plus various methods to make personnel and vehicles more visible. Most fire departments carry portable generators and floodlights to improve functions and operations at emergency scenes. By supplying illumination, efficiency and safety are improved. The crews can more easily see what they are doing and supervisory personnel have more effective oversight. Initially, the side effect approaching individuals was underestimated. Bright scene illumination has the added benefit of warning and advising oncoming motorists to the dangers ahead. It acts to alert them without entrancing them.

Three light sources are available: ground-level floodlights powered by extension cables from the apparatus or portable generators, scene lights mounted in the upper body of the apparatus, and extendible poles or towers. More and more departments have enhanced their portable and body-mounted floodlight capability by adding telescoping towers with banks of floodlights. Often these towers, which can be elevated, rotated, and tilted by remote control, are mounted on the pumpers or rescue trucks. The high position of the lights shining down on the scene acts to limit shadows and also are highly visible to approaching drivers. The immediate area is very highly illuminated and acts to inform civilians of an unusual occurrence on or near the roadway. With the highly lighted area of operation, some perceptual enhancements are provided for the oncoming driver. At night, some objects—even those highlighted with retroreflective material—have a tendency to appear to be disassociated from the land surface or the horizon and are difficult to localize for distance and location. With the increase brightness, judgments are less complicated, location and distance are better defined, and the resulting reaction time is reduced.

16.8 Traffic Control Device Set-Up Pattern

The establishment of the working distance from the first warning sign to the scene of operation has been discussed. After computing the safe warning distance, the placement of devices can be initiated. The first warning to an unsuspecting motorist should be as large as possible. On an interstate highway with heavy and fast traffic, the large blinking arrow pointing to the direction of the lane shift should be first. This device is usually mounted on a vehicle. If sufficient warning devices are available, the display vehicle should not be in the roadway, but if possible on the shoulder and on the same side as the event ahead. If a second vehicle with a large arrow is available, it should be parked about 400 feet downstream and block the lane. This may be more effective than an emergency vehicle parked along the shoulder with flashing lights because the former actually gives an instruction to the approaching driver. The flashing lights may do nothing more than distract that driver from looking at the road ahead. A pedestrian law enforcement officer can not be seen across the several lanes of active traffic and should not be the initial warning. Several spaced cones and sign boards can then be placed along the lane of travel starting on the roadway near the shoulder and progress in a diagonal line to ease traffic into the adjacent lane. After lane control has been established, a string of traffic cones can be placed all the way toward and past the incident. It is helpful to note the end of the restricted area and where normal traffic can resume. Fire, EMS, or law enforcement personnel can take positions along the board, flare and cone line to guide drivers away from the blocked lane or lanes. If an emergency vehicle is the first warning on the highway, it is best if there are no individuals sitting inside.

Chapter 17

Emergency Service Personnel as Pedestrians

Stephen S. Solomon, B.S., O.D.

Emergency services personnel usually arrive by vehicle. As soon as they exit that vehicle and proceed on foot, they are pedestrians. Among the most common incidents of pedestrian injuries are the law enforcement traffic officers standing near the driver's side door in the typical "ticket writing" posture when the vehicles are stopped along a highway shoulder. It is not surprising to have a law enforcement officer in this situation struck and seriously injured or killed. Remembering the information from Chapter 16 about flashing lights, recall that some drivers have a tendency to drive toward attention drawing stimuli, and the stronger the stimuli or the brighter the light, the longer vision is directed toward the source, and the more susceptible the driver. The warning lights on the law enforcement vehicle were thought to warn approaching motorists, but in actuality, they may draw vulnerable operators toward the vehicle by displaying very bright or complex flashing lights. If the flash has a long "off" interval, is very bright, or is included in a group of flashes, the risk is greater and increases still more if the approaching driver is tired, under the influence of drugs, or having recently consumed alcohol (even if the blood alcohol is below the legal limit of DUI).

To protect the pedestrian officer, a special garment should be supplied and used when exposed to traffic. Typically, this is a stripped fluorescent and retroreflective vest capable of being worn over the basic uniform. It can be removed easily for nontraffic functions. Additionally, all flashing warning lights should be extinguished except amber, nonstrobe flashers that face oncoming vehicles.

17.1 Traffic Control at Emergency Scenes

Directing traffic in routine municipal settings is a hazardous duty. The officer is almost always standing on the shoulder or in the middle of the roadway and exposed to traffic moving about him. This function is performed mainly on municipal streets and traffic is not traveling at highway speeds. Even so, the officer is very vulnerable to being struck by a vehicle. Yet this job responsibility and risk is not comparable to directing and controlling traffic at an emergency scene. The greatest differences are the distractions presented to approaching civilian drivers. Until they make visual contact and then interpret that contact, they do not suspect any hazards ahead and are driving in their normal manner. The emergency can involve any combination of provider groups, (fire, EMS, or police) accompanied by emergency vehicles, spectators, and the scene itself. With the intensity presented, it is understandable that approaching motorists may not be attending to driving tasks as they get closer and are attracted to the "carnival" atmosphere. A traffic control officer is highly vulnerable to injury or death in these circumstances because it is so difficult to contrast the officer from the many other unusual attractive stimuli are simultaneously available. Because the traffic officer can easily see the approaching vehicle, it does not mean the driver distinguishes the officer from the background. This situation forms the judgment basis for a tragic circumstance because the officer expects the motorist to comply and the motorist probably would if he sees, comprehends, and reacts in time. It may not be an issue of 20/20 vision, but rather the ability to filter out the information that is not immediately related to safely driving his car or truck. In a sense, the officer may be camouflaged against the rest of the background.

Three other factors significantly contribute to accidents involving pedestrian emergency service providers: weather, working at night, and driver conditions which include fatigue, alcohol consumption or drug consumption.

Weather examples are easily perceived in their relationship to decreased visibility and reaction time. Concurrently, the road surface conditions are likewise understood. However, what is not as evident is the response of the driver to the change in the environment. Reaction times are affected by the changes in visibility and braking distances are increased by less-than-optimum road surfaces. If the weather is in transition from clear to inclement, then the driver may not yet have made a mental adjustment to his driving attitude. Suddenly having to react to a pedestrian where one is not expected can initiate a panic reaction which may culminate in aggressive steering, forceful braking, and even starting a chain reaction with other trailing drivers. In the simplest form, the automobile may hit the civilian, while in more complicated scenarios collisions between vehicles as well as the pedestrian are possible. The first defense against being struck would be to avoid an active roadway if possible, and the second would be to wear high-visibility clothing.

Nighttime is extremely dangerous for more than the obvious visibility reasons. The pedestrian can see the lights of the approaching vehicle and assumes he can also be seen, yet logic tells us this is blatantly wrong. Darkness also removes many of the cues we use for depth and distance judgments. It is easier for a driver to misjudge locations at night. What may appear to be a pedestrian on the road shoulder can actually be that pedestrian in the roadway because the path may not be straight ahead, but rather on a curve. Unilluminated obstructions can be present and not detected until the headlight beams strike them. Stopping distances on highways usually exceed the distance lighted by the high beams. All these conditions add up to a very dangerous environment for the pedestrian emergency worker, especially one who is focusing on rendering assistance rather than concentrating on approaching traffic. As apparatus is placed at the night scene, illumination by the use of the towers or portable lights should be a high priority. Ample light provides a safer work area and will often highlight pedestrian activity in the immediate area. The higher the lights are above ground level, the more diminished the shadows, the less the disabling glare and the greater the perceptual enhancement.

In some instances the responder could arrive before the apparatus. That individual is extremely vulnerable because there is no emergency vehicle displaying warning signals, protective clothing may not be worn, and traffic control may be nonexistent. Again, the recommendation is for the use of fluorescent or retroreflective overwear. The trim material is more effective when mounted on the cuffs of pants and jackets. This is because of the high degree of motion imparted by walking and arm movements. The movements also provide a rhythm of human motion which is quickly recognized by observers as pedestrian activity. It should be noted the beam of a headlight at night will stimulate a response from retroreflective material hundreds of yards away.

Alcohol use has been well discussed in a widespread venue ranging from medical texts to beer advertising and the full realm in between. Reaction times are dulled, judgments are biased, and driving ability is impaired. However, when discussing drug use the issue takes on another dimension. The typical thought when discussing drugs is illicit material, yet prescribed medications or over-the-counter products can have a significant affect on alertness, reaction time, and driving ability. Regardless of the chemical used, the affect on driving is singular in that the ability to operate a vehicle is reduced, impaired, or incapacitated. While fatigued drivers are not impaired in the sense of alcohol or drugs, they are functioning at a diminished level of response and reaction. This is known as a *divided attention task.* As conditions deteriorate, fatigue increases, or if chemical impairment is present, the ability to shift from point to point decreases. Phrased another way, attention is not easily diverted to another point of view.

The death of emergency service pedestrians at the hands of compromised drivers is an all too common event. There are cases in which some of these individuals actually drove around traffic barriers and into equipment or personnel. Therefore providing scene protection when exposed to oncoming drivers must include a consideration of this possible occurrence. While there is no perfect solution, some agencies use their larger vehicles as a physical barrier to intrusion. Even so, some drivers may find their way around and into the scene. As has been discussed, the stronger the stimulus the greater the reaction to look toward it. Compromised operators have difficulty releasing the visual stimulus. This causes them to look toward the flashing light for too long a period. Eventually they drive in the direction of their gaze and into the placed or parked emergency vehicle. Driving always consists of looking at the roadway ahead and scanning the periphery for additional hazards and information.

One of the most vexing problems in attempting to prepare for safe operations is the uncanny ability of compromised drivers to ignore, bypass, or penetrate all manner of warnings, signals, barriers, and commands. There is a general awareness within the emergency services of the high risks involved with these unreliable drivers. Unfortunately, severe injury and death involving emergency service providers occur all to frequently on the roadway. Defensive measures must be in place to address safety problems.

7.2 Fire Police Officials

A firefighter who may have received special training in traffic control and traffic law may be used for scene control and directing traffic at emergencies. Often they work in conjunction with law enforcement and share duties at a call. Other times they may be the only trained personnel on the scene to render traffic control. In some states they may be sworn officers and may have limited police powers. Some fire departments allow older members who no longer actively fight fire to continue service in the department as a fire police officer.

As previously discussed, directing traffic at an emergency scene has increased risks. In order to provide for increased visibility for these individuals they should have ample fluorescent and retroreflective material in their protective garments. A substitute for a vest worn over regular clothing is a full length long coat of fluorescent quality and striped with retroreflective tape. Additional trim on the arm and leg cuffs provide increased and valuable visual cues. Gloves that are visible day or night are an asset. A handheld light is important but should not be covered with an orange wand as the plastic sheath further diminishes the weak light output. If possible, the area where the fire police or police stand should be externally illuminated, preferably from an elevated tower.

17.3 Scene Safety

It has been suggested by the National Fire Protection Association that fire departments supply safety officers to oversee all operations for safe and protective practices at any scene. This includes all functions on streets and highways. Often this officer has the authority to shut down or suspend operations until safe procedures can be implemented. He is directly responsible to the incident commander and does not have to follow the chain of command upward. Therefore at every scene there should be a trained official assigned to observe the safety of the area and implement plans to enhance any lapses of safe and secure operations. It may not be possible to create an absolutely secure environment, but all reasonable methods should be employed. Standard operating procedures are often prepared in advance for the activities of the units and they should include safety considerations.

There are two times of heightened vulnerability for the pedestrian emergency worker: when they first start their scene activity and appropriate warnings are not yet in place and supporting manpower may not be available, and second, when controls are removed and traffic is being released into a normal flow pattern at the end of the event.

The safest suggestion for pedestrian activity is to be one for the briefest period of time. During pedestrian exposure, the traffic must be controlled using the devices discussed in this chapter. Protection can include emergency vehicles placed before the scene and blocking the entire lane of activity. Every individual working in this environment should wear high visibility gear including fluorescent and retroreflective tape or material. For firefighters, the most recent NFPA Standard 1971 contains directions for the minimum amount of high visibility material to be displayed on firefighting gear. (However, an earlier version of the NFPA standard recommended a more comprehensive version of fluorescent and retroreflective material which promoted a higher and safer level of visibility.) NFPA Standard 1521 discusses the role of the fire department safety officer. Those recommendations at a minimum should also be in force and documented.

Within the fire service there are strict guidelines directing individuals to wear appropriate protective when exposing themselves to a fire or hazardous material. Even if civilians are in dire need of help the fire service is taught to protect themselves first, not to be selfish but to preserve their well-being so they can help others. If the rescuer is injured he or she can not help the victims, but becomes another victim needing limited resources and distracting efforts from the primary scene. While this safety approach is well established in firefighting situations, it may receive less attention at a highway incident in that traffic control and lane isolation may not occur as the primary protective step. The parallel to putting on full protective gear at a fire before entering a burning building is the control of traffic before immediate rescue at a highway accident.

Just the reverse may occur in that the first arriving emergency vehicle may be an ambulance or fire truck. These units usually are placed at or very near the incident with one of their first responsibilities to analyze of the event to determine the equipment and manpower needed, and communicating these needs. The next step could be laying a hose line, patient assessment, or disentanglement of victims. The personnel who are now pedestrians on the scene may start traffic control, or others may initiate this function as they arrive. However, traffic control must be given a higher priority and implemented more quickly in order to provide safety for all concerned at the scene.

Part III

Statutes and Case Law

Paul F. Hill, Esq.

Introductory Essay

Scope, organization, and purpose of Part III

Part III of this book deals with the laws pertaining to emergency vehicles on emergency runs. The drivers of police vehicles, fire trucks, and ambulances who are responding to emergency calls or in the pursuit of a fleeing driver have the right to ignore, with due caution or to act without reckless disregard, the rules of the road. These rules include speeding, running stop signs and signals, driving the wrong way, and entering the opposing traffic lane. All other drivers have the duty to yield the right of way regardless of the traffic law. While such emergency runs and pursuits must happen many times a day without incident (after all, how many times do we hear a wailing siren in the distance and hardly notice it?), accidents, injuries, and fatalities do happen as a result. Aggrieved persons may contact a lawyer about the feasibility of making a claim. Part III was researched and written to provide a thorough and current (as of 2001) study of the statutory and case law in the fifty states plus the District of Columbia.

There are common patterns among the states in emergency vehicle law, but there are many differences. An attorney or police officer in Arizona needs to know the law of Arizona, not New York or Arkansas. That includes identifying the relevant statutes and examining the more recent cases. A general study would not be very useful to that person. If all this book did was to act as a shortcut to research, it would be a bargain. More than that, it should help the lawyer in deciding whether to take the case, or if on the defense side, what defenses have been asserted.

Part III is organized alphabetically, Alabama through Wyoming. The reader will want to check the current statutes for any changes, and read the cases in full and search for the latest cases. Federal cases are placed in the state in which they arose.

Emergency vehicle statutes

Set out below are the relevant sections from the Uniform Vehicle Code that governs emergency vehicle operation and civilian drivers' duties. Note that the UVC is not the law of any state. It is a model code that the states have enacted, with whatever changes the state legislature desired to make. I have noted in each state section what changes and additions were enacted.

UVC 11-106. Authorized emergency vehicles

(a) The driver of an authorized emergency vehicle, when responding to an emergency call or when in pursuit of an actual or suspected violator of the law or when responding to but not returning from a fire alarm, may exercise the privileges set forth in this section, but subject to the conditions herein stated.

(b) The driver of an authorized emergency vehicle may:
- (1) Park or stand, irrespective of the provisions of this chapter;
- (2) Proceed past a red or stop signal or stop sign, but only after slowing down as may be necessary for safe operation;
- (3) Exceed the maximum speed limit so long as life or property are not thereby endangered;
- (4) Disregard regulations governing direction of movement or turning in specified directions.

(c) The exemptions herein granted to an authorized emergency vehicle shall apply only when such vehicle is making use of an audible signal meeting the requirements of §12-401(d) and visual signals meeting the requirements of §12-214 of this code, except that an authorized emergency vehicle operated as a police vehicle need not be equipped with or display a special visual signal from in front of the vehicle.

(d) The foregoing provisions shall not relieve the driver of an authorized emergency vehicle from the duty to drive with due regard for the safety of all persons, nor shall such provisions protect the driver from the consequences of the driver's reckless disregard for the safety of others.

UVC 11-405. Operation of vehicles upon approach of emergency vehicles

(a) Upon the immediate approach of an authorized emergency vehicle making use of [audible and visual signals], or of a police vehicle properly and lawfully making use of an audible signal only;

(l) The driver of every other vehicle shall yield the right of way and shall immediately drive to a position parallel to, and as close as possible to, the right-hand edge or curb of the roadway clear of any intersection and shall stop and remain in such position until the authorized emergency vehicle has passed, except when otherwise directed by a police officer.

(2) [Refers to a similar duty for streetcar operators.]

(b) This section shall not operate to relieve the driver of an authorized emergency vehicle from the duty to drive with due regard for the safety of all persons using the highway.

UVC 11-510. Pedestrian yield to authorized emergency vehicles

(a) Upon the immediate approach of an authorized emergency vehicle making use of [audible and visual signals], or of a police vehicle properly and lawfully making use of an audible signal only, every pedestrian shall yield the right of way to the authorized emergency vehicle.

(b) This section shall not relieve the driver of an authorized emergency vehicle from the duty to drive with due regard for the safety of all persons using the highway nor from the duty to exercise due care to avoid colliding with any pedestrian.

UVC 11-907. Fleeing or attempting to elude a police officer

(a) Any driver of a motor vehicle who willfully fails or refuses to bring his or her vehicle to a stop, or who otherwise flees or attempts to elude a pursuing police vehicle, when given a visual or audible signal to bring the vehicle to a stop, shall be guilty of a misdemeanor. The signal given by the police officer may be by hand, voice, emergency light or siren. The officer giving such signal shall be in uniform, prominently displaying officer's badge or office, and officer's vehicle shall be appropriately marked showing it to be an official police vehicle.

(b) Every person convicted of fleeing or attempting to elude a police officer shall be punished by imprisonment for not less than 30 days nor more than six months or by a fine of not less than $100 nor more than $500, or by both such fine and imprisonment.

Commentary on §11-106

(a) "When responding to an emergency call."

Normally there is little dispute of whether there is an emergency call. Normally, if an officer signals a driver to stop, and if the driver then flees, the officer is acting in an emergency and within the protection of emergency driving statutes if he decides to chase and is observing the pursuit policies of his department. He does not have to wait and notify his dispatcher first, but would be expected to notify the dispatcher quickly. A Virginia case pushed the envelope when the court held an officer to be protected by sovereign immunity from the moment the officer decided to pull over a red-light runner. The officer glanced at his right-hand outside mirror to check for traffic before getting into the right-hand lane, and rear-ended the vehicle in front. *Nationwide v. Hylton,* 530 S.E.2d 421. (See the Virginia section.) Note that police department pursuit regulations may prohibit an officer from joining in a pursuit in progress unless specifically authorized.

A Colorado case, *Corsentino v. Cordova*, 4 P.3d 1082, has a good discussion of what standard to apply to determine if there was a bona fide emergency. The standard is an objective standard from the perspective of a reasonable emergency vehicle operator. In the fire truck context, the definition of what is an emergency is broad. If a truck is dispatched to a fire, some departments send a replacement truck to that station, and several cases have held that the second truck may proceed as in an emergency and take the right of way. In ambulance cases, it is not uncommon for the plaintiff to allege that the operator was not really on an emergency run, but the benefit of the doubt will go to the ambulance operator. Of course, if the emergency vehicle is not on an emergency run, the operator has the same duties as other drivers.

(a) "When in the pursuit of an actual or suspected violator of the law."

Most states have repeated this language. Iowa has made a substantial change: "When in pursuit of an actual or suspected perpetrator of a felony, or in response to an incident dangerous to the public." The intent is discourage police pursuits for misdemeanor offenses. I am skeptical how much effect it has. If literally observed, an officer could not speed to catch up with a speeder to stop him to issue a ticket. A police officer would have to have a good grasp of what grade a suspected violation is, misdemeanor or felony, and make that decision quickly. And a fleeing driver may commit a number of offenses in rapid succession. Local pursuit agency or department regulations may prohibit pursuits for minor traffic violations, but in general officers are given wide discretion in enforcing the law.

(b) This section lists four specific types of traffic laws that may be disregarded. Note that there are two specific standards given, parts (2) and (3), which refer to driving through red lights or stop signals, and exceeding the speed limit. States typically enact this section with little change. The standard for running red signals and stop signs is frequently used by the courts, but the standard for exceeding the speed limit is less frequently cited. Some state statutes expand this section to permit an emergency vehicle operator to drive on the shoulder, or go left of center in a no-passing zone. Even if not in the list of permitted acts, courts read into the act that permission has been granted to break any traffic law if done with the specified standard of care.

(c) This section, which refers to audible and visual devices, is frequently changed in state statutes. Many states permit visual or audible signals, not visual *and* audible signals, at the operator's discretion. Some state statutes specify situations in which a police officer may approach a crime scene or a suspicious situation silently and without flashing lights to apprehend suspects and preserve evidence. Even if not in the statute, courts are likely to grant the privilege. An officer in this mode would be expected to be extra cautious in running lights and stop signs.

(d) Due care or reckless disregard: which is it? This section of 11-106 has contradictory standards. I presume the drafters of the Uniform Vehicle Code couldn't agree and included both. No doubt state legislatures generally enacted the Uniform Vehicle Code without agonizing over each section. The phrases do not carry the same meaning. Courts generally hold due care synonymous with ordinary care or that care a reasonably prudent person would exercise in the same situation. In our context, that means exercising due care while running red lights, stop signs, speeding, or pursuing a fleeing driver. Only a minority of states apply the due care standard in emergency vehicle cases. Look at your state statutes and leading cases carefully. A minority of states set out one standard or the other, not both, in their statutes. Two state courts do not like the due care standard because it sounds like an ordinary negligence standard, and do not like the reckless disregard language either. The Texas legislature has changed due care to appropriate regard in its statute, and the courts construe that as meaning a reckless conduct standard. That standard is below reckless disregard. Pennsylvania similarly holds that an emergency vehicle operator must observe due care but for him or her that includes a degree of recklessness permitted. Otherwise, what is the point of the emergency vehicle statute?

Reckless disregard is held to have the same meaning as gross negligence, reckless conduct, willful disregard, heedless of consequences, malicious, wanton, rash, conscious indifference (but perhaps not deliberate indifference), and so on. The term is a common one in many types of statutes, and courts may develop specific meanings for specific statutes. It is necessary to read the appropriate cases to see how reckless disregard is construed in one's jurisdiction. One state, Illinois, includes both due care and reckless disregard in its emergency vehicle statute, and under the principle that all words in a statute should be given effect if possible, holds that an emergency vehicle operator must exercise due regard not to act with reckless disregard. At least that is creative.

A skeptic would point out that these are just words that cannot be applied precisely to real life situations, and there is some truth to that. Suppose a deputy sheriff is chasing a suspected drunk driver in moderate traffic on a state highway at speeds up to 90 mph. The deputy knows that there is a traffic signal a mile ahead, and backs off, but the drunk runs a red light and hits an innocent bystander. That is a common scenario. The trier of fact could find the deputy at fault or innocent under either standard, depending on subjective and possibly emotional factors. Of the several hundred cases analyzed in this study, however, the majority find for the defendant, usually on summary judgment.

Official and governmental immunity

In many states, the emergency vehicle statutes set out above are on the books but do not control. Roughly half the states have immunity statutes, each with its own standard of care stated, that control claims arising out of police, fire, and emergency medical vehicles. Sovereign immunity is an old common law doctrine, and all states necessarily retain immunity for some discretionary acts. Government officers and employees must be given some room to perform their duties without being second-guessed by the courts. In the traffic context, government typically is not liable for such decisions as installing or not installing traffic control signs and signals, street and highway design, and so on. Negligent implementation and maintenance may be actionable, since that would be a ministerial duty. For example, if government installs a traffic signal, it may be liable for not maintaining it.

Regarding the operation of government vehicles, there typically is a statute that makes the government liable for negligent operation, the same as any other driver. This is an exception to governmental immunity. Claimants, depending on the state's tort claims act, may have to file with a board, or may be able to proceed directly to court. There will be a statutory cap on damages, typically ranging from $50,000 to $250,000 per claimant and another limit on the total for one incident. I have not set out the details of such tort claims acts in each section, and in many states the emergency vehicle statutes apply instead. Look at the state section in this book for your state.

Many states provide an "exception to the exception" for emergency vehicle operation. Such states thus recognize that the duty to provide police and fire protection is a high duty and that operators on emergency runs enjoy immunity. Typically there is a standard of care stated. Such operators are immune so long as they do not act with gross negligence, reckless disregard, actual malice, and so forth. There may not be a specific emergency vehicle exception, but a general exception for the performance of discretionary functions unless done with actual malice. Then the issue is whether, in a police pursuit context, the pursuit, or parts of it, was a discretionary function. It can get complicated. Almost everywhere, the decision to pursue is discretionary. The conduct of the pursuit may or may not be discretionary, and the same for the decision to terminate the pursuit. You have to study the cases.

Some states, such as Illinois, provide immunity for those executing or enforcing the law. Pennsylvania has an exception for the pursuit of an escaped prisoner or person fleeing the police. Virginia courts apply state immunity for emergency vehicle runs, subject to a gross negligence standard. So a plaintiff in Virginia who confidently relied on the state's version of UVC 11-106 would get a shock when he received and answer to his complaint. Some states apply immunity but derive the standard of care from their UVC 11-106 statutes, thus giving weight to both statutes. And sometimes UVC 11-106 is applied to fire and ambulance cases but not to law enforcement cases. There is a continuing interplay between court decisions and responses from the state legislature.

The law of Texas is particularly complicated. The Texas Supreme Court holds that a police officer in pursuit is not liable (and thus his employer is not liable) if he was within the scope of his employment, was exercising a discretionary function, *and* was acting in good faith. The good faith concept was borrowed from §1983 claims in federal courts, which a legal morass, in the author's opinion. It may be difficult for a litigant to judge with much confidence how his case may turn out. By contrast, the pursuit law of California is simple. If the law enforcement agency had adopted pursuit regulations that covers four areas as specified in the statute, the officer and the agency is not liable to an innocent person injured or killed by the fleeing driver. That's it. End of case. Or at least until the legislature or the courts say differently.

Police pursuit

More than half the cases in this book are pursuit cases. The public does not much notice when a fire truck or ambulance collides with another driver, but pursuit cases are controversial. The public understands that the fire truck it going to a fire, but may wonder whether a police pursuit was necessary. On the other hand, the public wants vigorous enforcement of the law. The public hardly notices when a pursuit is successful and the law violator, often charged with a felony, was caught, and to be fair such law enforcement benefits should be considered as well when compared to the pursuits that ended in injury or death.

Pursuit statistics are beyond the scope of this book. See, for example, Geoffrey P. Alpert et al., "The Constitutional Implications of High-Speed Police Pursuits . . ." 27 U. Mem. L.R. 599 (1997), for the results of studies in selected police departments. No one doubts that the police have a difficult task in deciding whether and how to pursue a driver who flouts their order to stop. Judges frequently comment on the unfairness of applying hindsight to the officer's decision. The officer must balance the offense and risk the fleeing driver is posing to the public against the risk of pursuit, and do so under pressure, make split-second decisions, and continue to evaluate the risks in ever-changing circumstances. From my evaluation of the law of the fifty states, I can state that it appears that the public in general wants vigorous law enforcement to the extent that it causes a conflict between fighting crime and endangering the public. There apparently is little public support for blanket no-pursuit policies.

A law enforcement agency may well adopt a no-pursuit policy or restrict it. For example, the City of Pharr, Texas, prohibits pursuit speeds to no more than ten miles over the speed limit, and prohibits any pursuit outside of city limits. This policy will not guarantee that no innocent person will be hurt. Suppose a Pharr officer signals a red light runner to stop, and instead the culprit speeds away and hits an innocent person? Pursuits run the gamut from very brief ones to others that last for miles and involve several police vehicles. One in North Dakota went on for 60 miles, but at night in rural North Dakota the suspect was more likely to hit a deer than another car. Many pursuits are trailing pursuits. The officers hope the suspect will stop, and may follow in part to warn other drivers of the danger. The suspect may or may not slow down when the pursuers drop back out of sight. Whether they are dangerous to others depends on many factors, and officers must continually evaluate the risk versus the law enforcement benefit.

Similarly, fleeing drivers are not all the same. One group flees for, what seems to us, no rational reason other than to avoid a traffic ticket. But for some people, one more ticket for speeding means loss of license. Another group is under the influence of alcohol or drugs, or just plain irrational, and whether it is more dangerous to try to stop them rather than letting them go poses a dilemma for the officer. The public generally would want the attempt made, I suspect. The third group has a very good reason to flee. If they are caught, they probably will face jail time for the felony they are currently committing or have committed in the past. This third group is substantial. In California in 1994, over 50 percent of the approximately 5,000 pursuits were begun for a traffic violation, but 73 percent resulted in felony arrests. (Amicus brief of the County of Riverside, *Lewis v. County of Sacramento,* U.S. Supreme Court docket 96-1337.)

The most fruitful approach in mitigating the dangers of pursuits has been in department regulations and training, especially in training the newer officers. Most police dread pursuits and do not undertake them lightly. Pursuit policies cover when not to pursue, what factors to continually evaluate in deciding whether to continue or to terminate, the role of supervisors, how many police vehicles may participate, the use of helicopters, and so on. An attorney with a pursuit case will need to see the department guidelines in evaluating the case. I was somewhat surprised at what little consideration appellate courts gave department pursuit polices in their opinions. Unfortunately, other than the use of helicopters and devices like tire-puncturing strips (if the police can get them where the suspect will run over them), it is still pretty much chasing and setting up roadblocks ahead. Devices that would harpoon the suspect's vehicle or shut down his engine are still in the fantasy stage.

The plaintiff and defendant should both be aware of the range of the opposing parties and their status. The law enforcement officer himself is rarely the plaintiff. The fleeing driver may be injured and bring suit. A few states have statutory bars on his claim, and in general such cases are difficult to win. Generally the passengers in his car or a passenger of a motorcycle are considered innocent bystanders. Note that Michigan, by a recent decision of its supreme court, holds that such passengers must prove that they are innocent bystanders. If they are knowingly riding in a stolen car, have encouraged the driver to flee, made obscene gestures to the pursuing officers, and so on, they are presumptively assumed to be of the same status as the fleeing driver. The next category of plaintiffs are those who collide with the fleeing driver. Assuming they were observing the traffic law, they are innocent bystanders and are without fault. Pedestrians would be in the same category. A final category are those who collide with a law enforcement vehicle that is in pursuit or on an emergency run. Here it is not enough to for them to have observed traffic lights, speed limits, and stop signs. If the emergency vehicle had its lights and siren on, they do have a duty to yield.

One must not overlook the role of insurance. Do not assume that all fleeing drivers do not have liability insurance, and at least collect that. If the culprit is not insured, or has low limits, bring in the plaintiff's own auto uninsured or underinsured coverage. This sometimes is applicable if the plaintiff collided with a government vehicle and there is no recovery because of immunity, or because due care or reckless disregard was not proven. The majority of cases in this book were resolved in favor of the defendant in summary judgment. It is very important to fully research the law and craft the complaint accordingly. Summary judgment was granted to defendants in some cases in this book because the plaintiff did not support his allegations in response with the affidavits of experts.

Federal cases

Because of the difficulty of prevailing in state court, beginning in the mid-1980s there was a trend to try these cases in federal court. The claims are quite different, although one may assert state law claims along with federal constitutional claims. Two constitutional violations have been asserted seriously. (Some other claims are frivolous, such as the claim that being forced to be part of a roadblock violates the Thirteenth Amendment against involuntary servitude.) The first is unlawful seizure in violation of the Fourth Amendment, and the second the deprivation of life, liberty, or property without due process of law (substantive due process) under the Fourteenth Amendment. I have placed federal court cases with the state section in which they arose.

There are two U.S. Supreme Court cases that apply. See the California section for my summaries. (It is better, of course, to read the cases.) *Brower v. County of Inyo,* 489 U.S. 593 (1989), is a roadblock case, and is the authority for the claim of unlawful seizure. *County of Sacramento v. Lewis,* 523 U.S. 833 (1998), is the authority for Fourth and Fourteenth Amendment claims. The court took *Lewis* because of a conflict among the circuits about what standard to apply, deliberate indifference or shocks-the-conscience. The pursuit in question was not particularly outrageous compared to many others in this study, and the court held that the highest standard, conscience-shocking, applies to pursuit cases. I noticed that several state cases and lower cases in the federal courts were delayed until that decision came down, and then they were decided in favor of the defendants. I predict that these cases will tend to reduce the number of federal claims in pursuit cases. Many judges have commented that pursuit cases are state tort claims that belong in state courts.

State-by-State Brief Summary

Alabama

UVC 11-106 enacted with insignificant change. It is not clear which standard the courts use, due care or reckless disregard. A 1984 state supreme court case specifically cited reckless disregard.

Alaska

The only significant change from the UVC is the requirement that emergency vehicles stop at stop signs and signals before proceeding. It is not clear which standard is applied.

Arizona

No significant change from the UVC. The courts are moving toward the due care standard.

Arkansas

No change from UVC. Due care is the standard. Governmental liability is capped at $25,000 per claimant, $50,000 per incident, which are on the low side.

California

UVC 11-106 is not enacted as such. Emergency vehicle operations are established from traffic code provisions, citing a range of section numbers. Law enforcement officials and agencies are not liable to anyone injured or killed by a fleeing driver as long as the agency has adopted regulations conforming to four requirements. It is not clear which standard applies when the emergency vehicle hits the plaintiff, because both are cited.

Colorado

Colorado has enacted UVC 11-106 with little change, but emergency vehicle operators are immune by statute, 24-10-106, but are subject to subsection (2) of their version of the UVC 11-106 (slowing as may be necessary for safety at red lights and stop signs, and not speeding so that life or property are endangered). The due care and reckless disregard subsection do not apply because they are not referenced in the immunity statute.

Connecticut

UVC 11-106 has been enacted, but the reckless disregard language has not been included. Due care is the standard.

Delaware

UVC 11-106 has been enacted, but instead of due care and reckless disregard, the statute protects an emergency vehicle driver except for "gross negligence or willful or wanton negligence."

District of Columbia

The District includes UVC 11-106 in its regulations. The district is immune except for gross negligence. This applies only to emergency vehicles owned by the District. One case held a Virginia officer pursuing into the District to a due care standard.

Florida

UVC 11-106 enacted without significant change. The standard the courts apply seems to be "negligent, careless or wanton." Sovereign immunity has been asserted.

Georgia

Has enacted UVC 11-106, but omits the reckless disregard standard. Section (d)(2) states that when a fleeing suspect hits plaintiff, officer will not be the proximate cause or a contributing cause unless he or she acted with reckless disregard. The tort claims act waives motor vehicle immunity except for "[t]he method of providing law enforcement, police, or fire protection." Some cases apply immunity for discretionary acts unless malice is shown; others do not. A 1995 case applied due regard. The emergency vehicle law in this state is unsettled.

Hawaii
UVC 11-106 has been enacted without change. No cases.

Idaho
UVC 11-106 has been enacted. No significant cases.

Illinois
Illinois has enacted UVC 11-106, but also provides immunity, except for willful or wanton conduct, when the public employee is "responding to an emergency call," or executing and enforcing the law.

Indiana
UVC 11-106 has been enacted without change, but until recently emergency vehicle cases were governed by immunity for discretionary acts or of enforcing the law, unless the government official's acts were outrageous. The Indiana Supreme Court held in a 1993 decision that the UVC-derived statute would control in emergency vehicle cases.

Iowa
UVC 11-106 has been enacted with one change. A law enforcement officer is protected only when he pursues a "suspected perpetrator of a felony." The Iowa Supreme Court has decided that reckless disregard, not due care, applies.

Kansas
UVC 11-106 has been enacted. The courts seem to apply the standard of due regard.

Kentucky
Kentucky's UVC 11-106 statute gives the standard of due care only.

Louisiana
Louisiana has enacted UVC 11-106 and applies due care.

Maine
Maine has enacted UVC 11-106 without change. The single case does not discuss the standard by which to judge the emergency vehicle operator's

acts, but focuses on the duty of the pursued to anticipate that the officer may well follow closely.

Maryland

Maryland omits the reckless disregard language in its UVC-derived statute, but cases are controlled by state governmental immunity, except for gross negligence or malice.

Massachusetts

The state did not enact UVC 11-106. Its emergency vehicle statute covers the essence of the privileges, and specifies the standard of emergency vehicle operators as that of due regard. Discretionary immunity plays a part, but there are too few state cases to clearly indicate the law. Several recent cases were brought in federal court.

Michigan

There are two emergency vehicle statutes, one which is the UVC-derived text and apparently an older statute. The standard is reckless disregard. The cases are controlled by governmental immunity, however. Governmental employees are immune in the discharge of a governmental function absent gross negligence (see the statute for the definition of gross negligence). A sweeping supreme court decision, *Robinson* (2000), amplifies on the current law of emergency vehicles and holds police not liable where the fleeing driver hits the plaintiff.

Minnesota

The emergency vehicle statutes set forth both due care and reckless disregard. Governmental immunity, not specific to law enforcement, applies. The leading case, *Pletan* (1992), holds that police pursuit is protected by immunity.

Mississippi

Mississippi has enacted the essence of UVC 11-106, but the immunity statute states that government employees shall not be liable for an act or omission relating to police or fire protection absent reckless disregard, and shall not be liable in any circumstances to one engaged in criminal activity.

Missouri

UVC 11-106 has been enacted. The sovereign immunity statute provides that such common law immunity that existed before Sept. 12, 1977, remains in effect. (This was in response to a Missouri Supreme Court decision abolishing sovereign immunity.) Cases indicate that police officers may have discretionary function immunity, and that to enjoy immunity the officer must comply with the emergency vehicle statute. Cases hold that the police are not liable when the pursued driver hits the plaintiff.

Montana

Montana has enacted UVC 11-106, and its one case holds that the due care standard applies.

Nebraska

Nebraska has enacted UVC 11-106 without change. The courts have not specifically decided between due care or reckless disregard in applying the statute. However, Nebraska is unique in its statute providing that an innocent person injured in a police pursuit accident that was "proximately caused" by the pursuit will recover damages, up to $1 million (the highest statutory cap among the states). The statute has been construed to hold that the officers do not have to be negligent, just a factor in the accident, for the statute to apply.

Nevada

Nevada has enacted UVC 11-106, and there is no significant case.

New Hampshire

New Hampshire has enacted UVC 11-106 without significant change, and there are no significant cases construing it.

New Jersey

New Jersey has enacted UVC 11-106, but immunity with its exceptions controls the law. Two immunity statutes can apply. One provides that a government employee is not liable for the good faith execution or enforcement of any law. The other provides, in part, that neither a public entity or public employee is liable for any injury resulting from an escaped or escaping person, or a person resisting or evading arrest. The latter statute was a 1997 amendment to restore immunity in police pursuits. The public employee is not protected if he was acting fraudulently, with malice, or with willful conduct.

New Mexico

New Mexico has enacted UVC 11-106 without significant change, and the single case in state court does not specifically indicate whether due care or reckless disregard is the proper standard.

New York

UVC 11-106 has been enacted. The court of appeals specifically held that reckless disregard, not due care, is the proper standard (in the leading case of *Saarinen,* 1994). No immunity is asserted, but the numerous cases show that New York strongly protects police pursuits.

North Carolina

North Carolina has enacted UVC 11-106, and court decisions hold that the police are held to a reckless disregard or gross negligence standard, not due care.

North Dakota

North Dakota has enacted the essence of UVC 11-106. The single case holding for the innocent passenger plaintiff was reversed for failure to apply comparative negligence against the fleeing driver. The court discussed the two standards without deciding if one or both applied.

Ohio

Like California, Ohio has not enacted UVC 11-106 as such, but rather states that some forty-four sections of the vehicle code do not apply to emergency vehicles responding to a call. The standard of care required is due care. However, immunity applies to claims arising from the operation of emergency vehicles unless the operator acted with willful or wanton conduct.

Oklahoma

Oklahoma has enacted UVC 11-106 with few changes, and has an immunity statute stating that the state or a political subdivision shall not be liable for a claim resulting from the method of providing police, law enforcement or fire protection. Only three relevant cases were found. Two cases applied the method of providing law enforcement statute, and one under the emergency vehicle statute found that due care was properly applied.

Oregon

Oregon has enacted the essence of UVC 11-106, and states a due care standard. The immunity statute makes public bodies and their employees immune in the performance of a discretionary function, even if the duty is abused. There are few cases. A 1996 case found for the plaintiff when a deputy sheriff ran a red light. A 1989 case in which the fleeing driver hit the plaintiff found the pursuing officer had discretion in deciding to initiate and to end a pursuit.

Pennsylvania

The state's emergency vehicle statute gives a due care standard only. Cases hold that this is not the same as ordinary care. The emergency vehicle operator must exhibit some degree of recklessness to be held liable. A statute provides that there shall be no liability to any persons fleeing or resisting arrest.

Rhode island

The state has enacted UVC 11-106. The few cases do not give a clear answer of what standard is applied to emergency vehicle operators. Rather, they turn on whether there was proximate cause.

South Carolina

South Carolina has enacted UVC 11-106, with the exception that it omits the reckless disregard language. No immunity is claimed.

South Dakota

UVC 11-106 has been enacted. There are no relevant cases.

Tennessee

Tennessee has enacted UVC 11-106 with two changes. It applies due care only, and specifically states that there shall be no liability to the fleeing driver, nor for injuries to third parties proximately caused by the fleeing driver, unless the pursuing officer was negligent and a proximate cause. The cases primarily deal with the issue of proximate cause.

Texas

Texas has enacted UVC 11-106 with some changes and additions. Uniquely, a physician responding to an emergency call may claim the privileges of the emergency vehicle statute. Texas statutes now refer to appropriate regard, not due regard, as the standard of care required of emergency vehicle operators. That is held to be somewhere between due care and reckless disregard. Three immunity statutes can apply. One provides immunity for the method of providing police or fire protection, the other is based on discretionary function immunity. The courts determine cases based on whether the defendant was within the scope of his employment, was performing a discretionary function, and acted in good faith. It is complicated. Read the cases with care.

Utah

Utah has enacted UVC 11-106 with some additions, such as the requirement that a law enforcement agency must have a written pursuit policy and provide training. The waiver-of-immunity statute has an exception for the operation of emergency vehicles, but the operators must observe the emergency vehicle statute (UVC). The cases make clear that due care is required by emergency vehicle operators.

Vermont

Vermont has enacted UVC 11-106 with minor changes. It requires "fresh pursuit" of the suspect, although it is difficult to see how significantly this differs from pursuit. The few cases tend to leave the issues to a jury rather than grant summary judgment, in contrast to the majority of the cases from other states.

Virginia

Virginia has enacted UVC 11-106 with some changes, and applies the standard of reasonable care. However, the cases are decided on common law sovereign immunity with a standard of gross negligence.

Washington

Washington has enacted UVC 11-106 with only minor changes. The courts apply the standard of due care. The discretionary function defense has been asserted, but the courts hold that only high-level decisions and planning are immune, not law enforcement and emergency vehicle operation.

West Virginia

West Virginia has enacted UVC 11-106 without change. The courts apply due care in nonpursuit cases, and reckless disregard otherwise, reflecting the realities of police pursuit.

Wisconsin

Wisconsin has enacted UVC 11-106, and spelled out in detail the circumstances in which a police officer may approach a suspected or actual crime scene quietly. Wisconsin also observes governmental immunity for the performance of a discretionary function. The leading case, *Polenska* (1988), held that some decisions and conduct relating to a pursuit are discretionary and some are ministerial.

Wyoming

Wyoming has enacted UVC 11-106 without change, but there are immunity issues. The legislature, in passage of the Governmental Tort Claims Act of 1979, abolished all judicially created categories such as governmental or proprietary functions and discretionary or ministerial acts. However, the Act did not abolish common law defenses. The two cases identified wrestled with these concepts.

State-by-State Statutes and Case Law

Alabama

Ala. Code §32-5A-7. Emergency vehicles

[Alabama has enacted UVC 11-106 with only slight changes in wording, but none in meaning.]

Ala. Code §6-5-338. Peace Officers; On or off duty

Every peace officer, except constables, who is employed or appointed . . . by the state or a county or municipality thereof . . . shall at all times be deemed to be officers of this state, and as such shall have immunity from tort liability arising out of his or her conduct in performance of any discretionary function within the line and scope of his or her law enforcement duties.

The most recent case is *Williams v. Crook,* 741 So. 2d 1074 (Ala. 1999). It dealt with the problem faced by officers who want to approach a possible crime scene without alerting suspects. Sergeant Crook, employed by the city of Bay Minette, received a report of a domestic disturbance. While driving to the address, he collided with two other vehicles at an intersection. There was evidence that he was exceeding the speed limit, and it was undisputed that he did not use his blue lights or siren. The trial court granted summary judgment for Crook and the city on the basis of §6-5-338. The state supreme court reversed for trial. There was no question that the officer was exercising a discretionary function, but if indeed he was speeding he was required to turn on his blue lights and siren.

A 1987 case was in the Alabama Supreme Court twice. It involved catch-up driving, not actual pursuit. A state trooper was traveling at night traveling south on a four-lane highway, checking for speeders in the northbound lanes. He detected a speeder, and made a U-turn to pursue. He did not turn on his flashing lights or siren; instead he sped up to 80 mph in the practice of "catch-up driving," as he had been taught. The purpose of catch-up driving is to prevent the speeder from initiating a high-speed chase by coming up without warning. He saw some reflectors ahead and struck and killed a bicyclist. The officer had braked and moved to the left lane, thought he had cleared whatever vehicle was ahead, and resumed speed. In the first trial, the officer was found not negligent. That result was reversed because the trial court refused to instruct on wantonness. The second jury found the officer's conduct not to be wanton, but the supreme court again reversed for a new trial. It was error to permit testimony on catch-up driving because such testimony tended to excuse the defendant. The court pointed out that catch-up driving is contrary to §32-5A-7(c), and was not relevant to the issue of wantonness. Two dissents disagreed. Two juries had already found the trooper's driving neither negligent nor wanton. The fourteen-year-old deceased bicyclist was to some degree negligent. The trooper had been taught that catch-up driving was in some circumstances the safest procedure. (In some jurisdictions, the emergency vehicle statute specifically permits police to approach crime scenes or speed to catch up and inspect drivers without emergency signals.) *Smith v. Bradford,* 512 So. 2d 50 (Ala. 1987).

A 1984 case involved hot pursuit. Two police officers in the city of Winfield noticed a vehicle driven by Adkins, speeding over 70 mph in a 50-mph-zone within the city limits. Adkins, who was drunk, later said he tried to outrun the pursuit. The officers pursued, with blue lights flashing, at speeds of up to 100 mph. The pursuit ended when Adkins ran a stop sign and hit a vehicle in which a passenger was injured. The issue was whether the officers acted with reckless disregard. The jury found for the plaintiff and awarded damages. The supreme court reversed for a new trial as to the officers and the city. The trial judge erred in not giving the defendant's Instruction No. 17. The jury should have been instructed that the officers were not obligated to allow Adkins to escape and that they had a duty to arrest Adkins as long as their actions were not negligent or wanton. They cannot be responsible for Adkins' actions even if the pursuit contributed to the acts. *Madison v. Weldon,* 446 So. 2d 21 (Ala. 1984).

A 1988 case followed the reasoning in *Madison* and granted summary judgment to the defendant officers. The facts were similar: a pursued driver ran a red light and injured his own passengers and occupants of the other vehicle. The defendant officers by affidavit showed that, while they exceeded the speed limits, they also slowed to as low as 10 mph to cross intersections against a red light. The officers' pursuit may have contributed to the accident but were not its legal cause. *Doran v. City of Madison,* 519 So. 2d 1308 (Ala. 1988).

A 1994 case arose from the death of an innocent bystander. A pursued driver hit the innocent bystander head-on and she was killed. Following *Madison* above, the trial court dismissed the complaint. The supreme court reversed for trial. The pursuit began when the officer attempted to stop a suspected drunk driver. The site was a state highway, and speeds reached 100 mph, according to a witness's deposition. The plaintiff's expert's affidavit opined that the officer should have broken off the pursuit by turning off his flashing lights and siren and resuming a legal speed when the pursued driver turned off his own lights, and when the officer knew a roadblock was set up ahead. Merely dropping back half a mile is not truly breaking off pursuit, because the pursued can see the flashing lights behind. The court held there was a triable issue. *Seals v. City of Columbia,* 641 So. 2d 1274 (Ala. 1994).

A different pattern is presented when the pursued driver himself is the injured party and there is no innocent bystander. A 1989 case involved a high-speed pursuit of a motorcyclist. The flight seems rather ill-advised. The motorcyclist was clocked at 64 mph in a 55-mph-zone on an interstate highway. When police attempted to stop him, he fled at speeds of up to 120 mph. He exited the highway and then lost control on a curve, crashed, and died from his injuries several days later. The supreme court affirmed summary judgment for the defendant, noting that if officers are not liable to an innocent bystander, they are even less likely to be liable for mishaps to the pursued driver himself. The court also affirmed that there was no deprivation of life or liberty within the meaning of the due process clause and dismissed the §1983 action. *Blair v. City of Rainbow City,* 542 So. 2d 275 (Ala. 1989).

A 1989 case was brought in federal court under a §1983 claim. The case arose after Little Rock police pursued a car thief and the thief apparently caused the death of an innocent occupant of another car. The case does not summarize the facts of the chase. The district court judge dismissed the claim. The court followed *Jones* (see Tennessee section) in pursuit cases in holding that gross negligence must be more than simple negligence "with the addition of a vituperative epithet." The brief chase in this case occurred in light traffic and was well short of outrageous conduct. The court notes that while it may be a good policy not to chase those who present no apparent danger, such a policy is not constitutionally required. *Britt v. Little Rock Police Dept.,* 721 F. Supp. 189 (E.D. Ark. 1989).

The only case found involving an ambulance or fire truck is an old one, *Echols v. Vinson,* 124 So. 510 (Ala. 1929). An ambulance on an emergency run collided with the plaintiff driver at an intersection. The court reversed a judgment for the plaintiff for a new trial. The trial judge must instruct on the special rights of emergency vehicles. The trial court focused only on which driver has the right of way at uncontrolled intersections.

Alaska

Alaska Admin. Code §13 t. 02-517. Authorized and other emergency vehicles

[Alaska places traffic rules in the administrative code rather than the Alaska Statutes. Here is a summary of the variations from UVC 11-106 (See the Introductory Essay for the text):

- (a) This subsection extends the privilege of disregarding traffic laws to a driver of an emergency vehicle who is "in the performance of his official duties," not only one who is on an emergency call, in pursuit of an actual or suspected law violator, or going to a fire.
- (e)(2) This subsection permits the emergency vehicle driver to exceed the speed limit "if he does not unreasonably endanger life or property, slows at each intersection, and exercises care under the circumstances."
- (2)(3) This subsection permits illegal turns or directions if the driver stays under the speed limit for that location.

(a) The driver must stop at each stop sign or red signal before proceeding. The UVC requires the driver to slow down and stop only if necessary for safety.]

The most significant case involved a police pursuit and injury to an innocent bystander. It was in the supreme court twice. The pursuit arose when an officer stopped a driver who turned out to have a revoked license. The driver fled at high speed, and the pursuit was on. The pursued entered an intersection on a green light, albeit in a dangerous manner. The officer, with flashing lights and siren on, entered the intersection without stopping on a red signal and collided with the plaintiff's car, which had the green light, and the plaintiff was seriously injured. There was evidence that others at the scene did notice the police vehicle, and that the plaintiff had an unobstructed view. The plaintiff was required by Admin. Code §13, t. 04.201(d) to yield to the emergency vehicle. At a bench trial, the court found all parties negligent and apportioned fault as 60 percent pursued driver ($187,937), 32 percent city ($80,503), and 8 percent plaintiff. The plaintiff's second claim was for a violation of 42 U.S.C. §1983 rights for inadequate training of police officers amounting to deliberate indifference of the rights of others. The trial court granted the city summary judgment on this claim.

The state supreme court affirmed on the negligence claim and reversed for further findings on the §1983 claim, 863 P.2d 240 (Alaska 1993). The trial judge was to determine if the pursuit-driving training was inadequate, whether any inadequate training could be said to represent city policy, and whether any inadequate training indicated that the city was deliberately indifferent to the public's safety. On the remand, the trial judge found no §1983 violation, and the supreme court affirmed, *Hildebrandt v. City of Fairbanks,* 957 P.2d 974 (Alaska 1998). The court pointed out that "[d]eliberate indifference on the part of the city in adopting inadequate police training policies, without more, cannot sustain a substantive due process claim. If we allowed such a result, courts would become embroiled in evaluating municipalities' resource allocation decisions—a task for which the courts are ill-equipped." The court did not find that the police officer's conduct in the chase was shocking to the conscience. It cited several police pursuit cases in which speeds where higher and innocent bystanders were killed but where such conduct similarly did not rise to the level of a constitutional violation.

Note that the plaintiff in *Hildebrandt* did succeed in winning damages in his claim for ordinary negligence, a result that is by no means typical in police pursuit cases, with only an 8 percent reduction for comparative negligence. He did not succeed in persuading the court to hold the pursued driver and the police officer to be a single party under Alaska Stat. §09.17.080. Their actions as causes of the accident were easily separable, and comparative negligence was properly applied to each separately.

There can be problems in collecting a judgment from the pursued driver: he may have no liability insurance or attachable assets, have low limits, or his insurer could attempt to deny coverage on the basis that the incident was more like an intentional tort than an accident.

A 1995 case of police pursuit resulted in the death of the pursued driver. An airport security officer noticed several cars illegally parked on the premises, and noticed a liquor bottle on top of one of the cars. One suspect, later identified as Day, an individual wanted for several law violations, ran into the woods. The officer left the woods, assuming that Day would come out. He did, driving his car at high speed. The officer says that he pursued the subject at 70–75 mph, but then lost contact and stopped the pursuit. Meanwhile, Day ran off the road and died in the crash. His blood alcohol level was .21 percent. The estate sued on the claim that the officer had a duty to arrest Day and prevent him from reaching his car. The trial court granted summary judgment to the state on that claim. Then the estate argued that the officer was negligent in the pursuit in that he turned off his siren as he entered a dip in the road, then came out of the dip and startled Day into a crash. The court again granted summary judgment to the state. While officers owe a duty to the innocent bystander in conducting pursuit, little is owed to the pursued driver himself. The blameworthiness of the fleeing driver's own conduct in illegally fleeing is paramount. Any other holding would encourage drivers to flee and make police reluctant to catch lawbreakers. *Day v. Willis,* 897 P.2d 78 (Alaska 1995).

The only nonpolice emergency vehicle case found involved a fire truck on a nonemergency run. The driver killed a pedestrian running across in front from a unfavored street. The fire truck driver accordingly has the duties of any other driver in that situation. The driver was going 30 mph, and the only real issue was whether the posted speed limit was 20 or 35 mph. The case includes a photograph of the area. The driver had entered the lane leading to a slight curving ramp into a business district. The dedicated lane was posted at 35 mph, while the business district, the accident site, a short distance beyond, was 20 mph "unless otherwise posted." The supreme court affirmed a jury defense verdict. The determination of the applicable speed limit was a question of fact for the jury. *West v. Municipality of Anchorage,* 754 P.2d 1120 (Alaska 1988).

Arizona

Ariz. Rev. Stat. 28-624. Authorized emergency vehicles

[The only significant change from UVC 11-106 (see the Introductory Essay for text) is that Arizona permits the driver of an emergency vehicle to exceed the prima facie speed limit, whereas the UVC refers to the maximum speed limit. The prima facie speed limit, referred to in UVC §11-801 as the "basic rule," is speed reasonable and prudent under the conditions, and may well be less than the maximum statutory speed in adverse conditions.]

An older case dealt with the validity of a city ordinance that conflicted with the state statute. The usual rule is that an ordinance that deals with a matter of statewide concern may not so conflict. A Phoenix ordinance required an emergency vehicle on a run to come to a full stop at stop signs and red signals before proceeding with caution. The state statute requires such vehicles only to slow as necessary for safe operation. The court reversed a verdict for a plaintiff who was injured while riding in an ambulance. Ariz. Stat. §28-626 itself states that such provisions be uniform throughout the state and in all political subdivisions. *Phoenix Respirator v. McWilliams,* 408 P.2d 951 (Ariz. App. 1970).

We deal first with cases in which the plaintiff is an innocent bystander. There are generally two views. One is that the pursued driver, who caused the pursuit and often is incredibly reckless, is solely responsible and is the proximate cause of the injury or death of the innocent bystander. The other is that the police owe a duty to the innocent bystander and can be negligent in the pursuit or attempted arrest. The most recent case arose in the early morning hours in Tucson. Three private security officers noticed a driver weaving and speeding, then running a red light, driving over a curb, sideswiping a car, and continuing. A police officer arrived and stopped the driver. Then the driver sped off, driving recklessly, and eventually ran a red light and killed another driver. Two officers at this point were pursuing and had been joined by a police helicopter. The estate sued the city and made a number of allegations: the officer who made the initial stop should have blocked the driver's car, the officers should have broken off pursuit and slowed to let the air crew coordinate the pursuit as required by regulations, and the air crew was negligent in illuminating the pursued driver.

The city moved for summary judgment, arguing that the officers had a duty to pursue, that they exercised reasonable care, and that their actions were not the proximate cause as a matter of law. The court granted the motion. The appellate court reversed for trial. It determined that courts are moving to the second view noted above. It cited, among others, a Connecticut case, *Tetro v. Town of Stratford,* 458 A.2d 5, holding that even in the face of negligent or reckless behavior of the pursued driver, that a negligent pursuit may be a proximate cause of an accident with an innocent bystander. The court also noted that ordinarily such questions are for a jury. The court remanded for trial. There is no further appellate record of this case. *Aten v. City of Tucson,* 817 P.2d 951 (Ariz. App. 1991).

The controlling statute, Ariz. Stat. §28-624(D), states that the driver of an emergency vehicle is not protected from the consequences of his reckless disregard for the safety of others. A 1970 case came about when a highway patrolman, chasing a speeder on a highway at 8:00 A.M., and apparently exceeding the 60-mph-speed limit, collided with a driver who pulled out from a roadside café. The case does not state whether the flashing lights or siren were on. The patrolman saw the driver 300 to 400 yards away, and presumed the driver would yield. There was a moderate collision, and the driver was able to continue until he could pull off the road safely. The trial judge refused to give the plaintiff's requested instruction on reckless disregard for the rights of others. The jury found for the defendant, and the supreme court affirmed. The patrolman was speeding for the purpose of performing his job, and the evidence that he noticed the plaintiff's vehicle and was able to take evasive action and almost avoid a collision did not justify the requested instruction. *Thompson v. State*, 466 P.2d 5 (Ariz. 1970).

Ariz. Stat. §28-624(C) requires the driver of an emergency vehicle to sound his siren "as reasonably necessary." In a 1977 case, an officer was driving to a report of an armed robbery in progress, and observing police regulation "Code Two." He was found not liable when he collided with a driver at an intersection. Code Two permits the officer to drive to the scene without the use of a red light or siren, but he must drive at a safe and lawful speed. The officer was going 45 to 55 mph in a 35-mph-zone. The jury found the plaintiff driver contributorily negligent (this was before comparative negligence in Arizona) and awarded him no damages. The appellate court affirmed. The plaintiff's requested instruction on negligence per se was properly rejected. Mere speeding is not negligence per se. Also, the statutory requirements for operating an emergency vehicle did not apply because under Code Two the vehicle is to be driven in accordance with all traffic laws. *Hurvitz v. Coburn,* 572 P.2d 128 (Ariz. App. 1977).

Law enforcement officers owe a greater duty of care to innocent bystanders than to the pursued himself and to any passengers in his vehicle. A 1975 case has some unusual facts, and deals with the beliefs of the pursued driver. A patrolman was parked across from a tavern at night and saw two men in a car argue with a man in the parking lot, who tried to enter the car. (It was discovered later that the men in the car had used the services of the other man's prostitute and had not paid. The pimp later said he would have killed them if he had caught them.) The driver and passenger sped away at high speed, and the officer pursued. The officer reached speeds in excess of 100 mph without closing nearer than one-third of a mile. He slowed for an S-curve, didn't see the car, went back to the curve and found the car had crashed. One of the men was dead and the other dying. The family of one sued the patrolman and the state. Their theory of liability was that since the officer did not turn on his flashing lights or use his siren, the pursued driver thought his pursuer was the pimp and drove in fear for his life. In other words, if the officer had identified himself by these means, they would have stopped.

The plaintiff relied on Ariz. Stat. §28-954(E). This section deals with horns and warning devices, and says that the driver of an emergency vehicle shall use a siren "when reasonably necessary to warn pedestrians and other drivers of the approach of the emergency vehicle." The trial court granted the defendants motion for summary judgment and the appellate court affirmed. Since there were no pedestrians or other drivers involved, the cited statute did not apply. The statute cited does not require an officer to identify himself or to warn a pursued driver. A duty to the deceased here could only arise only if the officer knew or should have known that the deceased thought that he was being pursued by the pimp. *Hederick v. State,* 530 P.2d 1144 (Ariz. App. 1975). Note that this case was the subject of a law review article, "Police Officers, Duties to Pursued Motorists," 18 Ariz. L. Rev. 834 (1976).

A 1952 case arose over the question of whether a driver of an ordinary vehicle can in the right circumstances assume the rights of an emergency vehicle. A patrolman noticed a speeding driver, whom he stopped. The driver was taking a knife wound victim to a hospital. The patrolman told the driver to "follow me" and to "stay close to me," and led the way toward the hospital at high speed but with appropriate lights and siren. The officer entered an intersection on a green signal. The signal turned red, and a driver, the defendant, not noticing the following car, started up and there was a collision. The trial court awarded damages to the passengers in the car carrying the wounded man. The supreme court reversed and dismissed the complaint. It noted that the following driver entered the intersection on a red signal on the wrong side of the road. The plaintiff's car was not an emergency vehicle simply because it was following a police officer on an emergency run. *West v. Cruz,* 251 P.2d 311 (Ariz. 1952).

Arkansas

Ark. Code Ann. §27-49-109. Drivers of authorized emergency vehicles

(a) The driver of any authorized emergency vehicle when responding to an emergency call upon approaching a red or stop signal or any stop sign shall slow down as necessary for safety but may proceed cautiously past the red or stop sign or signal. At other times, drivers of authorized emergency vehicles shall stop in obedience to a stop sign or signal.

(b) No driver of any authorized emergency vehicle shall assume any special privilege under this act except when the vehicle is operated in response to an emergency or in the immediate pursuit of an actual or suspected violator of the law.

Ark. Code Ann. §27-51-202. Restrictions not applicable to emergency vehicles

(a) The prima facie speed limitations set forth in this subchapter shall not apply to authorized emergency vehicles when responding to emergency calls and the drivers thereof shall sound an audible signal by bell, siren, or exhaust whistle.

(b) This section shall not relieve the driver of an authorized emergency vehicle from the duty to drive with due regard for the safety of all persons using the street, nor shall it protect the driver of any emergency vehicle from the consequences of a reckless disregard of the safety of others.

Ark. Code Ann. §21-9-301. Tort liability—immunity declared

It is declared to be the public policy of the State of Arkansas that all counties, municipal corporations, school districts, special improvement districts, and all other political subdivisions of the state shall be immune from liability and from suit for damages, except to the extent that they may be covered by liability insurance. No tort action shall lie against any such political came because of the acts of its agents and employees.

Ark. Code Ann. §21-9-303. Motor vehicle liability insurance required

(a) All political cames shall carry liability insurance on their motor vehicles or shall become self-insurers, individually or collectively, in the minimum amounts prescribed in the Motor Vehicle Safety Responsibility Act, §27-19-101 et seq.

[The act goes on to state that if a governmental entity does not purchase insurance, it is treated as a self-insurer and its liability is limited to the minimum coverage required by §27-19-101. Nothing prevents an entity from purchasing larger coverages, and any claimant can collect to that amount. The latter two sections set out above are typical state responses to the historic doctrine of governmental immunity. The motor vehicle exception is very commonly enacted. As of this writing, the minimum limits prescribed in Arkansas is $25,000 for one claimant, $50,000 for two or more claimants in the same accident, and $25,000 for damage to property. These low limits are designed to protect the financial resources of governmental units. The act applies to all government-owned motor vehicles, not just police, fire, and ambulance vehicles, as well as torts other than vehicle accidents.]

A 1989 case, *City of Little Rock v. Weber,* 767 S.W.2d 529 (Ark. 1989), brought another chapter in the attempt to end sovereign immunity. In a 1968 case involving an accident between a city-owned vehicle and a driver, the Arkansas Supreme Court abolished sovereign immunity. In response, the legislature passed an act the next year restoring immunity with the provision that government vehicles carry liability insurance. *Weber* arose out of a collision between a city police officer and another driver. The officer, with lights and siren turned on, entered an intersection on red and they collided. The trial court declined to grant the city summary judgment and at trial the jury awarded the injured driver $4,750. Despite this paltry award, the city appealed, arguing that it was absolutely immune for any negligence in the operation of an emergency vehicle. The supreme court held that the city was attempting to have the court apply a test made obsolete by later decisions. The test had been applied in the past to permit an injured person to sue a public official or employee even though the government was immune. Statutes carving out an exception for motor vehicle accidents and granting public employees immunity from suit for acts or omissions within the scope of their duties now governed. The court also held that an instruction stating that the driver of an emergency vehicle who is exercising special privileges, such as going through red signals without stopping as was done here, must exercise ordinary care. The city sought an instruction on reckless disregard instead, obviously a higher standard. The court held that, read together, the statutes supported the ordinary care standard for the officer.

A 2000 case illustrates how the insurance cap works. The case came to the supreme court as one of first impression. Two police officers in the city of Caddo Valley received a BOLO (be on the lookout) for a truck that was stolen from a gas station in a nearby town. Officer Whittle saw the thief (Sherman) drive through Caddo Valley and immediately switched on flashing lights and siren and pursued. Officer Kelloms heard Whittle's radio call and joined the pursuit. Speeds were between 75–90 mph. The officers heard on their radios that Arkadelphia police were setting up a roadblock on Highway 67. Kelloms, the superior officer, told Whittle to back off, in hopes of getting Sherman to slow down. Whittle eventually slowed, but Sherman did not slow. The roadblock ahead was formed by two police cars parked across the two-lane highway with just enough room between for a driver to get past slowly. The plaintiff, George, driving a jeep, was about to pass through the roadblock when Sherman crested the hill. Sherman ignored an officer with a drawn pistol, accelerated, and knocked the jeep off the road and George was thrown out and injured.

The plaintiff sued the two Caddo Valley police officers and the thief, Sherman. (Note that she did not sue the Arkadelphia officers. Later the court permitted Caddo Valley to be substituted for the two officers.) The theory of negligence was that Whittle and Kelloms were at fault for pursuing at high speed and for not breaking off pursuit sooner. The case went to trial and the jury found all defendants negligent. It apportioned liability as 90 percent Sherman and five percent each as to Whittle and Kelloms, and awarded $150,000. The trial judge then ruled in a later hearing that the city was jointly and severally liable, but its liability extended only to its policy limits of $25,000. Both parties appealed. The plaintiffs urged that since two Caddo Valley police vehicles were involved, the award should be $50,000. The city argued that the officers were not negligent nor a proximate cause, and that the city should have been granted a directed verdict because any negligence was cut off by the superseding negligence of Sherman. The supreme court found that the city's immunity from suit began only where its vehicle insurance policy limits ended. The officers were not immune from suit. When they made the discretionary decision to pursue, their subsequent actions are judged by an ordinary care standard. Continuing to pursue after they knew of the roadblock could be found to be negligence. Officer Whittle was pursuing 90 to 100 feet back at high speed. The pursuit was Whittle's first, and he had not been trained in high-speed pursuit. He did not immediately back off after being ordered to do so. The officers were only four or five car lengths behind Sherman at the site of the accident.

The court also found that the insurance policies of each police vehicle was payable to the plaintiff, for a total of $50,000. Section 21-9-303 plainly states that an owner must insure *each vehicle* for a minimum of $25,000. *City of Caddo Valley v. George,* 9 S.W.3d 481 (Ark. 2000).

A 1995 case involved an ambulance. The plaintiff filed suit in federal district court rather than state court, a recent trend. The facts were that the ambulance was on an early evening "cardiac call" in Bentonville, Ark. The ambulance driver slowed to about 40 mph as he approached a red light at an intersection. The driver and another paramedic sitting in the passenger seat looked both ways to see if all traffic had stopped and sounded all four of the siren's frequencies. All cars, north, south, and east were stopped, and they entered the intersection on the red signal. The plaintiff, apparently ignoring all the stopped traffic, also entered the intersection and they collided. The plaintiff claimed three acts of negligence by the defendant driver. First, a row of trees or shrubs obstructed the views. The court noted that the paramedics testified that they could easily see over the vegetation. Second, there were a number of cars "in or near" the intersection. The court pointed out that this, if anything, could indicate the plaintiff's own negligence. Might he not wonder why they had stopped on a green signal? Third, the plaintiff's expert stated that it would never be safe to enter an intersection in these circumstances at 40 mph. He taught his students that any speed over 10 mph is dangerous. The court noted that the controlling statute says that an emergency vehicle must slow as necessary for safety but may enter on red.

The plaintiff alleged a violation of the Due Process Clause of the Fourteenth Amendment (a state deprivation of life, liberty, or property without due process of law). The Eighth Circuit recognizes two scenarios: one in which the plaintiff is in custody, and one in which the state has affirmatively placed the plaintiff in danger, which is the claim in this case. In this second scenario, the plaintiff must show that the state acted with culpability beyond mere negligence. The court decided that the test of culpability shall be the "shocks-the-conscience" test. Tests of recklessness, gross negligence, and deliberate indifference are not appropriate concerning emergency vehicle drivers who are permitted to break the traffic rules. Such conduct for the typical driver would indeed be reckless, but here are permitted for the public goods of catching suspected law violators or responding to medical and fire emergencies. Any other test than shocks-the-conscience would convert ordinary tort claims into federal constitutional cases. The court cited a number of cases in several circuits in which the shocks the conscience test is used in this kind of claim.

The court found the negligence alleged here to be far from shocking the conscience. It cited especially *Temkin v. Frederick Co. Commissioners,* 945 F.2d 716 (4th Cir. 1991), a police chase case termed by an expert to be "reckless," "totally irresponsible," and "wanton." The court granted the city summary judgment, along with the comment that the case did not belong in federal court. Since there was no constitutional violation, there can be no liability. The court also dismissed the plaintiff's claim of unreasonable search and seizure as without merit. The court dismissed the state law claims of outrage and negligence without prejudice. *Parton v. City of Bentonville,* 901 F. Supp. 1440 (W.D. Ark. 1995).

California

Cal. Veh. Code §21055. Exemption of authorized emergency vehicles

[California has substantially adopted UVC §11-106 (see the Introductory Essay for the text) with these differences. Vehicle Code §21055 states that emergency vehicle drivers are exempt from the traffic code, by language such as, "Chapter 2, commencing with Section 21350," and so on. This practical approach differs from the UVC in that the UVC refers only to parking and standing, proceeding through stop signs or signals without stopping, exceeding the speed limit, and disregarding proper turning movements and directions. Section 21055 thus omits the language about "slowing as may be necessary for safe operation." These privileges are contingent on sounding a siren and "displaying a lighted red lamp visible from the front as a warning."]

Cal. Veh. Code §21056. Effect of exemption

Section 21055 does not relieve the driver of a vehicle from the duty to drive with due regard for the safety to all persons using the highway, nor protect him from the consequences of an arbitrary exercise of the privileges granted in that section.

Cal. Gov. Code §820.2. Discretionary acts

Except as otherwise provided by statute, a public employee is not liable for an injury resulting from his act or omission where the act of omission was the result of the exercise of the discretion vested in him, whether or not such discretion was abused.

Cal. Veh. Code §17004. Authorized emergency vehicles

A public employee is not liable for civil damages on account of personal injury to or death of any person or damage to property resulting from the operation, in the line of duty, of an authorized emergency vehicle while responding to an emergency call or when in the immediate pursuit of an actual or suspected violator of the law, or when responding to but not returning from a fire alarm or other emergency call.

Cal. Veh. Code §17004.7. Liability of public agency employing peace officers conducting vehicular pursuits

(a) The immunity provided by this section is in addition to any other immunity provided by law. The adoption of a policy by a public agency pursuant to this section is discretionary.

(b) A public agency employing peace officers which adopts a written policy on vehicular pursuits complying with subdivision (c) is immune from liability for civil damages for personal injury to or death of any person or damage to property resulting from the collision of a vehicle being operated by an actual or suspected violator of the law who is being, had been, or believes he or she is being or has been pursued by a peace officer employed by the public entity in a motor vehicle.

(c) If the public entity has adopted a policy for the safe conduct of vehicular pursuits by peace officers, it shall meet all of the following minimum standards:

    (1) It provides that, if available, there be supervisory control of the pursuit.

(2) It provides procedures for designating the primary pursuit vehicle and for determining the total number of vehicles to be permitted to participate at one time in the pursuit.
(3) It provides procedures for coordinating operations with other jurisdictions.
(4) It provides guidelines for determining when the interests of public safety and effective law enforcement justify a vehicular pursuit and when a vehicular pursuit should not be initiated or should be terminated.

(d) A determination of whether a policy adopted pursuant to subdivision (c) complies with that subdivision is a question of law for the court.

Cal. Penal Code §13519.8. High speed vehicle pursuits; Training courses and guidelines

(a) [This section directs the Peace Officer Commission on Standards and Training to develop a course on vehicular pursuits.]
(b) The course or courses of basic training for law enforcement officers and the guidelines shall include adequate consideration of each of the following subjects:
(1) When to initiate a pursuit.
(2) The number of involved law enforcement units permitted.
(3) Responsibilities of primary and secondary law enforcement units.
(4) Driving tactics.
(5) Helicopter assistance.
(6) Communications.
(7) Capture of suspects.
(8) Termination of a pursuit.
(9) Supervisory responsibilities.
(10) Blocking, ramming, boxing, and roadblock procedures.
(11) Speed limits.
(12) Interjurisdictional considerations.
(13) Conditions of the vehicle, driver, roadway, weather, and traffic.
(14) Hazards to uninvolved bystanders or motorists.
(15) Reporting and postpursuit analysis.

California is a large law state. Veh. Code 17004.7, effective Jan. 1, 1988, is unique. The statute has a dual purpose: to immunize public agencies employing peace officers (note that the officers are already statutorily immune) in pursuits in which the pursued vehicle collides with a bystander, and thus encourage peace officers to enforce the law without fear of governmental liability. I have set out Penal Code 13519.8 above to indicate the range of issues that should be covered in training officers in police vehicular pursuits. Note that whether an agency's written pursuit policy meets the requirements of the statute is to be determined by a court. That is not a jury question. California is also noteworthy because the leading case on the shocks-the-conscience test, *County of Sacramento v. Lewis,* 523 U.S. 833 (1998), arose in California. Because of the number of cases discussed below, I will try to keep the discussion of each brief. The reader, as usual, is encouraged to read the entire case.

One of the early cases construing §17004.7 was *Kishida v. State,* 280 Cal. Rptr. 62 (Cal. App. 1991). The case arose when a driver being pursued by two CHP officers hit another vehicle and caused a death and two injuries in that vehicle. The plaintiffs alleged that the highway patrol had failed to adopt and *implement* written pursuit policies in compliance with §17004.7. In other words, it is not enough to adopt such policies, the defendant must prove that the policies were actually being followed, and thus there are triable issues. The trial court did not agree and granted the defendant summary judgment. The appellate court, the Fourth District, affirmed. There is no logical reason why "implement," assuming it is implied in the statute, means proving that the policy was adhered to in each case. It could better be interpreted as requiring that pursuit training be conducted.

A 1991 case, *Weiner v. San Diego,* 280 Cal. Rprt. 818 (Cal. App. 1991), made a similar finding. Again, the plaintiff was an innocent bystander in another vehicle. The plaintiff argued that there must be a factual determination in each case of whether the pursuit policy was followed "in practice." The court affirmed a dismissal by the trial court judge. The legislature clearly meant for the courts to determine only if the written guidelines complied with the statute. The trial court's finding that the guidelines "substantially complied" was sufficient to support the dismissal of the plaintiff's case.

A 1998 pursuit case resulted in the death of an innocent bystander. A highway patrol officer attempted to stop a driver with an expired license plate. The driver fled, and the officer pursued for 45 minutes. The time was early afternoon. The pursuit was at times high speed, and was apparently in the county, not the city of Sacramento. The driver ran a red light and killed an occupant of another vehicle. The family brought suit (after obtaining a default judgment against the fleeing driver), and the trial court granted summary judgment to both county and state. The court of appeal, the Third District, affirmed. First, the court held that there was no due process violation. Similar immunity statutes have been held constitutional. Second, the CHP pursuit guidelines, contrary to what the plaintiff contended, did give sufficient guidance to the pursuing officers. Yes, the guidelines must be more specific than simply stating that officers must use their best judgment. However, pursuit guidelines cannot cover every situation that might arise; that would be impractical. Here, the guidelines covered when to initiate a pursuit, how to conduct it, and what factors to consider in breaking off pursuit. The court rejected the plaintiff's argument that the officer must consider the seriousness of the offense, and that a driver who fails to stop when ordered to do so should not be pursued for minor traffic offenses. The court stated that this view does not recognize that the flight itself may create a danger to others, and a pursuit may be proper. *Ketchum v. State,* 73 Cal. Rptr.2d 152 (Cal. App. 1998).

A 1993 case held a city's pursuit guidelines were not sufficient, and reversed summary judgment for the city. As above, an innocent bystander in another vehicle was killed by a driver being pursued. At issue was the requirements of Veh. Code 17004.7(c)(4), defining the circumstances under which an officer should initiate and break off a pursuit. The policy of the defendant, the city of Perris, was held by the appellate court to be flawed in leaving too much to the officer's discretion, and failing to set out adequate guidelines. The guidelines read:

> **When pursuits should be discontinued.** Officers should consider discontinuing a pursuit when it poses a serious and unreasonable risk of harm to the pursuing officer or to the public, balanced against the seriousness of the violations. Justification to continue a pursuit will be based on what reasonably appears to be the facts known or perceived by the officer.

The fact that the policies in question had been suggested as model guidelines by various peace officers' associations was not enough. The determination is for the court. The court cited with approval guidelines that would pass muster published by a Bellevue, Washington police department and by the National Law Enforcement Policy Center. In essence, the defendant's guidelines were simply too short. Listing the various factors to consider would have passed muster. *Payne v. City of Perris,* 16 Cal. Rptr.2d 143 (Cal. App. 1993).

A 1992 case also involved pursuit policy language in which the city of Gardena's policy was held not sufficient to grant the city immunity. A nine-year-old driver took his parents' car and fled when police attempted to arrest him. The pursuit lasted for three miles at speeds up to 100 mph. Then the driver ran into a third vehicle and killed its driver. The city argued that its pursuit guidelines were similar to those of San Diego, and the trial court granted summary judgment. The appellate court found that there were key differences in the two guidelines, and reversed for trial. Gardena's guidelines permit a pursuit in all cases if "there is reasonable cause to stop vehicle and the driver fails to stop." The court notes that this could be for a minor traffic or equipment violation. San Diego's policy gives further things to consider: is the driver a known felon, or have the vehicle occupants committed a crime in the officer's presence, among other factors? Gardena's guidelines on when to break off pursuit are "when the risk of harm to the officer or to the public outweighs the seriousness of the violations." The court found this did not offer sufficient guidance to police officers. Other cities' guidelines mention weather, vehicular and pedestrian traffic, speed, distance pursued, nature of the offense, and so on. In addition, the defendant's guidelines were lacking in designating primary and secondary pursuit vehicles. *Colvin v. City of Gardena,* 15 Cal. Rptr.2d 234 (Cal. App. 1992). See also *Berman v. Daly City,* 26 Cal. Rptr.2d 493 (Cal. App. 1993), for a case with similar facts and results as *Colvin.*

We now look at three cases involving innocent bystanders in which the guidelines passed muster. The facts of *Bryant v. County of Los Angeles,* 32 Cal. Rptr.2d 285 (Cal. App. 1994), are unusual. Finney kidnapped a man who owed him money, and tried to "ransom" the man to his girlfriend. She called the sheriff's department, the officer arrived, and chased Finney around the house. Finney managed to steal the officer's cruiser and was pursued by other officers. He caused a collision that severely injured the plaintiff. The plaintiff contended that the department pursuit guidelines failed to cover the situation of a driver fleeing in a stolen vehicle as opposed to one fleeing in his own vehicle. The appellate court affirmed summary judgment for the defendants. To require officers to leaf through a manual before taking any action is impractical, and in many instances the erring driver would escape while the officers were trying to find the right section. There is no reason to treat the driver of a stolen vehicle any differently from any other driver.

A pursuit case arose in the city of Corona in which a fleeing minor driver pursued by an officer went through a red signal and a stop sign before entering another controlled intersection at 55 to 65 mph. He broadsided the plaintiff's vehicle and the passenger was injured. The plaintiff contended that the city's pursuit guidelines were not in compliance with the statute, Veh. Code 17004.7. The appellate court affirmed summary judgment for the city. Unlike the cases above, the guidelines adequately met the requirements of the statute and put in the magic words left out above. (The case sets out in full the city's pursuit guidelines in an appendix.) Possibly the primary reason why the case was appealed was that plaintiff alleged an equal protection violation. Section 17004.7 makes several classifications. It distinguishes between those injured by the pursued driver and those injured by a police vehicle, and whether the agency had adopted guidelines or not. The court held that such classifications were rational in effectuating a legitimate legislative purpose. *Billester v. City of Corona,* 32 Cal. Rptr.2d 121 (Cal. App. 1994).

The appellate court in *Brumer v. City of Los Angeles,* 29 Cal. Rptr.2d 515 (Cal. App. 1994), had little difficulty in holding pursuit guidelines adequate and in affirming summary judgment for the defendant. The police were chasing a robber, and the latter caused a fatal crash. I will mention two points. The guidelines did not list various perils to consider in beginning or ending a pursuit, as was suggested in some of the cases above. The court here rejected the idea that such perils necessarily be listed. "Surely police officers recognize driving hazards, such as the slipperiness of a wet road, without being warned by their chief." The guidelines require pursuing officers to drive with due regard for the safety of others, and to continuously consider whether the pursuit should continue. As for supervisory control, the guidelines do require that a supervisor assumes management of a pursuit, but the ever-changing circumstances of a pursuit are best left to the discretion of the officers, not to one who cannot observe them.

The California Supreme Court took a case, decided in 2000, to clear up some confusion regarding Veh. Code §17004 (granting immunity to drivers of emergency and police vehicles on emergency runs or in immediate pursuit of an actual or suspected law violator), §21055 and 21056. Section 21055 permits emergency vehicle drivers to break traffic laws if they use the siren and lights, and 21056 requires them nevertheless to observe due care. The case arose when a police officer on radar duty noticed a speeder and began following without activating his siren or lights. The evidence was in conflict on whether the speeder was aware of the pursuit. He ran a red light and killed another person. The trial court granted summary judgment to both the county and the officer. The court of appeal reversed as to the officer, finding triable issues based on sections 21055 and 21056. The court of appeal relied on a 1978 California Supreme Court case, *Brummett v. County of Sacremento,* 582 P.2d 952, but the issue in that case was the immunity of the county, not the officer. The court held that the officer was already immune under §17004, and since 1988 the employing agency is immune under §17004.7. *Cruz v. Briseno,* 994 P.2d 986 (Cal. 2000).

Section 17004.7 does not apply when the plaintiff is the person pursued. A thorny case that was decided by the California Supreme Court arose from the following facts. Richmond police officers answered a report of armed men at a street corner at night. When they drove up; one fled and failed to stop on a command by the officer. He darted in front of a cruiser and was seriously injured, and he sued the city and the officer driving the cruiser. The trial court granted summary judgment to the defendants. The court of appeals reversed as to the city. The supreme court decided that the officer was without doubt immune under §17004. The plaintiff contended that the city could be liable under Gov. Code §815, the California Tort Claims Act, or Veh. Code §17001 (city liable for negligent acts of employees driving a motor vehicle). The city claimed immunity under Gov. Code §845.8 (public entity not liable for injury to a person resisting arrest). The majority found that §845.8 did not grant the defendant immunity. The case is far too long to summarize here. It is sufficient to say that the majority opinion discussed apparently conflicting statutes from different parts of the code, and a line of cases in trying to decide whether the legislature meant to overrule them in whole or in part in enacting §17004.7. The court affirmed in finding that the city was not immune. In concurring, Judge Mosk argued that §845.8 was the only relevant controlling statute, and opined that the plaintiff did not "cause" his injury. A long dissent said that this was the issue, and in my experience in researching many traffic cases, few dart-outs win damages. *Thomas v. City of Richmond,* 40 Cal. Rptr.2d 442 (Cal. 1995).

We now look at a few cases that were brought to federal courts, ending with the U.S. Supreme Court decisions of *Lewis v. Sacramento* and *Brower v. Inyo County.* The first three cases had the United States as the defendant. In a 1994 case, the U.S. Border Patrol was pursuing a driver later caught and charged with several felonies, and the driver hit and injured a truck driver. The Border Patrol claimed immunity under Veh. Code §17004.7. The district court judge granted the U.S. summary judgment, and the case came to the Ninth Circuit. The question on appeal was whether the state statute was preempted by the Federal Tort Claims and the U.S. Constitution. In a short opinion, the court reversed on a finding that several Ninth Circuit cases followed the Supreme Court in holding that FTCA claims may not be defeated by state immunity. The court noted a Nevada pursuit case in federal court was that bound by the state monetary cap, on damages, but that is not immunity. *Montes v. United States,* 37 F.3d 1347 (9th Cir. 1994). In a similar Border Patrol pursuit case, the trial judge at a bench trial erroneously applied §17004.7 immunity, but also made a finding that the agents were not negligent. The 9th Circuit affirmed, after stating that the immunity ruling was in error. The fact that the trial judge was openly skeptical, not to say sarcastic, with respect to the some of the plaintiff's witnesses did not merit a reversal. The judge correctly excluded the testimony of an expert witness, an academic who frequently testifies in pursuit cases, on probable cause to stop the suspect driver. The judge is qualified to decide that issue. *Stuart v. United States,* 23 F.3d 1483 (9th Cir. 1994).

Another 1994 case is similar, and also was decided on negligence principles. In the early morning hours, border patrol agents were watching a driver who pulled off I-15 and stopped for several minutes. They suspected alien smuggling, and followed the car. It bypassed the border patrol checkpoint, and came back to I-15. They stopped the driver, but he suddenly fled. The agents followed at high speeds for less than three miles. Traffic was sparse. The driver exited and hit the plaintiff's car. It turned out that the driver had stolen the car, and was not smuggling. The trial judge found the pursuit reasonable under the circumstances, and found that the agents were not negligent. He found it unnecessary to rule on immunity. The Ninth Circuit affirmed in all respects. *Cimo v. I.N.S.,* 16 F.3d 1039 (9th Cir. 1994).

We now turn to a police pursuit case that was accepted by the U.S. Supreme Court, *County of Sacramento v. Lewis,* 523 U.S. 833 (1998). My first comment is that from a plaintiff's perspective, this was not a particularly good case to go the highest court. The officer's conduct in this fatality was not nearly as outrageous as in some other cases I have annotated for this study. Also, the victim was a passenger in the pursued vehicle, and not quite an innocent bystander as often is the case. I will just cover the main points in keeping with the limited scope of this book. The reader will want to read the entire case. At least twelve law reviews have published articles on this case for those readers wanting exhaustive commentary.

While the facts can be shaded one way or the other by how they are described by witnesses and in the cases and the briefs filed, the following is what seems to have happened. At 8:30 P.M. in May, two Sacramento County deputy sheriffs in separate cars responded to a report of a fight. A motorcycle approached slowly. The driver, Willard, was eighteen and the passenger, Lewis, sixteen. (The Ninth Circuit case refers to them as minors, or boys, not young men. Both were on probation for theft charges, and one condition was that they not associate, so there was a motive to flee. Also, the motorcycle was stolen.) One officer yelled for them to stop, and attempted to block them by pulling closer to the other cruiser. The passenger, Lewis, said, "Let's get out of here." (Defendant's petition of certiorari, p. 42.) The motorcyclist slowed and drove between the cars, and then took off at high speed. The other officer, Smith, could not hear what was yelled, but he promptly turned around and pursued, turning on his siren to alert other traffic. The pursuit was short: 75 seconds and 1.3 miles. Willard drove recklessly, sometimes in the oncoming traffic lane, and almost hit a bicyclist. Officer Smith appeared to be more cautious. He slowed enough to make turns without screeching his tires, for example. His speed was at times high to maintain contact, at a distance of 200 feet, he testified. Several cars had to take evasive action. Willard crested a hill and attempted an impossible (considering his speed) left turn, and skidded to a stop. His brake lights were not working. Smith crested the hill, and saw that the motorcycle had stopped. He locked up his brakes, skidded 147 feet, struck Lewis and apparently the motorcycle, and stopped some 32 feet farther. Lewis was killed.

The parents filed a §1983 action in state court, alleging a depravation of Fourteenth Amendment rights under the Due Process Clause. The defendant moved the action to federal court. That court found that the plaintiff's rights under due process in the context of a police pursuit was not "clearly established" by case law and found the Smith had qualified immunity. The court granted summary judgment to all defendants. The court of appeals reversed as to Officer Smith. It found a triable issue of whether he acted in reckless disregard or deliberate indifference toward Lewis, and found that he was not entitled to qualified immunity despite state statute §17004. The court cited three police pursuit cases from other circuits. All found the officers' conduct not sufficiently egregious to trigger the Due Process Clause. In this case, Officer Smith did not radio the dispatcher, and it could be argued that this omission did not balance the offense that caused the pursuit with the possible dangers to others, as is required in pursuit policies. Here the court found that issue close enough to merit a trial.

The case went to the U.S. Supreme Court. The defendants argued that there was a conflict among the circuits. The Ninth Circuit by this case established the standard of police conduct in pursuit cases as "reckless disregard" or "deliberate indifference," while the First, Third, and Fourth Circuits required a higher standard, "shocking to the conscience." (The higher standard favors the police.) With Justice Souter writing the opinion, the Supreme Court reversed. All other justices either wrote or joined in concurring opinions. First, there was no "seizure" of Lewis and thus there is no Fourth Amendment violation, citing *California v. Hodari,* 499 U.S. 621 (1991). Second, only the most abusive, egregious official conduct can trigger the Due Process Clause. And the test of that conduct in pursuit cases is that which shocks the conscience. The Fourteenth Amendment is not a "fount of tort law to be superimposed upon whatever systems may already be administered by the States." The standard of "deliberate indifference" contemplates some time for deliberation, such as when a prisoner is in custody, and not a sudden prison riot. Police pursuits are somewhat like prison riots. Police officers "are often forced to make split-second judgments—in circumstances that are tense, uncertain, and rapidly evolving."

This is what the Supreme Court held. High-speed chases, with no intent to harm suspects physically or to worsen their legal plight, do not give rise to liability under a §1983 action. Officer Smith was faced with lawless, reckless behavior that he did nothing to cause, other than not to continue the chase. Smith acted instinctively, but his actions were not tainted by improper or malicious motives. Justice Scalia concurred and noted that the motorcycle driver, Willard, was at least as culpable in "depriving" Lewis of his life. Issues of negligence, contributory negligence, and assumption of the risk are tort issues, not constitutional ones. California has enacted §17004, and if the law is to be changed, California must do it, not the Supreme Court.

A 1990 case applied *Lewis.* A driver in a "muscle car" passed a deputy sheriff, weaving in and out of traffic and driving erratically. The officer turned on lights and siren and tried to make an arrest, but the driver fled. The pursuit lasted only a minute. The driver ran a red light and injured two others in a car. The officer's speed never exceeded 45 mph, while the driver was traveling 60–70 mph. Neither officer came closer than 300 feet to the pursued driver. The plaintiffs filed a §1983 claim. The trial court granted summary judgment to the defendants. Before the case was heard in the Ninth Circuit, the *Lewis* decision was handed down. The Ninth Circuit had to determine whether *Lewis* applied to third party innocent victims. The language in *Lewis* referred to suspects, and dealt with a plaintiff who was tainted by the fault of the driver of the motorcycle. The court found that *Lewis* applied to pursuit cases regardless of the class of the plaintiff. The court had little problem in determining that the conduct of the officers in this case was not nearly as shocking as in *Lewis,* and that case passed the shocks-the-conscience test. The court of appeals affirmed. *Onossian v. Block,* 175 F.3d 1169 (Cal. App. 1999).

The final case is the second case in the U.S. Supreme Court, *Brower v. Inyo County,* 489 U.S. 593 (1989). This is a roadblock case, and anyone with a pursuit roadblock case will want to study it. The case arose from a pursuit of a stolen car. The driver was Brower. The time was around 11:30 P.M., and the location was Highway 395 in Inyo County. Deputy Oyster was in pursuit. At some point in the twenty-mile pursuit, Oyster called ahead for a roadblock. Deputy Sides directed a semi driver to park his rig across the two-lane highway and block both lanes. Sides then parked his cruiser 200 feet ahead of the semi and pointed his headlights toward the direction of the pursuit. Brower came up at high speed, passed Sides, and crashed into the semi, abruptly ending his life.

The family sued in federal district court. They alleged negligence and wrongful death on their state claims, and alleged a 42 U.S.C. §1983 violation. The district court judge dismissed the private defendants, the truck driver and his employer, and dismissed the governmental defendants for the plaintiffs failure to state a claim. The case came to the Ninth Circuit.

The plaintiffs argued that the roadblock was illegal; a violation of substantive due process. The plaintiffs claimed that the roadblock was essentially a deathtrap in that the truck was placed past a curve that cut down the deceased's view, that it was not illuminated, and the cruiser's headlights shined in the deceased's eyes. The plaintiffs alleged that this was a "pattern and practice" of the defendants. The court held that the plaintiff stated a claim, but the plaintiffs probably need to produce more than "conclusory allegations" at a latter proceeding to survive summary judgment.

The plaintiffs claimed that there was a Fourth Amendment violation because there was an unreasonable seizure by means of deadly force. The court held that there was no unreasonable seizure because Brower chose to avoid being stopped, even though he had several opportunities. He was stopped by the roadblock but not seized, in the constitutional sense, by the police.

The U.S. Supreme Court accepted the case and reversed the Fourth Amendment finding for consideration in light of its opinion. The court of appeals erroneously made an analogy with a situation in which a fleeing suspect loses control and crashes. There would be no seizure in that case because a violation of the Fourth Amendment requires "an intentional acquisition of physical control." The same is true even when the arrest or apprehension of the individual is intended and desired. The deputy sheriffs indeed wanted to apprehend Brower, but there is a seizure only when it is effected *through means intentionally applied.* The court gives two hypothetical situations. One is the example the court of appeals stated, which would not be a seizure. The other would be if a pursuing police officer deliberately sideswiped the suspect's vehicle with the intention of causing it to crash. That would be a seizure, and then the question would be whether it was unreasonable.

No doubt the officers hoped that Brower would voluntarily stop, but the court declined to speculate on this subjective intent. Whether the seizure was unreasonable will depend on all the circumstances: where the roadblock was placed past a curve, and especially whether the police vehicle's headlights were used to blind the suspect. Was the roadblock designed to produce a collision? The court points out that if a roadblock was well illuminated and set up on a straight section of the road, that would be a different situation.

A concurring opinion, joined by three other justices, thought that the majority expressed dicta in its "intentional acquisition of physical control." There was an intended seizure; the question was whether it was reasonable. No more need be decided.

The final reported proceeding is *Brower v. Inyo County*, 884 F.2d 1316 (9th Cir. 1989). The court denied the defendant's motion to dismiss and remanded to the district court. The plaintiffs' pleading, responding to a motion to dismiss for failure to state a claim, must be taken as true and construed in the nonmoving party's favor. The issue of whether a nonlethal alternative to the roadblock existed is a question of fact.

Colorado

Col. Rev. Stat. §42-4-108. Public officers to obey provisions—exception for emergency vehicles

(1) The provisions of this article applicable to the drivers of vehicles upon the highways shall apply to the drivers of all vehicles owned or operated by the United States, this state, or any county, city, town, district, or other political subdivision of the state, subject to such specific exceptions as are set forth in this article with reference to authorized emergency vehicles.

(2) The driver of an authorized emergency vehicle, when responding to an emergency call, or when in pursuit of an actual or suspected violator of the law, or when responding to but not upon returning from a fire alarm, may exercise the privileges set forth in this section, but subject to the conditions stated. The driver of an authorized emergency vehicle may:
   - (a) Park or stand irrespective of the provisions of this title;
   - (b) Proceed past a red or stop signal or stop sign, but only after slowing down as may be necessary for safe operation;
   - (c) Exceed the lawful speed . . . so long as the driver does not endanger life or property;
   - (d) Disregard regulations governing directions of movement or turning in specified directions.

(3) [Requires the driver of an emergency vehicle to use flashing lights and siren when exercising the privileges set out in (2)(b) to (d)

above, and to use flashing lights when exercising the privilege authorized by (2)(a) above. Adds this language: "Except that an authorized emergency vehicle being operated as a police vehicle while in actual pursuit of a suspected violator of any provision of this title need not display or make use of audible or visual signals so long as such pursuit is being made to obtain verification or evidence of the guilt of the suspected violator."]

(4) The provisions of this section shall not relieve the driver of an authorized emergency vehicle from the duty to drive with due regard for the safety of all persons, nor shall such provisions protect the driver from the consequences of such driver's reckless disregard for the safety of others.

Col. Rev. State §24-10-106. Immunity and partial waiver

(1) A public entity shall be immune from liability in all claims for injury which lie in tort regardless of whether that may be the type of action or the form of relief chosen by the claimant except as provided otherwise in this section. Sovereign immunity is waived by a public entity in an action for injuries resulting from:
    (a) The operation of a motor vehicle, owned or leased by such public entity, by an employee while in the course of employment, except emergency vehicles operating within the provisions of section 42-4-108 (2) and (3).

[Colorado has enacted U.V.C. 11-106 with little change from the model code. The only significant change is noted in (3) above. It permits police in actual pursuit to pursue without emergency lights or siren if the purpose it to "obtain verification of or evidence of the guilt of the suspected violator." This is a commonsense provision to allow police to speed and run signals and signs (safely) to catch up with and inspect the vehicle and driver without alarming him. No doubt this is routinely done; Colorado has specifically authorized it by statute. The immunity statute cited above is a common one. Colorado courts have applied it along with 42-4-108 in police pursuit cases. The pattern in Colorado is for defendants to plead the immunity statute. Because the immunity statute does not refer to 42-4-108(4), only the standard of endangering life or property while speeding can be applied, not "reckless disregard" or "due regard for the safety of others." The Colorado Supreme Court specifically so stated in the *Fogg* case noted below. There are relatively few recent cases.]

A 2000 case in the Colorado Supreme Court arose from a burglary alarm. The sheriff of Pueblo County received a dispatch reporting a home burglary alarm. Deputy Cortese responded by speeding 50 to 60 mph in a 35-mph residential area, and with lights and siren activated. He approached an intersection just as another driver, Cordova, was making a left turn. There was a crash, and she died from her injuries. The sheriff (defendant) claimed immunity under the Government Immunity Act (GIA), set out above. At a hearing on defendant's motion, the trial court made two findings. First, the deputy was not responding to an emergency call because home burglary alarms are too unreliable to so qualify. Second, the deputy endangered life or property in violation of 42-4-108(2)(c), also set out above. The trial court denied the claimed immunity, and the court of appeals affirmed (without making a finding on the first issue). The Colorado Supreme Court took the case mainly to define the proper legal standard for determining the definition of an emergency in this context.

The court noted that one could employ a subjective standard, looking at the facts from the deputy's viewpoint. Another approach would be to look at the facts from the caller's perspective. Or one could let the courts to look at all the facts and circumstances and make the call. The court determined that the best standard would be an objective standard from the perspective of a reasonable emergency vehicle operator. Would such an operator reasonably believe there was an emergency, based on what he knew or should have known? The court hoped that such standard would strike a balance, and protect the reasonable operator acting in good faith, while still protecting the public from operators who unreasonably respond to calls in a dangerous way. From this perspective, the court rejected the trial court's finding that there was no emergency. That was an after-the-fact call by the trial judge. However, the court agreed that there was a triable issue of whether the deputy endangered life or property in speeding. The court did not agree with the defendant that this finding would negate the GIA. The court said that the decision of endangerment must not be based on the fact that the accident happened, but rather on the facts and surrounding conditions before the accident. Here, the deputy was speeding 60 mph in a 35-mph-zone and approaching an intersection with poor sight distance because of shrubbery and trees. The court reversed the first finding of the trial court and affirmed the second, and remanded for further proceedings. *Corsentino v. Cordova,* 4 P.2d 1082 (Colo. 2000).

The Colorado Supreme Court had a similar case earlier, *Fong v. Macaluso,* 892 P.2d 271 (Colo. 1995). One question was, what is an emergency? Deputy Macaluso parked his cruiser in the passing lane of I-25 to aid a motorist with a flat tire. The disabled car was in the median, and the deputy was helping change the tire. The cruiser's flashing lights were on. Fong moved to the left lane to pass traffic on long downhill curve and hit the cruiser. The trial court granted the defendants summary judgment on the finding that the deputy was responding to an emergency and was entitled to "park or stand" irrespective of the traffic code. The plaintiff appealed, arguing that the defendant had not operated his vehicle with due regard for the safety of others. The trial court specifically found that this provision was not incorporated in the Government Immunity Act. The court of appeals affirmed.

The primary issue for the supreme court was to determine what the legislature meant by "emergency." The statute refers only to pursuit of suspected criminals and responding to fire alarms. The court referred to an earlier case, and adopted the dictionary meaning of "an exigency or an unforeseen combination of circumstances that call for immediate action." Pursuit of criminals and going to fires require an immediate response, and some risks in driving are expected and permitted. On the other hand, returning from a fire alarm is specifically exempted from emergency driving. The court remanded and instructed the trial court to apply the definition above in determining the emergency issue. As noted above by the author, the court also refused to read 42-4-108(4) into the immunity statute. The court specifically overruled *Cline v. Rabson,* 862 P.2d 1035 (Colo. App. 1993), on this point. (Police officer on emergency run struck elbow of motorcyclist he was passing held not negligent.)

A 2000 case dealt with both witness and expert witness testimony. A police pursuit began in the city of Westminster when a police officer attempted to stop a driver whom he suspected of armed robbery. The pursuit speeds were high. The fleeing driver crashed into the plaintiff and caused injuries. The trial court held an evidentiary hearing and ruled that the officer's actions were legally justified, and dismissed the complaint. The court of appeals agreed that the trial judge applied the correct statute as a standard, 42-4-108(2)(c), however the judge failed to make a finding of whether the officer's speed improperly endangered life or property. The case was remanded for this omission. As for the witness who said the officer was going about 70 mph, the testimony was properly excluded for lack of proper foundation. The witness also said that the cars were going fast, but any specific speed would be a guess. The plaintiff's expert testified that the officer was reckless and disregarded the safety of others. This stated a legal conclusion and was properly excluded. *Quintana v. Westminister,* 8 P.2d 527 (Colo. App. 2000).

Note a 1992 case, *Zapp v. Kukuris,* 847 P.2d 150 (Colo. App. 1992) in which a fleeing driver of a stolen car struck and killed another driver. The court of appeals reversed a dismissal for the defendant (the pursing officer). The court based its decision on 42-106(4) (now codified as 42-4-108(4)), driving with due regard and not being protected from the consequences of the officer's reckless disregard. The Colorado Supreme Court has since, as noted above, held that this section, which is not specifically mentioned in the immunity statute, will not be read into the statute by the courts, so this case would not be a good precedent.

We conclude the Colorado section with two federal cases. *Medina v. Denver,* 960 F.2d 1493 (10th Cir. 1992) arose from a bicycle-police car accident. The police were pursuing the driver of a stolen car. The fleeing driver turned suddenly to elude police cars coming from opposite direction and injured a bicyclist. The bicyclist filed a §1983 claim. The federal district court granted summary judgment to the defendants on the finding that the defendants' conduct was not directed toward the bicyclist; he was a bystander. The Tenth Circuit disagreed with that finding but affirmed on other grounds. The court held the police officers not liable for the conduct of the pursued driver, commenting that this appeared to be the law in 1986, at the time of the accident (later cases may not be used to try to establish a different rule). The court also held the city and county not liable because all the plaintiff had to defeat the motion for summary judgment was documented incidents of police chase accidents and the statement of an expert terming the actions of the police as recklessly disregardful and deliberately indifferent to human life. Such allegations were conclusory, not evidence. The police manual on pursuits alone refuted such allegations.

A 1996 case, *Williams v. Denver,* 99 F.3d 1009 (10th Cir. 1996), involved an officer with a poor driving record. Around 4:00 A.M. on a Sunday morning an officer called for a backup in arresting a car thief. The officer did not indicate he was in danger. Officer Farr responded by driving 60 mph in a 35-mph-zone on a major street. He entered an intersection against the red and struck a vehicle entering the intersection at 20 mph. The cruiser had its flashing lights on but no siren. (This does not seem particularly outrageous to the author, considering the time of day, the day of the week, and the fact that the officer was on a major street. There was evidence that it was not uncommon for officers to not use their sirens and to run red lights in such circumstances.) What was more damning was the record of the officer. Farr had numerous moving violations in his record, three revocations of his operator's license, a recommendation from the Civil Service Commission that he not be hired, and so on. One wonders what the full story is behind his hiring. The district court granted summary judgment to all defendants. The appellate court found that the city could be liable on a failure-to-train claim. The court reversed as to Officer Farr on a finding that Farr's state of mind indicates that he was speeding for speeding's sake (a finding the dissent found preposterous).

Connecticut

Conn. Gen. Stat. Ann. §14-283. Rights of emergency vehicles

[Connecticut has adopted UVC 11-106 with only insignificant changes in wording (see the Introductory Essay for the UVC text). Thus police and other emergency vehicle operators may disregard usual traffic rules as long as they do not endanger life or property, and must drive with due regard for the safety of all persons and property. The statute does not include the reckless disregard language. As in California, Connecticut requires an official pursuit policy promulgated by the Commissioner of Public Safety and other officials conforming to the specifications set out in §14-283a. The statute states that the Commissioner and other officials "shall take whatever measures that are necessary to assure that each officer understands the pursuit policy established," in other words, see that officers are trained. Unlike California, doing so does not relieve officers and their employers of liability. The Act apparently was intended to reduce the danger arising from pursuits, not to grant immunity. Section 14-283a took effect Jan. 1, 2000.]

We begin with older cases and end with the most recent. In a 1954 case, a police officer sued a motorist when they collided at an intersection. The officer received a call to answer a fire call. He turned on lights and siren and collided with a truck. The truck driver approached the intersection slowly because he had a red signal. It turned green and he entered on a green. The officer entered on red at about 30 mph. A parked tractor-trailer obstructed the views. The officer sued for personal injury, and the defendant cross-claimed for property damage. The superior court held that each party was negligent and awarded no damages. The officer was negligent despite having the right of way under 14-283, and the truck driver was also at fault for failing to yield. Failure to hear the siren was no excuse. *Matcheski v. Gutkin,* 109 A.2d 879 (1954).

A 1983 case is more typical of police chase cases. Two Stratford police officers noticed three youths in a car at a shopping center. They thought the youths were too young to drive legally, and they suspected that the car may have been stolen. When they attempted to question the occupants, a chase ensued at speeds of 50 mph in a congested area, and heading the wrong way on a one-way street. The plaintiff driver heard the siren but before he could determine from what direction he was hit head-on by the fleeing youths. He sued the town of Stratford and the jury awarded him damages. The defendant appealed on two issues: the fault was the fleeing driver, and the statute contemplates police liability only when the plaintiff is injured by the emergency vehicle itself. The Connecticut Supreme Court affirmed the original decision. The defendant may not escape liability because of the intervening negligence of the pursued driver, according to later cases. The defendant had cited earlier cases now held to be a minority view. On the second issue, the statute, 14-283, clearly requires police drivers to exercise due care for the safety of all persons. There was evidence that the pursuit was not in conformance with the department's own policies, especially with regard to the seriousness of the offense, speeds in a congested area, and going the wrong way. *Tetro v. Stratford,* 458 A.2d 5 (Conn. 1983).

A 1988 case resulted in four fatalities, all in the pursued vehicle. Two Bethel, Connecticut detectives in an unmarked car noticed an El Camino (half car, half pickup truck) with teenage boys hanging over the sides in the back. The officers made a U-turn to issue a warning, and, after a detour in a parking lot, a chase ensued. The officers turned on a flashing red light on the dashboard and the siren. Speeds were well above the posted limits. The license number was observed and radioed in. The dispatcher then determined that the car was not stolen, but the officers never asked for the dispatcher to call back with this information. The El Camino crashed and the driver and three in the back were killed. Bethel police pursuit policy (set out in a footnote in the case) would seem to show that the pursuit was imprudent. No chase should be initiated except for dangerous felonies, only when "absolutely necessary," not for misdemeanors, and should be broken off when the risk of an accident outweighs possibility of an arrest. The jury returned a general verdict for the defendants. The supreme court found it necessary to affirm (in spite of apparent violation of the pursuit policy). The general verdict rule applied. The plaintiff failed to ask for interrogatories to ascertain the basis of the verdict. The verdict could have been supported by either the defendant's lack of negligence or the plaintiffs' greater negligence. On another issue, the court held it proper to introduce evidence of drinking by the pursued driver. *Staudinger v. Barrett,* 544 A.2d 164 (Conn. 1988).

A 2000 case is most unusual and makes an important finding. An assistant fire chief was on the way to a fire emergency in a vehicle that not described in the trial records. He was operating his siren and emergency lights. He swerved briefly into the oncoming lane to pass a motorcycle on a curve where visibility was limited. An oncoming driver had to pull off the road "to her distress," but there was no collision or injury. She had heard the siren and was on the alert, but did not pull over earlier in compliance with statute. Based on these facts, the chief was charged with several violations. He was convicted of improper passing, §14-232, and improper driving on the left side of the highway at a curve, §14-235, and was fined $150. The trial judge rejected the defendant's request to apply the emergency vehicle statute, §14-283. The judge found that the emergency vehicle statute did not permit the defendant to violate §14-232 and 235, both governing direction of movement. The appellate court reversed and ordered an acquittal. The court interprets the statutes that govern direction of movement. The state argued that §14-283(d), imposing due regard for safety, must be applied to subsection (b)(4), so the legislature must have so intended. The court declined to "engraft" the due regard standard of subsection (d) onto subsection (b)(4), especially in a no-accident case. This very close interpretation appears to be unique to Connecticut. *State v. Nesteriak,* 760 A.2d 984 (Conn. App. 2000).

A 1986 case was filed in federal court. It arose from a roadblock accident. A state trooper began pursuit of a speeding motorcyclist. The time was 8:45 P.M. in September. The rider sped, according to the officer, at 73 mph. The motorcyclist consciously tried to avoid arrest. Meanwhile, Salem Town Constable Cordell was informed of the chase heading southbound on Route 85. Cordell was northbound on a dirt path adjacent to the highway. He decided to make a U-turn and wait on the southbound shoulder. At this point guardrails on both sides reduced the width available, and Cordell not have room to complete the turn. He backed across the highway to try again. You guessed it. At the same time he saw a single headlight bearing down no him, and he thought the safest thing to do was not to move, but to let the motorcyclist get past. The rider saw the police car and said that he thought to pass on one side, but the officer backed into his path. The motorcyclist was severely injured.

The plaintiff alleged a violation of his liberty interest under the Fourteenth Amendment, and violation of the Fourth Amendment to be free from unreasonable seizures. The district court judge declined to grant summary judgment to Constable Cordell. The plaintiff alleged that Cordell intentionally backed to cause the plaintiff to crash, possibly an unconstitutional seizure of the plaintiff. The court said the evidence "arguably supports this inference." Such inferences must be drawn most favorably for the nonmoving party. (Frankly, the notion that the officer intended this result seems most unlikely to the author.) The court did grant summary judgment to the state trooper who began the pursuit. His actions were reasonable, as were those of his supervisor. The plaintiff alleged that the latter should have called off the pursuit as more risk than benefit. The court held that such decision was for the pursuit officer. The court also dismissed the claims against Cordell in his official capacity (essentially the claim against the Town of Salem), as well as any state law claims. *Stanulonis v. Marzec,* 649 F. Supp. 1536 (D. Conn. 1986).

Delaware

Del. Code Ann. tit. 21 §4106. Authorized emergency vehicles

[Delaware has enacted UVC 11-106 without significant change (see the Introductory Essay for text) except for subsection (d). That UVC subsection states that an emergency vehicle driver must driver with due regard for the safety of all persons and is not protected from the consequences of his own reckless disregard for the safety of others. Instead, Delaware's subsection (d) reads as follows.]

> The driver of an emergency vehicle is not liable for any damages to or loss of property or for any personal injury or death caused by the negligent or wrongful act or omission of such driver except acts or omissions amounting to gross negligence or willful or wanton negligence so long as the applicable portions of subsection (c) have been followed. The owner of such emergency vehicle may not assert the defense of governmental immunity.

Rather than use the standards of due regard and reckless disregard, the Delaware subsection uses gross negligence or willful or wanton negligence. The author has identified three cases. The primary issue in a 1996 case was whether there was a pursuit. Defendant Sutton was a Wilmington police officer. A fellow officer, Wyatt, attempted to stop a driver who failed to stop at a stop sign. The driver stopped momentarily and then sped away with Wyatt in pursuit. Wyatt apparently radioed Sutton to help, although the dispatcher did not authorize Sutton to join in the pursuit. Sutton stopped his cruiser facing the wrong way on a one-way street to intercept the fleeing driver. The plaintiff third-party driver turned onto the street and hit the parked cruiser at slow speed which allegedly caused personal injury. (This is one of the tamest pursuit cases in this book.) The jury found for the defendants, and there was an appeal. One claim was that the trial judge found that there was a pursuit as a matter of law. The superior court disagreed and remanded, finding this to be a jury issue. The court opined that since the dispatcher had not told Sutton to aid Wyatt, he could not come under the protection of 21-4106. *Garden v. Sutton*, 683 A.2d 1041 (Del. App. 1996).

I feel compelled to comment on *Gardner* above. While it was up to the jury on remand (if the case didn't settle in the meantime), the superior court's suggestion seems preposterous. If his fellow officer radios for aid, Sutton is involved in the pursuit. If an officer is ordered by the dispatcher to end the pursuit, he can be at fault for not doing so. But that didn't happen here. The court states that Sutton joined the pursuit voluntarily but was responding to a radio call from Wyatt. Perhaps all the court meant to do was to have the jury determine the facts rather than the trial judge.

A 1979 case arose from a collision between a fire truck on a fire call and a pickup. The pickup driver, Green, slowed to about 30 to 35 mph as he approached an intersection with traffic signals. It was a clear morning.

The fire truck approached on the other highway from Green's left, with its signals and siren operating. Green had the green light, and said he did not notice the fire truck. Smith, the fire truck driver, entered on red at 30 to 35 mph, and said he didn't see Green. There was a crash. The trial judge found both drivers negligent as a matter of law. The court reversed as to Green, the plaintiff, finding a jury question. Green claimed that because of the bright day he couldn't see the flashing lights. The superior court commented that before Green can be found negligent for not yielding to an emergency vehicle, he must be found to have seen or should have seen the fire truck. The jury also should determine whether Smith's negligence was gross or willful and wanton. *Green v. Millsboro Fire Co.,* 403 A.2d 286 (Del. Super. Ct. 1979).

A 1963 case involved a collision between private ambulance and a car driver. The car driver approached an intersection that had, for him, a yellow blinking caution light. The ambulance driver had a stop sign and a red blinking light. The trial judge found both drivers at fault. The Delaware Supreme Court reversed as to the ambulance driver's counterclaim. Unlike the case above, the car driver did not have a clear right of way. Rather, he had a caution light, and if he could not hear the siren because of his radio, that is no excuse. As for the ambulance operator, he was not necessarily negligent for not stopping, as long as he was driving with due regard for others. To hold otherwise would negate the emergency vehicle law. His contributory negligence is a jury question. *Matthews v. Bryerton,* 193 A.2d 83 (Del. 1963).

District of Columbia

D.C. Code Ann. §1-1212. Governmental immunity for negligent operation of vehicles

[Section 1-1212 is a typical waiver of immunity for claims arising from the operation of government vehicles, except "provided, that in the case of a claim arising out of the operation of an emergency vehicle on an emergency run the District shall be liable only for gross negligence." The D.C. Code does not include the typical provisions of the Uniform Vehicle Code. They are set out in Title 18 of the D.C. Municipal Regulations (not seen by the author), but the courts generally apply the statute above in emergency vehicle cases.]

The above statute is referred to as the Employee Non-Liability Act. The act states the standard of gross negligence, and a 1997 case amplified on the meaning of that standard. The case arose from a pursuit resulting in the fatality of an innocent bystander. A juvenile driver was noticed by District police because of his youth. A check revealed the car was stolen. The driver fled and was pursed by D.C. police into Maryland, where St. George's County police took over. The driver sped to 90 mph. The parkway turned into a two-lane road, and the driver crossed the double yellow line to pass three cars and hit the deceased's car head-on. The fleeing driver said he couldn't turn back to his lane because his passenger grabbed his arm. The entire chase distance was five miles, traffic was light, and the visibility was good. The plaintiff husband, a Maryland resident, sued the District police under Maryland's wrongful death act. (Two cases were filed, and the appellate court decided both in this opinion.) The plaintiff alleged both gross negligence and negligent training of the officers. The jury found for the plaintiffs on both claims. The District moved for judgment notwithstanding the verdict, the trial judge denied the motion, and the case was appealed.

The court of appeals stated that "gross negligence" requires "such an extreme deviation from the ordinary standard of care as to support a finding of wanton, willful and reckless disregard or conscious indifference for the rights and safety of others." Thus, the court includes most of the usual terms used to describe negligence beyond ordinary, and lumps them together as gross negligence as contemplated by the statute. The issue for the court was whether the D.C. police were grossly negligent. The plaintiffs' expert found the police conduct grossly negligent in several ways: running red lights against police policy, not balancing the dangerousness of the pursued against the danger of pursuit, the fact that the only apparent crime was driving a stolen car, and the fact that the driver did not flee and ignore traffic laws until police attempted to stop him. The court did not agree, and pointed out that it was not clear that D.C. police were a proximate cause of the accident. The court cited a number of cases from other states where speeds, weather, visibility, and the presence of other traffic and pedestrians would appear to make pursuits more unwise than here, and yet virtually all those cases held the police were not grossly negligent. The JNOV (judgment *non obstante veredicto*, or judgment notwithstanding the verdict) should have been granted. The court also held that the standard to judge negligent training must be gross negligence, and remanded for a new trial on that claim. (The court commented that it is difficult to see how the plaintiff could prevail.) *District of Columbia v. Walker,* 689 A.2d 40 (D.C. App. 1997).

A 1994 case is noteworthy primarily for errors made by the District's counsel. The facts were very similar to the above case. Again, a young driver in a stolen car was noticed by police, and a check confirmed that the car was stolen. A "be-on-the-lookout-for" call was broadcast and shortly afterward the original officer saw the car and was able to get within three blocks of it. The driver fled. Another officer spotted the suspect weaving to pass other cars, and turned on his lights and siren. The suspect fled at high speeds, running red lights, and hit the plaintiff. As usual, the suspect was not hurt and was caught on foot. The plaintiff alleged gross negligence in the pursuit and negligent supervision. Experts testified for each side, the one contending that the pursuit was against the police department's General Order 301.3, the other saying that the pursuit was consistent with the order and general police practice nationwide. The jury awarded $360,000, and the court of appeals affirmed. The court did not make a finding on the grossly negligent pursuit claim (the dissent thought it well short of gross negligence, to which the author would agree), but upheld the verdict of negligence in supervision of the pursuit. The jury was instructed on negligent supervision, not on gross negligence as required by law, but the defendant's counsel failed to object before the case went to the jury and thus may not raise the issue later. The court notes that it must construe the facts most favorably toward the plaintiff on reviewing a judgment notwithstanding the verdict. *District of Columbia v. Banks,* 646 A.2d 972 (D.C. App. 1994).

Metropolitan police on emergency runs are regulated by Title 18, the District's traffic regulations derived from the Uniform Vehicle Code and by department policies. Title 18 requires them to slow as much as necessary for safety when running red lights on emergency runs, while department policy requires them to stop before proceeding on red. The latter has the lowest status, but violation of either is not necessarily proof of gross negligence. In a 1998 case, two officers on a "Code One" run (drivers trapped in vehicles after an accident) ran a red light and collided with a 1987 Jaguar. The jury awarded $100,000 to the Jag driver and $10,000 each to the two child passengers. The trial court rejected the District's motion for judgment notwithstanding the verdict, but required the driver to accept $50,000 or a new trial. (Neither the driver not the children had objective injuries, but the driver alleged various aches and pains and stayed off work for six weeks.) In a short opinion, the court reversed and granted the JNOV. The police officers' conduct was far short of gross negligence. They entered the intersection five or ten miles over the speed limit with lights flashing and sirens screaming. As far as violating department regulations, the evidence would barely support a finding of simple negligence. *District of Columbia v. Henderson,* 710 A.2d 874 (D.C. App. 1998).

A 2000 case involving an ambulance questioned whether it was up to the jury or the judge to decide if the ambulance was on an emergency run at the time. An ambulance had picked up a woman with a head wound. The paramedics did not regard the injury as particularly serious, but as a matter of policy treated the trip to the hospital as an emergency with flashing lights and siren on. They attempted to pass a driver on the left as the driver was beginning a left turn at an intersection. The trial judge determined that the ambulance was on an emergency run, and gave the jury only the issue of gross negligence. Not surprisingly, the jury found for the fire department. The court of appeals affirmed. D.C. Code §1-1211(4) clearly permits the department to determine what is an emergency. *Browner v. District of Columbia,* 756 A.2d 927 (D.C. App. 2000).

We conclude the D.C. section with two cases brought under the Federal Tort Claims Act with the United States as defendant. We will see that there is a striking difference in the results. A 1991 case arose when a park police cruiser operated by an officer broadsided a car at an intersection, killing the driver. The officer was driving on Independence Avenue. He came to a red light and saw two vehicles in front run the light. He accelerated into the intersection to make arrests, and whether his siren was on is not clear from the case. There is little doubt that he did not enter with caution. The deceased slowed enough to begin a left turn. Apparently the officer was intent on catching the red light runners, as the judge opines. The standard applied was simple negligence, which is vastly different from gross negligence under District law. The award was $1.25 million. *Groves v. United States,* 778 F. Supp. 54 (D.D.C. 1991).

A 1995 case explained why D.C. Code 1-1212 does not apply. Agents of the Drug Enforcement Agency were attempting to arrest a drug suspect who managed to flee into traffic on E Street. The suspect reached speeds between 60 and 70 mph with an agent in pursuit. The suspect ran a red light and injured a motorcyclist. The pursuit was during morning rush hour. The district court found the pursuit utterly reckless and awarded $629,000. (The plaintiff had one leg amputated below the knee.) The court of appeals affirmed. Under the FTCA, the United States would be liable if a private person would be liable under the same circumstances. The court pointed out that no private person has the right to so pursue, so there is no private analogy. The defendant argued that the correct analogy therefore is the D.C. standard of gross negligence. The court cited a number of cases, primarily *Briscoe v. Arlington County,* 738 F.2d 1352 (D.D.C. 1984), in holding that 1-1212 immunity is limited to employees of the District. Federal legislative history showed that the primary purpose was to protect the financial resources of the District. *Hetzel v. United States,* 43 F.3d 1500 (D.D.C. 1995).

Briscoe arose from a pursuit begun in Arlington County, Virginia. Arlington officers were alerted to a bank robbery and given a car description. Nine minutes later Officer Kyle tentatively identified the vehicle and the suspect. Kyle stopped the driver on the E Street exit ramp of the Theodore Roosevelt Bridge, which is in the District. The robber, Brooks, got out and walked toward the officer with his hands in the air. Then the officer made a mistake. He walked Brooks back to his car and told him to keep his hands on the trunk. When Kyle walked to the passenger side to observe the passengers, Brooks dashed into the car (he had left the engine running and the door ajar) and fled at high speed. Kyle followed, and there was a high-speed chase. It ended when Brooks hit a pedestrian and inflicted injuries that resulted both legs being amputated. The plaintiff sued Officer Kyle and Arlington County, Virginia.

The court of appeals affirmed an award of $4 million to the pedestrian and reduced the award to his wife to $350,000. The court held that D.C. law applied, rather than Virginia's law. The latter would apply sovereign immunity and dismiss the claims. I will not go into the long discussion on the conflict of laws issue. (Under Virginia law, the defendants would be protected by sovereign immunity for discretionary acts. See the Virginia section.) This meant that Kyle's actions here could reasonably be ministerial, not discretionary, and that applied to the allegation of negligent supervision and training. On the all-important issue of the standard of care, it also meant that due care, not gross negligence is the proper standard.

The court held that the statutory standard of gross negligence applied only to District-owned vehicles. The court has a long discussion of why it found the due care standard applicable, since District regulations setting out UVC 11-106 gives both due care and reckless disregard. Whether it is persuasive, I'll leave to the reader. *Briscoe v. Arlington County,* 738 F.2d 1352 (D.C. Cir. 1984).

Florida

Fla. Stat. Ann. §316.126. Operation of vehicles . . . upon approach of an authorized emergency vehicle

(1) Upon the immediate approach of an authorized emergency vehicle, while en route to meet an existing emergency, the driver of every other vehicle shall, when such emergency vehicle is giving audible signals by siren, exhaust whistle, or other adequate device, or visible signals by use of displayed blue or red lights, yield the right-of-way to the emergency vehicle and shall immediately proceed to a position parallel to, and as close as reasonable to the closest edge of the curb of the roadway, clear any intersection and shall stop and remain in position until the authorized emergency vehicle has passed, unless otherwise directed by any law enforcement officer.

(2) Every pedestrian using the road right-of-way shall yield the right-of-way until the authorized emergency vehicle has passed, unless otherwise directed by any police office.

(3) Any authorized emergency vehicle, when en route to meet an existing emergency, shall warn all other vehicular traffic along the emergency route by an audible signal, siren, exhaust whistle, or other adequate device or by a visible signal by the use of displayed blue or red lights. While en route to such emergency, the emergency vehicle shall otherwise proceed in a manner consistent with the laws regulating vehicular traffic upon the highways of this state.

(4) Nothing herein contained shall diminish or enlarge any rules of evidence or liability in any case involving the operation of an emergency vehicle.

(5) This section shall not operate to relieve the driver of an authorized emergency vehicle from the duty to drive with due regard for the safety of all person using the highways.

Fla. Stat. Ann. 316.072(5). Authorized emergency vehicles

[This section sets out the text of UVC 11-106 without significant change. See the Introductory Essay for text.]

Note that Section 316.072(5) includes both due care and reckless disregard standards, as given in UVC 11-106. Section 316.126, the duty to yield to emergency vehicles, applies a due care standard to emergency vehicle operators. Notwithstanding these statutes, the cases are based mainly on the common law. We take the cases with the oldest first. In a 1961 case, the court of appeals reversed a trial court's finding that the pursuing officer and his employer were negligent as a matter of law. An officer attempted to stop a driver who sped through town. Another officer joined in, and the chase went on for 16 miles in the country at speeds up to 100 mph. At one point, one of the officers fired four shots in the air. The pursued driver, Gnann, swerved from side to side to prevent the officers from coming abreast or passing, and intermittently turned off his headlights trying to shake them. During one of these dark periods, and while Gnann was on the left side of the highway, an opposing vehicle approached. There was a head-on collision and both drivers were killed. The widow of the innocent driver alleged reckless and negligent pursuit. There was a trial on damages only and the jury made an award. The court of appeal reversed because generally causal negligence is a jury question. It was error for the trial judge to grant the plaintiff summary judgment on liability. *Mount Dora v. Bryant,* 128 So. 2d 4 (Fla. App. 1961).

A 1967 case is quite different. In *City of Miami v. Horne,* 198 So. 2d 10 (Fla. 1967), the Florida Supreme Court said that an officer has the duty to enforce the law, including the duty to pursue, but may not "exceed proper and rational bounds and not act in a negligent, careless or wanton manner." The Miami officer had stopped a driver for speeding in a 30-mph-zone and also cited him for not having an operator's license. The driver suddenly took off at 90 mph and the officer pursued. The suspect killed an innocent bystander. The court stated that the officer is under a duty to pursue in this instance, and even if the pursuit contributed to the recklessness of the pursued driver, the city is not liable to the third party. A later case (*Brown* below) called this case "hard to interpret," and interpreted it to mean that the officer properly initiated the pursuit, and did not have a good opportunity to terminate it. In other words, even if the pursuit is halted, the suspect is still a massive danger to the public unless he also slows and drives as he should. Which decision (pursue or break off) puts the public at greater risk is not easy to predict.

A short 1971 case, *Reed v. Winter Park,* 253 So. 2d 475 (Fla. App.), cited and followed *Horne.* The pursuit does not seem to make much sense. An officer en route to an accident scene with siren on noticed a driver speeding up instead of getting out of his way as required by law, so the officer tried to stop the driver. The driver sped away and the officer began a pursuit which ended when the pursued hit another vehicle that had the right of way. The court did not question the right to pursue, but remanded for trial on whether the officer was exercising proper care by not continuously sounding his siren. Had he done so, the plaintiff argued, the he might have been warned and could have avoided the fleeing driver.

A 1986 case explicitly followed *Horne.* A fleeing suspect struck the plaintiff's car. The trial court granted summary judgment to the defendant sheriff and county on the grounds that they enjoyed sovereign immunity. The court of appeal affirmed. The plaintiff argued that such immunity was waived by Fla. Stat. 768.28 (waiver of sovereign immunity for tort actions). The court held that the decision to pursue is a discretionary act which, absent a malicious purpose or willful and wanton conduct, is protected. Of course, there is still a common law duty of care in a police pursuit. However, the only negligence alleged is that of conducting the pursuit at high speed on busy streets. The pursuing officer was well away from the crash scene, and had lights and siren operating. According to *Horne,* as long the as pursuit is conducted with due care, the officer is not responsible for the acts of the pursued. *Rhodes v. Lamar,* 490 So. 2d 1061 (Fla. App. 1986). A 1989 case, *Miller v. Dept. of Highway Safety,* 548 So. 2d 880 (Fla. App. 1989) held it to be a jury question of whether the pursuit was negligently conducted, and reversed summary judgment for the defendant on the basis of sovereign immunity. The suspect stole a police car left with the engine running and unlocked while the patrol officer was checking an abandoned car.

The most important recent case is *Brown v. Pinellas Park,* 557 So. 2d 161 (Fla. App. 1990). Near 1:00 A.M. a driver, Deady, ran a red light in Pasadena. A deputy sheriff saw the traffic violation and pursued. Two more officers jumped in, all running red lights. Incredibly, before it ended, as many as twenty police and sheriff cars pursued at high speeds to catch a red light runner (not to minimize the danger of red light runners). Thirteen minutes, 34 traffic signals, and 25 miles later Corporal Rusher, ahead of the pursuit at a crossroad, waited for a green signal. Next to Rusher was another car also waiting. Rusher did not attempt to warn that driver of the approaching caravan. Then Rusher turned right on his green, intending to get ahead of Deady and try to stop him. The other car started across the intersection and was hit broadside by Deady. Both occupants in the car were killed. There was some evidence of an order to cease the pursuit, but no firm evidence of who gave it or who could have responded. (This pursuit is almost a caricature of everything that can go wrong in a pursuit situation: the illegal but apparently harmless traffic violation, the length of the pursuit, the number of officers who joined it, the failure to break off after weighing the danger with the original violation, and the inaction of the trooper who might have warned the deceased individuals. All these acts appeared to be in violation of both written and oral policies of the sheriff's department.)

The plaintiffs appealed a dismissal. The court of appeal made a thorough analysis of Florida police pursuit law, discussing most of the cases noted above, and the opinion should be carefully studied by anyone with a pursuit case in Florida. At the least, the plaintiffs raised jury questions. The court noted that two societal values conflicted: enforce the law but don't needlessly risk innocent lives. The court asked the Florida Supreme Court to clarify the law, specifically, "Is the continuation by law enforcement officers of a high speed vehicular pursuit of a traffic law violator which results in deaths of innocent bystanders an actionable breach of duty involving an operational level governmental function which is not immune from liability when it is alleged that under the circumstances the officers should have known that continuing the pursuit would create an unreasonably dangerous hazard to innocent bystanders, including those who were killed when the traffic violator's vehicle collided with their vehicle?"

The supreme court held that the defendants were not entitled to sovereign immunity. Their own pursuit policies forbade this kind of pursuit. Law enforcement officers are entitled to a high degree of judicial deference, and normally police actions in the face of emergencies will be discretionary acts and thus immune from liability. In this case there was no emergency, and the method of conducting the pursuit was operational and not discretionary. The court's majority held that the plaintiffs stated a case for the jury as to the proximate cause. The court found its decision consistent with *Horne,* in that the pursuit must not be conducted in a negligent or wanton manner. The court also applied the "zone of risk" created by the pursuit. This doctrine was identified primarily in *McCain v. Florida Power,* 593 So. 2d 500 (Fla. 1992), and is applied in Florida in all kinds of negligence cases. The three dissents are worth reading. One thought that the majority's view will result in more harm than good. *City of Pinellas Park v. Brown,* 604 So. 2d 1222 (Fla. 1992). On remand and a trial, the jury awarded $2.5 million. The court of appeals held the award excessive, but because it was subject to the $200,000 state-imposed damage cap ($100,000 per person, $200,000 per accident) no further action was required. Note that the plaintiffs settled with some defendants for $900,000. The deceased were two young women. *Rice v. Brown,* 645 So. 2d 1020 (Fla. App. 1994).

A 1997 case arose when two youthful offenders in some kind of detention facility operated by the Department of Agriculture stole a van and fled. A "be-on-the-lookout-for" bulletin was issued, and the van was located and a pursuit began. When the boys crossed a county line, at speeds of over 100 mph at times, the original pursuers broke off and local officers took over. Thirty-five miles farther, the van tried to get through a rolling roadblock, and there was a multiple vehicle crash. The officers who were sued were the ones who had broken off. The court of appeal affirmed summary judgment in their favor. They were not at fault for beginning the pursuit and, since the crash happened long after they broke off, they were not responsible for the injuries. *Porter v. Dept. of Agriculture,* 689 So. 2d 1152 (Fla. App. 1997).

We conclude the Florida section with two federal cases. The first case is not even a pursuit. A Volusia County deputy sheriff was speeding on a state highway for no particular reason (in other words, acting like all too many drivers). The plaintiffs were traveling in the opposite direction and attempted a left turn. They collided with the deputy whose speed was "approximately" 73 mph. There were serious injuries. The plaintiffs filed a §1983 action. The trial court granted summary judgment to the defendants, and the court of appeals affirmed. The plaintiffs argued that this is only a case of reckless driving, not police pursuit, and therefore the defendants may not rely on those decisions (which rarely find a constitutional deprivation in pursuit cases). If so, the court reasoned that simply going a few miles over the speed limit without an emergency is just negligence, and that kind of a claim is "grist for the state law mill." *Roney v. Watson,*

101 F.3d 1378 (11th Cir. 1996). A 2000 case applied the U.S. Supreme Court case *Sacremento v. Lewis* (see the California section) in holding that a high-speed pursuit, conducted for the legitimate purpose of arresting a carjacker and not with malicious intent to harm, does not arise to a constitutional deprivation of rights. *Levy v. City of Hollywood,* 90 F. Supp. 2d 1344 (S.D. Fla. 2000).

Georgia

Ga. Code Ann. §40-6-6. Authorized emergency vehicles

[Sections (a) to (c) sets out UVC 11-106. See the Introductory Essay for text.]

(d)(1) The foregoing provisions shall not relieve the driver of an authorized emergency vehicle from the duty to drive with due regard for the safety of all persons.

(d)(2) When a law enforcement officer in a law enforcement vehicle is pursuing a fleeing suspect in another vehicle and the fleeing suspect damages any property or injures or kills any person during the pursuit, the law enforcement officer's pursuit shall not be the proximate cause or a contributing proximate cause of the damage, injury, or death caused by the fleeing suspect unless the law enforcement officer acted with reckless disregard for proper law enforcement procedures in the officer's decision to initiate or continue the pursuit. Where such reckless disregard exists, the pursuit may be found to constitute a proximate cause of the damage, injury, or death caused by the fleeing suspect, but the existence of such reckless disregard shall not in and of itself establish causation.

(d)(3) The provisions of this subsection shall apply only to issues of causation and duty and shall not affect the existence or absence of immunity which shall be determined as otherwise provided by law.

It is clear from the above statute that an officer pursuing a suspect is not liable if he does not act with reckless disregard, or is otherwise immune. As is always true in negligence allegations, what is reckless disregard depends on the facts and prior case law, and, frankly, how the court or jury reacts to the facts. There are no magic words that can clearly distinguish between ordinary negligence and reckless disregard. One meaning has shades drifting into the other, and only the extremes are an easy call.

A 1954 case gave some of the history and rationale. It has long been recognized that law officers need some discretion in enforcing the law, while on the other hand "protection to person and property is the paramount duty of government." One of those rights is use of the public thoroughfares with reasonable safety. The case arose when a deputy sheriff, the defendant, received a call of a shooting in a nearby town. The plaintiff, another deputy, rode along. They proceeded along a densely populated road at 80 mph (plaintiff's figure) or 65 mph (defendant's figure), and crashed into a truck they were overtaking as it turned left. Their lights and siren were operating. The trial court granted summary judgment to the defendant. The court of appeals reversed, concluding that whether the deputy drove with reckless disregard was a jury question. *Archer v. Johnson,* 83 S.E.2d 314 (Ga. App. 1954).

Newer cases are more pertinent to the law as interpreted today. A 1990 case dealt with immunity. The facts are sketchy, but it appeared that a deputy sheriff was responding to a call of a fight, exceeded the speed limit without using his blue light or siren, and failed to yield the right of way to another. The employing county did not carry liability insurance. The trial court granted immunity to the county, but not to the deputy, who appealed. The supreme court held that the deputy was exercising a discretionary function when responding to the call, and that his failure to comply with the traffic law was only negligence. The important point was that the county had not elected to be covered by insurance. The state is required to purchase such insurance; counties may or may not. If not, they retain their sovereign immunity, and as such their employee, the defendant, is also immune. If the county did have insurance, the plaintiff would be entitled to a trial on negligence and reckless disregard for the safety of others. A dissent strongly urged that the plaintiff was so entitled. *Logue v. Wright,* 392 S.E.2d 235 (Ga. 1990).

The Georgia Supreme Court case, *Mixon v. City of Warner Robins,* 444 S.E.2d 761 (Ga. 1994), caused the legislature to make a change to the statutory law. An officer saw a driver fail to stop at a stop sign. The officer pursued. The driver ran another stop sign and killed another driver. The trial court granted the defendant summary judgment on a finding that the fleeing driver was, as a matter of law, the sole proximate cause. The court of appeals affirmed. The supreme court reversed for trial. The court pointed out that the statute, 40-6-6, neither specifically provides that the pursuing officer can be held liable for an accident caused by the pursued driver, nor does it specifically provide for immunity. The statute does, in §40-6-6(d), state that the officer must drive with due regard of the safety of all persons. The court of appeals held that such officer shall not be the proximate cause or a contributing cause unless it is shown that the pursuit was a "higher threat to public safety than is ordinarily incident to high speed police pursuits." *Mixon*, 434 S.E.2d at 71. The supreme court turned to a Texas case for guidance, *Travis v. City of Mesquite,* 830 S.W.2d 94 (Tex. 1992). That case said that a pursuing officer can be the proximate cause if the risk of injury to third parties is unreasonable as compared to the benefit of arresting the suspect. The court noted that the fleeing driver had slowed to 5 mph or less in going through the stop sign, and the pursuit reached speeds up to 60 mph, more than twice the speed limit.

A concurring opinion preferred the policy as stated by the lower court as striking the better balance between competing values. It noted that if the officer had not pursued, he could be sued "in the current legal climate," and cited a case in which this happened, *Landis v. Rockdale County,* 445 S.E.2d 264 (Ga. App. 1994). In that case, an officer failed to pursue and arrest a drunk driver who then caused a fatal accident. The court held the officer not negligent, and there was a dissent to that decision. And so it goes in this controversial area of law. In response to *Mixon*, the legislature amended 40-6-6 to add 40-6-6(d)(2), which is quoted at the beginning of this section. Law enforcement officers and representatives of municipal and county governments feared that police officers would be held liable just for doing their jobs, and that local governments, mostly self-insured, would face large civil damage awards. Originally the proposed bill called for a California-type statute that would grant immunity if law enforcement agencies adopted pursuit policies meeting specified guidelines. See 12 Ga. St. U.L. Rev. 295 (1995) for an account of this legislative action.

The usual rule is that sovereign immunity for the operation of government vehicles is waived to the extent of liability insurance coverage. As in most states, Georgia has so waived immunity by enacting a state tort claims act. A statute, the Georgia Interlocal Risk Management Agency (GIRMA), §36-85-1 et seq., authorized local governments to pool their risks with other agencies, including those risks arising from motor vehicle liability. That statute provided that a governmental agency joining the risk pool did not waive its immunity. The Georgia Supreme Court held that provision to be unconstitutional. An officer responding to a call collided with the plaintiff. The court of appeals held that the city was not immune. Disputed evidence suggested that the officer turned off his siren as he approached the crime, and veered to the plaintiff's side of the road. The officer's version was quite different. The court remanded for trial. *Harden v. Burdette,* 420 S.E.2d 626 (Ga. App. 1992).

The supreme court amplified on the immunity issue in a 1994 police case. A sheriff's deputy, Richardson, was responding to an emergency and collided with the plaintiff. The county had joined the risk pool, GIRMA. The court held that immunity was waived to the limits of the liability policy. Richardson herself was immune as correctly held by the trial court. She had official or qualified immunity. The basis of such personal immunity is that a government employee, acting within the scope of her employment, must be permitted some discretion in doing her job. It is manifestly unfair to place a person "in a position in which he is required to exercise his judgment and at the same time is held responsible according to the judgment of others, who may have no experience in the area and may be much less qualified than he to pass judgment . . . or who may now be acting largely on the basis of hindsight." (Quoting Restatement (Second) of Torts §895D, comment b.) Richardson could be liable only if performing her discretionary duty with malice. On the other hand, the county, as represented by the sheriff, enjoyed sovereign or governmental immunity (the court held the terms synonymous) but here it had been waived by the purchase of liability insurance. *Gilbert v. Richardson,* 452 S.E.2d 476 (Ga. 1994).

One city claimed that it had not waived immunity because its liability policy (covering claims in excess of $250,000) was a general one, not specifically pertaining to motor vehicles. The court held that the policy language was broad enough to encompass the claim in a 1994 case. An officer on an emergency run ran a red light to make a U-turn and collided with the plaintiff. On remand, the jury was to decide whether the officer's acts were discretionary. If so, he would be immune. The acts would be discretionary if the officer were responding to an emergency call. There was a factual issue of whether the emergency had been "cleared." If so, the officer would be obligated to observe traffic laws. *McLemore v. City Council of Augusta,* 443 S.E.2d 505 (Ga. App. 1994).

A 1996 case in federal court arose from a call an officer received about a driver who had eluded police in the adjoining county. The officer saw the driver run a stop sign, and pursued for five miles. It was then that the officer collided with the plaintiff, an innocent bystander. The federal district court granted summary judgment to the defendant. A 1991 amendment to the state constitution authorized the legislature to waive immunity by enacting a state tort claims act. Such immunity extended to governmental agencies and employees in the performance of discretionary acts unless they acted with actual malice or actual intent to cause injury. If the officer was on an emergency run, he was exercising discretion and would be immune. The plaintiff argued that the officer was not responding to a bona fide emergency, and that this was a triable issue. The court decided otherwise. The best person to decide if there was an emergency is the officer himself. True, the officer had the license number, but he could reasonably assume that when a driver has been fleeing other officers, he has committed a crime. It was 4:00 A.M. when little traffic was about. *Morgan v. Causey,* 910 F. Supp. 651 (M.D. Ga. 1996).

A 1986 case, also in federal district court, held that it was premature to grant summary judgment to the defendants. The court noted that applying §1983 to a pursuit case is somewhat novel (at that time), and with the facts hotly disputed and the law in this area "ever-changing," it would not dismiss the claim, but would entertain motions for directed verdict unless the plaintiff can support his case at trial. The facts were than police chased a speeding and reckless driver who passed between two police cars not yet set up as a roadblock, and crashed into a tree on a curve and killed himself. The court did not decide any issues other than not to grant the defendants summary judgment. There is a long discussion. *Easterling v. City of Glennville,* 694 F. Supp. 911 (S.D. Ga. 1986).

A 2000 case made a finding on the relationship between reckless disregard and malice. An officer on an emergency run (pursuing a suspected car thief) ran a stop sign without slowing, according to the plaintiff, and as required by Code §40-6-6. The trial court granted summary judgment to the officer. The court of appeals affirmed. The court notes that the plaintiff did not allege that the officer acted with malice, only that he acted with reckless disregard for the safety of others. Reckless disregard does not equate with malice under Georgia case law. Code §40-6-6 itself states that it is not intended to waive any immunity to which a defendant may be entitled. As to the codefendant city, since it had chosen not to buy liability insurance, it retained its immunity. Finally, the plaintiff alleged that the city created a nuisance by not properly training its officers in pursuit procedures. The court rejected that claim because the plaintiff produced no evidence to support it. *Williams v. Solomon,* 531 S.E.2d 734 (Ga. App. 2000).

See another 2000 case decided on reckless disregard with no mention of a deputy sheriff's immunity. A deputy was chasing a driver (reason not given) on a two-lane road. The fleeing driver crossed the center line and killed an oncoming driver. The trial court granted the deputy and the sheriff summary judgment. The court of appeals reversed for trial as to the deputy. Code §40-6-6(d)(2) states that an officer cannot be the proximate or a contributing cause unless the officer acted with reckless disregard for proper law enforcement procedures in the decision to begin or continue a pursuit. An expert for the plaintiff claimed by affidavit that the officer did so act. Summary judgment as to the sheriff stands because the plaintiff submitted nothing to oppose it. (Hey, no one says pursuit law is an exact science.) *Land v. Becham,* 530 S.E.2d 746 (Ga. App. 2000).

A 1990 case did credit state immunity over any language in Code §40-6-6. The state was sued after a state trooper chasing a speeder collided with the plaintiff. The court of appeals affirmed a dismissal for failure to state a claim. The Georgia Tort Claims Act, §50-21-20 et seq., waives immunity. One of the exceptions is that lack of immunity is waived for "the method of providing law enforcement, police, or fire protection." Absent any allegation that the trooper acted with malice, dismissal was proper. *Hilson v. Dept. of Public Safety,* 512 S.E.2d 910 (Ga. App. 1999).

A county police officer on a domestic dispute call attempted to pass a driver as the driver was turning left. The officer had not activated lights or siren. The court of appeals reversed a summary judgment for trial. There was no question that the officer was engaging in a discretionary act, but it was up to a jury to decide if the act was merely negligent (and thus excused) or was in reckless disregard for the safety of others. The court was unwilling to make that finding as a matter of law, noting that there were similar cases each way. *Johnson v. Gonzalez,* 478 S.E.2d 410 (Ga. App. 1997). See a 1996 case in which the court of appeals had little doubt that a sheriff and his deputy were immune. The deputy was chasing a suspected car thief and collided with a third driver. The decision to pursue in the early morning hours was clearly a discretionary one, and there was no evidence of reckless disregard. *Morgan v. Barnes,* 472 S.E.2d 480 (Ga. App. 1996).

We look at the remaining recent cases briefly. A 1996 case held that an officer acted properly when he got a "fight, snatch thief" report with the license number and description of the car and occupants. He saw the car twenty minutes later and pursued without lights, siren, or speeding. The suspect then sped dramatically and ignored traffic rules. The officer called for help, and when in his judgment the traffic density increased the officer resumed a normal speed and called streets and directions. The suspect crashed into the plaintiff. The court held there was no reckless disregard. *Wilson v. City of Atlanta,* 476 S.E.2d 892 (Ga. App. 1996). A 1995 case held that a county that did not purchase liability insurance but maintained a reserve fund for paying claims had not waived its sovereign immunity. The fatal accident happened when an officer responding to a Code 4 call struck the plaintiff's left-turning car. If the officer was not sounding his siren at the time, that would be at most simple negligence, and such negligence in the performance of a discretionary act does not negate the officer's immunity. *Tillman v. Mastin,* 453 S.E.2d 85 (Ga. App. 1995). See a 1992 case with an identical holding. Again, the officer had his flashing lights on but not his siren at the time. He was held immune, being at most merely negligent. *Banks v. Patton,* 413 S.E.2d 744 (Ga. App. 1992).

A 1995 case held that whether the officer's acts showed due regard for the safety of all persons was a jury question. There is no mention of immunity. The officer was pursuing a suspect on a rainy day and both vehicles (plaintiff's version) or only the suspect's vehicle (officer's version) crashed into plaintiff's car while it was parked on the edge of the highway. *Thompson v. Payne,* 453 S.E.2d 803 (Ga. App. 1995).

We conclude the cases with two involving fire trucks. A volunteer fireman received an emergency call and proceeded to it with flashing lights and the siren sounding, and driving above the posted limit. He looked away for a moment and rear-ended a pickup that had stopped in response to the siren. The pickup passenger sued. The jury found for the fireman. The court of a ppeals r eversed. The error was in holding the defendant to be the driver of an authorized emergency vehicle and thus permitted to disregard some traffic laws. The fireman had not yet been certified, and cannot come under the statute's protection. *Pryor v. Phillips,* 473 S.E.2d 535 (Ga. App. 1996).

In the second case, a fire truck using its warning signals entered an intersection against the red signal and collided with the plaintiff. The plaintiff heard the siren but nevertheless entered the intersection. The court of appeals affirmed summary judgment for the city and the fireman. Firefighting is a governmental function, and the city is immune from any liability, as prescribed in Code §36-33-1(b): "Municipal corporations shall not be liable for failure to perform or for errors in performing their legislative or judicial functions. For neglect to perform or improper or unskillful performance of their ministerial duties, they shall be liable." The city did not lose its immunity by purchasing non-motor-vehicle liability insurance, nor by joining a risk pool with other cities. The fire truck operator was performing a discretionary function, and even if he was negligent he does not lose his immunity. The court noted that both parties had misstated the nature of governmental immunity. A city is immune from liability for the negligent performance of its governmental duties, but may be liable for the negligent performance of its ministerial duties. *Sinkfield v. Pike,* 411 S.E.2d 889 (Ga. App. 1991).

Georgia cases appear to be grounded on a variety of statutes, including the typical UVC provision enacted by the states, as well as immunity statutes and Art. I, §2, §9 of the Georgia Constitution. Cases that seem similar factually do not cite all the related statutes. Some are decided on reckless disregard with no mention of immunity, some on immunity with the distinction between ministerial and discretionary functions, and some on malice and willful and wanton conduct. One with a pursuit case will need to carefully study the applicable law and craft his complaint accordingly.

Hawaii

Haw. Rev. Stat. 291C-26

[This statute sets out UVC §11-106 without any change in text (see the Introductory Essay for text).]

Hawaii appears to be in the happy position of having no police pursuit or other authorized emergency vehicle cases.

Idaho

Idaho Code §49-623. Authorized emergency or police vehicles

[Other than giving a required decibel rating for sirens and a visibility standard for emergency lights, Idaho has adopted UVC §11-106. See the Introductory Essay for text.]

There are no cases in which a plaintiff alleges damage or injury caused by emergency vehicles on emergency runs. Three criminal cases involve citations issued for failure to yield to authorized emergency vehicles. In *State v. Barsness,* 628 P.2d 1044 (Idaho 1981), the defendant was convicted of failure to yield. A Boise police officer responded to a report of a felony in progress. The officer switched on his warning lights but not his siren. The Idaho Supreme Court affirmed the conviction. Section 49-623(3) states that to come under the protection of the statute the officer must use his warning lights "and/or" his siren. He satisfied the statute by using his lights only. Whether he was driving with due care was for the jury. The court noted that the city ordinance required both lights and siren. The court held that state law must control. A long dissent argued that the ordinance was not in conflict, rather it went beyond the state statute in response to city traffic conditions.

A 1989 conviction for failure to yield to an emergency vehicle was overturned because the state did not prove that the sheriff's vehicle complied with the equipment requirements. *State v. Monaghan,* 783 P.2d 311 (Idaho 1989). Similarly, the amber deck lights of a police car did not meet the requirement of being visible in 360 degrees at 1,000 feet. *State v. Pick,* 861 P.2d 1266 (Idaho 1993).

Illinois

745 Ill. Comp. Stat. 5/1. Immunity of State as defendant or party in court

[States that except as otherwise provided, the State shall not be a party]

745 Ill. Comp. Stat. art. II. Local government tort immunity

10/2-109. A local public entity is not liable for an injury resulting from an act or omission of its employee where the employee is not liable.

10/2-201. Except as otherwise provided by Statute, a public employee serving in a position involving the determination of policy or the exercise of discretion is not liable for an injury resulting from his act or omission in determining policy when acting in the exercise of such discretion even though abused.

10/2-202. A public employee is not liable for his act or omission in the execution or enforcement of any law unless such act or omission constitutes willful and wanton conduct.

10/5-106. Except for willful or wanton conduct, neither a local public entity, nor a public employee acting within the scope of his employment, is liable for an injury caused by the negligent operation of a motor vehicle . . . when responding to an emergency call.

Illinois Compiled Statutes ch. 635. Rules of the road

5/11-205 [Sets out UVC §11-106 without significant change. See Introductory Essay for text.]

Emergency vehicle cases are governed by state or local government immunity, the rights and duties of emergency vehicle operators as standardized in the Uniform Vehicle Code §11-106, or both. Most of the cases annotated below have been decided by the lower appellate courts in Illinois. A 1992 case in the Illinois Supreme Court dealt with immunity. A state trooper on patrol on I-80 near Joliet either did or did not (the evidence is not clear) receive a call to come to a disturbance on Eastern Avenue in Joliet. Normally such matters would be handled by city police. The trooper entered the city and soon found himself going the wrong way on a one-way street. As he made a U-turn, he collided with the plaintiff driver. The jury found the trooper at fault and granted a modest award. The case came to the Illinois Supreme Court on the issue of official immunity. The court explained that sovereign immunity was abolished by constitutional amendment in 1970, and the court of claims was established with exclusive jurisdiction over claims against the state, with a $100,000 damage cap. After a study of cases, the court affirmed the verdict. The court gave two reasons. First, the trooper would be immune only if his acts were unique to his duty as a state officer. Because it was not clear that the officer was on an emergency run, his duty in driving is the same as any other driver. He is to observe the law and exercise due care. Second, the officer was not performing a governmental discretionary act and thus has the same duty as any other driver. The decision to turn onto one street rather than another is one made by all drivers. *Currie v. Lao*, 592 N.E.2d 977 (Ill. 1992).

In a similar case, a state trooper on patrol, Julian, heard on his radio that another trooper was chasing a motorcyclist for a traffic infraction. When that officer reached the Wisconsin line, he abandoned the chase. The officer who heard the report meanwhile headed toward the pursuit at high speed, and crashed into another car at an intersection. There was conflicting evidence about who had the green light and whether the lights and siren were activated on the officer's vehicle. There was evidence that Julian violated several pursuit policies, including one that authorized officers in hot pursuit only to cross state lines. The jury awarded damages against Julian. The court of appeals agreed that Julian was not protected by immunity because his pursuit was in violation of any authority. It remanded on unrelated grounds, but found Julian liable. (Julian's counsel improperly argued to the jury that it would be unfair for the trooper to have to pay for the next fifty years, when counsel knew that the state had agreed to cover any damage award. *Lenz v. Julian,* 657 N.E.2d 712 (Ill. App. 1995).

A 1994 case cited *Curie* in finding that the plaintiff's only remedy in a pursuit case is the court of claims. A state trooper received a report of an armed robbery, the vehicle's color (green), and the street and direction of the vehicle. The trooper waited at an intersection, saw a vehicle matching the description, and pursued with appropriate warnings. The officer entered an intersection on the green signal. The plaintiff apparently entered on red and there was a crash. The plaintiff argued in vain that the officer was speeding and that there was no green car being pursued, according to three witnesses. The appellate court affirmed a dismissal. The witness' affidavits merely showed that they did not notice a green car, not that there was none. The other evidence clearly showed that the officer was in performance of his duties and was immune. *Hickey v. Huber,* 635 N.E.2d 791 (Ill. App. 1994).

It is obvious that when there are two standards by which to judge an officer's conduct, official immunity or the traffic rules applicable to emergency vehicles, there will be questions. If they conflict, which will control? If only one is before the court, can the other be ignored? The standards have different origins. The immunity standard grew out of common law official immunity, now made constitutional and statutory. The other grew out of rules of the road as developed by the National Committee on Uniform Traffic Laws and Ordinances and published as the Uniform Vehicle Code. The states have been required to adopt the UVC with any modifications they desire. This conflict came up in a 1998 case. A city police officer responded to an emergency call of an imminent shooting and collided with a car in front of him that moved into the left lane to make a turn. The officer was momentarily distracted in locating his siren button. The trial court granted summary judgment to the officer because he was enforcing or executing the law and his conduct was not willful and wanton. The plaintiff argued on appeal that the more specific statute, 625 ILCS 5/11-205, must control. The majority agreed with the plaintiff and reversed. It found that a specific statute, relating to only one subject, controls over the more general one relating to a variety of fact situations. A dissent argued that a specific statute should not override general immunity. *Bradshaw v. City of Metropolis,* 688 N.E.2d 332 (Ill. App. 1998).

The dissent in *Bradshaw* above cited *Postich v. Henricks,* 641 N.E.2d 975 (Ill. App. 1994). That case involved an officer who received an ASAP report of "man with a gun" and in responding collided with another vehicle. The court held that immunity applied and the plaintiff's only recourse was the court of claims. The court held exactly the opposite from the holding in *Bradshaw.* If immunity applies, then 11-205 is irrelevant. Go figure.

Bradshaw was decided by the Fifth Appellate District Court. A 1999 police pursuit case, decided in the Appellate Court for the Second District, found the reasoning in *Bradshaw* flawed. An officer in pursuit of a suspect ran a red light and collided with the plaintiff. The trial court dismissed the complaints, and the plaintiff appealed. The plaintiff argued that *Bradshaw* should be followed; that the more specific statutes, §§11-205 and 907, must control over §2-202 of the Tort Immunity Act (a public employee is not liable for his act or omission in the execution or enforcement of any law unless such act or omission constitutes willful and wanton conduct). First, the court noted that it was not obligated to follow a case from another appellate district in the state. Second, the plaintiff had improperly quoted from the dissenting opinion as if that was the opinion of the court, and the court urged counsel never to do that again. Third, the court simply thought that *Bradshaw* was wrongly decided. Each statute stands "in its own sphere." Section 2-202 specifically immunizes government employees, while §§11-205 and 209 covers the duties of drivers of emergency vehicles, who may or may not be government employees. (Perhaps the court had in mind volunteer fire departments personnel and private ambulances.) The court remanded for trial of whether the officer's conduct was willful and wanton. (This is unlikely since the officer had on his lights and siren.) *Carter v. Du Page County,* 710 N.E.2d 1263 (Ill. App. 1999). Litigants in Illinois will have to pay attention to the development of the law in each appellate district and watch for different holdings.

Whether an officer is executing or enforcing the law is fairly broad. In a 1998 case, a sheriff's deputy was responding to a call of shots fired. He was not specifically dispatched to the scene. He did not have his flashing lights or siren on, and he did not think the report was an emergency. He struck a pedestrian en route. Nevertheless, he was granted summary judgment, and the Appellate Court for the Second District affirmed. The plaintiff argued in vain that no person was arrested or charged, and that no specific law was being violated. The court pointed out that immunity may not be determined by circumstances after the fact. *Bruecks v. County of Du Page,* 658 N.E.2d 538 (Ill. App. 1995). Now, see a 1992 case in which a Chicago police officer on parking violation duty received a report of a missing person. The officer did not regard the incident as an emergency, drove within the speed limit, and did not switch on emergency signals. He collided with a minor bicyclist. The appellate court remanded for trial on a finding that the officer was not executing and enforcing the law at the time. He had the same duty of all drivers to observe the traffic laws and maintain a proper lookout. *Simpson v. City of Chicago,* 599 N.E.2d 1043 (Ill. App. 1992).

A 1987 case held an officer's acts to be negligent but not willful and wanton. The officer was responding to a burglary alarm with his siren and lights. He collided with another driver at a three-way stop intersection. There was a fatality. The jury made an award. The appellate court reversed on a finding that the defendant city was immune under the Tort Immunity Act. Under the act, the defendant can be liable only if the officer's act was willful and wanton conduct. A dissent complained that the majority ignored §11-205. *Rosen v. City of Collinsville,* 520 N.E.2d 638 (Ill. App. 1987).

A 1995 chase resulted in the death of a motorcyclist. A city police officer noticed a motorcyclist with an unhelmeted passenger speeding and weaving in and out of traffic and running red lights. The motorcyclist had spent several hours drinking. The officer pursued and advised his dispatcher several times. When he reached the limits of his jurisdiction, he called the dispatcher for authorization to continue the chase. No supervisor was available, so the dispatcher told the officer to use his own discretion. The officer considered the department's pursuit policies that required him to balance the risks, and in view of the light traffic, he decided to continue. Eventually the officer had to drop back and lost the suspect. Then he came upon an accident scene and found that the motorcycle passenger had been killed. The parents sued. The trial court granted summary judgment to the defendants. The issue on appeal was whether the officer's conduct was willful and wanton as contemplated by §2-202 of the Local Government Tort Immunity Act. With no dissents, the court found the pursuit in fact to be careful and was not even the proximate cause. The motorcyclist's actions posed a danger to his passenger and to others. The court did not agree with the plaintiff's expert that the officer increased the danger to the passenger by pursuing. It cannot be the law of Illinois to prohibit officers from pursuing unhelmeted riders. *Urban v. Village of Lincolnshire,* 651 N.E.2d 683 (Ill. App. 1995).

In most appellate cases, police officers' conduct in emergencies have not been found to be willful and wanton. A 1998 case even quoted the U.S. Supreme Court case, *Sacramento v. Lewis* (see the California section). An officer was alerted of a possible drunk driver by the driver's flashing of his lights and driving recklessly. The officer pursued and did get the license number but continued the chase at high speeds. The officer stayed about six car lengths behind, and was considering ending the pursuit when the driver crashed into another car, with that driver injured and the suspect killed. The injured driver sued. The appellate court affirmed summary judgment for the defendant, and emphasized the danger posed to all by drunk drivers. The site was a four-lane highway, the traffic was light, and the officer had his lights and siren on, and the pursuit was brief. The court noted the conflict among the courts on which standard to apply, but found that the pursuit was justified under either standard. *Hall v. Village of Bartonville,* 699 N.E.2d 148 (Ill. App. 1998).

A smash-and-grab thief is not quite like a drunk driver. If not pursued, the thief is unlikely to be any particular threat to other drivers. The conflicting value, of course, is police departments' duty to catch thieves. A Chicago suburb had been the site of several such incidents: the driver of a gray Riviera would park in front of a store and his two passengers would smash a window, grab merchandise, and run to the car to escape. Police had a description, and following another incident located the car. The thief stopped at first, but then fled with the usual reckless driving. Officers from several jurisdictions stayed behind or tried to set up roadblocks, and the suspect drove around at least one roadblock. The officers generally drove little above the speed limit. The suspect ran a red light and killed two people. The thief was convicted of murder and was sentenced to life imprisonment. He had five different drugs in his body at the time. The appellate court affirmed a defense summary judgment. The police were not the proximate cause of the tragedy, and their conduct was well short of willful and wanton. The defendants made an interesting argument: if they were held liable here, all a fleeing suspect needed to do make an escape would be to driver recklessly, the more the better. *Nelson v. Thomas,* 668 N.E.2d 1109 (Ill. App. 1996).

When the pursued himself is the fatality, the incident does not register as high on the outrage scale as when the fatality is an innocent party. A seemingly simple 1992 case has some unusual facts. A small village hired an officer, the former mayor, as its sole, part-time police officer. The officer had no police training, but was given an unmarked car. The officer noticed a car with a dragging muffler that caused sparking, and attempted to stop the driver. The officer did flash his red light. The driver may have been a little over the speed limit. They circled a block three times and collided at one point. The driver stopped once but then sped off. The officer couldn't pursue because of a flat tire. The fleeing driver overturned on a sharp curve less than a mile away and was killed. An expert for the plaintiff argued that the officer's conduct was willful and wanton. He may have been wearing a black leather jacket and perhaps the deceased did not realize that he was a police officer. The offense, whatever it was, was trivial. The plaintiff also urged liability on the village for hiring an untrained officer who also had served time for a drug offense. The appellate court affirmed summary judgment for the defendants, finding little evidence of conduct that would defeat tort immunity. *Sank v. Poole,* 596 N.E.2d 1198 (Ill. App. 1992).

A 1997 case involved inconsistent verdicts. See the case for the full facts. Briefly, two plainclothes Chicago officers responded to a call of a red car possibly involved in gang activity. They came upon a Nissen, stopped it, and approached with drawn guns. The driver fled, running lights and so on. The officers pursued with their oscillating lights on but did not use the siren, which contravened pursuit policy. They thought that such use might cause drivers to stop suddenly and thus create a hazard, and they were not themselves running red lights. Eventually the red car crashed at a bridge, and the plaintiff was found severely injured on the sidewalk. Apparently she was hit by the car. The jury awarded $2 million to the plaintiff for willful and wanton conduct, with the city responsible for 20 percent and the fleeing driver 80 percent. The jury also found by special interrogatory that the officers' conduct was not willful and wanton. The trial judge resolved the contradiction by granting judgment notwithstanding the verdict to the city. The appellate court affirmed. The pursuit conduct was short of willful and wanton, and although department pursuit policy says to use the siren, there are situations when its use may not be advisable. The policy is evidence only, and cannot dictate the result in a case. *Morton v. City of Chicago,* 676 N.E.2d 985 (Ill. App. 1997).

One case was filed in federal district court with a claim of violation of the Fifth Amendment. An officer in East St. Louis attempted to stop a driver who made an illegal U-turn. The driver fled at high speed instead of stopping. The officer pursued with his spotlight and siren on. The pursued driver went airborne and crashed into a cement barrier. He had nine passengers in the car, including six children. One was killed and others injured. The plaintiffs filed §1983 claim alleging deprivation of life and liberty without due process. The district court granted summary judgment to the defendant officer and city, and the court of appeals affirmed. The court saw the plaintiffs' claim as alleging that the officer had a constitutional duty not to chase for a minor traffic offense. The court points out that the officer was trying to give notice, a traffic ticket, that would lead to a judicial hearing. (The driver may have been providing illegal taxi service and may have fled for that reason.) The court says that flight itself is a new offense and may well be more serious than the original reason for the attempted stop. The attempted stop was a cause of what happened, but not the culpable cause. The only person responsible here is the driver. The district court judge applied the shocks the conscience test, and a concurring opinion agreed that was the proper test. *Mays v. City of East St. Louis,* 123 F.3d 999 (7th Cir. 1997).

We conclude this section with a glance at fire and rescue cases. A 1994 case arose when a fire fighter on a call drove a fire truck through a construction zone. The truck hit a cone, and the cone then struck and injured a workman on the site. The plaintiff sued under the Fire Fighter Liability Act that makes fire departments liable for negligent operators of fire vehicles. The defendant asserted a later enactment, the Tort Immunity Act. That Act immunizes public employees from liability, except for willful and wanton conduct, while acting in the scope of their employment while responding to an emergency call. The question of which act applied was certified to the Illinois Supreme Court. The court agreed with the court of appeals in holding that the former act was repealed by implication. *Jahn v. Troy Fire Protection Dist.,* 644 N.E.2d 1159 (Ill. 1994).

A 1999 case held there was a triable issue on willful and wanton conduct. Three fire vehicles answered an automatic fire alarm in midafternoon in Springfield. The lead vehicle, driven by Forgas, approached the scene, the Urban League Building. Forgas noticed the occupants out of the building but saw no smoke. He entered the intersection on a red light, but with emergency lights and sirens sounding. His witness said he entered at 15 mph, while a disinterested witness guessed the speed at 40 mph. In other words, there was a typical range of estimates. The plaintiff did not notice the fire trucks and entered on green and was injured. The fire alarm turned out to be a false alarm, but the court held that there was a proper emergency. A dial-in call is an emergency call, and the mere fact that no smoke is visible does not convert the call to a nonemergency one. However, the court held that there was a genuine issue of willful and wanton conduct. (The court also held that the tort immunity statute applies, not §205 of the Traffic Code.) Forgas did proceed into a crowded intersection without stopping, perhaps as fast as 40 mph. There is a jury question. On another point, the plaintiff may introduce into evidence the fact that Forgas was cited for a traffic infraction, pleaded guilty, and paid a fine. *Young v. Forgas,* 720 N.E.2d 360 (Ill. App. 1999).

The Appellate Court for the Fourth District in *Young* above disagreed with a ruling by the Second District Court of Appeals which was decided in 1994. In that case, a passenger in an ambulance was injured in a collision. The ambulance was driven by a fireman, an employee of the City of Waukegan. The fireman entered an intersection on the red signal and was hit by a truck driver with the green signal. The emergency lights and siren were in use. A number of witnesses gave the ambulance's speed as in the 15 to 35 mph range. The court of Appeals affirmed summary for the defendants. The ambulance driver's conduct lacked the necessary reckless, deliberate indifference to the welfare of others to be held as willful and wanton conduct. (This and the case above have nearly identical facts, but a different holding. So the result depends on which judicial district in which the case is tried.) *Hampton v. Cashmore,* 637 N.E.2d 776 (Ill. App. 1994).

If an accident is caused by defective equipment, the fact that the equipment is being used on an emergency run is of no aid to the defendant. A 1993 case arose when a fire truck and another vehicle collided when the rear brakes of the fire truck failed. The court held that the immunity statute did not apply. Whether the fire department had actual or constructive notice of the leak was for the jury. *C.D.L., Inc. v. East Dundee Fire Protection Dist.,* 624 N.E.2d 5 (Ill. App. 1993).

Indiana

Ind. Code Ann. §34-13-3-3. Loses for which governmental entity or employee not liable

A governmental entity or an employee acting within the scope of the employee's employment is not liable if a loss results from:

(6) The performance of a discretionary function.

(7) The adoption and enforcement of or failure to adopt or enforce a law.

Ind. Code Ann. §9-21-1-8. Authorized driving procedures for person driving authorized emergency vehicle

[Sets out UVC §11-106 without change. See the Introductory Essay for text.]

[Section 34-13-3-3 is the Indiana Tort Claims Act, effective 1974. The Act marked a change in the standards used to evaluate and decide police emergency driving cases. Note that 34-13-3-3 does not specifically include fire vehicles on emergency runs. I have found no cases of fire trucks involved in accidents on the way to a fire, but several cases hold that a fire department has immunity in decisions made and methods used to fight fires. I cannot state that ambulances and fire trucks are or are not immune when on emergency runs. In any event, they would be subject to 9-21-1-8. Note also that, unlike Illinois above, the Tort Claims Act does not make an exception for willful and wanton conduct. Most states have a state claims board that aggrieved individuals may petition for payment of such claims.]

A 1966 case indicates how the law worked before the Tort Claims Act. A fireman on an emergency run collided with the plaintiff, with the latter having the green light. The court of appeals affirmed an award for the plaintiff. Section 9-21-1-8 required the fireman to slow as necessary for safety at a red or stop signal or sign and then proceed cautiously. Curiously, the court seemed to hold it reasonable for the plaintiff to proceed on green unless he heard the siren. Others heard the siren. *Clark v. Sterrett,* 220 N.E.2d 779 (Ind. App. 1966).

Fast forward to 1981. A state trooper patrolling on I-65 noticed a Nova in the opposing lanes speeding and also noticed that the car had a rear bumper higher than was lawful. The officer called an officer going the same direction as the suspect, and that officer stopped the driver. The driver failed to produce appropriate car documents and suddenly fled, almost hitting the officer. Now the officer had adequate reason to arrest the driver, and the chase was on. The pursuit continued on a state highway, and the officer struck an oncoming car whose driver suddenly attempted a left turn. The driver was killed. The issue was whether the officer was "enforcing" the law within the meaning of the Tort Claims Act. The trial court ruled he was, the court of Appeals reversed, and the supreme court reversed again. One dissent thought the decision was "astounding," and the other dissent asked if Indiana had returned to the rule of "the King can do no wrong." On rehearing, the majority stuck to its guns. The majority affirmed its finding that the statute was constitutional (I presume the limp "not unconstitutional" means constitutional). The majority did say that a state employee's acts *might* be so outrageous, or willful and wanton, as to be incompatible with the performance of the duty undertaken. *Seymour Nat'l Bank v. State,* 428 N.E.2d 203 (Ind. 1981) (Note that in a 1983 case, not involving police pursuit, the court of appeals held the language directly above describing conduct not within the scope of employment as dictum, and declined to follow it. *Jacobs v. City of Columbus,* 454 N.E.2d 1253.

One case alleging a §1983 violation was filed in federal court. A driver was killed in the early morning hours during a pursuit incident in which the pursuing officer reached speeds in excess of 100 mph. The deceased was an innocent bystander. The district court granted summary judgment to the defendant. The court noticed the "outrageous" language in *Seymour* but held that this was a simple tort case arising out of an automobile accident. The plaintiff had a remedy in state law. *Ellsworth v. Mockler,* 565 F. Supp. 110 (N.D. Ind. 1983).

A 1991 case, *Frye v. Town of Akron, 759 F. Supp. 1320* (N.D. Ind. 1991), was also brought in federal court. A police officer attempted to stop a motorcyclist who was carrying a passenger. The motorcyclist had crossed the center line several times, and this was the reason for the attempted stop. The motorcyclist fled at high speeds on winding roads at night. The pursuit ended when the officer's car hit the motorcycle. The passenger was killed in the crash. The trial judge held that the plaintiff had not shown that the officer made an unlawful seizure. However, the plaintiffs stated a claim for violation of substantive due process in alleging that the officer's acts were arbitrary, unreasonable, and reckless. The court also held that the plaintiff had stated a claim as to lack of training in pursuit policies, and denied the defendant's motion for summary judgment. *Frye v. Town of Akron,* 759 F. Supp. 1320 (N.D. Ind. 1991).

Crews v. Brockman, 510 N.E.2d 707 (Ind. App. 1987), arose from a "possibly violent" domestic disturbance. A Franklin police officer responded with lights and siren. There was an intersection accident, and the injured plaintiff sued. The allegation was that the officer was not enforcing the law at the time. (The trial judge held that there was a threshold question of whether the act was discretionary or ministerial. The court of Appeals did not address this point.) The plaintiff argued that the officer had not arrived at the disturbance, and thus was not enforcing the law. The appellate court found no merit in this argument, nor in the argument that §9-21-1-8 applied. Immunity presupposes liability. Also, the court had already rejected the willful and wanton concept in *Jacobs.*

A 1987 case decided whether an officer coming to assist another officer in a vehicle pursuit was enforcing the law. The trial judge thought not and declined to grant summary judgment. The trial judge also found a jury question of whether an officer's actions eleven blocks away from a chase, who is reaching speeds of 110 mph on busy streets in the early evening hours, might constitute conduct so willful and wanton as to take the officer out of his scope of employment. The court of appeals reversed and granted summary judgment. There is no question that the officer was enforcing the law in coming to join a chase. The court noted that the willful and wanton conduct language came from *Seymour*, and which is the case in which the supreme court held that immunity did apply in a police pursuit. *Indiana State Police Dept. v. Swaggerty,* 507 N.E.2d 649 (Ind. App. 1987).

A 1988 case arose over the immunity of Kentucky officials acting in an incident in Indiana. Kentucky police chased a suspect across the state line into Evansville, where Indiana police took over. The suspect rear-ended and killed an innocent bystander. The estate sued a host of Kentucky and Indiana entities. The defendants filed motions for dismissal for failure to state a claim on which relief could be granted. The trial judge granted all motions, and the plaintiff urged on appeal that his suit against Kentucky entities should be reinstated. The court of appeals affirmed the original decision. Kentucky was entitled to the benefits of the Indiana Tort Claims Act. Even if the police were negligent, they were immune. *Clement v. State,* 524 N.E.2d 36 (Ind. App. 1988).

A 1990 case sounds like simple negligence, if that. A county police officer responded to a call of a suspected suicide. He turned on his flashing lights and siren. (Question: what's the hurry?) The officer at some point turned off his siren to use his radio. He slowed to 25 to 30 mph to descend a hill in rainy conditions, but lost control at the bottom and crossed the center line. A passenger in an oncoming truck was injured. The plaintiff argued that, while the officer was enforcing a law, his conduct was outrageous. The court of Appeals listed the cases that have dealt with the troublesome language in *Seymour,* and held that the language did not alter the immunity granted in §34-13-3-3. There was nothing outrageous about the officer's conduct, and he was certainly within the scope of his employment. A concurring opinion hoped that in some future appropriate case, a jury could take up the *Seymour* language. *Carver v. Crawford,* 564 N.E.2d 330 (Ind. App. 1990).

In *Bevis v. City of Indianapolis,* 565 N.E.2d 772 (Ind. App. 1991), an officer going to a burglary scene was held to be enforcing the law and thus immune even though he was not using his lights and siren. The court of appeals held that this act was a reasonable one under the circumstances.

A 1991 case brought in state court alleged violations of the U.S. Constitution. The case involved a roadblock. The circumstances that provoked the chase are briefly given in the dissenting opinion. Apparently Roddel "drove through the park" around midnight in the town of Flora. The town marshal tried to stop Roddel, but he fled. Under state law, flight alone is a Class A misdemeanor, and flight that creates a substantial risk to others is a Class D felony. The police enlisted the support of county law enforcement officers, and a roadblock was set up. Roddel left the road to avoid hitting the police cars hit a tree. His left leg was severed, but nevertheless an officer handcuffed him to the steering wheel. The plaintiff alleged a §1983 violation, a claim based on tort liability, and a claim that several Indiana statutes were unconstitutional. The trial judge dismissed the claims, and the court of appeals affirmed. Citing the California case in the U.S. Supreme Court, *Brower v. County of Inyo,* 489 U.S. 593 (1989) (see the California section), as a "beacon of light," the appellate court found the §1983 dismissal to be proper. The plaintiff alleged only that the roadblock was unreasonable because the chase was unnecessary. Only if the roadblock had been designed to cause serious injury or death could a valid claim be stated. The court held the statutes cited by the plaintiff to be constitutional, including 9-21-1-8 (rights of operators of emergency vehicles). Finally, the comparative fault law does not apply, so the plaintiff's act of fleeing at speeds up to 95 mph was contributory negligence. *Roddel v. Town of Flora,* 580 N.E.2d 255 (Ind. App. 1991).

Guess what? The Indiana Supreme Court revisited the issue of law enforcement immunity in three cases decided October 25, 1993. The primary discussion is in *Quackenbush v. Lackey,* 622 N.E.2d 1284 (Ind. 1993). An officer received a report of a man hitting a woman in a parking lot. The officer did not use his emergency lights and siren. (That is a common police tactic used to approach a crime scene quietly, but then the officer must driver accordingly.) The officer ran a red light and collided with the plaintiff's vehicle. The court decided that the legislature did not intend to overrule, but rather to codify, the common law. Under the common law, even when eroded by governmental immunity, a public employee was not liable for all acts and omissions, but only for those involving the breach of a private duty owed to an individual. The court noted that 34-13-3-3(7) conflicted with 9-21-1-8, and that such conflicts should be harmonized, if possible. The court overruled *Seymour* by a 4-2 vote, and essentially held that 9-21-1-8 should be given priority when officers are responding to an emergency. A dissent said the court assumed improperly the duty of the legislature and that *Seymour* followed the clear and unambiguous language of the statute. The other two cases are *Fries v. Fincher,* 622 N.E.2d 1294, and *Belding v. Town of New Whiteland,* 622 N.E.2d 1291. These three cases are now the most authoritative in emergency vehicle cases in Indiana.

I mention an ordinary negligence case to show that police not operating in an emergency are in the same class as other drivers. A plaintiff attempted to convert a tragic accident into a federal constitutional violation. An off-duty police officer ran a red light and caused the death of another motorist. The estate settled with the officer for $250,000, but then sued the officer and the city of Indianapolis in federal court. The plaintiff alleged in part that the police attempted to cover up the true facts of the accident, and that they tended to ignore the more seriously injured or dying civilian driver to care for the officer, who had only minor injuries. That is, they denied medical care. The Seventh Circuit dismissed the case. The plaintiff failed to demonstrate that the officer was driving so recklessly (driving over the speed limit at night without lights, not being on an emergency run while running a red light) as to give rise to a §1983 action. The plaintiff would have to show not only reckless behavior, he would have to show that the officer knew the accident was imminent and that he consciously refused to prevent it. *Hill v. Shobe,* 93 F.3d 418 (7th Cir. 1996).

Iowa

Iowa Code Ann. §321-231. Authorized emergency vehicles and police bicycles

[Iowa has enacted UVC §11-106 (see Introductory Essay for text) with a few changes. The most significant change is in part (1). Instead of referring to "when in pursuit of an actual or suspected violator of the law," Iowa has changed the wording to "when in pursuit of an actual or suspected perpetrator of a felony, or in response to an incident dangerous to the public . . . " In other words, don't chase for a misdemeanor offense. But remember that if the flight itself is reckless, that may be a felony itself. See the Indiana section for a case on this point, *Roddel v. Town of Flora,* 580 N.E.2d 255 (Ind. App. 1991). If an actual or suspected violator of a misdemeanor simply refuses to stop, but continues driving and obeying all traffic laws, an officer should be able to get the tag number (assuming the vehicle has one), call for help to follow and eventually stop the suspect, or do both. But as soon as the suspect speeds recklessly or runs stop signs or signals, in a state with laws prohibiting fleeing, there is probably a felony offense and the police may pursue, even in Iowa.

The second change is in part (b). "Disregard laws or regulations governing direction of movement for the minimum distance necessary before an alternative route that conforms to the traffic laws and regulations is available." Don't drive the wrong way any longer than necessary. The third change makes it clear that a police bicycle is an emergency vehicle.]

There are few cases. We begin with a 1998 ambulance accident case. An experienced ambulance driver, Hinson, was taking a seriously ill elderly woman from Mediapolis to Burlington, Iowa. When Hinson arrived, he called the Burlington police dispatcher and requested traffic lights be set to flashing red in all directions on Roosevelt Ave. The dispatcher complied, and made the change in accordance with Hinson's location. Hinson had his warning lights and siren operating, and drove at 40 mph. The plaintiff, Bell, on Kirkwood Avenue, stopped and looked but could not see the ambulance because of traffic stopped in the left turn lane, and notwithstanding the fact that no one else was moving, drove into the intersection. She was hit broadside and was seriously injured. Hinson said that the intersection was clear as he entered it. The jury was instructed that it must find Hinson reckless for an award. The jury awarded over $800,000 less 25 percent for contributory negligence. The trial court then granted Hinson's motion for judgment notwithstanding the verdict, and the case went to the Iowa Supreme Court. The court held that the standard of reckless conduct is correct, and that the jury's verdict must be overruled. The court also held correct the trial court's exclusion of testimony from the plaintiff's expert, Randy Westfall, an expert in emergency vehicles. His testimony as to Hinson's conduct contained mixed questions of law and fact, and were properly inadmissible. *Bell v. Community Ambulance Service,* 579 N.W.2d 330 (Iowa 1998).

A 1998 ambulance case dealt with what might be called a poor drafting problem. Note that UVC §11-106(d), which is §321.231(5) in the Iowa Code, gives two standards applicable to emergency vehicle drivers: due care and reckless disregard. It is obvious that these standards are different. A look at *Words and Phrases* reveals, as seems obvious, that due care is simply an ordinary negligence standard of care. Reckless disregard is a conscious indifference to consequences. Many courts define reckless disregard as the same as gross negligence. The facts were similar to *Bell* above, with the ambulance on an emergency run and again entered an intersection on a red signal and hit the plaintiff's vehicle broadside. The trial judge did not find sufficient evidence of reckless conduct and refused to let that issue go to the jury. The judge allowed the case to proceed on the negligence count. The jury found each party 50 percent at fault, and the judge entered judgment in the amount of $63,598. The supreme court determined that §321.231 controlled.

The court had dealt with this problem in a police pursuit case, *Morris v. Leaf,* 534 N.W.2d 388 (Iowa 1995). In *Morris,* a suspect pursued by an officer collided with and injured another person. The Supreme Court held that neither the officer nor his employer owed a special duty to the plaintiff for the torts of the suspect. An officer can be liable only for his own reckless disregard for the safety of others. The "due care" language is controlled by the more specific reference of "reckless disregard." Any other holding would inhibit vigorous law enforcement. The court now applied this rationale to the ambulance case. It did not agree that the fact that the plaintiff in *Morris* was injured by a fleeing driver, not the officer, distinguished the case. The ambulance driver here must be given the same protection. The protection given emergency vehicle operators makes no distinction between ambulance and police vehicles. As for any conflict between "due care" and "reckless disregard," the court held that emergency vehicle operators are to drive with due care. However, "the threshold for recovery for violation for that duty is recklessness, not negligence." Since the plaintiff did not appeal the trial judge's ruling on recklessness, the court remanded for judgment for the defendants. The court overruled any cases to the contrary, particularly *Rush v. Sioux City,* 240 N.W.2d 431 (Iowa 1976). *Hoffert v. Luze,* 578 N.W.2d 681 (Iowa 1998).

There are a few older cases, but I believe they have been superseded by the holdings in the cases above. A short 1976 case dealt with whether a police car was on an emergency run. The officer noticed a driver who was speeding. He stopped her, and found that she was taking three sick children to the hospital. One child had a fever of 105 degrees; another had convulsed. The officer took the mother and children in his police car and received permission to take them to Cedar Rapids. He activated his lights and siren, and collided with the plaintiff. (Her claim was for a paltry $970.) The trial court erroneously determined that because an ambulance was available in Palo, the officer was not on an emergency run. The supreme court ordered a new trial. Whether the officer reasonably believed there was an emergency is a question of fact for a jury. *Hamilton v. Town of Palo,* 244 N.W.2d 329 (Iowa 1976).

A 1987 case arising in Sioux City was filed in federal court. The plaintiff was a motorcyclist. Around 2:00 A.M. a police officer watched the motorcyclist, Tagstrom, go through a stop sign without stopping and saw that the motorcycle's lights were not on. Tagstrom then lost control avoiding another vehicle and fell. The officer stopped and got out, but Tagstrom picked up the bike and took off. The officer pursued. The pursuit escalated, lasting eight miles through the streets on the west side of the city. Speeds were up to 80 mph. Six or seven cruisers were involved, and the motorcyclist evaded two roadblocks. The motorcyclist crashed into the back of a pickup and an officer crashed. Tagstrom appeared to be dead. He lived, and sued the officers and the city, alleging violation of the Fourth, Eighth, and Fourteenth Amendments. The district court judge questioned the wisdom of the pursuit, but noted that the issue was whether it was constitutional. The judge ruled that there was no seizure. The plaintiff voluntarily chose not to be restrained by fleeing when ordered to stop. The court ruled that there was no excessive force under the Due Process Clause. The officers were not attempting to run over the plaintiff to force him to stop; rather, they were matching his speed. The plaintiff could have eliminated the risk to himself at any time. The court denied the defendants' motion for summary judgment as to the officers on the final claim, that of deliberate indifference to the plaintiff's serious medical needs. The issue is weak, but it is a jury question. *Tagstrom v. Pottebaum,* 668 F. Supp. 1269 (N.D. Iowa 1987).

Kansas

Kan. Stat. Ann. §8-1506. Authorized emergency vehicles

[Kansas has enacted UVC §11-106 without significant change. Kansas added a section giving authorized emergency vehicle drivers permission to pass through tollgates without stopping, after slowing for safety. Kansas courts have decided the cases on the basis of §8-1506 only, so there are no issues of governmental immunity to complicate the matter. There are relatively few cases.]

The most recent case is a criminal one. A highway patrol officer at work on I-70 clocked a motorcyclist at 95 mph. The officer made a U-turn and, without activating his lights or siren, sped at 120 mph to catch up. There was no other traffic. The motorcyclist fled at an even higher speed, swerving from the inside shoulder to the center line to pass other traffic.

The motorcyclist exited to a hilly two-lane highway with a speed limit of 55 mph. The officer also exited to pursue. The motorcyclist pulled away at high speed through a residential area, and the officer lost sight of him. The officer came to an intersection, slowed but did not stop at his stop sign, looked both ways, and proceeded. (Another part of the case says the officer ran a red light.) He hit a car and that driver was killed. The motorcyclist later claimed that he was not aware that he was being pursued, and he was not under the influence of drugs or alcohol. Both the officer, Osland, and the motorcyclist were charged with involuntary manslaughter. Osland pleaded guilty for his failure to comply with §8-1506. (The case does not mention a civil settlement, but we probably can assume one was made.) The trial court dismissed the case against the motorcyclist on a finding that his conduct was too remote from the accident to be a cause. The Kansas Supreme Court reversed for trial. An officer has a duty to enforce the law, and that duty extends to hot pursuit of one attempting to evade arrest. In essence, the officer's actions were not so unexpected or abnormal as to insulate the motorcyclist from the causes of the accident. The court noted the usual rule that one need not foresee exactly how harm may come to another from his own acts to be liable, or here, guilty. *State v. Anderson,* 12 P.3d 883 (Kan. 2000).

The case above is unusual. No doubt there were sound reasons why the highway patrol officer pleaded guilty in this case. The court did note that because there was a guilty plea, it was not determined whether the officer's actions were merely negligent, reckless, cause for discipline, or discharge. Whether the officer had his lights and siren operating was not mentioned. If they were operating, the other driver has a duty to yield. In many states, the officer would not have been found at fault. Since there was a criminal charge, I presume the facts were in the outrageous category.

The Kansas Supreme Court distinguished a 1983 case, *Thronton v. Shore,* 666 P.2d 655 (Kan. 1983). In this case, two innocent bystanders were killed when the pursued driver collided with them. An officer on duty on the campus of Kansas University at 1:30 A.M. clocked a jeep traveling 35 mph in a 20-mph-zone. The Jeep driver committed numerous traffic infractions in his flight and ran a stop sign in downtown Lawrence. Both occupants of the other vehicle were killed. (The Jeep driver was later convicted of involuntary manslaughter.) An equally divided Kansas Supreme Court affirmed summary judgment for the defendant officer, but granted a rehearing. The court held 4-2, with two dissents, that the officer was not legally liable to the plaintiffs. The court cited a number of cases from other jurisdictions, and found that the majority would not impose liability in these circumstances. It noted that a common theme is that an officer has the right and the duty to enforce the law, and his conduct should not be judged only by hindsight. The fleeing driver who acted with recklessness is the cause. The dissents would reverse for trial. There simply were not enough facts on the record to show that the officer acted with "due regard" for the safety of others. Both the majority and dissenting opinion focused on the due regard standard rather than reckless disregard.

State v. Simpson, 732 P.2d 788 (Kan. App. 1987), found a St. George city police officer in pursuit of a fleeing vehicle guilty of reckless driving. The officer was in pursuit on Highway 24. It was dark, and the road was narrow and hilly. The suspect passed two cars in front. The officer then pulled out to pass in a no-passing zone on a hill and collided with an oncoming car. He was convicted of reckless driving. The appellate court affirmed. The officer conceded that a civilian who did the same thing would indeed be guilty of the charge, but asserted that as the operator of an emergency vehicle in a pursuit, he should not be held to an ordinary negligence standard. The appellate court found that he was not so held, but that his act of passing in the no-passing zone, on a hill, came within his duty to comply with the emergency vehicle statute. His reckless act fit the prohibition to drive with due care in the situation.

A 1993 case went to the Seventh Circuit Court of Appeals. An Overland Park officer noticed a driver, age sixteen, was weaving. The car had one headlight out. The driver fled at high speed instead of stopping. The pursuit lasted three minutes and ended when the driver hit a tree, killing himself and his passenger. The plaintiffs brought a diversity action against the police, alleging that the pursuit was in violation of several department policies on pursuits. The district court granted summary judgment to the defendants. The Seventh Circuit affirmed. The court noted that generally police officers do not owe a duty individually to members of the public. Two policies may have been violated: continuing the pursuit when the officer is entering an unfamiliar area, and not breaking off when the danger outweighed the immediate need to arrest the pursued driver. These decisions were at most negligent. The court followed *Thornton* above in finding the defendants were not the proximate cause of the deaths. A concurrence noted that the trend among the states is to find this to be a question for a jury. *Carl v. City of Overland Park,* 65 F.3d 866 (7th Cir. 1995).

A governmental entity has the right, of course, to bring suit against an erring driver. In a 1972 incident, a fire truck was on the way to a fire. The fire truck driver came to a stop sign (the intersecting street had the right of way). The driver looked east where he could see for 900 feet and saw no traffic. He looked west and proceeded into the intersection without again looking east. According to the driver and to several witnesses, he was going about 10 mph. His emergency lights and siren were operating. The defendant driver was heading east hit and hit the rear of the fire truck and rolled it. The fire district sued for property damage. The trial judge entered a verdict against both drivers. The Kansas Supreme Court reversed for a new trial. The trial judge erroneously instructed the jury according to the law as it existed before the 1970 amended emergency vehicle statute. The judge should not have instructed on ordinary negligence, but rather should have instructed on due care as given in the emergency vehicle statute. On another point, it was error not to instruct on the defendant's intoxication. *Shawnee Tp. Fire Dist. v. Morgan,* 559 P.2d 1141 (Kan. 1977). The plaintiff was a volunteer fire department and the driver was a volunteer fireman.

Note two cases in which the vehicle involved was held not to be an emergency vehicle entitled to statutory protection. A police motorcycle being operated on funeral procession duty was not in an authorized emergency and the operator is held to the same duties as civilian drivers. *Littell v. Malony,* 593 P.2d 11 (Kan. App. 1979). (This is the usual holding in other jurisdictions.) Similarly, a tow truck, normally a private vehicle, is not an emergency vehicle required to have its oscillating light operating when it is merely being used to tow a disabled car. If the tow truck were at an accident scene and was attaching the car, the light would be required. (A drunk driver hit the tow truck.) *Divine v. Groshong,* 679 P.2d 700 (Kan. 1984).

Kentucky

Ky. Rev. Stat. Ann. §189.930. Right-of-way to emergency vehicles

[Kentucky is the exception to the rule. It has substantially adopted UVC

§11-106. Subsections (1) to (5) of the above statute set out the duty of drivers to yield to emergency vehicles. Subsection (6) states: "This section does not operate to relieve the person who drives an emergency vehicle from the duty to operate the vehicle with due regard for the safety of all persons using the highway." Thus, a due care, not reckless disregard, standard is applied to emergency vehicle drivers. Section 189.920 has a rather detailed description of emergency lights and audible devices required on emergency vehicles, and one with a case would want to check that statute.]

There are hardly any cases in the courts. A 1995 case has scant facts. A state trooper apparently was proceeding on an emergency run, a burglary call, with his flashing blue light operating but not his siren. He came to a sharp curve and veered across the center line and collided with the plaintiff, injuring both him and his granddaughter. The trooper denied speeding, but the court comments that in any event his speed was too fast to negotiate the curve. The plaintiff first elected to file a claim with the Commonwealth Board of Claims, which is a remedy available in many states. The statutory limit for that board is $100,000. Apparently the plaintiff decided to trust his fate (as the court put it) to a jury rather than the administrative agency with the limitations it presents, and he abandoned that claim and filed in the courts. The jury awarded the driver $70,000 for lost wages, $260,000 for lost earnings, past and future, and $100,000 for pain and suffering. The appellate court made some minor modifications in no-fault benefit set-offs, which are not pertinent here. The court held that the trooper was exercising a ministerial, not a discretionary, function and cannot claim immunity. It was proper not to instruct on the plaintiff's duty to yield to an emergency vehicle because there was no evidence that the plaintiff could have seen the flashing blue light around the curve at a distance of 500 feet, as required by statute. In many jurisdictions, the trooper would be given some leeway for driving in an emergency, but here he was held essentially to an ordinary negligence standard. *Speck v. Bowling*, 892 S.W.2d 309 (Ky. App. 1995).

Another 1995 case was brought in federal court. An off-duty Fayette County deputy sheriff noticed a driver squealing his tires and speeding. The deputy suspected the driver was intoxicated. The deputy notified another police officer and a short chase ensued. The suspect ran a red light and collided with another vehicle. A pedestrian on the sidewalk was hit and killed. (The driver was found to have a blood alcohol of .12. He was convicted of wanton murder and given 30 years.) The deceased's family filed a §1983 action. Specifically, the plaintiff alleged that the officers conducted the pursuit without emergency lights or sirens, and that such failure to use the equipment "was an arbitrary abuse of police power which shocks the conscience." The district court analyzed a number of substantive due process cases, and primarily followed a Sixth Circuit case, *Nishiyama v. Dickson County,* 814 F.2d 277 (6th Cir. 1987). That case established actionable police conduct as a combination of reckless indifference and intentional conduct, subsumed under the heading "gross negligence." The court held that the police conduct here fell well short of that standard. The court did dismiss the plaintiff's state claims without prejudice to her right to file those claims in state court. *Smith v. Lexington Fayette Urban County Government,* 884 F. Supp. 1086 (E.D. Ky. 1995).

A 1992 case involved a police officer and resulted in a large award for the plaintiff, but it is not a true emergency vehicle case. The officer was patrolling on I-64 when his radio reported an accident ahead at the east end of Cochran Tunnel. The "run" was not specifically assigned to the officer, but he told the dispatcher he would go by the scene to help if necessary. He went through the tunnel and began to slow down and pull onto the left shoulder. The plaintiff driving behind did not notice his slower speed in time and rear-ended the police car. The jury awarded substantial damages and found the defendant 95 percent at fault. The appellate court affirmed. The defendant, in disregarding certain traffic regulations, must be especially aware that other drivers do not expect such disregard of regulations. There is no mention whether the officer activated his emergency lights. *City of Louisville v. Maresz,* 835 S.W.2d 889 (Ky. App. 1992). I would have to say that the jury was rather generous in finding the plaintiff only 5 percent at fault. A following driver noticed the officer's brake lights on as he slowed, but the plaintiff's only came on just before the crash. There appears to be violation of assured clear distance and proper lookout.

A 1972 case was determined based on a pedestrian's duties when crossing a street at other than a crosswalk, rather than by emergency vehicle privileges. A pedestrian crossing in midblock did not look the "wrong" way for traffic. The pedestrian had crossed the two northbound lanes in stopped traffic and apparently did not look to his left before stepping into the southbound lanes. He was killed by a police officer who had been dispatched to an accident scene and was going the wrong way to get around stopped cars. The officer was using his emergency lights. The appellate court affirmed a jury verdict for the defendant. Without deciding whether a pedestrian has a duty to look the wrong way, the court held that at most that was a jury question. The deceased breached his duty to use ordinary care. It was proper for the trial judge to refuse to instruct that the deceased had a duty only to look the direction from which traffic could be expected. *Wright v. City of Louisville,* 477 S.W.2d 146 (Ky. App. 1972).

I have encountered the scenario above several times in coauthoring another book, *Pedestrian Accident Reconstruction and Litigation* (Lawyers & Judges Publishing Co. 1994). No court is likely to rule that a pedestrian is or is not required to look the wrong way when crossing a street. Since writing the book, I have become much more aware of the dynamics of pedestrian accidents, and I always look the wrong way. Others may reasonably assume that the chances of a wrong-way driver are too remote. This question is a classic jury issue, and we cannot expect unanimous holdings.

Louisiana

La. Rev. Stat. Ann. §32:24. Emergency vehicles; exceptions

[Louisiana has adopted UVC §11-106 (see the Introductory Essay for text) with one change. Section 32:24(C) states that a driver of an emergency vehicle on an emergency run is entitled to the protection of the statute only when making use of audible *or* visual signals, not both as given in the UVC section. This is a common change made by state legislatures. It recognizes there are times when a police officer may want to approach a particular driver or a crime scene quietly. Louisiana has many cases, and we will focus on a dozen or so of the more recent ones. It is interesting that the courts rarely cite the statutory language, and primarily refer to due care . They do not mention the reckless disregard language, whereas Iowa, for example, makes a distinction between the two.]

A 1993 case involved a roadblock. A police officer noticed a motorcyclist leaving a bar's parking lot and in the process commit numerous traffic violations. The motorcyclist fled when the defendant officer attempted to stop him. The chase lasted thirteen miles, often at high speeds. The officer called ahead for assistance, and another officer parked his car across the traffic lane at an intersection. The chase ended when the suspect hit the rear of the car. He died from his injuries. The estate filed a diversity action in federal court, alleging wrongful death under state law as well as filed a §1983 claim. The jury reported a deadlock. The judge then submitted several interrogatories to the jury. It answered "no" as to whether the officer backed his car into the deceased's path. It answered "no" as to whether the deceased was negligent. The judge found no conflict and entered judgment for the defendants. The court of appeals affirmed. It found summary judgment proper on the federal claim, following the primary roadblock case, *Brower v. County of Inyo,* 489 U.S. 593 (1989). See the California section. The deceased was not "seized" because the defendant, the pursuing officer, did not even know of the roadblock—if that was what it was. The wrongful death claim depended on whether the pursuing officer used excessive force, and the court held that was not established. The court also affirmed the trial court's decision not to admit the testimony of the plaintiff's expert witness. He had last testified as an accident reconstructionist in 1965, and admitted himself that his scientific knowledge of physics and math was deficient. As for the jury interrogatories, they were not irreconcilable. There is no "rule" that anyone has to be found at fault in a negligence action. *Rosado v. Deters,* 5 F.3d 119 (5th Cir. 1993).

Sometimes whether there is a pursuit is not even relevant. A police officer responded to a disturbance call, and got a description of the individual and his car. The officer noticed a gray car ahead and tried to get close enough to see if the driver matched the description. The officer did not use lights or siren. The suspect rear-ended a car stopped for a red light. At that point the officer switched on his overhead lights and undertook traffic control, while a state police officer investigated the accident. (The suspect later pled guilty of DWI.) The officer was three car lengths behind when the crash happened. The occupants of the car hit in the rear sued the original officer. The appellate court affirmed summary judgment for the defendant. Whether the officer was pursuing was immaterial. The accident was caused by the drunk driver who was fumbling with a cassette tape. *McCarter v. Parker,* 701 So. 2d 1084 (La. App. 1997).

Another 1997 case involved a real pursuit. A police officer noticed a car being driven in a suspicious manner (it was a stolen car) and attempted stop the driver. Although surrounded on three sides by police vehicles, the driver fled by driving into the opposite traffic lane, striking another police car. The suspect drove around three roadblocks, using the sidewalk at the last one, and continued to travel at high speeds with his lights off. He got caught in a cul-de-sac and drove between two houses, hitting a pedestrian and then fled without stopping. The pedestrian had been on the porch and fled in the belief she would be safer. An officer then drove between the houses, thinking the four occupants of the car had abandoned the car and were now on foot. The officer did not see the pedestrian on the ground and also struck her. The pedestrian died from massive trauma. The estate sued the suspect and the police. The jury awarded damages and found the city liable for 40 percent, and the thief liable for 60 percent (lots of luck there). The court of appeal reversed as to the city. The thief is the only one responsible for the harm. The thief engaged in increasingly reckless and dangerous driving to elude the police that presented an extreme danger to the public. The police cannot be faulted for attempting to catch him. *Hayes v. City of Baton Rouge,* 683 So. 2d 899 (La. App. 1997). I will comment that this kind of case presents a problem for the principle, frequently set out in pursuit policies, to end the chase when it becomes too dangerous to continue. The suspect had run three roadblocks, nearly hit other pedestrians, and was a menace to public safety. In hindsight, they should have let him go. This time the chase had a tragic end. No one knows how many times a pursuit of a dangerous driver is successful and it is not possible to know what harm might have been prevented thereby.

In tort cases, the Louisiana Supreme Court has adopted a policy to analyze duty versus risk. To prevail, a plaintiff must prove that the conduct in question was a cause-in-fact of the resulting harm, that the defendant owed a duty of care to the plaintiff, that the duty was breached, and that the risk of harm was within the scope of protection afforded by the duty breached. A 1999 case involved a chase in which an innocent person was injured by the suspect. A citizen called the police about a suspected car thief, a man the officer had known all his life, and gave the location. The suspect was identified, and a chase followed. The police suspected the thief was armed. Speeds were high at times, but the officers crossed intersections more carefully than the thief. The thief was about .25–.5 mile ahead when he hit an innocent driver at an intersection. The case gives several pages of facts. The plaintiff sued a number of parties, including police officers, the supervising sergeant, and the city. The trial judge found the suspect solely at fault. The appellate court affirmed. The court said that generally an officer must act reasonably, and operate an emergency vehicle with due care. However, the officer's method need not be the "best" method among alternatives, as long as it is reasonable under the circumstances. Here, a roadblock left a path for the suspect to get through; an example of reasonable conduct. The plaintiff's expert suggested a number of better actions that the police may have taken, such as cordoning off an area. The court went through the allegations against each officer and found no liability. The expert who argued that pursuit training of the officers was deficient, himself a police officer, admitted that he was not qualified to teach it. *Lavine v. Jackson,* 730 So. 2d 958 (La. App. 1999).

Two bicycle cases had different outcomes. A bicyclist, of course, is subject to the traffic code (excepting those provisions that can have no application). This includes the duty to yield to an emergency vehicle. A bicyclist, riding at night with no lights, reflective clothing, or lighting, and intoxicated to boot, crossed an intersection in front of a police car on an emergency run with emergency lights and siren operating. He was found by the appellate court to be solely at fault for his frightful injuries. The officer was driving 58 mph and was on a major street. The court reversed a jury verdict finding the city 30 percent at fault. *Lawrence v. City of Kenner,* 761 So. 2d 796 (La. App. 2000). In the other case, a bicyclist left his job at a fast food restaurant in the early morning hours and was hit by a police car. The officer was responding to a suspicious person call (it was a burglary in progress at a pawnshop about four miles away and other officers made arrests). Thc officer was speeding 60 mph in a 40-mph-zone, did not use his siren, and had a blue strobe flashing light on his dashboard only, and rounded a curve just before the accident. The appellate court reversed judgment for the defendant, and awarded the plaintiff $300,000 for his frightful injuries. The court notes that an officer proceeding to a burglary scene silently must do so with due care and ensure that his way is clear. *Butcher v. City of Monroe,* 737 So. 2d 189 (La. App. 1999).

A common traffic accident pattern is when a driver enters an intersection on a green signal, attempts a left turn, and is hit by an oncoming vehicle. This was the pattern in a 1993 case. The oncoming vehicle was a police car going 50 mph in a 35-mph-zone, answering a burglary call, and having a flashing dash light. The appellate court affirmed dismissal of the complaint. The officer was entitled to exceed the speed limit under the circumstances, and his use of the dash-mounted red flashing light complied with the obligation to display an emergency visual signal. The court notes that the plaintiff (who died of a heart attack a few days after the accident) was 83 years old. She did not stop to see if the way was clear for a left turn, and failed to see what was in plain sight. *Smith v. Commercial Union Ins. Co.,* 609 So. 2d 1024 (La. App. 1993). I will comment that the outcome in tort cases depends largely on the facts. If the officer had been going 75 mph, for example, it would be difficult not to find him lacking in due care.

A 1978 case is one in which the officer was speeding 20 mph over the speed limit. That does not sound outrageous, but the speed limit was 70 mph. The accident happened at night in a rural area with little traffic. The officer was chasing two speeders, and did not turn on his dash-mounted flashing light or use his foot-operated siren. His practice was not to use these until he had closed with the speeders to signal them to stop. The plaintiff was an eighteen-year-old man who was crossing the favored highway. The driver pulled into the median, stopped, and then tried to cross the other two lanes. The officer said that the driver pulled out about 150 feet in front. The officer's only warning was to flash his high/low beams. The trial judge found that both drivers were negligent. The appellate court ordered judgment for the plaintiff, and awarded $20,000 to each parent for wrongful death. (These sums would be considered very low today for wrongful death, and before ballooning awards well beyond economic damages.) A dissent didn't agree. The officer, who braked hard, was slowing from 90 mph, and the civilian driver had stopped in the median and should have looked. When one sees a vehicle on the favored street coming fast and flashing the headlights, that should be enough notice. *Moore v. Travelers Indem. Co.,* 352 So. 2d 270 (La. App. 1978).

Failure of a police officer to exercise due care can be demonstrated several ways. In a 1964 case, an officer during the early morning hours noticed a driver run a red light and speed. The officer activated his siren pursued at speeds reaching 75 mph. He was following the suspect at only 15 feet, and ran into the rear of the suspect's car. The court notes that one purpose of a siren is to tell the pursued driver to stop, and tailgating then is doubly dangerous. The suspect driver was not even sure that he was the target of the siren. The appellate court affirmed a small award to the passenger in the car. *Fant v. Zurich Ins. Co.,* 160 So. 2d 443 (La. App. 1964).

A 1981 case similarly held that an officer responding to a call of a house fire did not exercise due care for the safety of all persons, as required in the emergency vehicle statute. The officer proceeded on the favored highway at, according to one witness, 50 to 60 mph. The plaintiff had stopped at his stop sign, looked, and then began to pull out to make a left turn. He heard the siren and saw the police car and stopped, somewhat in the traffic lane, and was hit. The appellate court affirmed an award for the plaintiff. The officer sped on a major road where heavy cross traffic could be expected. (A point not mentioned in the case is that a police car is not a fire vehicle. While the officer might be of help in traffic control at a fire, he is not rushing to put it out.) *Knecht v. Hartford Acc. & Indem.*, 408 So. 2d 1147 (La. App. 1981).

A 1986 case held an officer's conduct in an emergency to be reasonable. This is a pedestrian dart-out case, and as a rule dart-outs do not prevail in negligence actions. A state trooper responded to a call of an overturned truck. The truck driver, our plaintiff, was later found to be drunk. The officer approached the accident scene, along with a sheriff's car and an ambulance, and slowed below the 50-mph-speed limit. The plaintiff was running along the shoulder and suddenly darted in front of the trooper. In spite of the massive injuries to the plaintiff, the appellate court affirmed a dismissal. The officer's conduct could not be termed to be reckless disregard for the safety of others. *Puearry v. Dept. of Public Safety,* 496 So. 2d 1372 (La. App. 1986).

A 1989 case dealt with the issue, in part, of whether an ambulance was on an emergency run and thereby entitled to the protection of the statute. The court held that the ambulance driver was at fault, whether he was on an emergency run or not. The patient's relative called for an ambulance to take the patient to a hospital. Neither the doctor nor the relative believed there was an emergency. The jury was entitled to believe evidence that this was so. Even if there had been an emergency, the defendant did not exercise due care. He approached an intersection controlled by flashing yellow caution lights above the speed limit of 50 mph (obviously drivers are expected to slow below that speed at the intersection) and hit and killed a driver who was making a left turn. The ambulance had its emergency signals operating, and attempted to pass the deceased as they entered the intersection. The defendant argued that the deceased slowed and appeared to move to the shoulder, and at this point the defendant sped up. *Keating v. Holston's Ambulance Serv.*, 546 So. 2d 919 (La. App. 1989).

A 1998 case arose when a volunteer fireman responded to a fire in a rural area. He was an unpaid volunteer for the Fort Pike Volunteer Fire Department, a nonprofit corporation. He drove to the scene in his own vehicle. It was after dark, and there was smoke and a high wind. The volunteer struck a pedestrian, a resident of one of the "camps," when he attempted to park off the road. A bench trial found the fireman and Fort Pike liable. The plaintiff contended on appeal that the New Orleans Fire Department was also liable. The court agreed and amended the judgment to hold the city vicariously liable. Since the NOFD supplied Fort Pike with equipment and supplies, and trained its volunteers, there was a master-servant relationship between the City and the volunteer fireman. The court pointed out that the result would be different if Fort Pike supplied its own equipment and controlled its own actions. *Matlock v. Hankel*, 707 So. 2d 1016 (La. App. 1998).

I note three cases in which the vehicle is not an official public emergency vehicle. During a labor dispute, a union member threw a paint ball at the car of a union opponent, and the latter chased the union member to learn his identity but hit another car. Both drivers were responsible for an injury to the thrower. The former had the right to chase, but the duty to do so without harm. *Willis v. Letulle,* 597 So. 2d 456 (La. App. 1992). An individual in a bar came to the aid of another who fell unconscious and was thought to have suffered a heart attack. The rescuer was driving the person to the hospital and ran a red light while sounding his horn and holding a white handkerchief out the window. The court lauded him for his motives, but held him liable for damages on a finding that his behavior was not reasonable. *Herbert v. Perkins,* 200 So. 2d 15 (La. App. 1972). Finally, a tow truck was given the rights of an emergency vehicle in a 1974 case. The operator had been called to an accident scene by the police and was told to get there as soon as possible. The operator turned on his emergency lights and approached the site. Traffic was backed up, so the operator turned into the opposing lane and collided with the plaintiff. The plaintiff was found negligent himself for not yielding. The court found that in the circumstances the tow truck was an authorized emergency vehicle. *Prather v. Gautreaux,* 297 So. 2d 439 (La. App. 1974).

It is noteworthy that none of these cases went to the Louisiana Supreme Court. Overall, the Louisiana emergency law is uncomplicated, and the cases are decided on due care and reasonableness.

Maine

Me. Rev. Stat. tit. Ann. 29-A, §2054(5). Exercise of privileges

[Section 2054 includes the definition of emergency vehicles of all kinds and the equipment that is required. Subsection (5) sets out UVC §11-106 without change. This includes requiring the use of both visual and audible signals when on emergency runs and the duty to driver with due care. The operator will not be protected from the consequences of his reckless disregard for the safety of others.]

The only identified case is from 1968, *MacDonald v. Hall,* 244 A.2d 809 (Me. 1968). The case is very unusual. A deputy sheriff was killed in an accident while pursuing the defendant, and the estate sued for wrongful death. The sheriff's department had a felony warrant out on the defendant; something to do with serving papers in a divorce action. Twice before the defendant had successfully fled the officers, and he told someone that "he wasn't going to let them take him that evening." Two deputies saw the defendant, Hall, and gave chase. The pursuit lasted less than three miles. Hall ran a stop sign, went off the road, hit a sign post, regained the road and went even faster. At the next intersection, Hall made a left turn and then braked and fishtailed. This caused the officer to skid, lose control, overturn, and die. The jury found for Hall. The Supreme Judicial Court remanded for a new trial. At trial, the jury asked the judge if it should consider any evidence or events previous to the fatal pursuit, and the answer was no. The supreme court held that such evidence and events should have been considered. Hall had a duty to stop. Since he chose to flee, he must anticipate that his actions, considering the duty of the deceased to perform his duty, would put the deceased in danger. He should anticipate that a pursuing officer may very well "follow too closely" in attempting the arrest.

Maryland

Md. Code Ann., Transp. §21-106. Emergency vehicles

[Maryland has adopted UVC §11-106 without significant change. Use of audible and visual signals are required, except that an emergency vehicle operated as a police vehicle need not be equipped with emergency lights. (I presume this would not affect the use of portable emergency lights that often are mounted on rooftop or dash.) Subsection (d) follows the UVC in requiring emergency vehicle operators to drive with due regard for the safety of all persons, but omits the language about reckless disregard. Section 19-102(a) directs a police officer not to order a driver of a nonpolice vehicle to participate in a roadblock, and (b) states that if a police officer so orders, the state or political subdivision will be liable for any damages or injuries. I presume this means that an officer may accept volunteer help from the public, but any injury to them will be covered (to the limits set in §17-103, see below).

Section 5-639 of the Courts and Judicial Proceedings article is entitled, "Negligent operation of emergency vehicles." Subsection (b) states that an authorized operator of an emergency vehicle is immune from suit in his individual capacity except for malicious acts or gross negligence. The owner or lessee of the vehicle is not liable for the operator's malicious or grossly negligent acts. Subsection (3) states that a political subdivision may not raise the defense of governmental immunity in an action against it under this section. Subsection (d), limitation of liability, caps the maximum liability for the operation of an emergency to the limits of the state's compulsory insurance liability law, as set out in §17-103 of the Transportation Code. Currently the minimum required is $20,000 for one person, and $40,000 per accident. Normally a governmental unit is considered a deep pocket, but this kind of limitation or cap on damages is very common. This one is on the low side. If the entity carries a higher coverage, that of course is available.]

I understand that Maryland is unusual in that its largest city, Baltimore, has a no-chase policy for its police department. There are few cases, so we will look at all of them. A 1947 case seems unusual in today's terms, but it applied the emergency vehicle law. The emergency vehicle was "Big Bill," the "trouble truck" of the electric street railway system. The trouble truck was dispatched to problems in the system to make repairs. The case arose when the truck was sent on a call. The driver switched on his emergency lights and siren. He looked the "right" way at a red light protecting a one-way street, and entered the intersection at no more than 20 mph (the truck was rarely driven above this speed). The plaintiff's car entered on the green signal and they collided. The jury awarded the plaintiff passenger $950 for her injuries. The court of appeals reversed without ordering a new trial. The court held that the truck was a bona fide emergency vehicle. The operator could disregard certain traffic laws if he exercised due care, but would not be protected from his reckless disregard (note that this latter language is not in the current law). The court found that several witnesses heard the continuous sounding of the siren, and that the plaintiff's driver entered the intersection "completely oblivious" to the truck. The court noted that the driver sometimes used a hearing aid (the driver was 69) and had been irritated by slow drivers in front of him. Since he was solely responsible for the accident, his passenger could not recover damages. *Baltimore Transit Co. v. Young,* 56 A.2d 140 (Md. App. 1947).

A 1959 case involved fire truck versus a fire truck. It's bound to happen sometime: two fire trucks responding to an alarm collide. A Baltimore fire department truck on the way to a fire with all emergency signals operating, entered an intersection on a red light and collided with a Fire Insurance Salvage Corps truck entering on the green. Both had the privilege of the emergency vehicle statute. The court of appeals agreed that the city truck driver was the sole cause. The loser argued that the court should adopt the (then) "California rule," which supposedly holds that an emergency vehicle driver cannot be liable for ordinary negligence in disregarding rules of the road. The court held instead that the due care standard means the driver can be charged with negligence within the scope of the emergency vehicle law: run the red light but do it carefully. *City of Baltimore v. Fire Ins. Salvage Corp,* 148 A.2d 444 (Md. App. 1959).

A 1961 case is a man bites dog situation. One thing that will understandably arouse an officer's ire is the appearance that someone has deliberately injured a brother officer. Two police officers, at 10:30 P.M., had stopped two traffic law violators. A car came up at high speed and one officer attempted to get the driver to a stop by waving his flashlight. The car's speed increased and the officer leaped into the ditch to save himself. The other officer thought he had been hit, and took off after the speeder with a vengeance. (It turned out that the suspect was driving on a revoked license and wanted to escape.) The officer pursued up to 100 mph in a residential area, and thought he had the culprit trapped in a cul-de-sac. The suspect suddenly tried to make a left turn. Both cars skidded long distances, and there was a minor collision. The officer, however, was permanently injured and had to retire. The judge at a bench trial awarded damages to the officer, and the court of appeals affirmed. Without deciding whether the officer had a duty of care to the suspect, the court found that he had acted with due regard for the safety of others. The suspect had turned off his car lights and fled at high speed in a residential area. The officer could reasonably think that the suspect was a hit-and-run driver and posed a continuing danger to the public and should be stopped. The court declined to reverse the trial judge. *Martin v. Rossignol*, 174 A.2d 149 (Md. App. 1961).

In a 1968 case, an accident victim had a bad day: he was in two accidents. He had been picked up by an ambulance and was being taken to a hospital with the appropriate signals working. The ambulance approached a string of traffic, and all but one driver pulled off to the right. The one that didn't then stopped for a red light, and was rear-ended by the ambulance. The ambulance driver said he was braking gently so as not to shake up the injured passenger and planned to make a left turn. The driver in front said he made a normal stop and sat for some time before the collision. (The driver was the plaintiff in this case.) The appellate court ordered judgment notwithstanding the verdict for the plaintiff and ordered a trial for damages only. The defendant was negligent as a matter of law. *Altenburg v. Sears,* 239 A.2d 569 (Md. App. 1968).

A 1970 case brought up the situation of a tow truck operating as an emergency vehicle without using his air horn as an audible signal. The truck had been called to a four-car pileup. As the driver came to scene, he moved to the right shoulder to get around stopped traffic. His emergency lights were operating. The plaintiff, stuck in traffic, decided to call his office and suddenly pulled onto the shoulder to go to a nearby phone booth (no cell phones in 1970). The court of appeals reversed judgment for the plaintiff car driver. It was error to hold that the failure of the tow truck operator to sound his air horn was negligence per se. The court ordered a new trial. In addition, there was a question of whether the defendant violated the law prohibiting passing on the right under these circumstances. *Sullivan v. Costanza,* 267 A.2d 87 (Md. App. 1970)

A similar issue arose in a 1988 case. Is a police officer driving in an emergency situation still covered by the statute even though he is not sounding his siren? A Baltimore police officer, who received a report of "officer down," hit and killed a pedestrian who had been participating in a labor demonstration in the area. The widow sued the officer and the city for negligence. The trial court granted summary judgment to the defendants. The court of appeals affirmed. Citing *Sullivan* above, the court held that failure to sound the siren (defendants disputed this) does not take the officer out of the emergency vehicle statute, it merely means that he must drive in accordance with all the rules of the road, the same as any other driver. The defendant had also argued the siren issue was not relevant to causation. Since the police car was an emergency vehicle, the driver is immune under §19-103(b)(1). The police car is an emergency vehicle by statute. The officer was driving it in an emergency. Next case. *Taylor v. City of Baltimore,* 549 A.2d 749 (Md. 1988).

Moving up to 1991, we have another serious case. Around noon a Maryland state trooper noticed an erratic driver and suspected intoxication. Both drivers stopped at a red signal, and the officer got out and walked up to the suspect. He told the driver to pull off the road when the signal changed. Instead the driver fled at high speed on the green light. Deputy sheriffs and other troopers joined in the chase. The pursuit went on for seven miles. The plaintiffs claimed that the traffic was heavy. The pursuit ended when the suspect hit the rear of another car. One occupant of the rear-ended car died at the scene, the other several months later. The parents sued the state, the state trooper, Charles County, and its sheriff's department. The trial court granted summary judgment to the defendants. The court of appeals (Maryland's highest court), made a number of findings.

The state trooper, Officer Titus, who originally told the suspected intoxicated driver to pull off the road, did not violate §16-205.1(b)(2), which in part requires an officer to "detain the individual." Putting aside the question of whether telling the suspected intoxicated driver to pull over is not an attempt to detain, Maryland law does not make such officer liable to innocent third parties later injured by the suspect. (This is the typical rule among the states.) The court agreed with the trial judge that Titus is in any event immune under transportation code §19-103. Titus was not grossly negligent so that exception does not apply. His employer, the state, may be liable for Titus' negligent pursuit. The court held that the decision to pursue and the conduct of the pursuit need not be grossly negligent for state liability to apply. Maryland requires due care in such police actions. The court cautioned that the trier of fact must not apply a hindsight view, but must consider the situation the officers were evaluating at the time. "The officer is not to be held to the same coolness and accuracy of judgment of one not involved in emergency vehicle pursuit." The court commented that there are no hard and fast rules in this area, but that each case should be decided on its own facts. The court found that the plaintiffs failed to allege ownership of the county vehicles in their complaint and failed to raise the matter in oral argument. Nevertheless, the plaintiffs would be permitted to amend their complaint as to the county. *Boyer v. State,* 594 A.2d 121 (Md. 1991)

A 1991 case involved a Rockville police officer responding to an emergency call from another officer. The officer switched on her emergency lights, but not her siren, and entered an intersection on the red light. She collided with the plaintiffs who had the green light. The plaintiff alleged that the officer approached the intersection at 50 mph in a 25-mph-zone, and that the officer intentionally ran the light. The plaintiffs claimed that the officer saw or should have seen the plaintiffs' vehicle. The plaintiffs alleged that the officer acted with gross negligence in intentionally running the light and in not sounding her siren. The trial judge granted the officer's motion to dismiss for not stating a cause of action in gross negligence. Note that the plaintiffs had to allege gross negligence because Transp. Code 19-103 grants immunity to emergency vehicle operators except for gross negligence. The Court of Special Appeals, after examining several similar cases, affirmed the dismissal of the officer. *Taylor* above held that an emergency vehicle driven silently is still an emergency vehicle entitled to its statutory privileges. Gross negligence in the context equates to wanton or reckless disregard for human life. As to the liability of the city of Rockville, the court affirmed that its liability is limited to its self-insurance limit of $20,000 per claimant and $40,000 per event, not $200,000/500,000 as set out in the Local Government Tort Claims Act, §5-403. *Khawaja v. City of Rockville,* 598 A.2d 489 (Md. App. 1991).

One plaintiff filed an action in federal court in a 1991 case. A Frederick County deputy sheriff, on duty at 10:00 P.M., saw a driver peel out of a gas station, and gave chase. The station attendant called 911 to report that $17.00 worth of gas had been stolen. The chase took place on Route 355, which is a narrow, two-lane highway with varying population densities. Also, a carnival was taking place in the area, and there were cars parked along both sides of the road. Speeds were high. The suspect lost control on a curve, crossed the center line, and hit an oncoming driver. The officer braked hard, lost control, also went across the center line and hit the plaintiff broadside, inflicting serious and permanent injuries. The plaintiff filed a §1983 action. Such action must allege that (1), the officer was acting under color of state law, and (2), the plaintiff was deprived of rights, privileges, or immunities secured by the Constitution or laws of the United States. The action of the complaint must be so outrageous that no state remedy is sufficient to preserve the individual's constitutional rights. The district court granted summary judgment to the defendants. The Fourth Circuit affirmed. The proper test is whether the state act shocks the conscience. The court surveyed police pursuit cases from several other circuits. The court noted that various standards have been used: gross negligence, deliberate indifference, shocks the conscience. The court found that in the Fourth Circuit, at least, the proper test is shocks the conscience (see case for a review of the precedents). The court found that several other pursuit cases demonstrated more egregious police conduct and still passed muster. The court noted that the plaintiff's expert labeled Officer Selby's conduct "reckless," "totally irresponsible," and "wanton." The court found the conduct less flagrant than in some precedent cases. Finally, the plaintiff wanted more time for discovery, such as the allegation of the lack of pursuit training and whether the sheriff's department had in fact any pursuit policy. The court ruled that since the officer's acts were not a constitutional deprivation of rights, the allegations, even if true, would not change the result. *Temkin v. Frederick County,* 945 F.2d 716 (4th Cir. 1991).

A 1991 case filed in federal court is not, strictly speaking, a pursuit case. A state patrol officer received a report of a driver who did not pay his toll on I-95 and was driving recklessly. The driver was high on PCP, an illegal drug, and drove in a crazy manner, barely missing innocent parties a number of times. He drove against traffic on the interstate, and in general was a menace to the public. Eventually he was cornered in a field and drove over an embankment into a cornfield which concealed his Bronco. The police waited to see if the driver would emerge from the field. Meanwhile, Rucker, a civilian, involved himself in the melee against the orders of the officers. The drugged driver came barreling out of the cornfield in his Bronco. One officer fired twelve shots trying to flatten the Bronco's tires, and apparently one shot hit and killed Rucker, who was lying on the embankment. The Fourth Circuit affirmed a dismissal of all claims (the court did not express any opinion on the state law claims). Noting its decision in *Temkin,* the court said that the plaintiff might have a case if state officers acted irrationally and arbitrarily. They clearly did not so act in trying to stop "a madman running amok." *Rucker v. Harford County,* 946 F.2d 278 (4th Cir. 1991).

Massachusetts

Mass. Gen'l Laws Ann. ch. 89, §7B. Operation of emergency vehicles

The driver of a vehicle of a fire, police or recognized protective department and the driver of an ambulance shall be subject to the provisions of any statute, rule, regulation, ordinance or bylaw relating to the operation or parking of vehicles, except that a driver of fire apparatus while going to a fire or responding to an alarm, or the driver of a vehicle of a police or recognized protective department or the driver of an ambulance, in an emergency and while in performance of a public duty or while transporting a sick or injured person to a hospital or other destination where professional medical services are available, may drive such vehicle at a speed in excess of the applicable speed limit if he exercises caution and due regard under the circumstances for the safety of persons and property, and may drive such vehicle through an intersection of ways contrary to any traffic signs or signals regulating traffic at such intersection if he first brings such vehicle to a full stop and then proceeds with caution and due regard for the safety of persons and property, unless otherwise directed by a police officer regulating traffic at such intersection. The driver of any such approaching emergency vehicle shall comply with the provisions of section fourteen of chapter ninety when approaching a school bus which has stopped to allow passengers to alight or board from the same, and whose red lamps are flashing.

Other than being noteworthy for long sentences (198 words), the above provision covers the essence of UVC §11-106. Note that it makes no reference to use of audible or visual signals. No statutory duty is imposed on emergency vehicle operators on an emergency run to use such devices. Police vehicle drivers are not specifically granted the privileges "when in response to an emergency call or when in the pursuit of an actual or suspected violator of the law" as given in the UVC. The standard of care is caution and due regard for the safety of other people or property. Laws 89 §7 gives emergency vehicles "the right of way through any street, way, lane or alley." Laws 89 §7A directs other traffic to promptly pull off to the right and park there until the emergency vehicle has passed.

As in many other states, official immunity plays a part. The Massachusetts Tort Claims Act is set out at Laws 258. Section 9A promises to indemnify police officers for any damages awarded, and section 10(b) exempts liability under the Act for the acts or omissions of a public employee acting in a discretionary capacity while in the scope of his performance (again, a very common provision).

The pattern of Massachusetts cases is unusual in that there are some old cases and some new, with none identified in between. Let's begin with the old, as in 1888 old, to show that the common law required due care decades before it was codified in state traffic codes. A cab driver stopped his horse along the street, but not in a parallel park position as required by ordinance. A fire wagon going to a fire hit the cab, and the cab driver alleged injuries. The trial judge refused to instruct on illegal parking as evidence of negligence. The Supreme Judicial Court reversed a verdict for the plaintiff on a finding that the instruction should have been given. The fire wagon had the right of way and the right to proceed to the fire quickly. The plaintiff had the duty to pull to the right and get out of the way. If he is found negligent, he may not recover damages. *Newcomb v. Boston Protective Ass'n,* 16 N.E. 555 (Mass. 1888).

A 1932 case held that a verdict for the plaintiff must stand. The plaintiff came to an intersection in his car, stopped, looked, and pulled out to cross the "island" or square. He then saw a fire vehicle coming from his left at 50 to 60 mph. He shifted into second and tried to accelerate out of the way but was hit in the left rear of his car. Apparently the fire vehicle driver was not sounding his gong. The court held that it was not error to refuse to instruct on the plaintiff's duty to immediately pull to the left curb and stop. The plaintiff was hung up in the middle of the square. The statute simply did not apply to these facts. *Nestor v. Tewksbuty,* 182 N.E. 335 (Mass. 1932).

A 1952 case arose from a collision between a police car and city bus. The officer was on an emergency run and had switched on his siren. The officer slowed but did not stop at an intersection where he had the red light. The bus driver apparently did not hear or heed the siren, and entered on his green signal. There was a collision, and the officer sued for personal injury. The jury found for the officer. The trial judge then entered judgment for the defendant, and the officer appealed. The Supreme Judicial Court reinstated the verdict. There is no question that anyone other than an emergency vehicle driver who enters an intersection on a red signal is negligent. But since the legislature has carved out an exception here, whether the officer was negligent was properly a jury question, and the jury had spoken. *Neil v. Holyoke St. Ry. Co.,* 109 N.E.2d 831 (Mass. 1952).

A short 1996 case affirmed an award against the town of Milton for an accident during a police pursuit. The facts are not given other than the plaintiff was severely injured when a pursued driver crashed into the plaintiff. The plaintiff sued the town and the pursuing officer, Foley. The jury awarded $1.7 million against the town and $5,000 against Foley. (Note that the town's liability is capped at $100,000 by Laws 258 §2.) The officer and the plaintiff appealed. The plaintiff did not like the apportionment of fault. The town missed the deadline to file its appeal, and missed the extension permitted, so the court rejected its late filing. The plaintiff argued that Foley's conduct amounted to intentional tort. If so, that would take him out of the protection afforded him by Laws 258 §2 of the Tort Claims Act. Laws 258 §10(c) states that public employee immunity shall not apply to "intentional tort, including assault, battery . . . . " The appeals court held that while Foley's conduct may have been reckless, it was neither an intentional tort nor a battery. The court vacated the judgment as to Foley. *Jackson v. Town of Milton,* 669 N.E.2d 225 (Mass. App. 1996).

The remaining cases originated in federal court. A 1993 case arose out of a pursuit of a motorcyclist. Officer Meninno clocked him doing 59 mph in a 40-mph-zone. The operator fled instead of stopping. He had a passenger, the plaintiff, who told him to stop, but he refused. The chase took place at night, and there were no other vehicles involved. The pursuit covered 3.2 miles and lasted three or four minutes. Meninno alerted officers of the neighboring town of the pursuit, and one of them, Sullivan, stopped his cruiser in the opposing lane and waited, with flashing lights and "take-down" lights (temporary lights one can place on the top of a car) to illuminate the scene. The motorcyclist came around the curve at 60 mph. He would have had 480 feet in which to stop, if he wished. The trailing police car was 15 to 20 seconds behind. The motorcyclist appeared to slow, but lost control and the bike went down (the operator probably made the common mistake of locking up the rear wheel instead of applying the front brake hard) and wound up hitting the front of the blocking police vehicle. The operator was killed and the passenger was in a coma for a month and eventually had one leg amputated. The plaintiff sued seven defendants: three police officers, two towns, and their police chiefs. She sued in federal court alleging a violation of constitutional rights under U.S.C. 42 §§1983 and 1985, a violation of the same rights under the Massachusetts Constitution, and she also filed a claim under the Massachusetts Tort

Claims Act. The federal district court granted summary judgment to all defendants.

The court of appeals affirmed the dismissal of the §1983 claims against all three police officers. Officer Sadeck was not even at the scene. He arrived moments later. The plaintiff apparently relied on an erroneous newspaper account to the effect that Sadeck and Sullivan set up a staggered roadblock leaving room for the driver to slow and weave past. Officer Meninno had not effected a "seizure" of the plaintiff, relying on the primary roadblock case, *Brower v. County of Inyo,* 489 U.S. 593 (1989), so there was no violation of the Fourth Amendment. Officer Sullivan did not block both traffic lanes, and the cases cited by the plaintiff that best supports her position deal with complete or "deadman" roadblocks. Here the motorcyclist could have stayed in his lane and passed. The court held that at the time of the accident, it was not clear that a reasonable police officer would know that an illuminated, partial roadblock violated the Fourth Amendment. The court has an excellent discussion of the leading federal cases as of that date. The court also held that Sullivan was exercising a discretionary function. It was not persuaded in this murky area of law by the plaintiff's argument that the pursuit guidelines of Sullivan's department rendered his actions ministerial; that is, just follow the guidelines. There was substantial room for exactly how to conduct and terminate a pursuit.

The court took a pass on the final issue of whether the Massachusetts Tort Claim Act applied. Under the Act, the state is liable to the same extent a private person would be liable for a negligent act or omission of a public employee acting within the scope of his or her employment. One exception, §2(b), is the discretionary function. The plaintiff argued that the two towns were liable under the Act. The court notes that the discretionary function exception is necessary to prevent other branches of government from second-guessing the decisions of policymakers. After a long analysis of Massachusetts and federal cases, the court found that it could not state what the law is in Massachusetts and it certified the question to the Supreme Judicial Court. *Horta v. Sullivan,* 4 F.3d 2 (1st Cir. 1993).

The Supreme Judicial Court ruled that the exception did not apply to Officer Meninno. It noted that deciding whether certain discretionary conduct involves policymaking or planning is not simple, and that an "all-purpose guide to the limits of the exception is not likely." The court reasoned that the issue must be narrowly focused on the allegations of negligence, not on whether the conduct is part of a broader policy. The question in this case is whether Meninno, in his decision to pursue, made an ad hoc decision based on the situation, or on broader governmental objectives. The court found that any such pursuit must involve some discretion, but if it immunized the conduct, that wouldn't leave much left of the Tort Claims Act. *Horta v. Sullivan,* 638 N.E.2d 33 (Mass. 1994). I will comment that, as in many areas of law, there is simply not one right and one wrong position. Note the First Circuit's decision to vacate its original holding and remand the case for trial, *Horta v. Sullivan,* 36 F.3d 210 (1st Cir. 1994), in response to the above case.

The First Circuit decided a pedestrian accident case in 1996. Boston police officers noticed a person throw a bag through an open window of a car and leap inside. The driver took off and the police pursued. Persons in the car appeared to be swallowing potential evidence. The pursuit continued at 45 to 50 mph through heavy traffic, including pedestrians, near the Codman Square area. The suspect driver hit a ten-year-old girl who was trying to cross the street. The officers were familiar with the area. The pursuit lasted about two minutes. The plaintiff sued in federal court, alleging violation of federal constitutional rights and claimed under the Massachusetts Tort Claims Act. The trial judge dismissed the federal claims against the police officer. The court of appeals affirmed. After an analysis of similar pursuit cases, the court found the conduct here well short of shocking to the conscience. Some pursuit cases in federal court, with the same outcome, involved speeds of up to 100 mph. In one the offense was stealing $17 worth of gas. See *Temkin v. Frederick County* in the Maryland section above. The court did affirm awards of $100,000 to the plaintiff (the statutory limit), and $100,000 to the parents for loss of consortium. It found the trial judge could tell the jury what the statutory limit was, although normally it did not approve of this information being given. (There is no doubt that the jury generously rewarded the parents to compensate for the damage cap, and the court concedes as much.) *Evans v. City of Boston,* 100 F.3d 1033 (1st Cir. 1996).

The First Circuit applied the shocks-the-conscience test in a 1997 case, and again found the case failed the test. Around 9:00 P.M. on New Year's Eve in Saugus, Mass., officers got a call of a disturbance at a McDonald's. The restaurant is on U.S. 1, a six-lane highway, with a speed limit of 45 mph. There was a mixture of snow and rain on the roadway. While en route, the officers were radioed a description of the car, and tried to stop it as it was leaving the restaurant. The driver fled at high speed. The pursuit went on for three miles at speeds of up to 80 mph. The driver drove very recklessly, and at one time did a 360-degree spin and got back on the highway. He weaved through traffic, and tried to use other cars to block the officers. (Later the driver pleaded guilty to aggravated rape, burglary, and theft of a motor vehicle.) The driver exited the highway and hit a car in front of him waiting to turn right. The driver was seriously injured and sued the officer and the city in federal court. The district court granted summary judgment to the defendants. The First Circuit affirmed, noting that it had the benefit of *Evans*. The facts here were far short of conscience-shocking. The driver, potentially drunk, posed an increasing danger to others. The court notes that the plaintiff might be able to show negligence under state law, but here there was no constitutional violation. *Boveri v. Town of Saugus,* 113 F.3d 4 (1st Cir. 1997).

Michigan

Mich. Comp. Laws §257.603. Government vehicles, authorized emergency vehicles

[Michigan has adopted UVC 11-106 (see the Introductory Essay for text) with the two following changes. Section 257.603 allows police vehicles on emergency runs to proceed without an audible alarm if silence is required and still retain the benefit of the exceptions permitted. The section also omits subsection (d) of UVC 11-106 regarding due care and reckless disregard. That language is retained in §257.632 set out below. It is not clear why there are two controlling statutes with the same intent but with quite different style and somewhat different text. Both should be consulted in researching the law and identifying cases.]

Mich. Comp. Laws §257.632. Vehicles under police direction in emergencies or in pursuit of criminals, fire patrols, ambulances; exemption from speed limitations

[The speed limitation set forth in this chapter shall not apply to vehicles when operated with due regard for safety under the direction of the police when traveling in emergencies or in the chase of violators of law or of persons charged with or suspected of a violation, nor to fire department or fire patrol vehicles when traveling in response to a fire alarm, nor to public or private ambulances when traveling in emergencies. This exemption shall apply only when the driver of the vehicle while in motion sounds an audible signal by bell, siren or exhaust whistle as may be reasonably necessary or when the vehicle is equipped with at least 1 lighted lamp displaying a flashing, oscillating or rotating red or blue light visible under normal atmospheric conditions from a distance of 500 feet to the front of such vehicles, unless the nature of the mission requires that a law enforcement officer travel without giving warning to suspected law violators. This exemption shall not however protect the driver of the vehicle from the consequences of a reckless disregard of the safety of others.]

Mich. Comp. Laws §257.653. Authorized emergency vehicles; right of way yielded by other vehicles

(1) Upon the immediate approach of an authorized vehicle equipped with not less than 1 lighted flashing, rotating, or oscillating lamp exhibiting a red or blue light visible under normal atmospheric conditions from a distance of 500 feet to the front of the vehicle and when the driver is giving audible signal by siren, exhaust whistle, or bell: (a) The driver of another vehicle shall yield the right of way . . . .

Mich. Comp. Laws §691.1405(5). Government owned vehicles; liability for negligent operation

Governmental agencies shall be liable for bodily injury and property damage resulting from the negligent operation by any officer, agent, or employee of the governmental agency of a motor vehicle of which the agency is owner.

Mich. Comp. Laws §691.1407(2). Governmental immunity from tort liability

Each . . . employee of a governmental agency . . . shall be immune from tort liability for injuries to persons or damages to property caused by the employee . . . while in the course of employment . . . while acting on behalf of a governmental agency if all of the following are met:

(a) The . . . employee is acting or reasonably believes he or she is acting within the scope of his or her authority.

(b) The governmental agency is engaged in the exercise or discharge of a governmental function.

(c) The . . . employee's conduct does not amount to gross negligence that is the proximate cause of the injury or damage. As used in this subdivision, "gross negligence" means conduct so reckless as to demonstrate a substantial lack of concern for whether an injury results.

Section 257.653 requires other drivers to yield the right-of-way. This would include yielding where the civilian vehicle has the right of way by a green signal or the lack of a stop sign. The standard applicable to emergency vehicle operators in this section is due care. I have set the section out here because it is applicable to any case in which the emergency vehicle collided with another vehicle, and there are a number of case notes under it. Section 691.1405 waives governmental immunity for the negligent operation of governmental vehicles. Unlike some other states, there is no specific exception for emergency vehicles. Section 691.1407(2) immunizes a governmental employee who is not "*the* proximate cause" (note: not "*a* proximate cause") of the injury and who is not grossly negligent.

We have an unusual research situation in Michigan in that a year 2000 case, in a sweeping decision, overruled three lines of authority. Any new police pursuit case must be evaluated with this decision in mind. It answers most of the questions. The case is *Robinson v. City of Detroit,* 613 N.W.2d 307 (Mich. 2000.) The Michigan Supreme Court consolidated two conflicting cases from the courts of appeals. The first, *Cooper v. Wade,* 554 N.W.2d 919 (Mich. App. 1996), grew out of the pursuit of a stolen car. The driver of the stolen Jeep was fourteen years old, and he had three passengers. One police car pursued. The driver of the Jeep hit another car, lost control, and hit a house. He was killed and two passengers were injured. The court of appeals held that the city of Detroit was not entitled to summary judgment because the pursuit was operation of a government vehicle. If such operation was negligent, the city was not immune under §691.1405 (see above). As to the driver and passengers, the city did not owe a duty to the fleeing suspect (following several precedents), but owed a duty of due care to passengers and innocent bystanders.

The second, *Robinson v. City of Detroit*, 571 N.W.2d 34 (Mich. App. 1997), had similar facts. Again the police attempted to stop a driver of a car which later was found to be stolen. The driver fled, hit another car head-on, and his passenger was killed. (The passenger apparently had no part in the theft, but was offered a ride. He did actively encourage the driver to flee.) The court of appeals held that the city had no duty to protect this "voluntary" passenger. (An involuntary passenger would be a person forced into the car.) The court noted that the emergency vehicle statutes require police officers to exercise due care for life and property, and that officers are not protected from the consequences of their reckless disregard for the safety of others. Who are "others?" The statutes are silent. The court of appeals held that there is no duty to the passengers in the fleeing vehicle, and agreed that reasonable minds could disagree on this issue. The court noted that while a pursuing officer can exercise appropriate care in conducting the pursuit, the officer has no control over the pursued driver, other than abandoning the pursuit and thus "compromising law enforcement." The court suggested that this unsettled question required a solution by the Legislature.

The Michigan Supreme Court made several findings and overruled three cases. The court held that the police owe a duty to innocent passengers in fleeing vehicles, but not to passengers who are themselves "wrongdoers." An innocent passenger is one who neither helps bring about the pursuit nor encourages the driver to flee. The burden of proving innocence rests on the passenger or his estate, and this is not a question of law. (I will comment here that one can contemplate a range of a passenger's conduct. A passenger could be a participant in a crime, such as theft, drug-dealing, underage drinking, vandalism, and so on, and thus would have a good reason to flee. On the other hand, a driver may flee simply to avoid a speeding ticket, and the passenger would not be culpable. If the passenger encourages the flight, even if he did not initially tell the driver to flee, he is not innocent. In Michigan, the passenger is in the best position if he can show he tried to get the driver to stop.)

The city of Detroit is entitled to judgment as a matter of law. It is immune as a matter of law. The motor vehicle exception to immunity, §691.1405, refers to the negligent operation of a government vehicle and requires that the plaintiff's injuries "result from" that negligent operation. The court notes that Michigan strictly construes statutes in derogation of the common law, here common law sovereign immunity. The police vehicles in these two cases neither hit the fleeing vehicles nor caused them to hit another vehicle or other object, overruling *Fiser v. Ann Arbor,* 339 N.W.2d 413 (Mich. 1983). The injuries resulted from the acts of the fleeing driver.

The pursuing officers are immune because their actions were not the proximate cause in the context of governmental employee immunity, §691.1407(2). *The* proximate cause means the one cause most immediate, efficient, and direct preceding an injury, not "*a* proximate cause" (overruling *Dedes v. Asch,* 521 N.W.2d 488 (Mich. 1994). Fourth, it is not negligence for the police to decide to pursue a driver who flees (overruling *Rogers v. Detroit,* 579 N.W.2d 840 (Mich. 1998).

Now we look at each of the three cases the court overruled. The court noted that it had been asked many times to reexamine *Fiser.* The facts were that an Ann Arbor police officer attempted to stop a driver, Lehman, who ran a flashing red light in the early morning hours. A high-speed chase ensued. At one point Lehman was forced to stop, but sped off again, and the pursuing officer lost contact with the suspect. A short time later another officer, who had been following the chase on his radio, saw the suspect and pursued. Lehman then collided with the plaintiff driver at an intersection. The Michigan Supreme Court held that the officers could be found to have conducted the pursuit negligently, and could have been one of the proximate causes of plaintiff's injuries, and held that the plaintiff could go to a jury on the question.

The 2000 decision overruled this holding. It noted a 1989 case in the court of appeals, *Frohman v. Detroit,* 450 N.W.2d 59 (Mich. App. 1989), that grudgingly followed *Fiser.* The case arose when an officer walked up to an illegally (and stolen) parked van to tell the driver to move. The driver fled, running over the officer's foot while doing so. The officer pursued and then slowed and turned off his flashing lights and siren as they came to an intersection. The van driver hit and seriously injured another driver. The *Frohman* court objected to the "legal fiction" of *Fiser* that it is normally expected that a driver will flee when an officer attempts to stop him. A police officer has the sworn duty to enforce the law, and to hold him to be the cause of an injury when he has conducted the pursuit in a minimally negligent manner is a remarkable legal principle. In fact, most drivers ordered to stop do so. Those with another crime to conceal may not. The question should not be on how the officer was conducting the pursuit, but rather on what effect the manner in which the officer drove his vehicle had on the cause of the accident. The supreme court held that "the plaintiffs cannot satisfy the 'resulting from' language of the statute where the pursuing police vehicle did not hit the fleeing car or otherwise physically force it off the road or into another vehicle or object."

The second case overruled was *Rogers v. Detroit,* 579 N.W.2d 840 (Mich. 1998). Note that Justice Kelly wrote the majority opinion in this supreme court case, and Justice Taylor wrote the main dissent. In the 2000 case, they switched roles. Michigan has judges that are elected on a partisan basis and apparently campaign for office. Obviously there had been some electoral changes in the two years. (We can contrast this with the Nebraska system in which judges are appointed on a nonpartisan basis, and thereafter face only a ballot question every seven years in which voters simply vote to retain or dismiss them. Rarely do any judges lose. I have no doubt which is the better system, for what it's worth.) In *Rogers,* two pursuit cases were consolidated for review by the supreme court. In both, a driver fleeing the police collided with an innocent driver. The *Rogers* court held that a jury could find the police liable for the decision to pursue and for pursuing negligently within the context of the privileges granted to emergency vehicle drivers under the statute. The accident could result from the operation of a governmental vehicle even though there was no contact between the police vehicle and the two other vehicles. This holding was overruled in 2000, with Justice Taylor essentially rewriting his 1998 dissent as the now majority opinion.

The third case overruled was *Dedes v. Asch,* 521 N.W.2d 488 (Mich. 1994). *Dedes* is not a pursuit case. It arose when two school children were hit by a car while walking to their bus stop. They sued the school director of transportation for selecting an unsafe bus stop. At issue was M.C.L. 691.1407(s)(c). That statute permits a finding of liability against governmental employees who are "the proximate cause" of an injury or damage alleged by a plaintiff. The *Dedes* court did not believe that the legislature meant this to mean that there was only one proximate cause of an injury, or to prohibit the finding of concurring causes which might involve a governmental employee. The three dissenters thought that is exactly what the legislature did mean and that the meaning was clear. If the legislature meant otherwise, let them act to correct their "folly." The majority in the 2000 case overruled *Dedes* and found that the legislature indeed meant by using the word "the" that the proximate cause meant the one most immediate, efficient, and direct cause preceding an injury. Since the injury was caused most directly by the fleeing suspect, the defendants were statutorily immune.

The *Robinson* (2000) case was decided on a 4-3 vote, and there were three dissents. One commented that the law hadn't changed, the court had. That is a fair comment, but the case involved fundamental policy issues. The majority saw itself as restoring authority to the legislature that had been usurped by the courts. The minority argued that the majority, in overruling three distinct lines of cases, did not sufficiently consider the principle of stare decisis. The policy issues go well beyond the law of police chases, of course. By holding that "the" does not mean "a" and holding that a government vehicle operator is not liable for an accident unless he strikes another vehicle or forces it into another object, the case has authority under many kinds of facts.

Prior cases have to be viewed in the context of *Robinson.* An example is *Terry v. Detroit,* 573 N.W.2d 348 (Mich. App. 1998). Police pursued the driver of Cadillac who ran a red light. The officer suspected, correctly, that the Cadillac was stolen. The suspect ran a red light and injured an innocent driver and passengers. The court of appeals held that there was a triable issue of officer negligence. Since the officer did not strike the suspect or the innocent party, the case cannot be cited today as precedent. On another issue, the owner of the Cadillac, General Motors, was granted summary judgment. The car was parked in a secure and guarded company garage and the assigned driver left the keys under the floor mat to facilitate servicing the vehicle. The court found that General Motors did not act unreasonably.

The law which states that the police owe no duty to the pursued driver himself is unchanged. The matter was one of first impression in a 1994 case. State police clocked a motorcyclist at 85 mph on I-95, and gave chase. The motorcyclist exited the freeway and ran a red light. The pursuit began in earnest, traveling at high speeds through residential and commercial areas for eleven minutes. The officers were in communication with their supervisor during the pursuit. Another police car tried to cut off the motorcyclist by driving east as the motorcyclist headed west. As they approached, the motorcyclist turned into the eastbound land, sideswiped the cruiser, and was killed. The court held as a matter of law that the officers did not owe a duty to such wrongdoers, nor had not acted with gross negligence. *Jackson v. Oliver,* 514 N.W.2d 195 (Mich. App. 1994). I will comment that some would argue that under the circumstances, officers do owe a duty perhaps not to pursue or break off pursuit when the only apparent offense is speeding on the interstate. The court here is not persuaded. The police must sometimes allow fleeing suspects to get away because of concern for public safety, but it would be absurd to cease pursuit simply for the safety of the fleeing criminal suspect.

If the police officer on an emergency run collides with a vehicle he is not pursuing, the officer can be liable. A 1979 case is of this type, and is noteworthy mainly because in this case the Michigan Supreme Court established comparative negligence "in the interest of justice." The case arose when an officer on an emergency run with the appropriate warnings came to a stop sign. He slowed and entered the intersection and collided with the plaintiff. The plaintiff was following a car that turned right, and to some extent that car obstructed the view of the officer. The first trial found for the defendant, but was reversed and remanded for a new trial because the plaintiff was improperly questioned about whether she was wearing a seat belt. The second trial also found for the defendant. The court reversed again for a new trial because the jury instructions unfairly emphasized the plaintiff's duty to make sure the intersection was clear before entering it, although she was on the through street. Such driver has the right to assume others will stop at their stop sign unless in the exercise of ordinary care the driver learns otherwise; here, lights and sirens. The court noted that in the third trial, either or both litigants might be found negligent, and that comparative negligence would apply hereafter in Michigan. *Placek v. City of Sterling Heights,* 275 N.W.2d 511 (Mich. App. 1979).

Note that even before the *Robinson* case, officers could be found not at fault for the injuries to third parties who collided with the fleeing driver. A 1976 case had this result. The officers were pursuing a drunk driver with caution, slowing at intersections and operating sirens and flashing lights. That met the standard of proper regard for the safety of others. *Marion v. City of Flint,* 248 N.W.2d 580 (Mich. App. 1976).

Ambulance and fire truck drivers on emergency runs may have the benefit of the emergency vehicle statutes, but must exercise the care those statutes require. For example, an ambulance driver on an emergency run collided with a car. The ambulance operator was operating his lights and siren and came almost to a stop at an intersection. He looked right and then left, but did not again look right, and collided with the car. It was a jury question whether either or both were negligent. The plaintiff was the passenger in the ambulance. *Keevis v. Tookey,* 201 N.W.2d 661 (Mich. App. 1972).

Minnesota

Minn Stat. Ann. §169.03. Emergency vehicles; exemptions; applications

(1) Scope. The provisions of this chapter applicable to the drivers of vehicles upon the highways shall apply to the drivers of all vehicles owned or operated by the United States, this state, or any county, city, town, or any other political subdivision of the state, subject to such specific exemptions as are set forth . . . .

(2) Stops. The driver of any authorized emergency vehicle, when responding to an emergency call, upon approaching a red or stop signal or any stop sign shall slow down as necessary for safety, but may proceed cautiously past such red or stop sign or signal after sounding siren and displaying red lights.

(3) One-way roadway. The driver of any authorized emergency vehicle, when responding to any emergency call, may enter against the run of traffic on any one-way street or highway where there is authorized division of traffic to facilitate traveling to the area in which an emergency has been reported, and the provisions of this section shall not affect any cause of action arising to its passage.

(4) Parking. An authorized emergency when at the scene of a reported emergency may park or stand notwithstanding any law or ordinance to the contrary.

(5) Course of duty. No driver of any authorized emergency vehicle shall assume any special privilege under this chapter except when such vehicle is operated in response to any emergency or in the immediate pursuit of an actual or suspected lawbreaker.

Minn Stat. Ann. §169.17. Emergency vehicles

The speed limitations set forth . . . do not apply to an authorized emergency vehicle responding to an emergency call. Drivers of all emergency vehicles shall sound an audible signal by siren and display at least one lighted red light to the front, except that law enforcement vehicles shall sound an audible signal by siren or display at least one lighted red light to the front. This provision does not relieve the driver of an authorized emergency vehicle from the duty to driver with due regard for the safety of persons using the street, nor does it protect the driver of an authorized emergency vehicle from the consequences of a reckless disregard of the safety of others.

Minn Stat. Ann. §3.736. Tort claims

[This section sets out the state tort claims procedure. There is no specific exception for law officers. The act states the state will pay compensation for loss of property, personal injury, or death caused by an act or omission of a state employee who is acting within the scope of his employment if a private person would be liable under the circumstances. There is a $200,000 cap on damages for one claimant and $600,000 maximum for one incident. There is a 180-day limit on notice to the state of a claim. In short, this is a typical state tort claims act. The case notes indicate that most claims involving peace officers are arrest claims. See *Pletan* below for how the act applies to a pursuit case.]

Minn Stat. Ann. §466.01 et seq. Tort liability of political subdivisions

[This act is similar to the state act. Again, there is no particular law enforcement exception. The damage cap is $300,000 per individual and $750,000 for one incident. Both the state and local tort claims acts exempt claims for discretionary acts.]

A plaintiff in a police pursuit case attempted to impose strict liability on the city in a 1978 case. A St. Paul police officer identified a stolen car and attempted to arrest the driver. The driver fled at 75 mph. The police followed, at times quite close. The fleeing driver entered a four-way-stop intersection without stopping. He hit several parked cars, including the plaintiff's parked car, shoving it somewhat into the street. The police car then hit the plaintiff's car. The plaintiff sued for property damage. The trial court imposed strict liability on the city and awarded a whopping $1,150 in damages. The trial court's rationale for strict damages was that (1) innocent victims of police chases could collect damages without having to prove negligence, (2) the damages would be paid by the public that benefits from law enforcement, (3) and the effect would be to cause the police to question the benefit of police chases. The Minnesota Supreme Court reversed this finding. The trial court's use of the Restatement of Tort 2d §402A is clearly erroneous. That section deals with defective products. Similarly, the application of §520 is also wrong. That section deals with ultrahazardous activities, and there is an explicit exception for an activity imposed on the actor as a public duty. Law enforcement, including the pursuit and arrest of suspects, is clearly a public duty. (I am surprised that the trial judge accepted the plaintiff's position and applied two wholly inapplicable Restatement sections. There is nothing wrong with the position that the public ought to indemnify all victims of police chases, but it can't be supported by inapplicable authority.) The supreme court discussed the ramifications of strict liability as applied to police chases. One consequence would be its effect on law enforcement, and the unfairness of requiring the police to enforce the law and then impose liability when they conscientiously try to do so. The supreme court noted that the legislature already has imposed negligence standards on the driving of emergency vehicles. The court agreed that there was no negligence, and reversed the award. *Cairl v. City of St. Paul,* 268 N.W.2d 908 (Minn. 1978).

Minnesota recognizes official and discretionary immunity, and applied it to a 1992 pursuit case which ended in a fatal accident. (Note that this is a common law doctrine.) A child was walking home from school and was killed when struck by a fleeing felon. A Crystal police officer got a call about a "snatch and grab" shoplifting incident at a clothing store. The officer identified the car and driver and gave chase. Speeds were up to 75 mph in a 45-mph-zone at times, with the pursuit lasting about two minutes. After some dangerous driving, the felon came to a red light. A truck was stopped in the right lane. The pursuing officer saw a pair of legs running across in front of the truck. The officer slowed to five or ten mph, but the felon continued and killed the child. (The felon pleaded guilty to manslaughter.) The trial court granted summary judgment to the officer and city on the basis of discretionary function. The supreme court affirmed that finding. The court noted that an officer, in deciding whether to pursue, must weigh many factors, and do it in emergency conditions with little time for reflection. It would be hard to find a situation where more independent judgment and discretion are involved. The court held that the decision to begin and continue a police pursuit is protected by official immunity. *Pletan v. Gaines,* 494 N.W.2d 38 (Minn. 1992).

A 1998 case involved an ambulance. Note that a public official exercising his discretion and judgment is protected by official immunity unless the action constituted malicious conduct. A paramedic, Everson, received a report of an unconscious person. He proceeded toward the address with his emergency lights and siren on. He noticed a pedestrian, Kari, standing in a marked crosswalk at Lydia Street. Everson switched to the curb lane and changed his siren to the yelp mode. As he came up, Kari ran in front and was seriously injured. She had no recollection of the accident because of her injuries. The trial judge and the court of appeals held that the defendants had official immunity. The supreme court affirmed. The court noted *Pletan* and other cases in which an emergency vehicle operator must make split-second decisions, and found an analogy with operators of ambulances. Here there was no evidence that Everson's conduct was malicious. He had a right to fail to yield the right of way to the pedestrian, especially since he had made eye contact and reasonably thought that she would yield. Imposing liability would have the same chilling effect on the delivery of ambulances services that imposing liability on police would have on law enforcement. This was too much for a dissenting judge. He pointed out that the emergency vehicle statute required Everson to slow as needed for safety. The mere fact that there was a pedestrian in the crosswalk required, under the usual nonemergency vehicle statutes, Everson to yield. The dissent would not argue that Everson was or was not negligent, just that whether he complied with the statutes, and the standard of care provisions, was for a jury. The majority's position simply leaves no room for those statutes. *Kari v. City of Maplewood,* 582 N.W.2d 921 (Minn. 1998).

If the emergency vehicle operator is found immune by reason of exercising a discretionary function and is not acting with malice, that immunity extends to his employer as vicarious immunity. Any other holding would defeat the purpose of official immunity. So said the court in another ambulance case, *Nisbet v. Hennepin County,* 548 N.W.2d 314 (Minn. App. 1996). The case arose when an ambulance with all the bells and whistles operating, and driving in the left (oncoming traffic) lane hit a driver he was overtaking who was trying to turn left. The driver alleged the ambulance was going too fast for the sleet conditions at the time. The court affirmed a finding of official immunity for the ambulance operator and Hennepin County. In a footnote, the court did comment that if the question before it was whether the operator was negligent, that may well be an issue of fact. However, since official immunity applies, it simply does not come up. Like police officers, ambulance drivers on an emergency run must exercise discretion in how to drive, including breaking traffic laws, to get to the destination quickly. Some ambulance runs do involve life and death situations. Though such emergency driving can be dangerous, "it is thought that more good than harm" will result.

The court did comment that there are older Minnesota cases where official immunity was not recognized and the emergency operators were judged by a negligence standard of care. Apparently the defense of official immunity was not asserted in these cases. The court cited *Caril* (see above, *Van Tassel v. Hillerns,* 248 N.W.2d 3134 (Minn. 1976) (police in pursuit hit an innocent party), and *Markle v. Haase,* 73 N.W.2d 362 (Minn. 1955). (fire truck accident)). I will comment that this is not the only situation in which the law has shifted from judging cases solely by the emergency vehicle statutes related to immunity. Illinois is an example. Minnesota is somewhat unusual in not having a specific motor vehicle exception to immunity, and then having a law enforcement exception to that exception with, typically, a malicious action requirement for liability to be shown.

A 2000 Eight Circuit §1983 case is important because it was decided after the U.S. Supreme Court's California case, *County of Sacramento v. Lewis,* 523 U.S. 833 (1998). See the California section of this book. In *Lewis,* Justice Souter, writing for a unanimous court, held that the correct standard to judge police conduct in a pursuit case is whether it is shocking to the conscience. The court compared police pursuits to putting down a prison riot, when there is an emergency and the officers do not have the luxury of calm deliberation, but must make rapid decisions in ever-changing conditions without the luxury of hindsight. The court held that the officer, who actions were not tainted by an improper or malicious notice, and while perhaps not prudent, did not engage in conscience-shocking acts. The case below involves a longer pursuit.

The Minnesota case arose from a fatal accident chase on August 11, 1996. Minneapolis police officer Simonson, on duty, noticed a car that matched the description of a stolen vehicle. He pulled the driver over and approached with gun drawn and ordered the two occupants to put their hands up. The driver, Shannon, fled instead, and Simonson pursued, meanwhile advising the department. The driver ran stop signs and signals on city streets, and then entered I-94. Shannon eventually exited, turned around, and entered I-94 going the wrong way. Simonson made a U-turn and followed, with three other police cars somewhat farther back. The chase lasted six minutes and spanned about six miles. It ended when Feist, traveling on I-94 and apparently not noticing the slowing traffic as drivers saw and tried to avoid the fleeing driver and the officers, swerved into the right-hand shoulder and was killed when hit by Shannon head-on. The Eighth Circuit affirmed the denial of defendant summary judgment. The court noted that Simonson was an aggressive pursuer. In his ten-year career as a police officer, he had been involved in over 100 pursuits, and had never voluntarily terminated one. He was the officer involved in a 1990 pursuit that resulted in liability for the city.

The court found that, while Officer Simonson was perfectly justified in attempting to arrest Shannon, and thus begin pursuit, he had ample opportunity to evaluate the pursuit and call it off before the accident. Since he had time to deliberate, the test to be applied is not conscience-shocking, but rather deliberate indifference to the deceased's civil rights. The court found that Simonson is not entitled to qualified immunity because he knew or should have known that his actions potentially violated the plaintiff's Fourteenth Amendment rights. The court concludes that the case "is properly analyzed under the deliberate indifference standard of conscience shocking behavior." Is everybody confused now?

Note that this case was a 5-4 decision. Whether it will be followed or be an aberration remains to be seen. I will comment that the notion that if the officer had stopped pursuing when the suspect drove onto the off-ramp of I-94 the danger to others would be reduced is somewhat ingenuous. Surely a driver going the wrong way on the interstate is among the most dangerous of driving situations. It's an open question whether it is better to terminate the chase and hope the driver does not harm someone or to chase, displaying lights and sounding sirens, better serves the public's safety and the goal of law enforcement. Note that the case does not mention Minnesota's official immunity in performing a discretionary function doctrine.

Mississippi

Miss. Code Ann. §63-3-103. Types of vehicles

(d) "Authorized emergency vehicle" means every vehicle of the fire department (fire patrol), every police vehicle, and every such ambulance and emergency vehicle of municipal departments or public service corporations as is designated or authorized by the commission or the chief of police of an incorporated city. [An attorney general's opinion said that a private vehicle of a volunteer fireman is not an authorized emergency vehicle. Op. Atty. Gen. No. 97-0058.]

Miss. Code Ann. §63-3-205

No driver of any authorized emergency vehicle shall assume any special privilege under this chapter except when such vehicle is operated in response to an emergency call or in the immediate pursuit of an actual or suspected violator of the law.

Miss. Code Ann. §63-3-315
The driver of an authorized emergency vehicle when responding to an emergency call upon approaching a red or stop signal or any stop sign shall slow down as necessary for safety but may proceed cautiously past such red or stop sign or signal. At other times drivers of authorized emergency vehicles shall stop in obedience to a stop sign or signal.

Miss. Code Ann. §63-3-517
The speed limitations set forth in this article shall not apply to authorized emergency vehicles when responding to emergency calls and the drivers thereof sound audible signal by bell, siren, or exhaust whistle. This section shall not relieve the driver of an authorized emergency vehicle from the duty to drive with due regard for the safety of all persons using the street, nor shall it protect the driver of any such vehicle from the consequences of a reckless disregard for the safety of others.

Miss. Code Ann. §63-3-809
(1) Upon the approach of an authorized emergency vehicle, when the driver is giving an audible signal . . . the driver of every other vehicle shall yield the right-of-way . . . .
(2) This section shall not operate to relieve the driver of an authorized emergency vehicle from the duty to drive with due regard for the safety of all persons using the highway.

Miss. Code Ann. §63-7-19
[This section sets out the type of emergency lights required for emergency vehicles used to warn others to yield the right of way.]

Miss. Code Ann. §§11-46-1 to -23. Immunity from liability and suit.
[This article sets out the typical state and political subdivision immunity to tort liability. It creates the Tort Claims Fund and the Tort Claims Board. Section 11-46-9(c) states that a governmental entity and its employees shall not be liable for any act or omission relating to the police or fire protection unless the employee acted in reckless disregard of the safety and well-being of any person not engaged in criminal activity at the time of the injury. The cap on damages is $500,000 for a single occurrence (effective July 1, 2001)].

Mississippi has enacted UVC 11-106 with few changes, albeit broken down in several sections in the Mississippi code. There are few cases identified, and we take them with the oldest first. A 1959 case involved a police chase but the case was presented to the jury on ordinary negligence principles. A constable was chasing a speeding motorist on a gravel road. This raised a big cloud of dust. The plaintiff was approaching from the opposite direction and saw the dust. As he came off the blacktop, he slowed and turned to the right. He collided with the constable. The constable testified that the plaintiff came out of the dust cloud heading toward the oncoming lane, and they collided on the constable's side of the highway. There was testimony about the position of debris on the road. The supreme court affirmed an award for the constable, who had filed a cross claim. The jury was instructed that the constable was not excused from driving in a careful manner, even if he was chasing a suspected law violator. *Johnson v. Richardson,* 108 So. 2d 194 (Miss. 1959).

A Jackson police officer received a call of shots fired shortly after midnight. He proceeded toward the scene at 50 mph in a 35-mph-zone, with his blue flashing lights operating. He collided with a truck at an intersection and was seriously injured. Both drivers claimed the green signal. Buildings at the intersection cut down the visibility each driver. The officer sued the truck driver, and the jury found for the defendant. The supreme court affirmed. Apparently the jury found that the officer had a red light, and the statute requires him to slow and proceed cautiously, and to exercise due regard for the safety of others. True, the statutes require drivers to yield to an emergency vehicle displaying flashing lights, but here the lack of sight distance made that difficult. *Andrews v. Jitney Jungle Stores,* 537 So. 2d 447 (Miss. 1989).

We can dispose of a 1991 case quickly. A fire truck on an emergency run entered an intersection on the red light. The operator and another fireman in the truck testified that they slowed to 10 to 20 mph and thought the intersection was clear. They broadsided a car and the occupant died a few weeks later. The jury returned a defense verdict. The supreme court affirmed. It was not error to exclude pictures of the body at the morgue to show the severity of the injuries, and thus indicate the possible speed of the fire truck. The trial judge ruled that the pictures would be irrelevant and inflammatory. The plaintiff did not object to the instruction that the fire truck operator may proceed cautiously through an intersection when it is safe to do so, and may not raise the issue on appeal. The court did not agree that the verdict was against the weight of the evidence. *Walker v. Graham,* 582 So. 2d 431 (Miss. 1991).

A 1999 case is, not strictly speaking, an emergency vehicle case, but does involve a question of official liability for the failure of a police officer to prevent a drunk driver from driving. Usually such claims fail because of the discretion public officials have to do their jobs. The plaintiff was injured when she was a passenger in a car that hit the car operated by the intoxicated (.22 blood alcohol level) driver. He was on the highway without lights at night, and was stopped or driving in reverse. The plaintiff alleged that Officer Bradshaw had stopped the drunk driver shortly before the accident for erratic driving and for not using his headlights, and in order to prevent him from continuing to drive. The officer's employer, the city of Ruleville, claimed the protection of immunity, specifically Code 11-46-9(1)(c). That section grants public employees immunity except for acts of reckless disregard for the safety of others (except as to the wrongdoer). The trial court held that reckless disregard is the same as wanton disregard, which requires an element of malice, and did not find malice and intent to harm the plaintiff. Originally the supreme court affirmed this finding, but upon rehearing changed its mind and reversed for trial. It found that the officer's conduct was tantamount to an intention to harm others. A dissent would find that the record did not show that the officer intended to harm the plaintiff. *Turner v. City of Ruleville,* 735 So. 2d 226 (Miss. 1999).

I have looked at similar cases in other jurisdictions and other contexts, and would note that this decision is definitely a minority finding. See a Kansas case, *Spotts v. City of Kansas City,* 728 S.W.2d 242 (Mo. App. 1987). (Patrolman stopped an obviously intoxicated driver to issue a ticket for an expired plate, and was not responsible to a person whom the drunk hit fifteen minutes later. The patrolman did not owe a duty to that particular person, only a duty to the general public.)

A 2000 case shows that police on nonemergency runs enjoy no special privilege. An officer was going to dinner and was driving 57 mph in a 35-mph-zone at 11:00 P.M. he was not operating flashing lights or siren. He crashed into a car driven by a minor who was pulling out of an apartment parking lot to turn left. The officer testified that he customarily drove without noticing his speed. He was aware that he should have observed the speed limit. The city of Jackson claimed the immunity of Code 11-46-9, specifically claiming that the plaintiff, who did not have an operator's license, fit the criminal activity exception. The supreme court affirmed an award for the plaintiff, less 10 percent for his own negligence. Without determining whether the lack of a license was a felony or a misdemeanor, the court held that it had nothing to do with the accident. The statutory language was intended to cover criminal activity related to the accident; speeding, robbery, or whatever. The city's liability was limited to $50,000 by statute. The plaintiff was also entitled to collect uninsured motorists benefits of $50,000 carried on the car he was driving from Nationwide. The city was self-insured, and this fact did not defeat uninsured motorists coverage. *City of Jackson v. Perry,* 764 So. 2d 373 (Miss. 2000). This case should remind plaintiffs to seek all the insurance that may be available. When there is a government defendant, it is not unusual for UMI to apply even thought the claimant has recovered damages from the government.

Missouri

Mo. Ann. Stat. §304.022. Emergency vehicles defined—regulations

[Missouri has enacted UVC 11-106 (see Intorductory Essay for the text) without significant change. One change is to use the term "prima facie speed limit" rather than "maximum speed limit." The prima facie speed limit is the same as the basic speed limit, that is, speed appropriate for the conditions. Thus an emergency vehicle could be observing the statutory speed limit indicated by sign but exceed the basic limit in poor driving conditions. Code 300.100 also sets out UVC 11-106, this time without any changes, so it includes the language on due care and reckless disregard.]

Mo. Ann. Stat. §573.600. Sovereign immunity in effect—exceptions

[This section states that such governmental tort immunity as existed at common law prior to Sept. 12, 1977, remains in effect. This legislation was in response to a Missouri Supreme Court 1977 decision that ended sovereign immunity. The legislature promptly responded. One exception is for negligent operation of motor vehicles by public employees. There is no specific mention of immunity for discretionary acts, but Note 7, entitled "Police motor vehicles," indicates that emergency vehicle operators may have official immunity when performing a discretionary function.]

A 1993 case presented a novel question. A police officer received a call of an officer in distress. He notified the dispatcher that he was on his way and he collided with another driver at an intersection. (The case does not mention his use of emergency signals.) The trial court found for the plaintiff. The court of appeals reversed on a finding that the officer was entitled to official immunity. The plaintiff argued that the officer chose the wrong route and was speeding. The court noted that case law provides that public officers are not civilly liable to members of the public for the performance of discretionary duties. Officials need this protection to permit them to make decisions without fear of being sued. This was a case of first impression in deciding whether the official immunity doctrine applied to police emergency runs. The court noted that such immunity was found in police acts such as arrest, kicking down a door, deciding whether to arrest, and drawing a weapon. Here the decision on what streets to use, speed, and whether to stop at signs and signals is clearly discretionary. Any other result would inhibit law enforcement and aid to one in need. The court commented that perhaps the plaintiff might have a claim against the city (St. Louis) because sovereign immunity for the negligent operation of a motor vehicle has been waived. Official immunity and sovereign immunity are separate concepts. *Bachman v. Welby,* 860 S.W.2d 31 (Mo. App. 1993).

That may not be a viable suggestion, according to a 1997 case. A police officer was dispatched to an accident scene, and collided with a car. The officer claimed official immunity, and the trial judge so granted with prejudice. Then the officer's employer, the city of Fulton, moved for summary judgment because the only theory pleaded was *respondeat superior*. If the officer was acting within the scope of her employment, then the city cannot be held liable. The court of appeals agreed with the city. Since the officer was exonerated from liability, so must the employer be dismissed. *City of Fulton v. Hamilton,* 941 S.W.2d 785 (N. App. 1997).

The plaintiff in *Fulton* sought to apply the rule of *McGucklin v. City of St. Louis,* 910 S.W.2d 842 (Mo. App. 1995). A police officer on an emergency run ran a red light and collided with the plaintiff. The plaintiff alleged that while the officer's emergency lights were flashing, he did not sound his siren as required by Code 304.022(4). The officer testified that he had the green light and in fact was sounding his siren. The jury accepted the plaintiff's version and awarded $150,000. (The trial court reduced it to $100,000, the statutory limit.) The court of appeals affirmed. It ruled that the officer may not claim official immunity until he has complied with the emergency driving statute. The court distinguished *Bachman* above by noting that route and speed are discretionary acts, while the statute is clear in requiring that both lights and siren be used when running red lights. The court did note that its holding could have long-reaching effects on law enforcement, but found itself bound by the clear language of the statute. Many states do not require both lights and sirens, and some explicitly permit officers to go to a crime scene quietly when necessary. The officer claimed that he did use his siren. It is common, indeed expected, that participant and witness accounts of whether flashing lights, siren ,or horn were in use will conflict.

A 1996 case is almost identical to *Fulton.* In *Creighton v. Conway,* 937 S.W.2d 247 (Mo. App. 1996), a St. Louis police officer received a call of an accident with injuries, involving an overturned Jeep. Police regulations are that such priority one calls require that the officer respond with flashing lights and siren. The officer ran a red light and hit the plaintiff driver. The trial court granted summary judgment to both officer and city. There was no genuine issue of material fact that required a trial. The officer was complying with the statute while performing a discretionary act. The court of appeals affirmed. The plaintiff vainly argued that there was no emergency; the accident victim was already dead, so what's the hurry? The court noted that the officer did not learn this fact until hours later. (The plaintiff's argument was that if there was no hurry, the officer should have been more cautious.) The plaintiff had a more serious objection. Code 304.022 permits an officer on an emergency run to run red lights after slowing down as may be necessary for safe operation and not to exceed the prima facie speed limit if that endangers others. The court noted that this argument had already been handled in *Costello.*

Costello arose from a fatal accident. In the early morning hours, a police officer responded to backup call of a suspected burglary with his flashing lights and his siren set on "wail." He was slightly over the speed limit of 45 mph when he approached a green light from 450 feet. He did not see any other traffic. At 150 feet he saw the another car at the intersection from his right. He swerved but hit the driver's door and the driver was killed. At trial the jury found the officer 70 percent at fault and awarded damages, but then the trial judge dismissed the officer on the basis of official immunity. The city satisfied the award of $100,000, the statutory cap. (Note that this payment is in conflict with the decision in *Fulton*, a 1995 case above. Go figure.) The parents appealed the dismissal of the officer, alleging that he failed to slow when entering an intersection and sped when it endangered others. The court explained that the statutory language suggests how an emergency vehicle operator should drive depending on the particular circumstances at the time. I would comment that appropriate speed and requirements to slow are not precise instructions, whereas the command to use flashing lights and siren is clear. I would also comment that it appears that the deceased ran the red light, so the finding that the officer was 70 percent at fault seems unusual. *Costello v. City of Ellisville,* 921 S.W.2d 134 (Mo. App. 1996).

A 1995 case held that the mere fact that an officer switched on his flashing lights does not of itself prove an emergency situation. A police officer was on radar duty and he said that he clocked a speeder in the other direction. He flipped on his lights and made a U-turn, sped in pursuit and collided with another driver, killing the latter. A witness saw the U-turn but not a speeding driver. The trial court granted the officer summary judgment based on official immunity. The court of appeals reversed for trial. It found a question of fact whether the officer was actually chasing a speeder, and even if he were, whether so chasing was an emergency justifying immunity. Merely turning on his lights did not of itself so prove. *Anderson v. Jones*, 902 S.W.2d 889 (Mo. App. 1995).

A similar case is *Brown v. Tate,* 888 S.W.2d 413 (Mo. App. 1994). A police officer driving a paddy wagon at 2:00 A.M. came to an intersection with a flashing yellow light. She collided with and killed another driver. The plaintiffs alleged a number of negligent acts: driving too fast for conditions, failure to lookout, and failure to reduce her speed. The codefendant Kansas City did not claim that the officer was on an emergency run. It did contend that official immunity applied. The court of appeals held that summary judgment for the defendants was error. The officer had the same duty as any other driver. Neither did the public duty doctrine apply. The public duty doctrine applies to the public in general. Here the officer owed a duty to others that might have been on the road. *Brown v. Tate,* 888 S.W.2d 413 (Mo. App. 1994).

Lets turn to innocent third parties injured by a driver whom the police are pursuing. A 1996 case is unusual in that it alleged a personal vendetta.

Patzius and the son of a police officer, Nichols, got in an altercation over a girl. Later, two unidentified officers allegedly took Patzius to the station and beat him, and told him to stay away from the girl. Patzius didn't, and when he left her house two officers ordered him to pull over. He didn't, and a pursuit ensued, extending to beyond the city limits and without the knowledge of the police dispatcher. Patzius crashed into the plaintiff's car and she was seriously injured. The plaintiff sued the pursued, pursuers, and the city. The usual rule in Missouri is that the police are not liable to third persons injured by a fleeing driver. The trial court granted the city and the officer summary judgment. The court of appeals reversed as to the officers. If they were acting outside the scope of their employment, here engaged in a personal vendetta, and were negligent, they may be found liable to the plaintiff. The city itself is protected by sovereign immunity. They are not liable under *respondeat superior*, and the accident did not arise out of the operation of a motor vehicle within the scope of the officers' employment. *Bittner v. City of St. Louis,* 925 S.W.2d 495 (Mo. App. 1996).

In a 1986 decision the Missouri Supreme Court took up the issue of the liability of pursuing police for an accident in which the pursued driver causes the harm. The case arose when an Ellisville police officer noticed a car driven by a juvenile and attempted to stop it. The driver fled at high speed. Officers from two other towns joined the pursuit, and one officer, who learned of the pursuit over his radio, parked his cruiser as a roadblock on the side of a divided highway. The officer stood 100 feet from his vehicle on the other side of the median. The suspect, who had driven dangerously, swerved into the oncoming lanes and killed the officer. The widow sued the pursuing officers. The trial court dismissed the claim, the court of appeals reversed, and the supreme court took the case. The court cited a number of cases from other jurisdictions and found the defendants not negligent or liable. The pursued driver was solely at fault. The court noted that this is the majority position among the states. The defendant officers have two duties. One is to apprehend law violators and prevent them from harming others, the second is to do this in a manner that is neither careless, reckless, nor wanton. The court did not agree with the plaintiff that in attempting to stop the driver, the officers negligently caused the accident. They did not conduct the pursuit negligently. *Oberkramer v. City of Ellisville,* 706 S.W.2d 440 (Mo. 1986). Note that this is a Missouri Supreme Court case. That court rarely accepts such cases.

We look at two more recent cases similar to *Oberkramer.* The court of appeals made the same finding in a 1998 case. The St. Joseph police department received a report of a man trying to break into a residence. They investigated, and found a man sitting in an old Chevy Caprice (the owner had bought it to use in a demolition derby) and another man nearby. The man behind the wheel took off and ran a stop sign. The officers activated their siren and lights and pursued. The suspect, later found to be drunk, sped to 80 mph, lost control, and wound up in the plaintiffs' living room. The court affirmed that the officers were not negligent. The plaintiffs failed to plead any acts of negligence in the pursuit, nor that the pursuit violated any statute or regulation of the department. The decision to pursue is not a negligent act. The proximate cause was the pursued driver. The court also found for two other defendants, an insurance company and the owner of the car. *Sansonetti v. City of St. Joseph,* 976 S.W.2d 572 (Mo. App. 1998). For the same result, see *Peoples v. Conway,* 897 S.W.2d 206 (Mo. App. 1995).

A 1997 case arose from a fire truck parked on the shoulder of I-44. Two trucks responded to the call of a trailer fire. One truck's crew determined that the fire was in their district, and after borrowing the help of one of the other truck's crew, told the rest of the second crew that their help was not needed. The driver parked on the 7'10" truck on the 7'2" shoulder and waited. Ten minutes later the plaintiff driver, following two other vehicles, hit the truck when the lead vehicle, a tractor-trailer, suddenly slowed and swerved. The case does not say whether it was day or night. There was conflicting evidence on whether the fire truck had its emergency lights on. The court of appeals affirmed an award for $100,000 less 65 percent for the plaintiff's negligence. The jury apparently believed that either the fire truck was not in an emergency mode or that its emergency lights were not operating. In either case, the fire department was not entitled to emergency vehicle protection. The defendant claimed immunity under both the public duty doctrine and official immunity. These defenses were not applicable because the emergency vehicle statutes were not followed. *Pace v. Pacific Fire Protection Dist.,* 945 S.W.2d 7 (Mo. App. 1997). I will comment that inasmuch as the fire truck was at the scene and not returning from a fire run, it probably was in an emergency situation. If the jury had believed that its flashing lights were on, as the crew claimed, it should not be liable.

The facts in a 1993 motorcycle pursuit case are very similar to the U.S. Supreme Court case of *Lewis* (see California section). A Lee's Summit police officer saw a motorcyclist speeding 60 mph in a 30-mph-zone. There was a passenger. The officer pursued with siren and lights on winding and hilly streets. The officer had to accelerate to 100 mph to catch up, and the motorcyclist drove at 80 mph. The pursued driver stopped on the crest of a hill, the officer hit the motorcycle, and the passenger was injured. (The reader should know that larger motorcycles have a much greater power-to-weight ratio than cars, and can both accelerate and slow more quickly than cars.) The jury awarded the passenger $20,000, and both parties appealed. The passenger argued that an instruction to the jury—to consider whether the passenger was negligent if he knew he was in danger and failed to warn the driver if he had the time to do so—was improper. The court held this instruction to be proper. Any passenger has this obligation, and if breached, that is evidence of negligence. (Note that in pursuit cases, Illinois specifically holds that the burden of proof is on the passenger to prove that he is an innocent party.) Here the motorcyclist stopped momentarily in a driveway after pursuit had begun, and the passenger was at liberty to get off. An instruction stating that the jury could consider whether the passenger knew the operator had been drinking and whether such drinking impaired the operator's skills was proper. In fact, the passenger knew that the operator had had five beers, and should have known he was impaired from the erratic driving. The court notes that erratic driving goes beyond mere carelessness. A tendered plaintiff's instruction on whether the officer pursued negligently was properly refused. The evidence did not support the claim that the officer followed too closely under the hilly circumstances.

The trial court correctly granted a verdict in favor of the officer as enjoying official immunity. The city's pursuit policy states that there are three critical factors to consider: when to pursue, how to pursue, and when to terminate pursuit. All of these are "fraught with discretion." It is not possible in a department's policy to set out specific, detailed procedures for the nuances of how to conduct a pursuit. The court did note that stipulated medical expenses were $34,000, and amended the award to that figure. Nothing was collectible from the motorcycle operator, so the city is liable for the amount under the collateral source rule. *Hagedorn v. Adams,* 854 S.W.2d 470 (Mo. App. 1993).

We end the Missouri section with three cases filed in federal court. A 1993 case involved a major league chase. An adult and a minor stole a car (which one was driving was not given) and entered I-35 shortly before 7:00 A.M. They drove past a police officer. The officer gave chase, and was joined by two highway patrol officers and two deputy sheriffs (the chase defendants). Speeds were up to 100 mph. At one point the officers tried a rolling roadblock without success. Two city of Liberty police officers (the roadblock defendants) learned of the incident, and got ahead of the pursuit and parked their cruisers on both sides of the interstate. The roadblock defendants then drove onto southbound I-35 and stopped all traffic, which caused a mile-long traffic stop. As the chase came up, the pursuing officers drove on each shoulder, but the pursued driver drove between cars, killing one person, injuring others, and causing property damage. This case arose when a class of plaintiffs in the stopped cars, in effect involuntary roadblocks, brought a number of claims, including relief under state law and §1983 action plus other federal constitutional claims.

There is no substitute for reading the case, but I will try to summarize the decision of the district court judge. The judge refused to enjoin the defendants from forming such involuntary roadblocks in the future, as there appears to be an adequate remedy at law. Nor have the plaintiffs shown a probability of being so treated in the future.

There were several state law claims. The defendants claimed the protection of the public duty doctrine. State officials are not liable for injuries or damages to individuals when exercising a duty owed to the general public. Here, the court doubted whether the duties created by the emergency vehicle statutes created specific duties to individuals. However, it could not say the same about standards, policies, and procedures without reviewing them, and refused to dismiss the plaintiffs' claims to the extent they rely on these documents.

The court was dubious about whether the plaintiffs could defeat the defense of immunity under official immunity for discretionary acts, but would not dismiss the claims on this ground. The plaintiffs may try to show that the "arrogant disregard" exception for the safety of the public that would overcome official immunity.

Turning to federal constitutional claims, the court held that the Thirteenth Amendment was not violated. That amendment ended slavery; here requiring the assistance of a citizen by the police is a reasonable exercise of police power. The freedom of employment claim (forcing the plaintiffs to act as agents of the government by being a part of the roadblock) was dismissed as being without merit. Ditto for the violation of the freedom to travel claim. Likewise for the false imprisonment claim. The court refused to dismiss other claims, holding that more facts have to be developed before those claims can be disposed of by summary judgment or proceeding to trial. These include whether a reasonable police officer under the circumstances would know that he may have been violating the plaintiffs' Fourth Amendment rights, and the §1983 claims against the supervising officer. The plaintiffs alleged that Officer Houston was aware of the plans to set up the roadblock and was deliberately indifferent to the rights of the plaintiffs. *Boyle v. City of Liberty,* 833 F. Supp. 1436 (W.D. Mo. 1993).

A 1996 case is much easier to summarize. It is not truly an emergency vehicle case. An Overland Park police officer allegedly ran a red light without activating his warning lights and siren. He also tested positive for a controlled substance, marijuana. The plaintiffs alleged that the city was aware of the officer's previous drug problems and did not take appropriate measures. The plaintiffs filed a §1983 action. The district court judge dismissed the claim. There was no violation of substantive due process, citing primarily *Rochin v. People of California,* 342 U.S. 165 (1952). The bad acts alleged are not shocking to the conscience. This is just an ordinary tort claim. *Friedman v. City of Overland Park,* 935 F. Supp. 1015 (E.D. Mo. 1996).

An Eighth Circuit case arose out of a motorcycle incident. A Farmington police officer overheard on his cruiser radio that a highway patrol officer was attempting to catch a motorcyclist traveling on Route 67. Shortly afterward the police dispatcher notified the officer that the patrol had made a "generalized" request of help. The officer drove to Route 67 and parked in the median, waiting for southbound traffic, as he overheard that the fleeing biker was heading south. After two minutes, a southbound motorcycle appeared around a curve. There were no other vehicles, and what happened was in dispute. There was no evidence in the court's record that the motorcyclist was in fact the one being chased. The motorcyclist, Hawkins, said that he saw the cruiser's emergency lights come on and that the car began crossing the median. He thought the officer would turn into the left southbound lane, so he moved to the right lane. The officer continued across the lanes and hit his left side. There were serious injuries, including a broken back and an amputated leg.

The officer's account was that Hawkins was going very fast. (The officer got his figures wrong. He estimated that the motorcyclist traveled one half mile in seven seconds, which would be 250 mph.) The officer said that he intended to slow down or stop Hawkins, and crept into the southbound land with his lights and siren operating. He lost view of Hawkins because of the roof pillars and then the collision happened. The district court judge granted summary judgment to the city and police department, and the trial proceeded against the officer. The claim was that the officer made an unconstitutional seizure under the Fourth Amendment. After the plaintiff's evidence was presented, the judge dismissed the claims against the officer. There was no seizure and the officer enjoyed official immunity under Missouri law. The Eighth Circuit affirmed the finding that the officer's allegedly negligent acts were discretionary acts, and thus immune under official immunity. The court reversed summary judgment for the city. The city may be liable for negligent acts of its employee who was acting in the scope of employment. *Hawkins v. City of Farmington,* 189 F.3d 695 (8th Cir. 1999).

A 1989 case involved a Fredericktown police officer who tried to stop a suspicious driver whose car plates did not match the vehicle. The driver fled, lost control, and collided with an oncoming car. The officer hit the crash debris, and in the two collisions the fleeing suspect was killed, and both his passenger (Roach) and two innocent bystanders (the Kellers) were injured. Their claims against the officer and the city were dismissed, and the Eighth Circuit consolidated both cases in this appeal. The court affirmed the dismissal. Citing the U.S. Supreme Court case, *Brower* (see California section), the court held that Roach was not illegally seized. True, the officer sought to stop the vehicle, but here the stop was not attempted "through means intentionally applied." In other words, the officer did not intend the pursuit to end in the collision with another vehicle. As to the Kellers, the court characterized their claim of "unreasonable interference" as a Fourth Amendment violation of an unreasonable search or seizure. (Section 1983 is not of itself a source of substantive rights, but provides a method of vindicating rights elsewhere conferred.) The court also found against the plaintiffs on their claims of excessive force, negligence, and inadequate training. *Roach v. City of Fredericktown,* 882 F.2d 294 (8th Cir. 1989)

Montana

Mont. Code Ann. 61-8-107. Police vehicles and authorized emergency vehicles

[This section sets out UVC 11-106 with one change. An emergency vehicle operator on an emergency run has the protection of the statute when "making use of an audible or visual signal, or both." The UVC provision requires both.]

Only one case has construed §61-8-107. The case arose from an accident between a Missoula police officer and another driver. The officer was on parking violation duty when he received a report of a disturbance at a store. He activated his lights and siren and proceeded toward the scene. He entered an intersection on a red signal and in the opposing lane, hit the passenger's side of the plaintiff's car, and the plaintiff sued for personal injury. The main issue for our purposes is the standard of care instruction. The trial judge instructed the jury that the standard of care for all drivers is the conduct of an ordinarily prudent person in the same or similar circumstances. The jury found for the plaintiff. On appeal, the officer argued that it was error to impose the ordinary standard on him in view of §61-8-107. The Montana Supreme Court affirmed. The court noted that the statute requires an emergency vehicle operator to exercise due care, which can be termed ordinary care. The court declined to construe the "reckless disregard" language in the statute (an emergency vehicle operator is not protected from the consequences of his reckless disregard for the safety of others) as a higher standard to control over due care. Having no Montana precedent, the court cited a Vermont and a Kansas case in support of its decision. *Stenberg v. Neel,* 613 P.2d 1007 (Mont. 1980).

Nebraska

Neb. Rev. Stat. Ann. §13-911. Vehicular pursuit by law enforcement officer; liability

In case of death, injury, or property damage to any innocent third party proximately caused by the action of a law enforcement officer employed by a political subdivision during vehicular pursuit, damages shall be paid to such third party by the political subdivision employing the officer.

(1) [This section provides that a political subdivision paying damages, whether in voluntary settlement or in satisfaction of a judgment, shall be entitled to reimbursement from all applicable liability insurance, whether liability, disability, or medical. The damage cap is $1 million for one claim, and $5 million for all claims arising from one event.]

(6) For the purposes of this section, vehicular pursuit means an active attempt by a law enforcement officer operating a motor vehicle to apprehend one or more occupants of another motor vehicle, whether the driver of the fleeing vehicle is or should be aware of such attempt and is resisting apprehension by maintaining or increasing his or her speed, ignoring the officer, or attempting to elude the officer while driving at speeds in excess of those reasonable and proper under the circumstances.

Neb. Rev. Stat. Ann. §81-8,215

[This section, in part, is the same as 13-911, except it applies to the State of Nebraska.]

Neb. Rev. Stat. Ann. §29-211

[This section requires every law enforcement agency to adopt and implement a pursuit policy. It gives the standards required and requires that each officer be given pursuit training.]

Neb. Rev. Stat. Ann. §60-6,114. Authorized emergency vehicles; privileges; conditions

[This section sets out UVC 11-106 without change. As such, it requires an emergency vehicle operator to use both siren and flashing or rotating lights when taking advantage of the privileges granted to such drivers. Many states make some provision in the statute for silent approach to crime scenes.]

[Nebraska has unique statutes, §§13-911 and 81-8,215, applying to political subdivisions and to the state. The other states do not have an equivalent statute. Note that the statutes are not a blanket order to pay. Whether the claimant is an innocent party or whether the officer or officers proximately caused the damages must still be determined by negotiation or in court. In that respect, the statutes are not as far-reaching as seems on first glance. It is noteworthy that the maximum payable is much higher than in most other jurisdictions. Also, no kind of immunity defense is recognized.]

Omaha is the principal city in Nebraska. The *Omaha World-Herald* reported that from 1993 to Oct. 16, 2001, the city has paid more than $7 million in settlements or judgments in pursuit claims. *Omaha World-Herald,* Oct. 17, 2001, p. 1–2. Most of the payments are a result of five large payments rather than numerous smaller amounts. The number of pursuits has dropped in the last few years, especially 2000 and 2001. Police officers are advised to think and keep thinking whether to begin a pursuit and when to terminate one. As everywhere, the officers must weigh the importance of catching the fleeing driver against the dangers of the chase.

A Dec. 28, 2001 *Omaha World-Herald* account of a $565,000 settlement brought up a point frequently made by plaintiffs in such cases. The culprit was armed and had threatened to kill any police officer who tried to stop him. The police pursued, but as I recall from the newspaper accounts, the officers had backed off before the driver crashed into another vehicle and killed one person and left the other semicomatose. The police did know the identity of the driver. The plaintiff told the reporter that she did not sue for money, but rather to make the state patrol change its pursuit policy. An attorney for the state said, "No policy change has been made, no policy change will be made, and no policy change should be made. It was a pursuit that had to happen."

A plaintiff can allege violation of both statutes, that of innocent party and of negligent pursuit. A plaintiff in a 1993 motorcycle pursuit case so alleged. The innocent party was a passenger on the motorcycle. An Omaha police officer noticed a motorcyclist who he thought had eluded him four days earlier. The officer wanted to check the vehicle's registration, which had an expired plate. The passenger was killed in the two-minute pursuit through residential neighborhoods, when, after running several stop signs, the motorcycle operator ran into another vehicle. The passenger was thrown headfirst into a lamp pole. The parents alleged that the victim was an innocent passenger whose death was proximately caused by the chase, and that in the alternative the chase was wrongfully begun and conducted. A major issue was whether the victim had an opportunity to get off the motorcycle and declined to take it. If so, that would be relevant to her status as innocent victim. The supreme court affirmed an award of $356,932 against the city. There was evidence that the victim had two chances to get off: when the pursuit was about to begin and the motorcycle was stopped, and at a later point when the operator stopped again.

The court held that it was not clear that either were clear opportunities to get off, and in any event the officer could be found to have conducted the pursuit in violation of department's pursuit policy. Either finding justified the judgment. (The court notes that its policy is, as is usual, to construe the evidence in favor of the prevailing party.) *Stewart v. City of Omaha,* 494 N.W.2d 130 (Neb. 1993).

A claimant under §13-911 must satisfy three things to prevail. There must be a pursuit, the pursuit must be the proximate cause of the injury or damage, and the claimant must be an innocent party. In a 1992 case, two teenagers were killed by a speeding driver at an intersection. The wrongdoer had no liability insurance, but there was uninsured motorists coverage on the victims' vehicle, and the insurer paid $100,000 to each of the estates. The insurer then sued the city of Omaha for reimbursement, alleging that the wrongdoer was being pursued by a police officer. (Uninsured motorists coverage is not automatically paid. The insurer is entitled to claim against any party it thinks is the cause of the accident. Whether the estates claimed against the city is not revealed.) The case was tried before a district court judge, and he found that the sole cause was the speeding driver, and that the pursuit was not a proximate cause. The supreme court affirmed. The case has a long description of the events leading to the crash. Officer Vacarro responded to a call of a disturbance at a bar, and the bleeding bartender told Vacarro to stop a Buick that had just left. He did that, and learned that the driver, Williams, had a criminal record and was considered dangerous. He took Williams' operator's license, and the dispatcher told him to bring Williams to the scene of an assault and reported hit-and-run accident. Williams heard this on the radio, got back in his car and took off. Vacarro did not pursue because he thought he could locate the suspect later with the license number. Another officer, knowing of the events, saw Williams going very fast on 72nd Street, and got close enough to try to ram him (spin him out of control), but just missed. The officer watched Williams speed up to 90 mph (the speed limit is 40 mph), but did not attempt to pursue. He saw the fatal crash. Here there was no pursuit, or if there was one, it was not the cause. *Mid Century Ins. Co. v. Omaha,* 494 N.W.2d 320 (Neb. 1992).

I will comment that this case shows the ambiguities in pursuit cases. Here the second officer had the opportunity to try to physically stop the wrongdoer, and was right to try. Such cases indicate the decisions officers have to make quickly. Williams' dangerous speeds on a city street were not caused by the pursuit. The police in such cases are faced with several choices, all of which can be criticized in hindsight: try to catch and stop a driver who is posing a continuing danger to the public, but don't endanger the public in so doing, or don't try.

An older case came up in a criminal context. The defendant was attempting a left turn at an intersection in Lincoln and collided with a motorcycle police officer, killing him. The officer was responding to a report of "man down" and entered the intersection on a red signal at 40 to 45 mph. The speed limit was 35 mph. The officer had his siren sounding, but the defendant and family members in his car did not notice it. The defendant was convicted of vehicular homicide and sentenced to six months probation. He contended on appeal that it was error to instruct that the officer had the right of way, and that the defendant had the duty to yield. He contended that the officer was not pursuing a law violator as required by the statute. The supreme court affirmed the conviction. It would not read the statute so narrowly. The officer was entitled to emergency vehicle status. As to the siren, the fact that the defendant did not hear it is not a defense. The statute requires an "audible" signal, meaning capable of being heard. The siren could have been heard in the exercise of ordinary care. A dissent thought that the defendant did not exhibit a reckless disregard for the safety of others; he just didn't hear the siren. (The author would agree.) *State v. Otto,* 169 N.E.2d 612 (Neb. 1969).

A 1986 case involved a motorcyclist, but not in a pursuit context. Maple, the motorcyclist, was riding in the curb lane of Pacific Street. He approached an intersection five lanes wide, including the left turn lanes. He saw that his traffic signal was green. His view was blocked by school buses in the adjoining lanes. He heard a siren, and presumed that its source was a fire truck that had just made a right turn ahead of him. Even though adjoining traffic did not move, Maple entered the intersection at 35–40 mph. He collided with a police officer making a left turn at 2–3 mph, and suffered a broken leg. He sued the city of Omaha, and the city counterclaimed for property damage. The trial judge denied recovery for both parties. The supreme court reversed as to the city, and found it should recover property damages. The officer was properly dispatched to a personal injury accident. He was entitled to claim emergency status, with proper signals, and turn cautiously on red. He was responding to an accident call, with the information that there were two injured persons and was told that the injuries were not serious, according to a bystander. The city argued that the officer should be held to the reckless disregard standard, rather than due regard. Both are in the statute, with no indication that one or the other should control. The court did not see a major difference, and held that in any event the officer met the due regard standard. As to the plaintiff, his argument essentially was that one is not under a duty to yield to emergency vehicles until one is aware of them. The court held that the plaintiff's duty to yield does not depend on whether he accurately perceives the emergency vehicle. *Maple v. City of Omaha,* 384 N.W.2d 254 (Neb. 1986).

A 1981 case indicates the standard by which to judge the actions of police. The accident happened in 1976, so §13-911 was not in effect. The case held that both the due regard and the reckless disregard standards are proper. Two Omaha officers in a fast food restaurant noticed a driver whom they recognized as still under a thirty-day license suspension. The driver fled rather than stop. The driver, Recek, had a propensity to bait and then flee from the police, both on foot and behind the wheel, and in fact would not stop if pursued. The chase lasted seven minutes. It ended when Recek ran a stop sign and killed an innocent person. The supreme court affirmed judgment for the city. The plaintiff pointed out that Recek's violation was not even a moving one, and that the officers knew the identity of the driver. The court found that the pursuit was not negligently conducted. The officers went through intersections carefully and exceeded the 35 mph-speed limit by 10 mph. The officers exercised due care. "It must be remembered that foresight, not hindsight, is the standard by which negligence is determined, and that even an action which in retrospect turns out to have been ill-advised may still have been reasonable under the circumstances." *Lee v. City of Omaha,* 307 N.W.2d 800 (Neb. 1981).

If there is no pursuit, then the statutes covering the rights of emergency vehicles and the duty of others to yield control. Statute 60-6,151 requires drivers to yield immediately to emergency vehicles that are using the proper audible and visual signals. The emergency vehicle operator must exercise due regard for the safety of others. A 1992 case involved an ambulance on a Code Three run. (Code Three indicates a life-threatening problem.) The driver was on the way to the hospital with his patient and came to a large intersection with a state highway, speed limit 55 mph. The ambulance driver slowed to 5 mph and saw that all traffic had stopped. He did not notice that his signal turned red as he came up. A 95,000 pound dump truck struck him when he had almost cleared. That driver saw his signal turn green and did not see the ambulance. Witnesses were surprised that the ambulance did not stop. The ambulance owner, a city, sued the trucker for about $12,000 property damages, and the truck owner made a similar claim. The district court judge found the ambulance driver negligent and awarded property damages. The supreme court agreed as to the ambulance driver's fault. Considering the truck operator's duty to yield, it could not agree that the driver's negligence was only slight. Under comparative negligence principles (recently adopted in the state), the case was remanded to reduce damages accordingly. *City of LaVista v. Andersen*, 480 N.W.2d 185 (Neb. 1992).

Nevada

Nev. Rev. Stat. Ann. 484.261. Authorized emergency vehicles

[This section enacts UVC 11-106 without change except for including funeral escort vehicles, and limits those vehicles to 15 mph over the speed limit. Both visual and audible signals are required on emergency runs.]

We need not spend much time on the single case, *Avery v. Gilliam*, 625 P.2d 1166 (Nev. 1981). An ambulance operator was proceeding on an emergency run and he switched on his red light and placed his siren in the manual position. This required him to push the horn button to sound the siren. He entered the intersection on a red signal and collided with the plaintiff. The jury was instructed on the statutory requirement that both audible and visual signals were required. A police officer, who happened to be standing at the intersection, testified that the siren did not sound. The plaintiff driver testified that he did not recall. The jury found for the defendant, and the trial judge declined to grant judgment notwithstanding the verdict. The Nevada Supreme Court remanded for a new trial on its finding that the verdict was "manifestly and palpably" contrary to the evidence. The evidence established that the plaintiff had the right of way, since the siren was not sounding. Normally the court would not disturb a verdict in which there is a conflict in the evidence, but here there was no conflict.

New Hampshire

N.H. Rev. Stat. Ann. §265:8. Emergency vehicles

[This section enacts UVC 11-106 without substantial change, and includes the duty to yield to emergency vehicles on emergency runs by other drivers and by pedestrians. Section 265:8 specifically requires an emergency vehicle driver to operate both audible and visual signals when proceeding past a red signal or a stop sign without stopping, and when exceeding the speed limit. Section 265:8a requires all law enforcement agencies to adopt pursuit policies that conform to national accreditation standards. Such policies must be available for inspection by the attorney general and the police standards and training council. Note that Section 265:61, entitled Speed Exception, also permits emergency vehicles to exceed the speed limit. The section appears to be redundant.]

Only four cases were found, none of which are recent. Taking the oldest first, a 1946 case held that any negligence of the operator of a fire truck on a fire run could not be imputed to his passenger, a fireman superior in rank. The general rule is that the negligence of a driver is not imputed to a passenger. The law of joint enterprise did not apply. The passenger sued a truck driver who did not yield upon hearing the siren. The New Hampshire Supreme Court affirmed a verdict for the passenger. A driver must yield to an emergency vehicle even though the latter is not on the same highway. *Clough v. Schwartz,* 48 A.2d 921 (N.H. 1946).

A 1958 case is bordering on the frivolous, but we need a little light relief from all these tragic accident accounts. A deer hunter was chased, stopped, and cited by a state trooper for illegally hunting from a motor vehicle. The hunter initiated a private prosecution of the trooper for exceeding the speed limit of 65 mph in the chase. The state attorney general defended the trooper, who was acquitted. The officer had a right to speed if he exercised reasonable care. In addition, he did not break the law of speed being reasonable and prudent under the conditions (the basic speed law). The supreme court took the case to determine whether it was lawful for the attorney general to be in the case, and it found such representation was proper. *State v. Swift,* 143 A.2d 114 (N.H. 1958).

Another odd case arose in 1978. The city of Rochester enacted a noise ordinance that had the effect of outlawing sirens, and an ambulance operator was convicted of a violation. The supreme court reversed the conviction. State law permits such sirens on emergency runs, and state law preempts ordinances. *State v. Driscoll,* 385 A.2d 218 (N.H. 1978).

A 1984 case involved a fire truck and a motorist. The fire station in Kingston received a fire call on a very cold night just after the Blizzard of '78, so four-foot snow banks lined the streets. The firemen neglected to flip the switch at the fire station to turn on a flashing light at an intersection 500 feet away. The fire truck proceeded to Hunt Road, which has a stop sign. The cross street, Route 125, does not. The fire truck operator intended to turn left. He entered the intersection without stopping and collided with a couple in a car. Both were knocked unconscious. The jury returned a plaintiffs' verdict. The supreme court remanded for a new trial. The trial judge instructed on the rights of an emergency vehicle by reading the statute, but he omitted subsection VI setting out the duty of other drivers to yield. The court pointed out that such duty to yield applies when a driver sees or hears, or should have seen or heard, the visual and audible signals. There was evidence that the truck's warning lights were on. Also, the trial judge instructed on *either* audible or visual signals, not *and* as given in the statute. It makes a difference. No eyewitness claimed to have heard a siren. (The plaintiffs did not recall the accident.) *Saltzman v. Saltzman,* 475 A.2d 1 (N.H. 1984).

New Jersey

N.J. Stat. Ann. §39:4-91. Right of way of emergency vehicles

a. The driver of a vehicle upon a highway shall yield the right of way to any authorized emergency vehicle when it is operated on official business, or in the exercise of the driver's profession or calling, in response to an emergency call or in the pursuit of an actual or suspected violator of the law and when an audible signal by bell, siren, exhaust whistle or other means is sounded from the authorized emergency vehicle and when the authorized emergency vehicle, except a police vehicle, is equipped with at least one lighted lamp displaying a red light visible under normal atmospheric conditions from a distance of at least five hundred feet to the front of the vehicle.

b. This section shall not relieve the driver of any authorized emergency vehicle from the duty to drive with due regard for the safety of all persons, nor shall it protect the driver from the consequences of his reckless disregard for the safety of others. Nothing in this section shall be construed to limit any immunity or defense otherwise provided by law.

N.J. Stat. Ann. §59:5-2. Immunity

Neither a public entity nor a public employee is liable for:

a. An injury resulting from the parole or release of prisoner or from the terms and conditions of his parole or release or from the revocation of his parole or release.

b. Any injury caused by:
   (1) an escaping or escaped prisoner;
   (2) an escaping or escaped person;
   (3) a person resisting arrest or evading arrest;
   (4) a prisoner to any other prisoner; or

c. Any injury resulting from or caused by a law enforcement officer's pursuit of a person.

N.J. Stat. Ann. §59:3-14. Public employee immunity—exception

a. Nothing in this act [Tort Claims Act] shall exonerate a public employee from liability if it is established that his conduct was outside the scope of his employment or constituted a crime, actual fraud, actual malice or willful conduct.

b. Nothing in this act shall exonerate a public employee from the full measure of recovery applicable to a person in the private sector if it is established that his conduct was outside the scope of his employment or constituted a crime, actual fraud, actual malice or willful misconduct.

N.J. Stat. Ann. §59:3—3. Execution or enforcement of laws

A public employee is not liable if he acts in good faith in the execution or enforcement of any law. Nothing in this section exonerates a public employee from liability for false arrest or false imprisonment.

[New Jersey has a very interesting statutory scheme. Note that the last sentence of the first two statutes above was added by a 1997 amendment, effective Jan. 19, 1998. What the legislature did was to reassert sovereign immunity in pursuit cases. The various states fall into two basic categories. The first decides police pursuit cases under the state's version of UVC 11-106. The second employs some version of governmental immunity. New Jersey is clearly in the second category. In fact, although I have set out §39:4-91 above, the last sentence and the courts indicate that it is simply not applicable to police pursuit cases. It is applicable to other emergency vehicle cases, so the statute is not without effect. Note that it covers the essence of UVC 11-106 but omits some of the details, such as specifying that emergency vehicle operators may disregard parking, stop signs and red signals, speed, and turning and direction statutes.

Section 59:5-2 originally dealt with immunity for acts involving prisoners and parolees. The 1997 amendment added the words "or evading arrest" to subsection (b)(3), which could apply to a fleeing driver. Subsection (c) is entirely from the 1997 amendment and points only to police pursuit situations. So it is clear that the legislature intended to provide immunity here.

Section 59:3-14 sets out the exceptions to immunity. The case notes under this statute indicate that willful misconduct must go well beyond the absence of good faith or mere negligence. It must be the commission of a forbidden act with knowledge that it is forbidden, and the knowledge that there are no exceptions. In the context of a police pursuit, for example, if the department's pursuit regulations state that no more than two police vehicles may pursue unless more are specifically authorized by a superior officer, then an officer who joined in the pursuit without authorization might be denied immunity. The manner in which to conduct a pursuit, with the ever-changing circumstances and the rapid judgments that an officer must make, would normally not be willful misconduct.]

We examine first a recent (2000) case which indicates that immunity applies to a pursuit situation. One should know that police officers sometimes must engage in catch-up driving. They need to catch up with a driver to check his speed or get a better look at the car and its occupants. They are not in pursuit as such, since they are not attempting to stop the driver, but often must speed. Unless they switch on lights and siren (it depends on the exact state statute), they are not entitled to the protection of emergency vehicle status. This case arose when a police officer noticed an apparently speeding driver. The officer made a U-turn and sped to catch up. The officer hit a pedestrian running across the street, and the pedestrian sued. Each party claimed the signal light. The pedestrian saw no lights and heard no siren, while the officer said that he had activated both. There was no evidence that the speeder, who of course was never identified, knew he was pursued. The trial judge granted summary judgment to the officer and the city. The court found that the officer was immune because he was pursuing "an escaping person." (Subsection (c) of §59:5-2 was added after the accident, and so did not apply.) The superior court remanded for further proceedings. It did find that the officer was not engaged in a pursuit so §59:5-2 could not apply. The court suggested that the trial court determine if immunity applied under §59:3-3. (See above.) *Torres v. City of Perth Amboy*, 748 A.2d 125 (N.J. Super. A.D. 2000).

Torres above referred to the case of *Canico v. Hurtado,* 676 A.2d 1083 (N.J. 1996), a New Jersey Supreme Court case. Newark police officer Hurtado was dispatched to respond to a bank alarm. He had all signals flashing and sounding, and approached an intersection with the green light at 30 mph. The car in front, going the same direction, attempted a left turn as the officer passed, and the civilian driver was injured. She sued the officer. The supreme court affirmed that the officer had good faith immunity under §59:3-3 of the Tort Claims Act. The legislature has established immunity as the general rule and liability as the exception regarding the acts of public officials. The basic duty of a police officer is to enforce the law, and that is what he was doing. Responding to a possible bank robbery, a crime-in-progress, is objectively reasonable and the officer is entitled to good faith immunity. Even if the officer were negligent, and his use of emergency signals and moderate speed show an attempt to proceed with some care, he and his employer are entitled to immunity.

A 1993 case decided by the New Jersey Supreme Court, *Tice v. Cramer,* 627 A.2d 1090 (N.J. 1993), is significant. It was decided before the 1997 amendments to §59:5-2 were added, and it dealt with the meaning of an escaping person or a person resisting arrest in a pursuit context. A Wildwood police officer was dispatched to a reported melee at a tavern. He alleged that as he approached, a driver in a parking lot pulled out and almost hit him. The driver left without turning on his lights and the officer pursued. The pursuit lasted for five minutes with both drivers speeding and breaking traffic laws. (The fleeing driver claimed that a passenger threatened to kill him if he stopped.) The pursuit ended when the pursued vehicle crashed into a third driver and two people were killed. Some claims settled, but one against the officer and the city went to the trial court. That judge granted the defendants summary judgment, and the intermediate appellate court affirmed. The supreme court affirmed.

The court first noted that the evidence would support a jury finding of officer negligence. However, the issue is whether the officer was immune, not whether he was negligent. The court rejected liability based on §34:4-91. That statute states that an officer will not be protected from the consequences of his reckless disregard for the safety of others. The court reaffirmed the holding that immunity under the Tort Claims Act trumps the emergency vehicle statute. The question is not whether an emergency vehicle has the right of way. The court also noted but did not rule on good faith immunity under §59:3-3. Such immunity has been applied to police pursuits. Rather, the court focused on §59:5-2. Attempting to apply "resisting arrest" to pursuit cases has problems. A traffic stop for an infraction such as speeding is not an arrest. The individual normally stops, is issued a warning or a ticket, and then is free to leave. He is not arrested. In the course of a pursuit, however, the driver might commit other acts that justify an arrest and charges. The language in §59:5-2(b)(2) of an escaping or escaped person takes care of this problem. A fleeing driver is clearly an escaping person and thus immunity under that section applies. One does not have to determine whether the officer intended an arrest, or at what point he formed that intention.

The plaintiff also argued that the statute did not rule out liability for concurring officer negligence, as the statute does not grant immunity for any injury *solely* caused by an escaping person. The court points out that such wording is no immunity at all: no police officer is liable unless he is also negligent. Immunity thus is appropriate in police chases regardless of whether an officer's conduct is a concurring cause. The only exception is if there is willful misconduct. The court has an extensive discussion of California law because that jurisdiction has the same statute and substantial case law pertaining thereto.

The New Jersey Supreme Court revisited the issue in a 1995 case in which the police vehicle was actually involved in the collision. A Tinton Falls police officer stopped a motorcyclist for speeding. While the officer was back in the patrol car checking the operator's address (he did not have a license with him), the motorcyclist sped off and a pursuit began. The culprit and three police cars went through a major intersection on the green light. A following police officer was somewhat behind, and he entered, with appropriate signals, on a red light, and severely injured a driver with who had the green light. The trial court declined to grant the defendants summary judgment, and the court of appeals affirmed. The issue was whether *Tice* applies to cases in which the police car itself is involved in the crash. The case went to the New Jersey Supreme Court. The court reversed. The Tort Claims Act is not an insurance policy; it was intended to restore sovereign immunity. It simply makes no sense to hold that immunity (absent willful misconduct) applies under §59:5-2b(2) (an escaping person) only if the officer's vehicle does not collide with the innocent party's car. This result would leave the protection of immunity to mere chance, and that cannot have been the legislature's intention. The court finds the legislative purpose to be to encourage the vigorous enforcement of the law, unfettered by the possibility of liability. Even if the purpose was to save money by not paying compensation, as the plaintiff suggested, the court will not indulge in second-guessing.

The plaintiff argued that there was willful misconduct. Previous case law defined willful misconduct as the commission of a forbidden act with actual knowledge that the act is forbidden. The court found that there was an issue of material fact as to willful misconduct, and remanded on this point. The court affirmed summary judgment for the employing city because it cannot be liable for an employee's willful misconduct, pursuant to statute. The officer at some point heard on his radio an order to terminate the pursuit when the situation became dangerous to himself and others. If the officer in fact knew the situation was dangerous and he continued the pursuit, the issue of willful misconduct is a jury question. The court notes that there can be liability in a federal court regardless of immunity under the state tort act. *Fiedler v. Stonack,* 661 A.2d 231 (N.J. 1995).

A 1996 case in the superior court also involved a motorcyclist. There was also a claim of willful misconduct. The soon-to-be-deceased motorcyclist participated in a softball game with coworkers. The group then went to a bar for the next four hours. The motorcyclist left around midnight and his erratic driving was soon noticed by city police. He fled a pursuit, ran a red light, and sped to 80 mph. He then failed to make a curve, crashed into a guardrail, and died. The trial court granted summary judgment to all parties. (The bar was a defendant and settled.) The pursuing officers would be immune in the pursuit of an escaping or escaped person absent willful misconduct. The plaintiff argued that the officers violated General Order 83-3 of the Princeton Police Department. That order says not to begin a high-speed pursuit for a nonmoving violation. Here the officer witnessed the deceased weaving, overtaking dangerously, speeding, and running a red light. The pursuit was justified. On another point, the court held that whether the correct standard in a pursuit case of a federal constitutional claim is reckless indifference or shocks-the-conscience, neither standard was met. Finally, it was correct to dismiss the deceased's employer for encouraging attendance at the softball game and the after-game beer and pizza party. The employer had no control over alcohol sold at the bar. It was not an employer's function. *Kollar v. Lozier,* 669 A.2d 845 (N.J. Super. A.D. 1996).

A 1996 case does not involve an emergency vehicle, but deals with the issue of the right of way of a funeral procession in New Jersey. Other state laws or cases may have a different result. The plaintiff was driving in the procession to the cemetery. He had been told by funeral home employees to drive with his headlights on, but had no instruction on traffic controls. It was not the policy of the funeral home to instruct in any manner on traffic controls. He entered an intersection on red and collided with driver who had the green. He was following the car in front at two car lengths and was going 20 mph. Each driver sued the other. The trial court found each driver to be negligent but found that the negligence of the driver with the green signal was not the proximate cause. The superior court affirmed in part. The funeral home was not negligent. No evidence was presented to show that it created a hazardous condition. No statute, including 39:4-91, grants the right of way to a funeral procession. The fact that most drivers do yield is a result of custom and practice. The court remanded for a new trial to reconsider the negligence of the driver with the green signal. The driver testified that she looked but did not see the plaintiff. *Giantonnio v. Taccard,* 676 A.2d 1110 (N.J. Super. A.D. 1996).

We end the New Jersey section by briefly noting two cases brought in federal court. As the New Jersey Supreme Court noted above, state or local immunity in a pursuit case does not mean that a plaintiff cannot recover damages in a federal constitutional claim. (Unlikely, but not impossible.) The first case is *Fagan v. City of Vineland,* 22 F.3d 1283 (3rd. Cir. 1994), and the same case reargued en banc, 22 F.3d *Fagan v. City of Vineland,* 22 F.3d 1297 (3rd. Cir. 1994). Let me say at the outset that these two decisions run to 35 pages. In the early morning hours, a Vineland city police officer noticed a Camaro with a passenger standing up through the open T-top roof waving his arms. The officer followed the Camaro, intending to tell the driver not to permit his passengers to do that. The driver did not stop, and the officer turned on his emergency lights. The driver did not speed or commit traffic violations as first, but the pursuit eventually escalated into a high speed chase. Up to four police cars were involved. (Later, it was found that the driver was drunk, so there was a motive to flee.) The pursuit ended in a crash when the fleeing driver ran a red light at high speed and hit an innocent driver. Three people in the vehicles were killed and others seriously injured. (Note that because the court is reviewing summary judgment, the facts are given in the light most favorable to the plaintiffs.) The plaintiffs alleged that the pursuit was in violation of the Attorney General's guidelines, and that the city failed to train and supervise its officers in pursuit situations. Another claim was that the city followed a policy of not enforcing its pursuit guidelines. The district court judge granted the city and officers summary judgment.

The Third Circuit reversed on a finding that the city can be liable for a Fourteenth Amendment violation, here an allegation of deliberate indifference for failure to train, even if the officers themselves are not liable under the shocks-the-conscience standard. The somewhat longer en banc second opinion amplified on the finding that shocks-the-conscience is the correct standard. The opinion reviews and discusses the leading (at that time) cases in other circuit courts and the U.S. Supreme Court.

A 1999 case arose when a pursuing police car tried to pull over a suspicious driver to make a vehicle plate check. (It turned out that the car was stolen.) A pursuit ensued. It ended when the officer bumped the rear of the car. The driver hit his head and the car spun out of control and seriously injured a pedestrian, the plaintiff. The defendants were granted summary judgment, and the Third Circuit affirmed. The issue was whether *Sacramento County v. Lewis*, 523 U.S. 833 (1998), controlled. The plaintiff alleged that the officers were not acting on a report of a possible crime, that they willfully violated pursuit regulations, and that they used deadly force to stop the fleeing vehicle. The court found that *Lewis* applied and that summary judgment was appropriate. The officers were faced with the driver's flouting of the law, their reaction was instantaneous, and their motives were to do their job, not inflict injury. Even if they acted recklessly, there was no evidence that their actions "were tainted by an improper or malicious motive." A concurrence hoped that the *Lewis* case would not be applied over-broadly, noting that here the ramming of the car was intentional. *Davis v. Township of Hillside,* 190 F.3d 167 (3rd Cir. 1999).

New Mexico

N.M. Stat. Ann. §66-7-6. Authorized emergency vehicles

[This section sets out UVC 11-106 without significant change. The statute requires an emergency vehicle on an emergency run to sound an audible signal and display at least one lighted lamp displaying a red light visible 500 feet from the front. A police car need not display a red lamp to remain within the protection of the statute. Section 41-4-12, liability of law officers under the Tort Claims Act, does not specifically protect law enforcement officers when enforcing the law. The case notes indicate that an officer may be liable to harm to others if he fails to apprehend or detain an intoxicated driver. There are few cases.]

A 1990 case was in federal court. A deputy sheriff responded to a silent burglar alarm at an auto dealership. The officer allegedly drove 55–60 mph around a blind curve on a street with a 35-mph-speed limit without using his siren or emergency lights. It was after midnight and had been raining and sleeting. He collided with a driver pulling out from a restaurant and killed her. The plaintiff alleged that the deputy had seven traffic-offense convictions in four years and had drunk alcohol during his shift. The plaintiffs alleged a number of federal constitutional violations, primarily §1983 in depriving the plaintiff of her life without due process of law, and an unreasonable seizure of her person. The Tenth Circuit affirmed summary judgment for the defendants. If there was a seizure here, it was unintentional. An intentional seizure in this context would be one in which one's vehicle was used to set up a roadblock, for instance.

At most the plaintiff alleged negligence, essentially driving too fast for conditions, and the court held that the negligent operation of a police vehicle does not arise to the level of a constitutional violation. The court noted that this finding is consistent with decisions on other circuits. In a footnote (#3), the court points out that, "As a practical matter, any risks created for bystanders like Theresa are reciprocally created for himself. As such it is hard to imagine Romero being truly indifferent to the risks he was creating, which is a necessary characteristic of recklessness." As for failing to train or supervise, since there was no underlying constitutional violation by a county officer, there cannot be an action for failing to train or supervise. *Apodaca v. Rio Arriba County Sheriff's Dept.*, 905 F.2d 1445 (10th Cir. 1990).

A 1951 case held that a fire truck being used to take an inhalator to a residence was not on an emergency run even if its emergency signals were operating. The fire truck was not responding to a fire alarm as indicated by statute. Nor is it a public ambulance merely because it was transporting a medical device. The court held that the accident was caused by the fire truck driver's running a red light, and affirmed an award for the plaintiff driver. *Tiedebohl v. Springer,* 232 P.2d 694 (N.M. 1951).

New York

Veh. & Traf. Law 1104. Authorized emergency vehicles

(a) The driver of an authorized emergency vehicle, when involved in an emergency operation, may exercise the privileges set forth in this section, but subject to the conditions herein stated.

(b) The driver of an authorized emergency vehicle may:

1. Stop, stand or park irrespective of the provisions of this title;
2. Proceed past a steady red signal, a flashing red signal or a stop sign, but only after slowing down as may be necessary for safe operation;
3. Exceed the maximum speed limits so long as he does not endanger life or property;
4. Disregard regulations governing directions of movement or turning in specified directions.

(c) Except for an authorized emergency vehicle operated as a police vehicle or bicycle, the exemptions herein granted to an authorized emergency vehicle shall apply only when [emergency audible and visual signals are operated].

(d) An authorized emergency vehicle operated as a police, sheriff or deputy sheriff may exceed the maximum speed limits for the purpose of calibrating such vehicles' speedometers. Notwithstanding any other law, rule or regulation to the contrary, a police, sheriff or deputy sheriff bicycle operated as an authorized emergency vehicle shall not be prohibited from using any sidewalk, highway, street or roadway during an emergency.

(e) The foregoing provisions shall not relieve the driver of an authorized emergency vehicle from the duty to drive with due regard for the safety of all persons, nor shall such provisions protect the driver from the consequences of his reckless disregard for the safety of others.

(f) [States that ambulances operated in an emergency may use any street, sidewalk, etc. unless prohibited by local authority in the interests of the safety of persons so transported.]

The main difference from the UVC model law on emergency vehicles is that a police vehicle in an emergency does not have to sound a siren or switch on flashing lights to remain under the protection of the statute. Their use would normally be covered by department guidelines. The intent of the statute would be to permit an officer to approach a crime scene or to catch up with a suspicious vehicle without warning. Extending the protection to bicycle police officers is a laudable step. The standard of care generally applied is reckless disregard for the safety of others, not due regard for the safety of others. Note that no state or local tort immunity is claimed. Consequently, the cases are relatively unambiguous in applying only §1104.

A recent case may be regarded as having a typical result. In fact, it is a court of appeals case and has been followed by the lower courts. A Massena police officer noticed a driver who squealed his tires in making a turn. The officer decided to stop the driver so he followed, and saw the driver run a stop sign at 30 mph. He began to pursue using his flashing lights. The driver fled and ran red light. Realizing that the driver was not going to stop, the officer began to radio for help and heard a crash. The fleeing driver, who had been drinking, had crashed into another driver and injured her seriously. The pursuit lasted less than two minutes. The trial court dismissed the complaint. The appellate division reversed on a finding that there was sufficient evidence of reckless disregard for trial. The court found that at the relatively early hour of 10 P.M. some traffic would be expected in the area. The officer's speed reached 60 mph in a 35-mph-zone. There was a question of lack of proper pursuit training. The court of appeals reversed and reinstated the dismissal. The purpose of §1104 is to allow emergency vehicle drivers to perform their duties unhampered by the rules of road imposed on drivers. Police have to make split-second decisions in enforcing the law and protecting life and property. If the police were to be held liable or a mere failure of judgment, that could deter them from acting decisively in doing their job.

The court noted that the statute includes both due care and reckless disregard language. Courts have been uncertain which one controls. The matter was one of first impression. The court held that the correct standard to apply is reckless disregard. The problem with the due care standard is that it is associated with ordinary negligence. The court notes that the reckless disregard language appears twice in the statute, in stating also that the emergency vehicle driver is not protected from the consequences of his reckless disregard. "Due care" can be understood to apply to the more specific reckless disregard standard. An officer in initiating or continuing a pursuit must take due care not to act with reckless disregard. The court notes that emergency vehicle drivers must routinely make conscious choices that necessarily escalate the overall risk to the public at large in the service of an immediate, specific law enforcement or public safety goal. Here, the officer was faced with a reckless, obviously dangerous and probably intoxicated driver whose actions threatened the public. Plus, there is a legitimate law enforcement goal. The officer's speed of 60 mph on relatively empty, nonresidential streets was hardly a "breakneck" pace. On the issue of whether the village gave its officers too much pursuit leeway, the court held that the decision of the village to use the discretionary rather than the more restrictive or discouraging model is a matter of governmental policy. If not irrational, it may not be reviewed in a personal injury action founded on negligence. The plaintiff's experts had submitted various studies and articles indicating that the village's policy is generally disfavored. *Saarinen v. Kerr,* 644 N.E.2d 988 (N.Y. 1994).

I have started with the above case because it is both recent and is from New York's highest court. The problem with UVC 11-106 in giving both standards is handled variously among the states that apply the UVC-derived statute. Refer to the Introductory Essay for more commentary.

A 1999 case held that §1104 is an affirmative defense that should be pleaded. The case arose from an accident in which a state trooper en route to an accident, with warning lights on, turned into the path of the plaintiff. The state answered the complaint by arguing that "the defendant was not negligent in any manner, nor guilty of culpable conduct." The plaintiff sought to prevent the state from amending its answer to include an affirmative defense under §1104, which would include the language, "reckless disregard for the safety of others" as the hurdle the plaintiff would have to overcome. The plaintiff urged that the defendant must plead affirmative defense in the original answer, pursuant to C.P.L.R. §3018(G). (An affirmative defense is one that places the opposing party on notice that, if applicable, the defense will defeat the claim even if the plaintiff's allegations are true. The purpose is to alert the plaintiff to reevaluate his case before much time or money is expended.) The court agreed that the defense should have been included in the answer, but held that the plaintiff was well aware of the emergency status of the police car, and by implication should have known of §1104. The plaintiff's motion was denied. *Culhane v. State,* 687 N.Y.S.2d 542 (Ct. Cl. 1999). (Obviously the defendant should have known of the statute, too, and should have pleaded it in the original answer.)

We look at an older case that was decided without regard to §1104. It indicates that the common law protects an officer in a pursuit case. A state trooper patrolling near Utica noticed a (stolen) car clocked at 70 mph on a state highway at midnight. The officer activated lights and siren and pulled abreast of the driver five times, and each time the driver swerved left to force the officer back. The driver then rounded a curve and collided with plaintiff's oncoming car. The court of claims dismissed the claim. The court found the plaintiff's claim to be novel: an officer in pursuit of a reckless driver who is creating danger to others is himself increasing the danger and should be liable to innocent parties. The court agreed that an officer in a pursuit case can be negligent or reckless. However, here the officer should be commended in risking his own safety in his attempt to protect the public. The defendant simply did not cause or contribute to the accident. A police officer, in an emergency, must do his duty promptly, without hesitation, resolutely, and yet not be careless. The court admits that this is a harsh result for the plaintiff, but no other conclusion is possible. (This kind of language probably would not be common today, and the focus would be on a statutory defense.) *Wrubel v. State,* 174 N.Y.S.2d 687 (Ct. Cl. 1958).

It is a fact that most of the reported cases find for the defendant. We note now several cases in which the plaintiff recovered some damages or avoided a summary judgment for the defense. A 1999 action was brought in federal court (the plaintiff was a New Jersey resident). Police officers in Herkimer at 1:30 A.M. saw a driver speeding and running a red light. The driver was a seventy-two-year-old woman. The officers, each in a cruiser, tried to box in the driver. The driver forced both officers off the road by ramming them, lost control, crashed into a pole and died. The husband brought suit on federal constitutional and state negligence claims. The district court dismissed the federal claims and retained jurisdiction of the state claims in the interests of judicial economy. The judge declined to grant the defense summary judgment. The judge held that the rule of *Sacramento* (see California section) did not apply. That rule of conscience-shocking behavior applied only to due process claims. Section 1104 applies, with its standard of reckless disregard. The case may be tried. The officers pursued for eight minutes. Departmental policy prohibited a pursuit for "violations which do not create immediate danger, such as running a red light." The deceased was included in the category of reckless disregard for the safety of others. (The court did not comment on possible cause of the deceased's bizarre actions in forcing each officer off of the road.) *Czarecki v. Scherer,* 64 F. Supp.2d 92 (N.D. N.Y. 1999).

Ten percent is better than nothing. A 1998 case affirmed a verdict holding a village liable to that percentage for a verdict for the plaintiff. An officer pursued a suspect that, to the officer's knowledge at the time, had committed two traffic violations. The suspect hit another driver. The officer pursued at high speed, in the hours of darkness, in a residential area during rush hour. The officer was aware of regulations that he should slow if a pursuit "gets out of hand," and did in fact think about the dangers but continued. *Fioriello v. Sasson,* 680 N.Y.S.2d 662 (A.D. 1998).

A short 2000 case affirmed a verdict for the plaintiff on the evidence that the officer entered an intersection on a red signal without sounding his siren or slowing to a safer speed, and suddenly stopped and blocked the plaintiff's lane. The judge's charge erroneously referred to the duty to drive with due care for the safety of others (see *Saarinen* above), but the verdict sheet showed that the jury applied the reckless disregard standard. *Baines v. City of New York,* 703 N.Y.S.2d 463 (A.D. 2000).

A plaintiff can be a police officer. Two New York police officers were responding to an emergency call from another officer. The plaintiff-officer stopped for a red signal, then both officers tried to turn left. The second moved to the inside and hit the first vehicle with enough force to send it into a lawn. It was foggy and neither vehicle had emergency lights on. The appellate division reversed a dismissal and found enough evidence of reckless disregard to go to trial. *McCarthy v. City of New York,* 673 N.Y.S.2d 160 (A.D. 1998).

A 1997 case involved triple-parked vehicles. Two New York police officers stopped next to a double-parked vehicle that they had pulled over as a suspected stolen car. They exited the cruiser. A passing bus hit the cruiser's open door and injured the plaintiff-officer. The officer sued the city for the negligence of the other officer who was driving, and sued the bus company. The jury awarded damages against both defendants. The appellate division reversed as to the city. The officers were engaged in an emergency operation, which insulated the police vehicle driver from ordinary negligence. There was no evidence of reckless disregard. The court affirmed as to the bus company, but reduced the damages (or a new trial would be ordered) from $450,000 to $350,000. *Williams v. City of New York,* 659 N.Y.S.2d 302 (A.D. 1997).

For a similar result, see *Mulligan v. City of New York,* 664 N.Y.S.2d 484 (A.D. 1997). A police officer was a passenger in a police car that the driver was operating with siren and flashing lights activated, and in fact entered an intersection on a green light. Other traffic had stopped, but a van apparently entered the intersection and caused the accident. The codefendant police officer's conduct was in no way evidence of reckless disregard.

A controversial decision, *Campbell v. City of Elmira,* 644 N.E.2d 993 (N.Y. 1994), was decided by the court of appeals on the same day as *Saarinen* (see above). In *Saarinen*, the court by a 6-0 vote held that a pursuing police officer did not recklessly disregard the safety of the plaintiff, who was hit by the culprit. In *Campbell,* the court by a 4-3 vote held that an Elmira fire truck operator did recklessly disregard the safety of a motorcyclist who collided with the truck. Anyone with a New York case will want to examine both of these cases. The case arose when the fire truck responded to a general alarm at the county jail. The truck operator entered an intersection on a red signal at a speed of 10 to 15 mph. The motorcyclist was coasting up to a red light, but when he saw it turn green he accelerated despite the fact that no other drivers with the green light on both sides of the intersections moved. (See a case in the Nebraska section with the same facts where the court held that the motorcyclist should have been put on notice to watch out, and denied him recovery when he collided with a police on an emergency run.) The motorcyclist braked when he saw the fire truck in the intersection but hit the rear wheel well of the truck. The court notes that the evidence of whether the truck had visual and audible emergency signals operating was in conflict, but the fact that all others did not move on their green light is pretty conclusive that such signals were seen and heard. The injury caused the effective loss of use of one arm. The plaintiff was twenty-five years old. The jury awarded $500,000. The appellate division affirmed after a reduction of $55,000. The court of appeals affirmed, 4-3.

The court noted some differences from *Saarinen,* which did not go to a jury, so the standard of review is that a verdict should be affirmed unless it is "utterly irrational," citing *Cohen v. Hallmark Cards,* 382 N.E.2d 1145 (N.Y. 1978). (If this principle were truly followed, it's unlikely that any verdict would be disturbed.) In *Campbell*, the truck collided with the plaintiff. The majority made much of the truck driver's admission that he did not know whether he had a red or green when he entered the intersection, or that he may not have looked toward the motorcyclist. The majority found that the fire truck operator "intentionally violated the statutory mandate when he recklessly flaunted the risks of proceeding in an emergency setting into an intersection against a red light, indifferently and in disregard of any modicum of statutorily required attentiveness." The dissent simply found the defendant's action to be far short of reckless disregard. Red or green light notwithstanding, the defendant had the right to enter the intersection without stopping, and upon giving a blast from his air horn and seeing that traffic had stopped. (I would comment that the majority opinion's language about recklessly flaunting indicates a straining to affirm the verdict.) The moral of this case for the plaintiff is to survive summary judgment.

A short opinion in a 1993 case ordered a new trial. The facts are brief. The appellate division stated that evidence indicated that a fire truck entered an intersection on a red signal and hit a car broadside. The passenger in the car was killed. The jury found the car driver 70 percent at fault and the fire truck operator 30 percent, and awarded $2.5 million. The court ordered a new trial, both on the apportionment of fault against the civilian driver and on excessive damages, except that the plaintiff had the option of accepting a reduced sum of $500,000 from the city within thirty days. It was error to admit evidence that the car driver's blood alcohol level was .0439 when under state law a level under .05 indicates no impaired driving ability. The instruction on speeding was erroneous when there was no evidence of speeding by either driver. The trial judge also failed to properly explain the significance of the emergency vehicle law. *Marigliano v. City of New York,* 601 N.Y.S.2d 161 (A.D. 1993).

I will mention the rest of the recent cases briefly. All hold for the defendants. Looking first at those in which the police vehicle directly causes injuries to the plaintiff, a motorcyclist sued the village of Cobleskill following an accident. An officer attempted to stop a motorcyclist in the wee hours who exhibited signs of intoxication when he left a bar. The motorcyclist, whose blood alcohol level was 0.17, fled at speeds of over 100 mph and crashed on a curve after driving around a roadblock. (The plaintiff allegedly said, "You want to play? Let's go," when ordered to stop.) The court held that summary judgment for the defendant was proper. The officer's conduct may have been negligent but was not reckless. *King v. Village of Cobleskill,* 654 N.Y.S.2d 439 (A.D. 1997).

A 1997 case held that a momentary lapse of judgment is not reckless disregard for the safety of others. An officer responding to a report of a fight in a graveyard glanced down to switch on his flashing lights and hit and killed a bicyclist who was crossing the street in the middle of the block. The time was 5:00 P.M. in December, so it was approaching darkness. The court of appeals held that such action was not sufficient to satisfy the requirement of reckless disregard. The officer's speed was not more than 39 mph. The deceased was age sixteen. *Szczerbiak v. Pilat,* 686 N.E.2d 1346 (N.Y. 1997).

A driver was on Route 9A near Poughkeepsie and intended a to make a left turn. An officer behind her was responding to a report of a motorcycle accident and approached from the rear. Apparently the plaintiff did not signal a turn. The officer had his emergency lights and the siren on. The officer thought the plaintiff was slowing to yield to the right, as the law requires, but she attempted the turn and they hit. The case was tried in the court of claims, and both parties were found to be equally at fault. The appellate division reversed to find no liability on the officer's part. His conduct did not display reckless disregard. At the most, he was negligent. *Notorangelo v. State of New York,* 659 N.Y.S.2d 312 (A.D. 1997).

A 1998 case held that a police officer who used her "wiggle-waggle" lights (the front headlights flash alternatively) fell far short of showing a reckless disregard for the safety of others. The officer received a report of small white car being driven recklessly in the village of Gouverneur. She identified the vehicle and tried to stop the driver as indicated above. The driver was not even aware of being pursued. He said he was blinded by bright lights to the rear. He sped up, lost control on a curve, and hit a parked car, a pole, a house, and finally a tree. The plaintiff in the case was asleep in her house. The court reversed the lower court and ordered summary judgment for the defendant officer. *Martin v. Miller,* 680 N.Y.S.2d 300 (A.D. 1998).

A brief opinion in a 2000 case affirmed summary judgment for the defendant officer and city. An innocent bystander was injured by a suspect the police were chasing. Mere allegations that the officer was speeding do not meet the standard of reckless disregard. That standard requires intentional, unreasonable acts in disregard of an obvious risk that would probably lead to harm, and to do so with conscious indifference (citing *Saarinen).* Nor was there any evidence that a police supervisor was negligent in not calling off the pursuit. *Dorsey v. City of Poughkeepsie,* 712 N.Y.S.2d 604 (A.D. 2000).

A two-page opinion had the same result in a Putnam County case. The court ruled that a pursuit that lasted seven seconds in a quiet neighborhood at 2:35 A.M., in which the deputy did not exceed 35 to 40 mph in a 30-mph-zone, and did not activate his emergency signals, did not rise to the level of reckless disregard for the safety of others. The lower court had consolidated three actions, and had granted the officer and summary judgment. *Naue v. Higgins*, 662 N.Y.S.2d 527 (A.D. 1997)

An officer who pursued a motorist who refused to stop following a traffic infraction did not act with reckless disregard. The officer activated his emergency signals. The plaintiff was an innocent bystander. The court cited *Saarinen* as authority in this one-paragraph opinion. *Young v. Village of Lynbrook,* 650 N.Y.S.2d 804 (A.D. 1996).

A 1996 case is more extensive. It dealt with the issues facing police when they suspect that a driver is intoxicated. The plaintiff was in a car that the suspect hit head-on. The suspect, Juan Lopez, had been to a birthday party at the Eagles club in Rotterdam. When he left in his car, he spun in the gravel and drove into a yard. He backed out and sped off. A police officer witnessed this act, and closed with his lights and siren operating. He got the plate number and saw the suspect run a red light, weave, and almost hit another car. The officer backed off and turned off his siren so as not to further "agitate" the driver. Other officers were alerted. The officer realized that the suspect was entering a busier area, and turned on his siren to try to alert other drivers of the danger. Nevertheless, the suspect tried to negotiate a curve at 65 mph in a 35-mph-zone and collided with the plaintiff. (The suspect had a blood alcohol level of 0.20.) The court affirmed summary judgment for the Town. Nothing in the record demonstrated reckless disregard. In fact, the officer was obligated to use all reasonable means to stop Lopez. *Dibble v. Town of Rotterdam,* 650 N.Y.S.2d 897 (A.D. 1996). Note that the plaintiff also filed a dram shop action against the Eagles.

A 1995 case is very similar to *Dibble.* Again, an officer saw a driver pull out from a business (McDonald's) recklessly. Two pickups appeared to be racing each other 20 mph over the speed limit. The officer turned on his emergency signals and pursued. One driver did not stop. The officer received permission to pursue outside the city limits, and the pursuit in a rural area was at times up to 90 mph. The fleeing driver, drunk, failed to make a curve and hit the innocent driver head-on. Again, the court held that summary judgment was proper. The cause of the accident was the drunk driver, not the police officer. The officer was entitled to try to stop a driver who is posing a clear and present threat to public safety. *Mullane v. City of Amsterdam,* 622 N.Y.S.2d 346 (A.D. 1995).

A 1994 case is unusual. The plaintiff was injured when a riderless motorcycle collided with her car. (The facts given do not explain when and how the motorcyclist separated from the bike.) She alleged that the town police were at fault for pursuing the motorcyclist. The trial court judge denied the defendants' motion for summary judgment. The appellate division reversed. The record did not show that there was a chase, and even if there had been a pursuit, it was not conducted with reckless disregard. *Friel v. Town of Brighton,* 615 N.Y.S.2d 207 (A.D. 1994).

A 1995 case arose when a police officer tried to pull over a driver, Rocco, for an offense not given in the case. The driver ignored the officer, and the officer then got in front. Rocco struck the rear of the police car and caused it to spin off to the shoulder. Rocco then sped away at up to 90 mph. He soon crashed into an overpass and a tree, and his passenger, the plaintiff, was made a quadriplegic. At trial, the jury found Rocco 80 percent at fault and the officer 20 percent. The appellate division reversed the fault allocation against the police. It found Rocco to be the sole cause as a matter of law. *Paige v. Rocco,* 625 N.Y.S.2d 577 (A.D. 1995).

A 1996 case in the court of appeals dealt with the status of police department pursuit guidelines. The plaintiff police officer was a passenger in the police car and was injured in a pursuit. The officers saw a driver of a Cadillac cut through a gas station in an apparent attempt to avoid a red light. The police attempted to stop the Cadillac, but the driver fled. The chase escalated into the running of traffic signals and stops signs, swerving into the opposing lane and causing at least one uninvolved driver to take to the sidewalk, driving in a high school football field, back onto the street, and so forth. Once the Cadillac rammed the police car. After a second collision, the chase ended. The injured officer sued under General Municipal Law 205-e. That act gives police injured in the line of duty right to sue those responsible. The plaintiff alleged that the police car driver violated a number of pursuit guidelines. Memo No. 3 in particular advises officers to weigh the seriousness of the offense against the danger to himself and others, and to terminate when the risks outweigh the danger to the community and himself. The plaintiff pointed out that the offense, cutting through a gas station, did not justify the pursuit. A pursuit should be undertaken only when there is "a compelling need for an immediate apprehension."

The trial court found that, while the officer did not act with reckless disregard under the statute, he did violate Memo No. 3. The case came to the court of appeals. The issue was whether the Memo was a "requirement" as in the class of "statutes, ordinances, rules, orders and requirements" whose violation can support an action under §205-e. The court decided that it was not and dismissed the complaint. Memo No. 3 notwithstanding, whether and how to conduct a pursuit contemplates discretionary judgments. *Desmond v. City of New York,* 646 N.Y.S.2d 492 (N.Y. 1996).

A 1996 case arose when Mount Vernon police pursued a driver. The driver entered an intersection at high speed and killed one person and injured two others. The trial court found enough to go to trial, but the appellate division reversed and ordered the actions be dismissed. The pursued driver posed a threat to the public. That alone authorized the police to take whatever means necessary to stop the driver, as long as they are short of reckless disregard. The court found no material questions of fact to indicate such disregard. Any alleged violations of departmental rules did not establish recklessness. (It is an indication of how poorly plaintiffs fare in pursuit cases in New York that this serious case merited a one-page opinion.) *Powell v. City of New York,* 644 N.Y.S.2d 766 (A.D. 1996).

Two 1998 cases involve the same claims: the police failed to detain an intoxicated driver and then pursued recklessly. In the first, Niagara Falls police encountered an intoxicated person exiting a car and told him not to drive. Fifteen minutes later he nevertheless drove erratically on busy city streets. The police tried to get close enough to get a plate reading, and at one point turned on the cruiser's siren. The suspect sped away, and the officers stopped active pursuit. The suspect ran a red light and injured the plaintiff in a mobile home. The police were far enough back not to reach the scene for another 15 to 30 seconds. The appellate division affirmed summary judgment for the defendants. There was no special relationship between the police and the plaintiff and therefore there can be no liability for failure to detain the suspect. (This is the typical holding regardless of jurisdiction.) There was no evidence of reckless pursuit. The police have a duty to investigate erratic driving and to use all reasonable means to stop it. If the suspect is drunk, the duty to stop him is increased. *Jessop v. City of Niagara Falls,* 669 N.Y.S.2d 110 (A.D. 1998). The second case, *Cavigliano v. County of Livingston,* 678 N.Y.S.2d 186 (A.D. 1998), is practically identical.

We conclude this section with three fire truck cases. A North Babylon volunteer fire fighter was responding to a fire call and proceeded with rotating lights, siren, and air horn blaring. A driver in front heard the truck but thought she could not pull off to the right because of traffic conditions. Instead, she attempted to move into the left lane to make a left turn as the truck came up at slow speed and skidded into the car. The plaintiff was a passenger in the car. The court held that the driver of the car was properly not granted summary judgment. The court held that summary judgment for the fire fighter and town was proper. The fire fighter did not act with reckless disregard, nor with "willful negligence or malfeasance" as specified in General Municipal Law 205-b. *Tobacco v. North Babylon Fire Department,* 674 N.Y.S.2d 125 (A.D. 1998). A 1992 case held that a fire truck operator came to a red light with his emergency signals operating, slowed to 20 mph and looked both ways, had the right of way when he collided with a driver. The driver had the duty to yield. *Davis v. City of Mount Vernon,* 583 N.Y.S.2d 869 (Sup. 1992). A 1995 case extended the protection of the emergency vehicle statute to a volunteer fireman who was proceeding to a fire call in his private vehicle. *Skinner v. Scobbo,* 633 N.Y.S.2d 208 (A.D. 1995).

In conclusion, it is clear that New York strongly protects police pursuits. The language in *Saarinen* generally authorizes vigorous law enforcement. The cases make little mention of the effect of departmental pursuit policies. New York does not depend on any kind of official immunity. The cases are decided by applying the standard of reckless disregard for the safety of others. In New York, currently the standard is quite high. It is also noteworthy that few cases are filed in federal court alleging constitutional violations.

North Carolina

N.C. Gen. Stat. §20-156(b). Exceptions to the right-of-way rule

The driver of a vehicle upon the highway shall yield the right-of-way to police and fire department vehicles and public and private ambulances . . . when the operators are giving a warning signal by appropriate light and by bell, siren or exhaust whistle audible under normal conditions from a distance of not less than 1,000 feet. When appropriate warning signals are being given, an emergency vehicle may proceed through an intersection or other place when the emergency vehicle is facing a stop sign, a yield sign, or a traffic light which is emitting a strobe signal or a beam of steady or flashing red light. This provision shall not operate to relieve the driver . . . from the duty to driver with due regard for the safety of all persons using the highway, nor shall it protect the driver . . . from the consequence of any arbitrary exercise of such right-of-way.

N.C. Gen. Stat. §20-145. When speed limit not applicable
The speed limitations set forth shall not apply to vehicles when operated with due regard for safety under the direction of the police in the chase or apprehension of violators of the law or of persons suspected of any such violation, nor to fire department or fire patrol vehicles when traveling in response to a fire alarm, nor to public or private ambulances and rescue squad emergency service vehicles when traveling in emergencies.

N.C. Gen. Stat.§143.291. Tort claims against state
[This and following sections sets out the State Tort Claims Act. The Industrial Commission constitutes a court to administer the Act. The claimant can prevail only if a state employee is found negligent in a situation in which a private person would be found negligent, and the claimant must be free from any negligence himself. The maximum payable for any one claimant is $150,000.]

North Carolina has enacted the basics of UVC 11-106, albeit not all in one section. We begin with an older case because it adopted a reasonable person standard of care that was discussed in a recent case. The case arose when a deputy sheriff began to pursue a man who had not stopped at a stop sign. The deputy was going 70 mph (his account), or up to 90 mph (witnesses' estimates) in a residential area. The speed limit was either 35 or 55 mph, depending on the density of housing, and that issue was a jury question. The plaintiff was attempting a left turn into a private business and the officer hit the plaintiff. The deputy filed a claim in a cross action, and the jury awarded damages to the deputy. The supreme court reversed for a new trial. The trial judge read the emergency vehicle right-of-way statute to the jury, including the due care and reckless disregard language, and instructed the jury that if the deputy was in pursuit of a suspected law violator, mere speeding would not alone make his conduct culpable. The supreme court held that the deputy is to be held to the standard of a reasonably prudent person acting in the same circumstances (that is, ordinary negligence.) *Goddard v. Williams,* 110 S.E.2d 820 (N.C. 1959).

In a 1988 case, the supreme court confronted for the first time a case in which the pursuing officer's car was not involved in the accident. A Greensboro police officer in the early morning hours noticed a weaving driver, and tried to stop him. The driver fled, drove around a roadblock, and the pursuit entered the country. The officer was in contact with various police, sheriff, and highway patrol departments. The pursuing officers were advised to use their discretion in whether to continue the pursuit. The pursued driver, using only parking lights, tried to pass in a no-passing zone and killed two people in an oncoming car. The two pursuing police vehicles had dropped back somewhat at this point to lessen the risk. The trial judge denied the defendants' motion for a directed verdict, and the jury found for the plaintiffs. The supreme court reversed and ordered a directed verdict for the defendants. It held that the right standard to apply in this case is one of gross negligence, which is derived from the statutory language of reckless disregard. The trial judge erred in charging on reasonable care. The evidence in fact showed that the officers were not even negligent. They had a duty to try to apprehend. The court noted the toll taken each year by drunk drivers. *Bullins v. Schmidt,* 369 N.E.2d 601 (N.C. 1988).

The supreme court discussed both *Goddard* and *Bullins* in a 1996 case. The case arose when a Winston-Salem police officer noticed a car with one headlight out. The officer turned around and sped up without turning on his blue lights or siren. He wanted to catch up before alerting the driver. When he caught up, he intended to turn on his equipment and stop the driver. How much he was exceeding the 45-mph-speed limit was in dispute. The plaintiff, an innocent bystander, was approaching from the opposite direction. The officer came to an intersection that had a flashing yellow light. The plaintiff attempted a left turn and was hit by the officer. The city did not have liability insurance for claims up to $2 million. The court of appeals granted the officer and the city summary judgment based on sovereign immunity for claims under $2 million. (This is governed by

Code 160A-485, except for any negligence under Code 20-145.) The plaintiff argued that because the legislature had not acted to make any change in §20-145, the legislature intended that the ordinary negligence rule of *Goddard* be applied to police vehicle-plaintiff vehicle accidents. Usually this argument of legislative acquiescence is not persuasive, and it was not here. The court did not find a good reason to apply two standards depending on whether the police vehicle is in the accident. It held that gross negligence is the correct standard, and found none here. The court notes the brief of the North Carolina Association of Police Attorneys giving arguments for police immunity. There were two dissenting opinions. One dissent agreed that a single standard of gross negligence is correct, but thought that there was enough evidence of gross negligence (based on reckless disregard) for a jury to decide. *Young v. Woodall,* 471 S.E.2d 357 (N.C. 1996).

A 1994 case was appealed in front of the Industrial Commission. The commission decides state tort claims for the department of transportation. A highway patrol officer, patrolling on Highway 70, a four-lane road with a median, was notified of a possible drunk driver ahead. He sped up to 90 mph to catch up and identify the driver. He saw a car stopped in the median when he was 500 feet away. The driver intended to cross the road. She looked and saw oncoming traffic "at least 300 yards away" and started across. The officer hit the rear of her car. The Commission, applying the standard of *Goddard* (this is pre-1996), found the officer not negligent. The appellate court reversed on a finding that the officer was negligent as a matter of law. While finding the "pursuit" somewhat questionable, the officer was not free from any negligence. *Minks v. North Carolina Highway Patrol,* 449 S.E.2d 483 (N.C. App. 1994).

Another 1994 case involved no contact with the police vehicle. A deputy sheriff in the early morning hours responded to a report of a disturbance at an arcade. When he arrived, a hysterical female told him a man was shooting inside the arcade. Another person told him the gunman got in a car and was about to flee. The officer saw a car pull out, and believing that it was the gunman, pursued. (The case does indicate whether he was or not.) The officer pursued with lights and siren, staying two to five car lengths behind. The fleeing driver failed to negotiate a curve at 75 mph and all three occupants were killed. The trial court granted summary judgment to all defendants, and the appellate court affirmed. It applied the gross negligence standard of *Bullins.* The court found no gross negligence in alleged lack of guidance in departmental pursuit policy. The allegation that a superior should have called off the pursuit was not credible, given the brevity of the pursuit. *Clark v. Burke County,* 450 S.E.2d 747 (N.C. App. 1994).

A 1999 pursuit found that Durham police officers were not grossly negligent. A city police officer was on routine patrol at 1:00 A.M. and spotted Zambito driving a pickup. The officer knew that he had arrested Zambito a few months earlier for driving while impaired, and he suspected that Zambito was driving on a suspended license. Another officer was notified to be on the lookout. That officer spotted the suspect and followed, turning on his flashing lights when Zambito sped up. Shortly afterward Zambito entered the intersection of University Drive and Hope Valley Road on a red light at 70 mph and killed another driver. The pursuing officer did not see the crash but arrived shortly. Zambito's blood alcohol level was 0.13. Three days earlier Zambito had been arrested for driving while impaired, and he had said that he would run every time. There was no evidence that the officer in this case knew of the threat. The trial court granted the defendants summary judgment. The appellate court affirmed. It also found no gross negligence that, according to *Bullins,* is defined as "wanton conduct done with conscious or reckless disregard for the rights and safety of others."

The court, in a thorough analysis, identified some factors to consider in pursuit cases. Was there a reason for the pursuit? Was the suspect known to the police and could he be apprehended later? Was the suspect's conduct a clear danger to the public that could be abated only by a pursuit? Were the conditions (time of day, nature of the area, other traffic) likely to make the pursuit dangerous? The court found that all of these conditions made a decision to pursue valid. Yes, the suspect could have been arrested later (although in driving-while-impaired cases, a few hours makes a difference), but any suspected intoxicated driver is a present danger. On specific points, the allegation to pursue knowing the suspect's threat to flee (if the officers in fact knew of the statement) does not indicate reckless indifference. Police officers are not required to guess a suspect's state of mind. The plaintiff argued that his expert's affidavit should have been admitted. The expert, John Gormley, stated his opinion that the pursuit was grossly negligent and was a violation of city pursuit rules. The court may disallow such evidence as giving an opinion as to whether a legal standard has been met. That usurps the function of both judge and jury. *Norris v. Zambito,* 520 S.E.2d 113 (N.C. App. 1999).

A 1998 case has two noteworthy points. A Raleigh police officer received a radio call while driving, reached down to switch channels to monitor the situation and rear-ended the plaintiff's car. The plaintiff sued the city and the plaintiff's own automobile liability insurer under uninsured motorist coverage. The trial court granted all defendants summary judgment. The appellate court reversed as to the UM carrier. The city and officer enjoy official immunity if the alleged tort is committed in the performance of a governmental function, which policing is. The city had purchased liability insurance for claims over $1 million. The plaintiff did not allege damage to this amount. The officer is personally immune. He was in the scope of his employment and did not act with malice and was not tainted by corruption. On the insurance issue, the court reversed. Although the plaintiff's policy specifically excluded vehicles owned by a political subdivision in the definition of an uninsured motor vehicle, the court held for coverage. The legislature intended that the uninsured motorists coverage statute be liberally construed. The legislature cannot have meant the recovery of damages would hinge on whether the vehicle was privately owned. *Williams v. Holsclaw,* 495 S.E.2d 166 (N.C. App. 1998).

A short 1989 case arose from a high-speed pursuit in a rural area. A state trooper around midnight estimated a vehicle's speed at 80 mph. The officer followed, not to pursue and stop, but to try to see if this was the same car he had seen earlier. He clocked the speed at 115 mph eventually. The officer closed to within one half mile without using his blue lights or siren. He radioed for a roadblock to be set up at the Troy city limit and turned on light and siren to pursue. He then saw a dull orange flash on the horizon. The car had crossed the center line and all occupants of the vehicles were killed. The plaintiff argued that the rule of *Bullins* did not apply to this case because of the date. The court of appeals held that the rule of gross negligence did apply, and that in any event the plaintiff had not even shown ordinary negligence. *Fowler v. Dept. of Crime Control,* 376 S.E.2d 11 (N.C. App. 1989).

A 1999 case arose from a very high-speed pursuit that resulted in the death of an innocent person. It involves in part the state of mind of the pursued. A Hillsborough police officer at 2:00 A.M. stopped at a rural intersection, looked for traffic, and made a left turn after noticing a car approaching. He pulled onto NC-86, and ten seconds later the car passed him at high speed and in violation of a double yellow line. The civilian driver, Hill, saw that he had passed an officer and sped to 90 mph to discourage pursuit. The officer followed, notified the dispatcher of his intention to stop the car, and turned on his blue lights and siren. Hill turned onto I-85 and was confident that his borrowed BMW (from his sister) could outrun the Crown Victoria. (Nope, police Crown Vics are not quite like the one grandpa drives.) The pursuit went on for five miles on I-85 at speeds of 120 to 130 mph at times. The officer stayed in the left lane to minimize the danger to the few cars he passed, while Hill at times turned off his lights to confuse the officer. Hill exited, went south on I-85, crossed the medium and headed north. Hill then exited I-85 and headed south on U.S. 70 toward Durham. The officer had alerted Durham police, and Officer Dean of Durham waited along the highway. Hill streaked by at 90 to 100 mph. Both officers were cut off by a truck entering the roadway and lost sight of Hill. Hill thought he had outrun the pursuit but maintained his high speed to make sure. Another Durham officer was positioned to help and saw Hill pass by. She moved out to follow and saw the BMW go off the road. It ultimately crashed into a house. The body of Hill's passenger was discovered sometime later.

The estate sued in state court, alleging that the officers were grossly negligent. The claim against the chief and the city was for failure to train and supervise. The claim was brought under common law and federal constitutional law, including §1983. The trial court granted summary judgment to the officers and the city. The court of appeals reversed as to the individual officers, finding a question of fact as to their gross negligence. The North Carolina Supreme Court took the case under discretionary review, pursuant to §7A-31. One of the reasons for a discretionary review is whether a court of appeals decision appears likely to be in conflict with a supreme court decision. The supreme court reversed. It did not find gross negligence in the pursuit. The court did not find a difference here and in *Fowler* above. The court of appeals thought that since the pursuit in *Fowler* was very short, it did not apply in this case. The court noted that at no time did the officers try to overtake Hill, and when he crashed they were in fact some distance behind and Hill thought he had gotten away. The court especially held that the state of mind of the fleeing suspect is not to be considered a factor in judging the actions of the pursuers. The pursuers are not "forcing the accident." The court quoted from a Seventh Circuit case, *Mays v. City of East St. Louis,* 123 F.3d 999 (7th Cir. 1997):

> Death and disability haunt law enforcement. Lax law enforcement emboldens criminals and leads to more crime. Zealous pursuit of suspects jeopardizes bystanders and persons accompanying the offender. Easy solutions rarely work, and *ex post* assessments—based on sympathy for those the criminal has injured, while disregarding the risks to society at large from new restrictions on how the police work—are unlikely to promote aggregate social welfare. *Parish v. Hill,* 513 S.E.2d 547 (N.C. 1999).

A 1997 case was filed in federal court. The facts are complicated, and the federal district court declined to grant summary judgment to the several defendants, holding that there were triable issues. (There are no further reported proceedings in this case.) Marion city police noticed a Mustang stopped in traffic with its hazard lights flashing. The driver, Williams, got out of the car and told the police that a passenger, Westall, would not leave the car. Westall got behind the wheel and fled. Williams told the police that her ten-month-old baby was strapped in baby seat in the Mustang. Westall apparently was agitated about an impending stint in prison. (The court found that Williams was negligent in leaving the keys in the Mustang in these circumstances.) Shortly afterwards a police officer found the Mustang parked at a car wash, and reached in the car and tried to restrain Westall, but was dragged before letting go. There was an off-and-on chase involving several police vehicles from both the city of Marion and the McDowell County Sheriff's Department. Some of the officers were not fully informed about the facts of the chase, including whether the chase was called off, owing to poor radio communication, which was the result of obsolete equipment. The chase ended when Westall hit the innocent plaintiff's vehicle. The baby in the child seat was found 100 feet away from the crash. (The case does not indicate whether the baby was injured.)

The court found that a number of pursuit policies apparently were violated: evidence of a serious felony not established; failure to terminate when the driver's identity was established, danger the public; continually evaluating the danger to the public (including the baby passenger) against the benefit of catching the driver; caravaning; and officers entering the chase when not specifically authorized. A reasonable person could find that at least some of the law enforcement defendants acted with reckless disregard. *D'Alessandro v. Westall*, 972 F. Supp. 965 (W.D. N.C. 1997).

North Dakota

N.D. Cent. Code §29-06-13. Force may be used to make an arrest

If, after notice of intention to arrest the defendant, he either flees or forcibly resists, the officer may use all necessary means to effect the arrest.

N.D. Cent. Code §39-10-03. Class A authorized emergency vehicles (police, fire, ambulance)

[This section enacts UVC 11-106 with these exceptions: it does not include a standard of care, and it requires an audible signal or flashing lights to claim the privilege, not both. It includes subsection (1)(c), which permits an emergency vehicle operator to speed so long as the driver does not endanger life or property.]

N.D. Cent. Code §39-09-06. Liability of exempt driver for reckless driving

The speed limitations provided in this chapter do not apply to Class A authorized emergency vehicles. The exemptions provided for in this section do not protect the driver of any such vehicle from the consequences of a reckless disregard of the safety of others.

N.D. Cent. Code §39-10-26. Operation of vehicle on approach of an authorized emergency vehicle

3. This section does not operate to relieve the driver of an authorized emergency vehicle from the duty to driver with due regard for the safety of all persons using the highway. [Parts 1 and 2 require other drivers to yield to emergency vehicles.]

There appears to be but one case, *Jones v. Ahlberg*, 489 N.W.2d 576 (N.D. 1992). The case arose from the pursuit of a drunk driver, which ended when his passenger was killed. The parents are the plaintiffs. A Langdon city police officer, in the early morning hours, stopped a driver for weaving on the highway. A second officer stopped her car behind the suspect. The driver failed sobriety tests and was placed under arrest. His passenger appeared to be sleeping in the front seat. The arresting officer moved to the passenger's side to talk to her, and the suspect took that opportunity get back in his car and speed off, and the chase was on. Apparently the chase took place in rural area and no other drivers were involved. Up to eleven law enforcement units were called in to help. The chase went on for 60 miles (this is North Dakota, after all) through three counties and one town. Speeds were up to 100 mph, but slowed as the vehicles went through the town, Cando. Eventually a state trooper got in front and braked in an attempt to slow the suspect. The suspect tried to pass on the right shoulder and the cars bumped. The suspect went into the ditch, rolled, and his passenger was killed. The plaintiffs settled with the pursued driver and sued several law enforcement defendants. The case was tried before a judge and damages of $232,831 were awarded. The supreme court reversed for a new trial before a new judge.

First, the trial court must give precedence to the comparative negligence statute, even if one party has settled, and assign comparative fault. Second, the trial court did not clearly indicate the standard of liability used. The judge's memorandum opinion seemed to indicate simple negligence, or failure to use reasonable care. Another part referred to the duty of the police, even when exempted from speed limitations, not to drive with reckless disregard. The court cited a number of cases from other jurisdictions in holding that gross negligence, as evidenced by a reckless disregard for the safety of others, is the correct standard to apply. The court declined to rule as a matter of law for the defendants, and commented that reasonable persons could find reckless disregard. The court noted that the police knew the identity of the suspect. A concurring judge found these last remarks to be dicta and was concerned that the remarks might unduly influence the judge at the new trial.

Ohio

Ohio Rev. Code 4511.041. Emergency vehicles or public safety vehicles exempt

Sections [listing some 44 sections] do not apply to the driver of an authorized emergency vehicle or public safety vehicle if the emergency vehicle is responding to an emergency call, is equipped with and displaying at least one flashing, rotating, or oscillating light visible under normal atmospheric conditions from a distance of five hundred feet to the front of the vehicle and if the driver is giving an audible signal by siren, exhaust whistle, or bell. This section does not relieve the driver . . . from the duty to drive with due regard for the safety of all persons and property upon the highway.

Section 4511.03

The driver of any emergency vehicle or public safety vehicle, when responding to an emergency call, upon approaching a red or stop signal or sign shall slow down as necessary for safety to traffic, but may proceed cautiously past such red or stop sign or signal with due regard for the safety of all persons using the street or highway.

Section 4511.14

The prima-facie speed limitations set forth in section 4511.21 do not apply to emergency vehicles or public safety vehicles when they are responding to emergency calls [and are displaying the required visual and audible signals].

Section 2744.02. Political subdivision not liable for injury, death, or loss; exceptions

(A)(1) For the purposes of this chapter, the functions of political subdivisions are hereby classified as governmental functions and proprietary functions. Except as provided in division (B) of this section, a political subdivision is not liable in damages in a civil action for injury, death, or loss to person or property allegedly caused by any act or omission of the political subdivision or an employee of the subdivision in connection with a governmental or proprietary function.

(B)(1) Except as otherwise provided in this division, political subdivisions are liable for injury, death, or loss to persons or property caused by the negligent operation of any motor vehicle by their employees upon the public roads, highways, or streets when the employees are engaged within the scope of their employment and authority. The following are full defenses to such liability:

- (a) A member of a municipal corporation police department or any other police agency was operating a motor vehicle while responding to an emergency call and the operation of the vehicle did not constitute willful or wanton misconduct. [Parts (b) and (c) extend the same protection to fire vehicles and ambulances.]

Section 2744.01. Definitions
"Emergency call" means a call to duty, including, but not limited to, communications from citizens, police dispatches, and personal observations by peace officers of inherently dangerous situations that demand an immediate response on the part of a peace officer.

[We see that Ohio has not enacted UVC 11-106 as such, but simply exempts emergency vehicles on emergency runs from some 44 traffic provisions in the code. Two sections specifically exempt such emergency vehicle operators from observing speed limits and stop sign and signals. The standard of care required is due care, which normally is an ordinary negligence standard. The standard to be applied to emergency vehicle accidents is the one provided in §2744.03, the political subdivisons tort immunity act, which is willful or wanton misconduct. Claims involving state-owned vehicles must be brought in the court of claims under the waiver of state immunity, §2743.02. Note that §2743.02 does not specifically exempt the operation of emergency vehicles from a claimant's right to sue. The state would be liable if a private party would be liable in the same circumstances, but of course a private citizen is not exempt from traffic laws. Section 2743.02 does state that a claimant may sue a state employee if he or she acted with malicious purpose, in bad faith, or in a wanton or reckless manner. There are many cases applying the above statutes, and we examine some of the more recent ones.]

A 1998 case held that a police officer who allegedly pursued a motorcyclist after he was instructed to terminate the pursuit was not liable for willful or wanton misconduct. A Canton police officer, in the early evening hours in June, saw a group of motorcyclists. As the group cleared an intersection, one rider popped a wheelie (rode on the back wheel). This caused the officer to pursue that rider specifically. (As a former motorcyclist myself, I am not sure what law was violated. The court explains in its opinion that one cannot steer with the front wheel off the ground, but wheelies are of short duration normally and no sane or sober rider would pop one when he had to steer.) In any event, the officer chased with signals on and the motorcyclist got on the expressway and got away. The officer was unable to get the license number. Half an hour later the officer saw two bikes, one of which he thought was the original pursued rider. Both riders fled. The dispatcher told the officer not to chase if the cyclists did not stop, so the pursuit was again stopped. The officer saw the rider a third time and again chased. Again, he was told to stop. He said he slowed and turned off his emergency signals and resumed his usual patrol. He then came upon a motorcycle accident and realized it was the same motorcycle. The rider failed to make a curve and was killed. Witnesses said the officer arrived 30 to 40 seconds after the accident. The trial court granted the defendants summary judgment.

The court of appeals affirmed. The plaintiff argued that there were issues of whether the officer was the proximate cause, and whether the defendants were protected by sovereign immunity. The first question was whether the officer was responding to an emergency call. The court held that an emergency call is not limited to "an inherently dangerous situation." That conclusion would have a chilling effect on law enforcement. Any attempt to enforce the law qualifies. The court also held that the officer was too far behind the accident to be a proximate cause, and besides the decedent had a duty to exercise ordinary care for his own safety. Willful conduct implies some intent to harm, and the officer's actions were well short of willful or wanton conduct. A dissent thought that whether the officer actually stopped pursuing when he said he did was an issue for a jury. *Vince v. Canton,* 1998 WL 525570 (Ohio App. 1998).

The dissent in *Vince* above cited a 1991 case, *York v. Ohio State Highway Patrol,* 573 N.E.2d 1063 (Ohio 1991) that rejected a motion to dismiss for failure to state a claim. The case arose from a pursuit that ended in the death of the pursued driver. No other facts are given. The plaintiff contended that the deceased had committed no offense and was pursued in a willful and wanton manner. (Note that the court applied that standard as stated in the political subdivisions tort immunity act, even thought the defendants were state employees.) The supreme court held only that the allegations stated a claim, and would not speculate further. Two dissents argued that the mere fact of fleeing is an offense, and the mere fact that officers pursued is not on its face a willful or wanton act.

A 1995 case in the Ohio Supreme Court revisited the question of which standard applied to state employees. A fourteen-year-old driver and a friend took a car from the garage of their neighbor to drive—that is, joyriding, not theft of a vehicle. They were spotted by a township police officer, panicked, and fled. The pursuit moved to I-275, and two state patrol officers (the defendants) took control. Two Hamilton County deputy sheriffs learned of the chase and of their own volition set up a roadblock that brought the interstate traffic to a halt. (The time was near midnight.) The plaintiff was an innocent bystander stopped in the roadblock. He was rear-ended by one of the patrol cars. The issue of the case was what standard to apply to the defendant. The court of appeals applied a negligence standard in reversing the court of claims, which applied the willful and wanton conduct standard. The supreme court held that the standard set out for political subdivisions would apply. True, the "state" is not a "political subdivison," but it would not be fair to use different standards depending on the employing agency. In effect, the court assumes that the legislature did not intend this result. In addition, state police officers have functions that private citizens do not, those of law enforcement. Holding state patrol officers to an ordinary negligence standard would have a chilling effect. The court reinstated the judgment of the court of claims that the defendants were not liable. A strong and emotional dissent pointed out that §45.11 requires due regard in speeding. (I would point out that states like Ohio that employ an immunity defense hold that defense to control over a UVC-derived provision.) The dissent thought the court should not bend over backwards to encourage this kind of seemingly unnecessary high-speed pursuit that took on a life of its own. There was ample evidence of negligence: the officer drove at speeds up to 115 mph and followed too closely, and the internal report found that his acts were not within pursuit guidelines. *Baum v. Ohio State Highway Patrol,* 650 N.E.2d 1347 (Ohio 1995).

A 1991 case was properly disposed of by summary judgment for the city police officers. An Akron police officer saw a driver speeding on a city street. The officer made a U-turn, turned on his flashing lights, and pursued. He saw the driver go left of center on a curve. The pursuit continued until the driver ran a stop sign and hit another car, injuring both occupants. The city moved for summary judgment, contending that its officer was not the proximate cause and did not act with willful or wanton conduct. The court of Appeals affirmed on a finding that there was no genuine issue to try. A dissent would let the case be tried. There was a question of whether the officer was responding to an emergency call. In general, questions of recklessness and proximate cause are for the jury. (This is a distinct minority view in pursuit cases. Many are concluded with summary judgment.) *Lewis v. Bland,* 599 N.E.2d 814 (Ohio App. 1991).

A driver who was speeding, drunk, and driving without lights at night was properly chased by a state patrol officer, according to a 1993 case.

This tragic case was caused by teenage drinking. The driver, aged 16, and two friends took her Ford Probe, a sixteenth birthday gift, to Sandusky, Ohio. There they drank alcohol. At 10:12 P.M. they hurried back toward Port Clinton to make the strictly-enforced parental curfew. A patrolman observed them speeding on U.S. 2 at 97 mph. The officer crossed the median and turned on his red lights and chased. Both of the Probe passengers pleaded with the driver to stop without avail. The pursuit lasted five minutes, the latter part of which the driver turned off her headlights and passed vehicles on the shoulder. She spun out, hit a tree, and was killed along with one passenger. Her blood alcohol tested at .13. The plaintiff's main argument was that state patrol pursuit guidelines were not followed. The court sets out the guidelines in the case, and found that the plaintiff failed to so demonstrate by a preponderance of the evidence. The court particularly points out that when the decedent turned off her headlights, the officer had a duty to follow and try to alert other traffic. (I will point out that I am very familiar with this resort area, and there is heavy traffic in season.) *Stone v. Ohio State Highway Patrol,* 629 N.E.2d 1116 (Ct. Cl. 1993).

Officers other than city, county, and state peace officers are protected by the statutes. A police officer employed by the Toledo Mental Health Center was approached by a patient who said that she had been raped, and pointed to the rapist standing nearby. The latter struck the female officer and fled in his car. She called for another officer and asked him to get the license number. He followed in a Jeep Wagoneer (hardly a good pursuit vehicle) and observed the suspect speeding, running red lights, and turning off his lights. The Wagoneer was equipped with overhead flashing lights that he activated. The officer was five-eights of a mile behind the suspect when the suspect ran a red light and hit two cars. The plaintiff was in one of the cars. In court, the plaintiff argued that the officer was negligent in not sounding his siren, with which the vehicle was equipped, as he was driving on an emergency call. The court found that the officer was not "in response to an emergency call or was in the immediate pursuit" of the suspect. If he were, he was required to use both lights and siren. He was just trying to get the plate number. In any event, the only cause of the accident was the rapist. *Brasher v. Toledo Mental Health Center,* 584 N.E.2d 1374 (Ct. Cl. 1990).

Whether an officer is responding to an emergency call is not always clear. All the facts must be considered. In one case, a state trooper was proceeding to the site of a commuter plane crash in Gahanna, a suburb of Columbus. It was a fatal accident, and several police officers and ambulances were en route or at the scene. The officer stayed within the speed limit and did not switch on any emergency signals. The officer turned onto a small road that was icy. Ahead a driver had pulled over and almost stopped in response to an oncoming ambulance, and the officer skidded and struck the car. The court held that the officer's negligence was the sole cause and awarded damages. The officer was responding to an emergency but not driving on an emergency basis. That was a wise decision on the officer's part, but the case is one of simple negligence. *Quappe v. Ohio Dept. of Public Safety,* 679 N.E.2d 755 (Ct. Cl. 1997). The court in *Quappe* noted that the nonuse of emergency signals of itself does not determine the outcome. In *Moore v. Columbus,* 649 N.E.2d 850, police responding to a reported burglary at a school did not turn on their siren or flashing lights. A silent approach to a crime in progress makes sense. The police were still protected by the emergency vehicle statutes. Code 2744.02 specifically does not require audible and visual signals. Again, the immunity statute trumps the UVC-derived statute.

A 1998 case involved an off-duty deputy sheriff who was driving to work in a cruiser, a vehicle he was permitted to take home. He claimed that he overheard a report on his radio of a burglary in progress, and radioed the dispatcher that he was available to help. He continued and eventually ran a red light and hit a driver with the green. The defendant claimed immunity pursuant to §2744.02 as a matter of law, and the court granted him summary judgment. The court of appeals reversed for a trial. To be on an emergency call, the officer must "be called to duty." There was no evidence before the judge that the officer was himself called, and it is not clear that an officer can himself decide that he is on an emergency call. A long and thoughtful dissent pointed out that no Ohio case required such a "call to duty," and the deputy had in fact, according to witnesses, switched on his lights and siren. The other issue for the jury was to determine if running the red light was willful or wanton misconduct. *McGuire v. Lovell,* 715 N.E.2d 587 (Ohio App. 1998).

A 1991 case had a predictable outcome. A South Russell (Geauga County) police officer, at 1:00 A.M., was run off the road by an oncoming car that was speeding and driving on the center line. The officer turned around and pursued with lights and siren. He clocked the suspect at 80 mph. The officer maintained contact with his department and kept it advised of his position and progress. The pursuit continued over deserted rural roads for fifteen miles. It ended when the suspect crashed off the road. The pursued driver sued the officer. The court of appeals affirmed summary judgment for the officer. The officer was on an emergency call. He stayed well behind the suspect, far enough to keep it in sight, slowed or stopped at intersections, and his conduct could not be considered willful or wanton. *Rogers v. DeRue,* 598 N.E.2d 1312 (Ohio App. 1991).

We contrast *Rogers* with a 1987 case in which the court denied summary judgment to a city. The case arose when an Oakwood police officer at the city garage heard a radio call about a domestic disturbance. The officer jumped in a cruiser, turned on his lights and siren, and set out on a street (a boulevard with a tree-lined median), sped to 70 mph in a 25-mph-zone, entered an intersection on a red signal, and collided with the plaintiff. The case went to trial, alleging willful and wanton conduct on the part of the officer, as well as negligent training. At the close of the plaintiff's case, the city moved for a directed verdict, which was granted. The court of appeals reversed. There was a clear possibility that a jury could have found willful and wanton conduct. On the issue of negligent training, the court noted that the officer was age 22, had been on the force two months, and had had two weeks of training. There was also an issue of whether the officer was on a valid emergency call. *Reynolds v. City of Oakwood,* 528 N.E.2d 578 (Ohio App. 1987).

A 2000 case filed in federal court found two issues that could not be disposed of by summary judgment. The case arose from a fatal pursuit. Around midnight, a Carrollton police officer saw a driver exceeding the speed limit, doing 43 mph in a 25-mph-zone. No other vehicles were involved, nor were there any witnesses to the pursuit. The contentions of each side differed considerably. The officer says he followed at a moderate speed, stopped at stop signs, stayed back and lost sight of the vehicle at times, and did not witness the accident. (The driver hit a utility pole just outside the city limits.) The plaintiff's accident reconstructionist, depending primarily on paint markings, said that the officer must have hit the vehicle during the pursuit. The federal district court judge found that this was enough evidence of an unreasonable seizure under the Forth Amendment to deny summary judgment. The court relied on 1989 California case, *Brower v. County of Inyo,* 489 U.S. 593. Brower involved a twenty-mile chase that ended when police set up a tractor-trailer roadblock across the road, and had headlights shining toward the soon to be deceased as he came up and hit the truck. The U.S. Supreme Court held that whether there was an unlawful seizure was a triable issue in that case. Here, there is a question of whether the officer used his vehicle with an intention to stop an individual who would not voluntarily stop when chased. Whether the seizure, if there was one, was unreasonable depends on whether the offense justified the action. The plaintiff urged such justification improper when the offense was a traffic infraction and no one was in danger.

The court held that the second issue, that of municipal liability, also could not be disposed of by a summary judgment for the defense. The plaintiff contended that the city proximately caused the accident by inadequate training, which amounted to deliberate indifference. The court held that this issue could go to a jury. The plaintiff had shown sufficient evidence. The court dismissed all other constitutional claims. *Hockenberry v. Village of Carrollton,* 110 F. Supp.2d 597 (N.D. Ohio 2000.) Note that the officer was charged before a grand jury. The grand jury did not return an indictment. Obviously there was something more going on with this case than was brought out in the opinion. It appears that local passions were aroused by this accident.

We have covered enough police cases to indicate the current state of Ohio law. We conclude with ambulance and fire vehicle cases. While it seems a little late in the day to litigate the constitutionality of the immunity statute, that was the issue in a 1995 fire truck case. The plaintiff was sitting at a stop sign and was hit by a fire truck making an "improper" left turn. Probably the driver cut the corner and impinged on the plaintiff's lane. Obviously the plaintiff was completely free from fault. The case came up when the plaintiff's insurance company refused to pay damages under uninsured motorist coverage because the tortfeasors were immune. Uninsured motorist coverage stands in the place of the uninsured driver who must be found liable. (The plaintiff later settled with his insurer, so persistence can pay off.) The Ohio Supreme Court held that the statute, §2744.02, is constitutional. The legislature could properly find that quick response in an emergency outweighs the plaintiff's rights here. Also, protecting the financial soundness of the government is a legitimate purpose.

Whether these are the best ways to achieve the goals is for the legislature. A dissent found "The rationalizations and legal fictions employed to prop up this monarchial throwback are bewildering," referring to sovereign immunity. That is a distinct minority view, and I would point out that the statute in question merely sets out a small, in terms of number of claims, exception. *Fahnbulleh v. City of Columbus*, 653 N.E.2d 1186 (Ohio 1995).

Note a 1993 case in the court of appeals that similarly upheld the constitutionality of the statute. The plaintiff was in an accident with a fire truck, and alleged state and federal constitutional violations. *Lewis v. City of Cleveland,* 623 N.E.2d 1233 (Ohio App.1993).

A 1996 ambulance case held that summary judgment for the defendant was proper. The plaintiff said that he was driving on Superior Avenue in Cleveland on a winter day. He came to a T intersection where he had the right of way. He saw a flashing light to his left and slowed, but was hit by an ambulance that skidded on ice into the intersection. The court held that the plaintiff had not presented any genuine issue of material fact in dispute. The ambulance operator said that he did slow to 5–10 mph. This satisfied the statute permitting such drivers to slow as necessary for safety. The fact that he skidded on ice did not constitute willful or wanton conduct, and in fact showed due care for the safety of others. The plaintiff argued that he did not hear a siren, while the defendant said he used the siren. With no other witnesses, this is not enough to overcome summary judgment. *Harris v. Kennedy,* 689 N.E.2d 53 (Ohio App. 1996).

A 1992 ambulance case had a different result, with the court of appeals reversing summary judgment. A firefighter-paramedic responded to 911 fire call, which turned out to be all smoke and no fire. The only action taken at the scene was investigation. The accident happened on the way when the plaintiff became aware of the ambulance behind him. The plaintiff stopped in the curb lane at a green light. The ambulance operator passed and attempted a right turn, with moderate damage to both vehicles. The trial court granted summary judgment to the defendant. The court of appeals found that an emergency existed even though there was no fire. The court did reverse as to whether the defendant exhibited reckless or willful conduct beyond mere negligence. Maybe the plaintiff moved forward into the ambulance's side, or maybe the ambulance operator made a very careless right turn. *Brockman v. Bell,* 605 N.E.2d 445 (Ohio App. 1992).

A 1993 ambulance case held that even if the operator was not displaying visual and audible signals (he said that he did), that alone does not indicate willful or reckless conduct. The ambulance operator entered an intersection past a stop sign without stopping. He slowed to about 10 mph and collided with the plaintiff. The trial court granted summary judgment to the defendants. The court of appeals found that the plaintiff had presented no evidence that the defendant was not responding to an emergency alarm. The court noted that the immunity statute does not require emergency vehicles on emergency runs to display warnings. Even if the defendant should have come to a stop or slowed even more, as the plaintiff argued, his conduct was not reckless, willful or wanton. *Lipscomb v. Lewis,* 619 N.E.2d 102 (Ohio App. 1993).

Two fatalities were the result of an ambulance accident in a 1998 case. A driver of an EMS (emergency medical service) vehicle on an emergency run came to a red light at an intersection. She said she slowed to 15 mph and looked both ways twice before entering the intersection. She was hit broadside by a car and its two occupants were killed. The car's speed was estimated to be 55 mph. (The highway was U.S. 20A and I think that is the speed limit. The case does not state the limit.) The case went to trial and the jury awarded damages. The trial court then granted judgment notwithstanding the verdict on the defendant's argument that it was statutorily immune. The parties agreed that there was no evidence of willful or wanton conduct. The only real issue was the status of §4511.03, a part of the traffic code. It permits emergency vehicle operators to proceed cautiously past red signals and stop sign while exercising due care. The jury found that the operator did not exercise due care. The plaintiff pointed out that the immunity statute, §2744.02(B)(2) grants immunity for the operation of a fire department vehicle on an emergency run, absent willful or wanton misconduct. But subsection (c) immediately following grants immunity to "a member of an emergency medical service" vehicle if the operator complied with §4511.03, absent willful or wanton misconduct. (I must compliment the plaintiff for noticing this discrepancy.) Which of these applies? The vehicle is an emergency medical service vehicle. The court held that subsection (b) applied, and affirmed the judgment notwithstanding the verdict. The operator was a volunteer working for and trained by the Wauseon Fire Department, and the vehicle was owned by the department. Thus subsection (b) applied. A dissent would hold that the operator's function as an EMS driver should control, not her ancillary role as a volunteer member of the fire department. Also, the legislature should make up its mind about §4511.03. *Schwarzbek v. City of Wauseon,* 709 N.E.2d 579 (Ohio App. 1998).

Oklahoma

Okla. Stat. tit. 47, §155. Exemptions from liability

The state or a political subdivision shall not be liable if a loss or claim results from:

6. Civil disobedience, riot, insurrection or rebellion or the failure to provide, or the method of providing, police, law enforcement or fire protection.

Okla. Stat. tit. 47, §11-106. Authorized emergency vehicles

A. [This section sets out UVC 11-106 with two changes. An emergency vehicle on an emergency run must use an audible signal *or* visual signals to remain within the scope of the act. The use of both are not required. Part C states in part: "This subsection shall not be construed as requiring a peace officer operating a police vehicle properly and lawfully in response to a crime scene in progress to use audible signals."]

Only a few cases have been identified. A 1998 case construed 51 O.S. §155(6) in a case in which Tulsa police arrested a person, handcuffed him, and put him in a patrol car with the engine running. The officers were standing outside the car. The suspect managed to unlatch his seatbelt, and drove away and caused an injury to the plaintiff. The Court of Civil Appeals held that the city was immune because the acts of the police were related to the method of providing law enforcement. *Vaughn v. City of Tulsa,* 974 P.2d 188 (Okla. Civ. App. 1998).

A 1990 case arose from a pursuit. A Tulsa police officer stopped at a red light and noticed a car that screeched to a halt beside him. The officer noticed that the car's headlights were not on (it was 9:45 P.M. in January) and that the driver appeared either drunk or sleepy. The officer attempted to stop the driver, eventually using a continuous siren and flashing lights. The driver sped off at up to 65 mph on city streets. The pursuit lasted one minute and a distance of a little over one mile. It ended when the driver hit the back of a car in front of him, lost control, swerved into the oncoming lane and killed and innocent person. The estate alleged that the city failed to properly train and supervise the officer, that the officer failed to comply with pursuit policies, and that he drove with reckless disregard and want of due care for the safety of others. The trial court granted summary judgment to the city on the ground that the due care requirement of §11-106 does not apply to the decision to pursue a law violator.

The court of appeals affirmed. The plaintiff did not contend that the officer's operation of his police car was negligent. The sole reason why the suspect sped up was to evade a police stop. Section 11-106 does not address the decision to make a stop or to pursue if the driver will not stop. The court cited several cases from other jurisdictions in finding that if an officer in a pursuit is himself observing the law, he will not be liable to an innocent victim of the fleeing driver. The court also found as a matter of law that the officer was not the proximate cause of the accident. The officer's pursuit was not so outrageous as to outweigh the public benefit of enforcing the law. *Kelly v. City of Tulsa,* 791 P.2d 826 (Okla. App. 1990).

The third case arose from an accident between a police car and the plaintiff. An Oklahoma City police officer was responding to an emergency call (Code Three; officer needs assistance) with flashing lights in operation. The plaintiff ahead noticed the lights, and attempted to move into the right lane. The officer attempted to pass on the right at the same time and there was a collision. The plaintiff argued that she committed no traffic violation, while the officer did not use due caution as required by §11-106. The trial court granted the city summary judgment under the immunity statute. The court of appeals reversed. It found that the immunity statute did not apply. An emergency vehicle in response to an emergency call does not constitute a *method* of providing police protection. The operation of the emergency vehicle still must conform to §11-106. If such officer failed to observe the duty of care set out in the statute, citizens injured by such failure must not be left without a remedy. The plaintiff had presented sufficient documentation for a prima facie case. *Taylor v. City of Oklahoma City,* 914 P.2d 1073 (Okla. App. 1996).

I will comment that *Taylor* and *Kelly* are not in conflict. There was no allegation of any violation of §11-106 in *Kelly.* Note also that the plaintiff in *Taylor* was attempting to observe the law by trying to move to the right and stop. The statutory traffic law cannot cover every situation. If a driver is on the inside lane of a four-lane street and the emergency vehicle is approaching from the rear, the driver should move to the right only if it is safe. In the *Taylor* case, obviously the officer, who must still be considered more at fault, thought he could pass safely on the right. I also think that the way that Oklahoma courts give consideration—to both the immunity and the traffic statute—is an excellent policy.

A 2001 case was filed in federal court and has been in the Tenth Circuit Court of Appeals twice (so far). The case arose (May 1990) when Tulsa police began pursuit of a stolen car. The area was primarily residential. The fleeing driver ran at least eight stop signs at speeds between 40–50 mph, and speeds were around 65 mph otherwise. The chase ended when the driver ran a red light and hit a Ford Escort, killing all three occupants. The plaintiff, the estate of one of the deceased, alleged that the officers were aware of the likelihood of encountering traffic, and that pursuit regulations of the department were not followed. No supervisor was notified of the pursuit, a violation of the policy then in force. The policy permitted officers to pursue for any traffic offense, and the officer had the discretion of whether to pursue and when or if to terminate unless otherwise directed by a supervisor. Officers were trained to use caution and common sense and to terminate a pursuit when the risks outweighed the benefit.

The trial judge granted summary judgment to the defendants in 1993. The court of appeals, after some extended proceedings involving the district court, affirmed summary judgment for the three officers in finding that they had qualified immunity. It was not clear that in 1990 that they could be liable under a §1983 action. *Trigalet v. Young*, 54 F.3d 645 (10th Cir. 1995). (The plaintiffs revived the action against the city of Tulsa, and all proceedings were stayed pending the U.S. Supreme Court's decision in *Sacramento v. Lewis* (see the California section).)

The plaintiffs alleged a number of policies and practices involving police pursuits that violated the plaintiffs' constitutional rights, that were arbitrary or conscience-shocking acts or omissions. After considering *Lewis* and similar cases, the court held that the officers' acts in pursuing the driver of a stolen car did not violate the constitutional rights of the plaintiffs. There was no intent to harm them. If the officers did not commit a constitutional violation, neither can the city be held liable as to that claim. But note that the court expressed no opinion as to whether the city can be liable on the claim of indirect liability. *Trigalet v. City of Tulsa,* 239 F.3d 1150 (10th Cir. 2001).

A 1994 case also was filed in federal court. A Sapulpa police officer received notice that Gibbs had escaped from a correctional institution, was suspected of one rape and a rape attempt, was armed, and that his arrest was sought. The officer noticed a parked car matching the description and saw that a person was slumped over the wheel. The officer approached with gun drawn and tapped on the window. The door was locked. The officer recognized the suspect when he woke. Gibbs started the engine and drove off, and the officer pursued on Highway 66 toward Tulsa. Both vehicles were over the speed limit in the pursuit. Three minutes and 2.6 miles later Gibbs veered into oncoming traffic. Gibbs died and the occupants of the other car were severely injured. The occupants sued the officer and the city. The district court granted summary judgment to the defendants, and the Tenth Circuit affirmed. The court of appeals repeated an oft-made comment that §1983 "must not be used to duplicate state tort law on the federal level." The court agreed that the officer's conduct was not reckless and did not indicate a conscious indifference to the safety of others. The plaintiff's argument primarily was that the officer should not have awakened Gibbs, but should have called for backup and not given the suspect a chance to flee. The court finds that to be hindsight. Police officers often must make split-second decisions "in circumstances that are tense, uncertain, and rapidly evolving." *Webber v. Mefford,* 43 F.2d 1340 (10th Cir. 1994).

Oregon

Or. Rev. Stat. §820.300. Exemptions from traffic laws

(1) Subject to conditions, limitations, prohibitions and penalties established for emergency vehicles and ambulance drivers under ORS 820.320, the driver of an emergency vehicle or ambulance may do any of the following:
   - (a) Park or stand in disregard of statute, regulation or ordinance prohibiting that parking or standing.
   - (b) Proceed past a red signal or stop sign.
   - (c) Exceed the designated speed limits.
   - (d) Disregard regulations governing direction of movement or turning in specified directions.
   - (e) Proceed past the flashing bus safety lights without violating ORS 811.155 if the driver first stops the vehicle and then proceeds only when the driver:

(A) Determines that no passengers of the bus remain on the roadway; and

(B) Proceeds with caution.

(2) The provisions of this section:

(a) Do not relieve the driver of an authorized emergency vehicle from the duty to drive with due regard for the safety of all other persons.

(b) Are not a defense to the driver of an emergency vehicle or ambulance in an action brought for criminal negligence or reckless conduct.

(c) Except as specifically provided in this section, do not relieve the driver of an authorized emergency vehicle or ambulance from the duty to comply with ORS 811.155.

Or. Rev. Stat. §820.320. Illegal operation of emergency vehicle or ambulance; penalties

(1) A person commits the offense of illegal operation an emergency vehicle or ambulance if the person is the driver of an emergency vehicle and violated any of the following:

(a) The driver of an emergency vehicle or ambulance may only exercise privileges granted under ORS 820.300 when responding to an emergency call or when responding to, but not returning from, an emergency. The driver of an emergency vehicle may exercise privileges under ORS 820.300 when in pursuit of an actual or suspected violator of the law.

(b) The driver of an emergency vehicle or ambulance must use a visual signal with appropriate warning lights when the driver is exercising privileges under ORS 820.300.

(c) In addition to any required visual signal, the driver of an emergency vehicle or ambulance must make use of an audible signal meeting the requirements under ORS 820.370 when the driver is proceeding past a stop sign . . . .

(d) A driver of an emergency vehicle of ambulance who is exercising privileges granted by ORS 820.300 by parking or standing an emergency vehicle in disregard of a regulation or ordinance prohibiting that parking, stopping or standing, shall not use the audible signal.

(e) In exercising the privileges under ORS 820.300(1)(e) relating to buses and bus safety lights, the driver of an emergency vehicle or ambulance must first stop the vehicle and then must:

(A) Determine that no passengers or the bus remain on the roadway; and

(B) Proceed with caution.

(f) In proceeding past any stop light or stop sign under the privileges granted by ORS 820.300, the driver of an emergency vehicle or ambulance must slow down as may be necessary for safe operation.

(g) The driver of an emergency vehicle or ambulance must not exceed any designated speed limit to an extent which endangers persons or property.

(2) The driver of an emergency vehicle that is operated as an emergency police vehicle is not required to use either visual signal or the audible signal as described in this section in order to exercise the privileges granted in ORS 820.300 when it reasonably appears to the driver that the use of either or both would prevent or hamper the apprehension or detection of a violator of a statute, ordinance or regulation.

(3) The offense described in this section, illegal operation of an emergency vehicle or ambulance, is a Class B traffic violation.

Or. Rev. Stat. §30.265. Scope of liability of a public body, officers, employees and agents

(1) Subject to the limitations of ORS 30.260 to 30.300 [tort actions against public bodies], every public body is subject to action or suit for its torts . . . .

(2) Every public body and its officers, employees and agents action within their scope of employment or duties . . . are immune from liability for:

(c) Any claim based upon the performance of or the failure to exercise or perform a discretionary function or duty, whether or not the duty is abused.

[The statutes based on the UVC version are rather wordy, but enact the essence of UVC 11-106. The standard of care given in the first is due regard and reckless conduct, rather than the UVC and most states' use of the term reckless disregard. The second statute set out specifically defines the traffic violation of illegal operation of an emergency vehicle. One with a case would want to find out if an emergency vehicle driver had been given a ticket for this offense.]

There are not many identified cases. A 1996 case involved a deputy sheriff who responded to a 911 call. He entered the intersection of two state highways on a red signal, with his lights and siren on. He collided with the plaintiff driver who had the green signal. The Marion County

Sheriff's Board of Review investigated the accident. The board submitted a report to the sheriff's office, and apparently found some fault because the deputy was reprimanded. The board found that the deputy was not responding to an emergency, and that he went around stopped traffic at the intersection without stopping or using due care. The board's conclusion was that the accident could have been prevented if the deputy had observed the applicable statutes. The issue was whether the board's report should have been admitted in the civil trial. It was admitted, and the jury returned a plaintiff's verdict. The court of appeals affirmed. Normally a party may not bring in evidence of remedial action. An example would be the installation of a stop sign after an accident. Without deciding whether the report constituted remedial action, the court found that its admission did not affect a substantial right of the defendant. At trial the deputy admitted that he had been given a letter of reprimand and had not protested. *Ensign v. Marion County,* 914 P.2d 5 (Ore. App. 1996).

A 1991 pursuit case in which occupants of the pursued vehicle were injured has a very important issue: If the plaintiffs (passengers) settles with a private party, in this case the owner of the pursued vehicle, and that sum exceeds the statutory limit on governmental liability, is the government liable for any damages? The facts, viewed in the light most favorable to the plaintiffs, showed that at 10:50 P.M. a Gearhart police officer and a state trooper saw two cars "of the sort usually driven by teenagers" speeding on Highway 101. The officer pursued and reached 75 mph without gaining on the fleeing drivers. The pursued drivers split, and an officer pursued each. The Gearhart police officer had the accident. The officer lost sight of the car when he had to make a U-turn to follow, and then saw a car turn onto Pacific Way in Gearhart. The street is .8 miles long and ends at the beach. The officer got close enough that the occupants in the pursued car could see the officer's overhead lights. The driver ran into boulders that blocked traffic to the beach, and passengers in the car were injured. The plaintiff passengers settled with the vehicle's owner and sued the officer and the city. Without deciding the plaintiff's allegations of negligence, the trial court judge granted the city a directed verdict on the ground that the plaintiffs had already recovered more damages than the statutory cap of $100,000 per claimant and $500,000 per incident.

The court of appeals reversed on the statutory cap issue. The court considered the case law and Oregon's version of the Uniform Contribution Among Tortfeasors Act, ORS 18-460. The correct rule is that the plaintiffs may recover their full damages, however determined. If the damages exceed the amount for which they settled, the plaintiffs may turn to the defendant city for any amount up to the statutory cap. Note that there is a setoff for an amount received, so that if a plaintiff in this case is found to be entitled to $150,000, and he settled for $50,000 from the private party, he will get $50,000 from the city, not $100,000.

On the issue of negligence, the court found that to be a triable issue. There was evidence that the officer knew of the boulders at the end of the street. (I would have to say that this case is one of the tamest chases in the book. In most states, the defendants would be entitled to summary judgment.) The court also found issues regarding lack of training in pursuit, and the policy that determines the circumstances to terminate a pursuit was lacking in guidance. The court noted that the city intended to use the discretionary function defense based on ORS 30.265(3)(1) at the trial as an affirmative defense. *Dee v. Pomeroy,* 818 P.2d 523 (Ore. App. 1991).

A 1989 case held that ORS 30.265 applied to a pursuit in which the fleeing driver hit the plaintiff. The plaintiff alleged that the officer, who tried to stop a driver with an expired plate, was negligent. The court of appeals affirmed summary judgment for the county. Citing an earlier case, *Jackson v. Olson,* 712 P.2d 128 (Ore. App. 1985), the court held that a police officer has discretion and is immune for the decision to initiate a pursuit and the decision to end it. The manner of the pursuit may be negligent. *Dee* above applied this rule. *Lowrimore v. Dimmitt,* 781 P.2d 411 (Ore. App. 1989).

I would comment that Oregon is in a minority of states that applies ordinary negligence to police pursuit cases, even though ORS 820.300 includes the terms criminal negligence and reckless conduct. UVC 11-106 itself is ambiguous because it gives both a due care standard, which is usually the same as ordinary negligence, and reckless disregard. This gives room for diverse applications of the statute among the states.

Pennsylvania

75 Pa. Cons. Stat. Ann. §3105. Drivers of emergency vehicles

(a) General rule—The driver of an emergency vehicle, when responding to an emergency call or when in pursuit of an actual or suspected violator of the law or when responding to but not returning from a fire alarm, may exercise the privileges set forth in this section, but subject to the conditions state in this section.

(b) [This subsection gives the operator the right to go through a red signal or stop sign after slowing as necessary for safety, exceed speed limits as long as that does not endanger life or property, and disregard regulations directing movement or turning.]

(c) Audible and visual signals required—The privileges granted in this section to an emergency vehicle apply only when the vehicle is making use of an audible signal and visual signals meeting the requirements and standards . . . , except that an emergency vehicle operated as a police vehicle need not be equipped with or display the visual signals. An ambulance which is transporting a patient may use either the lights or the audible warning system, or both, as determined by the driver of the ambulance.

(d) Ambulances and blood-delivery vehicles—The driver of an ambulance or blood-delivery vehicle shall comply with maximum speed limits, red signals indications and stop signs. After ascertaining that the ambulance or blood-delivery vehicle will be given the right of way, the driver may proceed through a red signal or stop sign.

(e) Exercise of care—This section does not relieve the driver of an emergency vehicle from the duty to drive with due regard for the safety of all persons.

42 Pa. Cons. Stat. Ann. §8542. Exemptions to governmental liability

(a) Liability imposed—A local agency shall be liable for damages . . . if both of the following conditions are satisfied and the injury occurs as a result of one of the acts set forth in subsection (b):

(1) The damages would be recoverable under common law or a statute creating a cause of action if the injury were caused by [a private person].

(2) The injury was caused by the negligent acts of the local agency or employee . . . .

(3) As used in this paragraph, "negligent acts" shall not include acts or conduct which constitutes a crime, actual fraud, actual malice or willful misconduct.

(b) Acts which may impose liability—The following acts by a local agency or any of its employees may result in the imposition of liability on a local agency:

(b)(1) Vehicle liability—[For] the operation of any motor vehicle [which is] in the possession or control of the local agency, provided that the local agency shall not be liable to any plaintiff that claims liability under this subsection if the plaintiff was, during the course of the

> alleged negligence, in flight or fleeing apprehension or resisting arrest by a police officer or knowingly aided a group, one or more of whose members were in flight or fleeing apprehension or resisting arrest by a police officer. As used in this paragraph, "motor vehicle" means any vehicle which is self-propelled and any attachment thereto, including vehicles operated by rail, through the water or in the air.

Note that §3105 above gives the standard of due regard for the safety of others. Older cases referred to the standard of reckless disregard for the safety of others. Apparently that language was formerly in the statute. According to an older case, *Junk v. East End Fire Dept.*, 396 A.2d 1269 (Pa. Super. 1978), the "new" vehicle code of 1959 deleted the reckless disregard language to make it easier for plaintiffs in emergency vehicle cases to prevail. The court explained that under the due regard standard, a plaintiff still must prove more than simple negligence. He must prove recklessness. What would be reckless conduct by an ordinary driver may not be reckless conduct on the part of an emergency vehicle driver. If a fire truck driver with lights flashing and siren screaming comes to a red light, sees traffic that has yielded, and enters the intersection, that may not be reckless conduct.

Most recent cases are determined by the immunity doctrine if they involve police pursuit or apprehension. A 1994 case involved a twenty-year-old man who celebrated his birthday with a friend by driving around for four hours and drinking "an unknown number" of cans of beer. The two continued to misbehave by deliberately sideswiping a car and urinating in the parking lot of a convenience store. A customer of the store notified police, and an officer tried to make a stop. The driver sped away and got on Route 2, where his passenger noticed that the speedometer read 80 mph and that they missed utility poles by inches. The officer dropped back and notified another jurisdiction of the suspect. The suspect drove on the center line and the left lane at times. He then failed to negotiate a curve and hit a utility pole, which ended his life. The passenger somehow survived (the car broke in half and each part wound up over 100 feet apart). The plaintiff (estate) sued the officer, the borough, and the utility company for the placement of its pole. The trial court granted summary judgment to all defendants. The appellate court held that the borough was not liable because a previous decision held that such defendant would not be liable for injuries to a third person if that person was injured by a fleeing individual. Therefore, it would be strange indeed if the borough would be liable to the culprit himself. The court also affirmed as to the utility company. While its pole was only eight inches from the roadway, this accident was unforeseeable. *Beck v. Zabrowski*, 650 A.2d 1152 (Pa. Comwlth. 1994). An almost identical case in 1995 also held the city not liable to a driver who, while trying to elude police, hit a utility pole and killed himself. *Tyree v. City of Pittsburgh,* 669 A.2d 487 (Pa. Comwlth. 1995).

A 1992 case came to the Pennsylvania Supreme Court. A police officer in the early morning hours attempted to stop a driver he suspected of having an invalid license and being under the influence of drugs. The driver fled at high speed and hit and injured the plaintiff. The trial court refused to grant the officer and the township summary judgment. The supreme court reversed. It noted the general rule that the criminal and negligent acts of third parties are superseding causes that absolve the original actor from fault. The court noted that the immunity statute, §8542, absolves the government of liability for the acts of its employees who are operating within the scope of their employment. It would be incongruous to hold the government liable in this case, because the injuries were caused by the criminal acts of the fleeing driver. *Dickens v. Horner*, 611 A.2d 693 (Pa. 1992).

A 1993 case also involved injury to the pursued driver. A sixteen-year-old girl who did not have an operator's license fled when an officer tried to stop her. He pursued, and she went off the road at a curve and was severely injured. Her complaint alleged that the officer did not have probable cause to stop her, or that if he did he was negligent for following her several miles to a deserted road, and that he should have known she would not be able to negotiate the curve at her speed. Following the reasons given in *Dickens* above, the appellate court held the defendants immune. *Hawks v. Livermore,* 629 A.2d 270 (Pa. Comwlth. 1993).

The Pennsylvania Supreme Court revisited *Dickins* in a 1997 case. Again, the plaintiff was a bystander. The case arose when a Philadelphia police officer saw three cars in a row disregard a stop sign and then a red light. The officer began to follow. He then heard gunfire from the second car directed toward the first. The officer's siren failed to work, but he turned on his dome lights. He chased for thirteen blocks at speeds of 70 to 80 mph, according to the plaintiff's estimate. The pursued drivers entered an intersection against a red light and collided with the plaintiff, injuring him and killing his wife. The trial court granted summary judgment to the defendants, the city and officers, on the basis that the fleeing driver's acts relieved the police of liability as a matter of law. The commonwealth court reversed for trial, and the supreme court affirmed. Plainly put, *Dickens* was wrongly decided. It will be up to a jury to decide if the pursuit, and the way it was conducted, was a substantial factor. The plaintiff argued that if the siren had been working, he could have been warned. He said the fleeing driver was not using his headlights. Also, it was against police policy to pursue without a siren. If the siren was out of order, perhaps the pursuit should have been terminated. A concurring opinion thought that *Dickens* was consistent with this case. The claim in *Dickens* was that the police wrongly initiated a pursuit. A dissent thought that when a fleeing lawbreaker causes harm to another, both the immunity statute and case law protects the police and their employers. *Jones v. Chieffo,* 700 A.2d 417 (Pa. 1997). Note that the plaintiff also brought an action in federal district court claiming a §1983 violation. The district court judge granted summary judgment to the defendants. The plaintiff had not shown that the defendants acted with deliberate indifference. *Jones v. Chieffo*, 498 F/ Supp. 498 (E.D. Pa. 1993).

A short 1999 case held the city of Williamsport immune in a pedestrian pursuit. A city police officer, patrolling in the early morning hours, noticed two young men on a sidewalk. One was carrying what appeared to be a bottle of beer. The officer, in his cruiser, chased the men. The suspects tried to hide in a hedge, and the officer hit one of them, causing injuries. The trial judge granted the defendants summary judgment on immunity pursuant to §8542(b)(1) on its finding that the plaintiff was "in flight or fleeing apprehension." The appellate court affirmed. The plaintiff argued that whether he was in flight or fleeing apprehension was a factual issue that should be decided by the trier of fact. The court found no factual controversy here. The plaintiff himself said he was running because he was afraid of getting into trouble. *Forgione v. Heck,* 736 A.2d 759 (Pa. Comwlth. 1999).

We now look at several cases in which the plaintiff at least survived a defense summary judgment. A 2000 case arose from a high-speed pursuit in which a bystander was injured. Officers learned of a case of shoplifting. The suspect's car was identified and was found to have been stolen. An officer spotted the car, a high-speed pursuit ensued, and the plaintiff was injured when the thief crashed into his car. The plaintiff sued several boroughs. The trial court granted summary judgment to the defendants on the basis of immunity. The Commonwealth Court reversed summary judgment for the defendants. Whether the defendants' alleged negligence was a cause is a jury question. The court cited *Jones* above as authority, and noted that the immunity statute specifically states that there shall be no liability to those in flight or resisting arrest. It does not include innocent bystanders. *Aiken v. Borough of Blawnox*, 747 A.2d 1282 (Pa. Comwlth. 2000).

I comment on a 1992 case because it shows why §8542 was amended in 1995. The amendment said that there would be no liability to persons in flight or fleeing apprehension or resisting arrest, and made admissible into evidence information showing that the plaintiff was "engaged or participating in willful misconduct, including but not limited to, the illegal possession of controlled substances, firearms or ammunition." The accident in the 1992 case happened in 1979. Police on patrol at 4:00 A.M. say the plaintiff, riding a motorcycle, ran a stop sign. He fled when the police tried to stop him. He turned down a narrow, downhill street and saw that his way was blocked by a police car. He tried to steer around the car and was injured. (One leg was partially amputated.) There was dispute over whether the police car was moving, with the plaintiff arguing that the officer tried to block his path. The trial court instructed the jury that the plaintiff was negligent as a matter of law in not stopping, but left it to the jury to determine whether the plaintiff's negligence was a substantial factor in the injury. The jury awarded $750,000 on its finding that the city was 82 percent at fault. (The court reduced the award to the statutory limit of $500,000, but awarded $419,232 in delay damages.)

The Commonwealth Court affirmed most of outcome. It was proper to exclude evidence that the plaintiff was in possession of a handgun stolen from a government office. It was proper to exclude an assumption of the risk defense urged by the defendants. The court affirmed the admissibility of a plaintiff's expert witness testimony to the effect that nationwide standards require that in setting up a roadblock room must be left for the person in flight to get around. A dissent thought that there could hardly be a more clear example of assumption of the risk. *City of Pittsburgh v. Jodzis,* 607 A.2d 339 (Pa. Comwlth. 1992). Note that the court in *Jodzis* did mention the emergency vehicle statute, 75 Pa.C.S.A. §3105, noting that the standard of care required of emergency vehicle drivers is due regard for the safety of all persons.

A 1997 case does not add to our knowledge of Pennsylvania emergency law because it involves the apparent incompetence of counsel for the city of Philadelphia. One innocent person was killed and another injured when they were hit by a vehicle being pursued by city police. The city ignored discovery orders, was assessed monetary sanctions several times, and failed to show up for a hearing. The trial judge entered default judgment for the plaintiffs and ordered a trial without a jury on stipulated facts rather than on a case-stated basis. The damages were determined to be $800,000 including $300,000 in delay damages. The appellate court held that the default judgment was proper as imposed as a sanction. Thus the city did not have an opportunity to raise an immunity defense. Immunity would not necessarily mean that the city might not be liable to a bystander in a pursuit case. See *Jones* above. It would be up to the trier of fact as to whether police acts were substantial factors in causing the accident. *Taylor v. City of Philadelphia,* 692 A.2d 308 (Pa. Comwlth. 1997).

Several recent cases have been brought in federal court. The primary case is *Fulkersons v. City of Lancaster,* 801 F. Supp. 1476 (E.D. Pa. 1992), aff'd 993 F.2d 876 (3rd Cir. 1993) (table). An officer began pursuit of a motorcyclist who was breaking every rule in the book, including running red lights, speeding, and driving between lanes of traffic. Given the suspect's erratic driving, the officer assumed the driver was inexperienced, drunk, or on a stolen machine. Even his flight was inept. The officer continued the pursuit because he thought there was a good chance of arresting the suspect and that the suspect was a continuing danger to the public. The suspect crossed the center line one time too many and collided with the plaintiff. The plaintiff alleged violations of the pursuit policy. The case is very long, but the court concluded that the defendants did not act with deliberate indifference and dismissed the claim. The court did not dismiss the state law claims with prejudice, and noted that this tort case belonged in state court anyhow.

In a 1986 case, a Hazelton police officer pursued a suspicious driver who was parked behind a closed supermarket with his lights on and engine running. The driver fled and killed a man when the suspect ran a red light. The court granted the defendants' motion to dismiss, including the state law claims. At most the officer's actions were negligent. The court noted that this incident was the first of this type for the police department. Apparently the city did not have a pursuit policy as such, but the officer had the usual police training. *Timko v. City of Hazelton,* 665 F. Supp. 1130 (M.D. Pa. 1986).

Carroll v. Borough of State College, 854 F. Supp. 1184 (M.D. Pa. 1994) aff'd 47 F.3d 1160 (3rd Cir. 1995) (table), involved a motorcyclist. The rider passed a car on the right, I assume on a two-lane road, while speeding. The officer followed and then the rider ran a red light. The officer switched on lights and siren to stop the motorcyclist, but the latter sped up. The chase went on for three miles at speeds of up to 85 mph. It ended when the rider crashed on a curve. The trial judge granted summary judgment to the defendants on both state and federal claims. The officer testified that he was terminating the pursuit some 500 feet before the crash, and had radioed in the plate number. The court held that there was no unlawful seizure or an attempt to do that, and no conscience-shocking behavior. This case isn't even close.

A 1996 case, *Ploucher v. Nilan*, 1996 WL 677513, construed pursuit regulations. A police officer pursued a fleeing robber in a congested area. The robber drove recklessly, and the officer did consider breaking off the pursuit. He lost sight of the driver and then regained it. The officer did not have radio contact with his supervisor, which was a violation of regulations. The fleeing driver hit and killed two innocent persons. The appellate court held that two pursuit policy violations—not breaking off when losing sight and not maintaining radio contact—were not conscience-shocking, and found for the defendants.

A 2000 case was filed in federal court. It arose out of an Ardmore police officer responding to a nonemergency call, that of "a suspicious vehicle parked unoccupied." The officer was considerably exceeding the 25-mph-speed limit and he collided with the plaintiff's car. The plaintiff filed a §1983 action, alleging that the officer's negligent and reckless actions were a constitutional tort that would make the city liable. Specifically, the plaintiff alleged that his Fourteenth Amendment, substantive due process rights and the "liberty interest in his bodily integrity" was violated. The district court judge granted the defendant's motion for summary judgment. The court noted the U.S. Supreme Court case, *Sacremento v. Lewis,* 523 U.S. 833 (1998) holds that the Due Process Clause is violated only if the executive action is arbitrary or conscience-shocking. The court held that, while the officer had an obligation to respond to the call quickly, the act of speeding on a crowded roadway without use of siren or emergency lights may well have been negligence or possibly reckless. However, it did not rise to the level of shocking to the conscience. If the officer did not commit a constitutional violation, his employer cannot be liable for allegations of inadequate or ineffective training. *Leddy v. Township of Lower Merion,* 114 F. Supp.2d 372 (E.D. Pa. 2000).

We conclude with three fire truck or ambulance cases. A 1998 case involved a fire truck that was on its way to the fire station, apparently not on an emergency run. A diverter valve fell off the truck and landed on a two-lane road. A driver hit the valve and crossed the center line and hit the plaintiff's vehicle. The plaintiff sued the driver and the fire department. The facts were stipulated, and the trial court found damages to be $290,000. The trial court refused to grant the fire department judgment notwithstanding the verdict. The appellate court reversed. Section 8542 waives liability for the operation of government vehicles. "Operation" means movement of the vehicle. Here there was no movement. There was simply a valve that had fallen off the truck. It was a stationary object that had been on the road for some time. The court distinguished another case, *Mickle v. City of Philadelphia,* 707 A.2d 1124 (Pa. 1998), in which a passenger riding in an emergency vehicle was injured when the wheels fell off. That vehicle was moving. *Swartz v. Hilltown Township Volunteer Fire Co.,* 721 A.2d 819 (Pa. Cmwlth. 1998).

A 1992 fire vehicle case arose when the plaintiff entered an intersection at 1:50 A.M. with a green signal and collided with a fire truck on the way to a fire. The plaintiff suffered permanent brain injury. There was evidence that the fire chief at the scene of the fire (from Delaware County) radioed the Chester County fire department and asked that department to radio the involved Longwood fire truck to "slow down." The chief at the scene could not directly radio the Longwood truck which was in the collision. If such request had been received, it was probable that the truck would have stopped at its red signal. The plaintiff settled with the truck operator. At a later date, the trial court granted summary judgment to Delaware and Chester Counties. The appellate court held that the counties were not "operating" the fire truck within the exception to liability. Also since the plaintiff had already released the truck driver, he also released any vicarious liability that could be attributed to the counties. The court notes that it may not allow sympathy or outrage to control its decision on immunity. *Keesey v. Longwood Volunteer Fire Co.,* 601 A.2d 921 (Pa. Comwlth. 1992).

A 1991 case held that a volunteer, nonprofit ambulance service does enjoy governmental immunity. A pedestrian was injured in some manner by the ambulance. The defendants admitted liability, and paid $485,000. The parties agreed that an additional $400,000 plus interest would be payable if the defendant was not covered by governmental immunity. (If covered, the defendants still would be liable for negligence, but would be covered by the cap of $500,000. If so, the plaintiff would receive only $15,000 more.) The appellate court held that there was no immunity in spite of the defendants claims that there were various ways in which such services, such as volunteer fire departments, have been covered. The court pointed to a specific statute, 42 Pa.C.S. §8332. That section grants immunity but has a motor vehicle exception. The defendant argued in vain that the court should follow a case, *Zern,* (516 A.2d 799) that granted immunity to a volunteer fire department. *Scrima v. Swissvale Area Emergency Service,* 599 A.2d 301 (Pa. Comwlth. 1991).

Rhode Island

R.I. Gen. Laws §§31-12-6 to 31-12-9. Emergency vehicles

[These sections enact UVC 11-106 without change. Section 31-12-6(b) requires that uniform rules governing high-speed pursuits be adopted by the law enforcement agency accreditation council (when it is established) that are in accordance with the national law enforcement accreditation agency. Until that time, all police departments shall submit to the attorney general a copy of their high-speed pursuit policies, which shall be available to the public.]

There are only three cases to cover. Beginning with the oldest, we note a 1956 case. An International Travel-All, operated by a seventeen-year-old driver, and carrying seven people, was traveling on I-95. The road at the place of the accident was four lanes wide. The left two lanes were exit lanes and the right two were for through traffic. The driver was in the third lane from the right, and she turned into the fourth lane (the left lane) to pass a slower driver. At the same time, she saw a police car with flashing red lights apparently stopped in front. She struck the car and overturned. One of her passenger's was killed and another severely injured. At the same time of this tragic event, "a noted underworld figure and his companion" had been murdered by someone in a white Cadillac. The police, including the officer in the involved car, had received a be-on-the-lookout-for report to prevent the murderer's escape. The officer decided to exit at a so-called official turnoff and watch. The speed limit on I-95 was 60 mph maximum and 45 mph minimum. The officer was a short distance from this unofficial turnoff, going no more than 40 mph (his estimate) or stopped (others' estimates). The estate of the deceased passenger and the injured passenger sued both drivers, and the jury awarded damages. On appeal, the defendant officer argued that he was not liable because of his status as an emergency vehicle operator. The court noted that he still had the duty not to act with reckless disregard for the safety of others. The court also found that the officer was not operating in an emergency, so his duty was the same as all other drivers. The officer was driving well below the minimum speed limit and was not fully off the roadway. The officer was liable along with the other driver. *Roberts v. Kettelle,* 356 A.2d 207 (R.I. 1976).

A 1983 case held that a police officer was not liable to a passenger in the pursued car. A North Providence police officer noticed a driver, just after midnight, who appeared to be lost. The officer thought the occupants needed some assistance and turned on his flashing lights and beeped his horn to require the driver to stop. Then he turned on his siren. The driver stopped and the officer reported his location to the dispatcher and got out of his car. The driver suddenly sped off. It was learned later that the car was stolen. Since the flight was evidence of reckless operation, the officer pursued. The pursuit went on for two to three miles. The officer reached 80 mph without closing the distance. The suspect ran a red light, swerved to avoid a car, hit another car, and then a utility pole. A passenger was killed. The trial court granted summary judgment to the defendants. The supreme court affirmed. The officer's acts were not the proximate cause. The officer may have been negligent, as the trial court found, in skidding into the same car the suspect hit. However, the acts of the suspect driver began the initial chain of causation. The reckless decision of the fleeing driver showed a blatant disregard for his own life and those of his passengers. Such conduct was not reasonably foreseeable, given the fact that the driver had not broken any laws known to the officer at the time. *Almeida v. Town of North Providence,* 468 A.2d 915 (R.I. 1983).

The third case was brought in federal court. Coincidentally, the trial judge in *Almeida* is the same judge who wrote the opinion in federal district court in this 1993 case. The facts are similar. Two South Kensington police officers noticed a car lacking an inspection sticker. Otherwise the driver was not violating any laws. The officer attempted to stop the driver, but he refused and the officers pursued. A mile down the road, two officers who had learned of the pursuit set up a roadblock but left room for the driver squeeze between. The driver slowed, but then accelerated past the roadblock. One officer fired at the suspect and then joined the pursuit. Speeds reached 100 mph. The driver lost control, rolled, and he was killed and his passenger was permanently injured. The court declined to dismiss for failure to state a claim upon which relief can be granted. First, the court found that there was no seizure of the plaintiff (the passenger). A roadblock is intended to force the driver to stop if he does not do so voluntarily. The roadblock here left an escape that the driver in fact took. Also, the police had no intent of stopping the passenger. Their intent to get the driver to stop. There was no constructive seizure of the passenger.

The court did find that the plaintiff may proceed with his substantive due process claim of undue force. The court notes that the standard for this claim is very high. Police response to crime will at times create some risk to innocent bystanders, and the mere fact of risk of itself does not prove callous indifference. The court also allowed the plaintiff to proceed on the issue of proximate cause. The court was aware of *Almeida,* since the judge wrote that opinion, but held that here the plaintiff had not had his opportunity to present all his evidence. Also, in general, in the First Circuit, questions of intervening cause and so on are usually jury issues. *Medeiros v. Town of South Kensington,* 821 F. Supp. 823 (D.R.I. 1993).

South Carolina

S.C. Code Ann. §56-5-760. Operation of authorized emergency vehicles

(A) The driver of an authorized emergency vehicle, when responding to an emergency call or when in the pursuit of an actual or suspected violator of the law or when responding to but not returning from a fire alarm, may exercise the privileges set forth in this section, but subject to the conditions of this section.

(B) The driver of an authorized emergency vehicle may:
   (1) Park or stand, notwithstanding any other provision of this chapter;
   (2) Proceed past a red or stop signal or sign but only after slowing down as may be necessary for safe operation;
   (3) Exceed the maximum speed limit if he does not endanger life or property;
   (4) Disregard regulations governing direction of movement or turning in specified directions.

(C) The exemptions in this section granted to an authorized emergency vehicle apply only when the vehicle [is making use of both audible and visual signals), except that an authorized emergency vehicle operated as a police vehicle need not use an audible signal nor display a visual signal when the vehicle is being used to:
   (1) Obtain evidence of a speeding violation;
   (2) Respond to a suspected crime in progress when use of an audible or visual signal, or both, could reasonably result in the destruction of evidence or escape of its occupants who are suspected of involvement in a crime.

(D) The provisions of this section do not relieve the driver of an authorized emergency vehicle from the duty to drive with due regard for the safety of all persons.

(E) The Criminal Justice Academy shall promulgate regulations pursuant to the Administrative Procedures Act so as to provide uniform guidelines and training programs for law enforcement agencies which use emergency vehicles. Law enforcement agencies authorized to use emergency vehicles shall use the regulations developed by the Criminal Justice Academy to provide written guidelines and to provide training programs for its officers and employees regarding the operation of emergency vehicles. [Note that these regulations are codified in the South Carolina Code of Regulations at 38-600 to 38- 670.]

S.C. Code Ann. §56-5-750. Failure to stop motor vehicle when signaled by law enforcement vehicle

[This section states that, absent mitigating circumstances, it is unlawful for a driver to continue driving when signaled to stop by means of flashing lights or siren. Speeding up is prima facie evidence of failure to stop. Failure to see or hear the signals is not an excuse when the driver should have seen or heard. A first offense, in which no great bodily injury or death results, carries a penalty of thirty days' license suspension and no more than ninety days' imprisonment. A second offense carries a one-year license suspension and not more than five years imprisonment. If great bodily injury results from a failure to stop, the convicted driver must be imprisoned for not more than ten years, and if a death results, not more than twenty-five years. There is a provision for a restricted driver's license to permit one under suspension to drive to work or college if he lives more than one mile away. I have commented on this section because it is the most detailed and restrictive such statute among the states. Apparently the legislature decided this is another way to attack the problem of pursuit cases.]

South Carolina uses the statutory standard of due care as stated in §56-5-760. There is an especially detailed provision for police to have the privileges of the emergency vehicle statute without using lights or siren. There are not many cases. A 1966 case arose when a police officer had a suspected speeder under surveillance but said he was not chasing the car. The officer was going 35 mph (his version) or 60 mph (witnesses versions) in a 25-mph-zone. He collided with a driver who had a stop sign. That driver saw the officer but thought he was going slow enough to pull out safely. The officer did not have his flashing lights or siren on. The jury found for the plaintiff and awarded actual and punitive damages. The supreme court affirmed. It noted the violation of the officer in not using his warning signals. The plaintiff was entitled to proceed after stopping if he had ample time to clear favored traffic. The court referred to no previous case law. *Carter v. Beals,* 151 S.E.2d 671 (S.C. 1966).

A 1968 case is related to police emergency driving. A bank employee allegedly set off a burglar alarm in a negligent, willful and wanton manner. An officer was dispatched to the scene and ran a red light, and this caused a multiple car accident. The plaintiff was one of the drivers. He sued the bank as well as other defendants. The bank demurred to the complaint on the ground that its act was not the proximate cause, and that it could not have reasonably anticipated the accident at a result. The trial judge granted the demurrer. The supreme court reversed, finding that this novel question must be litigated. A judge concurring in the result had concerns that the courts should not discourage people from giving alarms, even if negligently done or it turns out that there was no emergency. South Carolina statutory law already makes the willful and malicious giving of a false alarm a criminal act. *Gossett v. Burnett,* 164 S.C. 1968).

A short 1982 case is unusual in that the plaintiff was the police officer who was injured while pursuing a speeder. The trial court awarded the officer $25,000. The supreme court affirmed. The defendant driver argued that the officer was negligent in that he sped. The court refused to hold such speeding as negligence per se. The officer is specifically authorized by statute to speed in such pursuits. The defendant failed to stop when ordered, and he himself drove in excess of 100 mph. Such questions were properly before the jury. *Jones v. Way,* 294 S.E.2d 432 (S.C. 1982).

The most recent case found is dated 1992. The plaintiff, age sixteen, was a passenger on a motorcycle that was pursued by the police. The motorcycle operator, age eighteen, was speeding at 80 mph to 85 mph, and neither person was wearing a helmet, which is required in South Carolina. The motorcyclist fled the pursuit and crashed, and the plaintiff was injured. The plaintiff sued both officers in their official and individual capacities, alleging that his constitutional rights under 42 U.S.C. §1983 were violated. The trial judge granted summary judgment to the defendants. The plaintiffs appealed with respect to §1983. The supreme court affirmed. Citing several U.S. Supreme Court cases, the court found that there was no seizure of the plaintiff in violation of the Fourth and Fourteenth Amendments. The officers violated no other constitutional rights. *Cone v. Nettles,* 417 S.E.2d 523 (S.C. 1992).

South Dakota

S.D. Codified Laws §§32-31-1 to -5. Emergency vehicles

[These sections set out UVC 11-106 (see the Introductory Essay for the text) without significant change. Consequently, both audible and visual signals are required for emergency operation. There are two exceptions. Police may exceed speed limits when checking the speed of other vehicles. Whether the police may do so without using lights and siren while so doing is not stated, but I think we can presume that this was the intention of the legislature. Also, an ambulance driver may speed outside the city limits without audible signals.]

S.D. Codified Laws §§32-33-10 to -14. Temporary roadblocks

[These sections authorize and regulate police roadblocks. Such roadblocks may be set up "for the purpose of identifying drivers, and apprehending persons wanted for violation of the laws of this state, or any other state, of the United States, who are using the highways of this state." The roadblock must be clearly visible from not less than 100 yards, and must be identified by a flashing red and a flashing yellow light. However, nothing in these sections shall limit the existing authority of law enforcement officers.]

S.D. Codified Laws §32-33-18. Failure to stop on signal or eluding [police]

[South Dakota, like South Carolina above, has made eluding or failure to stop a misdemeanor that carries license revocation as a penalty. In the proper circumstances, the individual may be permitted limited driving privileges. A violation which results in bodily injury, death, or substantial damages is a felony.]

No cases directly on point were identified. A year 2000 case involved a police officer from the city of Belle Fourche who was involved in an intersection accident while "pacing" two motorcyclists suspected of drag racing. The plaintiff was another car driver. The supreme court affirmed a verdict for the plaintiff. The case is of little value to us because the emergency vehicle statute was not an issue. The case was decided as if the defendant were a civilian driver. The officer pulled out from a stop sign and collided with the plaintiff. *Carpenter v. City of Belle Fourche*, 609 N.W.2d 751 (S.D. 2000).

Tennessee

Tenn. Code Ann. §55-8-108. Authorized emergency vehicles

[Sections (a)–(d) set out UVC 11-106 without significant change. See the Introductory Essay for text.)

(e) Notwithstanding the requirement of this section that drivers of authorized emergency vehicles exercise due regard for the safety of all persons, no municipality or county nor the state or any of its political subdivisions, nor their officers or employees, shall be liable for any injury proximately or indirectly caused by an actual or suspected violator of a law or ordinance who is fleeing pursuit by law enforcement personnel. The fact that the law enforcement personnel pursue the actual or suspected violator of a law or ordinance who flees from such pursuit shall not render the law enforcement personnel, or the employers of such personnel, liable for injuries to a third party proximately caused by the fleeing party unless the conduct of the law enforcement personnel was negligent and such negligence was a proximate cause of the injuries to the third party.

Tenn. Code Ann. §22-20-310 Determination to be made by court

(a) The court, before holding a governmental entity liable for damages, must first determine that the employee's or employer's acts were negligent and the proximate cause of plaintiff's injury, that the employee or employees acted within the scope of their employment and that none of the exception listed in §29-20-205 are applicable.

(b) No claim may be brought against an employee . . . unless the act or omission was willful, malicious, [or] criminal . . . .

Tenn. Code Ann. §29-20-202. Negligent operation of vehicles

(a) Immunity from suit of all governmental entities is removed for injuries resulting from the negligent operation by any employee of a motor vehicle or other equipment while in the scope of employment.

(b) This section shall not act as a repeal of §§55-8-101, 55-8-108, or 55-8-132, and the immunities provided by these sections are hereby expressly continued.

[Tennessee has, as elsewhere, set maximum payments to plaintiffs in tort claims against governmental entities, and these are set out at §29-20-403. The limits vary depending on whether the entity has purchased liability insurance, is self-insured, or was unable to purchase insurance.

Section 55-8-108, after adopting the model emergency vehicle act, UVC 11-106, specifically provides that there will be no liability to the fleeing driver, whether directly or indirectly caused. The pursuing officer may be liable to a third person if the officer's conduct was negligent and a proximate cause. Note that the legislature recognizes there can be more than one cause, and that the standard is simply negligence. Obviously, how the courts have applied the statute is all-important.]

In a court of appeals case in 1993, a judge, in affirming the dismissal of the plaintiff's suit, wrote that, "Under Tennessee case law, there are no meaningful restraints placed on officers who engage in high-speed chases to protect innocent third parties from injury or death." *Garren v. City of Englewood,* 854 S.W.2d 892 (Tenn. App. 1993). The judge thought that the "emergent majority view," as shown by a 1992 Florida case and a 1992 Texas case, offered some protection to innocent bystanders. The judge invited the General Assembly to examine the current law, §55-8-108.

The judge in *Garren* cited two Tennessee Supreme Court cases to support his comments. We will look at them first, and then at a 1994 case in which the supreme court revisited the issue. The first case, *Nevill v. City of Tullahoma,* 756 S.W.2d 226 (Tenn. 1988), arose when a passenger was killed when a car pursued by the police hit a tree. The police had chased the driver for about three miles when they turned off their sirens, backed off and followed at a reduced speed. The fleeing driver disappeared over a hill and hit a tree. The case was subject to the Governmental Tort Liability Act. At a bench trial, the judge awarded damages to the estate that, after a reduction of $25,000 received from the driver of the car, left a settlement of $105,628 against the city. (The Act limited recovery to $130,000.) The supreme court noted that this case presented a novel issue in Tennessee: may police officers be held liable for beginning a pursuit? The court of appeals had affirmed the trial court on its finding that the officers negligently decided to pursue a driver "cutting doughnuts" at a mall parking lot. The officers might have foreseen that a fleeing driver will not necessarily slow down when he has gone over a hill, but will continue to assure his escape. The supreme court reversed. It found that the sole proximate cause was the recklessness of the pursued driver. Any other conclusion would require law enforcement officers to pursue at a lawful speed or ignore law violators. The court noted that most jurisdictions review the manner of the pursuit, not the decision to pursue. A dissent thought that the findings of fact of the trial and the court of appeals should be respected.

The second case, *Kennedy v. City of Spring City,* 780 S.W.2d 164 (Tenn. 1989), arose from a stolen car incident. The thief had stolen the car in Georgia and was clocked for speeding in Tennessee. An officer attempted to stop the car, but the driver fled, wrecked the car, and escaped on foot. The thief then stole another car, and that car was identified. The driver was operating lawfully, but fled when signaled to stop. No other officers were within twenty-five miles, and the officer told the dispatcher that he would pursue. The fleeing driver ran a red light, and a mile later crossed the center line on a slight curve and killed an oncoming driver. The trial judge awarded $100,000 and the court of appeals affirmed. Both courts found the officer negligent for initiating a pursuit without assistance and without, if possible (as the plaintiff's expert said), planning the location and time where the catch would be made. If the suspect had fled once, the officer should have known he would flee again. Again, the supreme court reversed and held that the decision to pursue is not negligence. The court noted that part (e) of §55-8-108 became effective one day before the accident. There was the same dissent as in the above case.

(I would comment that in most states, the officer would not be found negligent for initiating pursuit under these circumstances.)

The saga continues. In 1994 the supreme court took another pursuit case very similar to *Kennedy* above. The court wanted to revisit the issue. The case arose when a police officer saw a Corvette (later found to have been stolen) with no taillights. The driver passed a car in the left turn lane, fishtailed, and was clocked at 55 mph in a 40-mph-zone. The officer turned on his blue lights to stop the driver, but the driver fled. His speed reached 100 mph. The pursuit was two to three miles long. The officer dropped back when he reached heavier traffic, but keep his emergency signals on. The chase ended when the fleeing driver crossed the center line and killed three people in that car. Obviously, outrage was high. Relying on the two above decisions, the court of appeals affirmed dismissals. The supreme court reversed for trial. The court reversed its previous holdings that the decision to pursue cannot be "conduct" within the meaning of §55-8-108(e). If that conduct was negligent and a proximate cause, the officer could be found liable. The court also noted the standards of due care and reckless disregard in the other parts of the statute. The court agreed with some other jurisdictions in not seeing any reason why the decision to pursue is not conduct. Whether that conduct was a proximate cause in this accident is for the jury. The court comments that whether the conduct was negligent depends on a number of factors, and these factors are generally covered by pursuit regulations. However, the conduct should "not be judged by the perfect vision afforded by hindsight." The court cited a number of cases from other states, and found its new position to be the majority. (I will comment that putting New York in this "majority" group is suspect. New York is one of the toughest states for pursuit plaintiffs. See the New York section.) *Haynes v. Hamilton County,* 883 S.W.2d 606 (Tenn. 1994).

Note a 1992 pre-*Haynes* case. Two officers received a "be on the lookout for" a stolen Pontiac. The driver was described and was said to have a loaded shotgun. The officers saw the driver and chased at 50 to 55 mph in city traffic. The pursued driver then sped to 70 to 80 mph just before an intersection. He ran a red light and killed three people. Following the case law of the time, the court of appeals affirmed a dismissal. *McClenahan v. Lawhorne,* 849 S.W.2d 773 (Tenn. App. 1992).

A 1998 case arose when a deputy sheriff saw a driver pass in a no-passing zone on a highway. The deputy turned on his blue lights and siren, and the driver sped off. Some three miles down the road, the deputy lost control on a curve and hit a car head-on, killing the driver. The deputy's car apparently was in poor condition. It had bad shocks and brakes, and the frame had been welded together by a jail trustee. The trial court held two trials, first trying the case against the deputy and the county before the judge. The court apportioned fault as 50 percent deputy, 50 percent fleeing driver. The court found the damages to be over $1 million, but limited the county's damages to $130,000 as capped by the Tort Liability Act, and dismissed Deputy Rose pursuant to that Act. The trial judge then gave the case against the fleeing driver to the jury, and it awarded $1 million, finding the driver liable for 84 percent and the deputy 16 percent. (The driver's insurer had already paid its $25,000 liability policy to the plaintiffs.)

The court of appeals affirmed. The Tort Liability Act clearly gives a public employee immunity where the agency itself is not immune. The employee is immune whether his acts are willful or not. The court also affirmed the dismissal of the §1983 claim (42 U.S.C. §1983). The worst thing to be said about Deputy Rose's conduct is that he might have known of his car's poor condition and driven accordingly. This is hardly an arbitrary, deliberate deprivation of constitutional rights. But note that the plaintiff did recover damages. It's just that in the context of $1 million in damages, $130,000 plus $25,000 isn't much. *Erwin v. Rose,* 980 S.W.2d 203 (Tenn. App. 1998).

In a 1995 case, an officer on an emergency run collided with another driver. A Knoxville police officer driving on Kingston Pike received a call of an accident with injuries. She turned on her siren and blue lights. She came to a large intersection and had the red light. She stopped behind stopped traffic, and then crossed into the oncoming lanes to get around the traffic. The plaintiff driver heard the siren but could not locate the emergency vehicle, and turned upon his left turn arrow. The vehicles collided. There was a bench trial. The judge found the officer to be 75 percent at fault and the plaintiff driver 25 percent (there were plaintiff passengers in his car). The court of appeals reversed and found that the plaintiff driver was solely at fault. It found the plaintiff driver to be in violation of the statutes governing the duty to yield to emergency vehicles and of another statute governing left turns. The supreme court modified the awards allocation to find the officer 25 percent at fault. It found no reason to reverse the trial judge when the preponderance of the evidence supported his decision. *Wright v. City of Knoxville,* 898 S.W.2d 177 (Tenn. 1995).

A 1987 case was filed in federal court. The culprit was a prisoner, Sherrill, who was incarcerated in the Nashville jail. He was allowed to go on furlough by the sheriff. He allegedly got drunk and began driving, although his license was revoked. Two White House, Tennessee, police officers responded to a report of a hit-and-run property damage accident. They located the vehicle and its driver, Sherrill, and pursued at high speeds outside the city limits. The plaintiff's husband was killed when Sherrill crossed the center line. The plaintiff filed a §1983 action, alleging a deprivation of rights without due process of law. The Sixth Circuit affirmed a dismissal of all federal claims. A police pursuit per se is constitutional. There can be a no constitutional violation unless the police exhibited gross negligence or wanton disregard. *Jones v. Sherrill,* 827 F.2d 1102 (6th Cir. 1987).

A 1986 case also was brought in federal court. A Nashville police officer attempted to stop a thirteen-year-old (no, that is not a misprint) driver for speeding. The driver fled and the chase reached speeds up to 100 mph. The pursuit ended when the driver ran off the roadway and was permanently injured. The parents filed a §1983 action, alleging an unlawful seizure under the Fourth Amendment. The district court judge granted summary judgment to the defendants because there was no seizure nor use of deadly force. The Sixth Circuit noted that while the plaintiff's argument that the officer's decision not to stop the pursuit increased the possibility of a serious accident might be pertinent to state law claims, it is inapposite as applied to a Fourth Amendment claim. The issue is whether there was a seizure and, if so, whether it was unreasonable. After giving the rule of the primary seizure case, *Terry v. Ohio,* 392 U.S. 1 (1968), the court held that there was no seizure. There was no physical restraint on the plaintiff's freedom to leave, nor was there a show of authority that restrained the plaintiff. He was restrained when he chose to flee and crashed his vehicle. The court also held that police pursuits of traffic violators is reasonable and only a minimal intrusion on Fourth Amendment rights.

The court affirmed summary judgment. It also held that the pursuit of law violators, actual or suspected, is not the exercise of deadly force. Allowing police officers to follow individuals who refuse to stop when signaled is not a violation of the Fourteenth Amendment and the right to life. (I would comment that people who oppose police pursuits often call use of the police vehicle deadly force, but it's a car, not a gun. The officer is following the person, not shooting at him.) *Galas v. McKee,* 801 F.2d 200 (6th Cir. 1986).

In summary, the Tennessee cases show that the law can change, both statutory and case.

Texas

Transp. Code. Ann. §546.001. Permissible conduct

In operating an authorized emergency vehicle the operator may:

(1) Park or stand, irrespective of another provision of this subtitle;
(2) Proceed past a red or stop signal or stop sign, after slowing as necessary for safe operation;
(3) Exceed a maximum speed limit, except as provided by an ordinance adopted under Section 545.365, as long as the operator does not endanger life or property; and
(4) Disregard a regulation governing the direction of movement or turning in specified directions.

Transp. Code Ann. §545.365. Speed limit exception for emergencies; municipal regulations

(a) The regulation of the speed of a vehicle under this subchapter [Subchapter 345; operation and movement of vehicles] does not apply to:
    (1) An authorized emergency vehicle responding to a call;
    (2) A police patrol; or
    (3) A physician or ambulance responding to an emergency call.
(b) A municipality by ordinance may regulate the speed of:
    (1) An ambulance;
    (2) An emergency medical services vehicle; or
    (3) An authorized vehicle operated by a blood or tissue bank.

Transp. Code Ann. §546.002. When conduct permissible

Section 546.001 applies only when the operator is responding to an emergency call, pursuing an actual or suspected violator of the law, or responding to but not returning from a fire alarm.

Transp. Code Ann. §546.003. Audible or visual signals required
Except as required by Section 546.004, the operator of an authorized emergency vehicle engaging in conduct permitted by Section 546.001 shall use, at the discretion of the operator in accordance with policies of the department or the local government that employs the operator, audible or visual signals that meet the requirements of Sections 547.305 and 547.702.

Transp. Code Ann. §546.004. Exceptions to signal requirements
(a) A volunteer fire fighter who operates a private vehicle as an authorized emergency vehicle may engage in conduct permitted by Section 546.001 only when the fire fighter is using visual signals meeting the pertinent requirements of Section 547.305 and 547.702.
(b) An authorized emergency vehicle that is operated as a police vehicle is not required to be equipped with or display a red light visible from the front of the vehicle.
(c) A police officer may operate an authorized emergency vehicle for a law enforcement purpose without using the audible or visual signals required by Section 546.003 if the officer is:
  (1) Responding to an emergency call or pursuing a suspected violator of the law with probable cause to believe that:
    (A) Knowledge of the presence of the officer will cause the suspect to destroy or lose evidence of a suspected felony, end a suspected continuing felony before the officer has obtained sufficient evidence to establish grounds for arrest, or evade apprehension or identification of the suspect or the suspect's vehicle; or
    (B) Because of traffic conditions on a multilaned roadway, vehicles moving in response to the audible or visual signals may increase the potential for a collision or unreasonably extend the duration of the pursuit; or
  (2) Complying with a written regulation relating to the use of audible or visual signals adopted by the local government that employs the officer or by the department.

Transp. Code Ann. §546.005. Duty of care
This chapter does not relieve the operator of an authorized emergency vehicle from the duty to operate the vehicle with appropriate regard for the safety of all persons or the consequences of reckless disregard for the safety of others.

Civ. Prac. & Rem. Code Ann. §101.021. Governmental liability
A governmental unit in the state is liable for:

(1) Property damage, personal injury, and death proximately caused by the wrongful act or omission or the negligence of an employee acting within his scope of employment if:
  (A) The property damage, personal injury, or death arises from the operation or use of a motor-driven vehicle or motor-driven equipment; and
  (B) The employee would be personally liable to the claimant according to Texas law; and

(2) Personal injury and death so caused by a condition or use of tangible personal or real property if the governmental unit, were it a private person, be liable to the claimant according to Texas law.

Civ. Prac. & Rem. Code Ann. §101.055. Certain governmental functions
This chapter does not apply to a claim arising:

(1) [Tax collection]
(2) From the action of an employee while responding to an emergency call or reacting to an emergency situation if the action is in compliance with the laws and ordinances applicable to emergency action, or in the absence of such a law or ordinance, if the action is not taken with conscious indifference or reckless disregard for the safety of others; or
(3) From the failure to provide or the method of providing police or fire protection.

Civ. Prac. & Rem. Code Ann. §101.056. Discretionary powers
This chapter does not apply to a claim based on:

(1) The failure of a governmental unit to perform an act that the unit is not required by law to perform; or
(2) A governmental unit's decision not to perform an act or its failure to make a decision on the performance or nonperformance of an act if the law leaves the performance or nonperformance of the act to the discretion of the governmental unit.

[Commenting on §§546.001–546.005 first, we see that Texas has enacted UVC 11-106 with some changes and additions. Audible or visual signals shall be used at the discretion of the operator in accordance with local ordinances or policies. Section 546.004 gives especially detailed guidance on when police officers may use or not use such signals.

Section 546.005 states that the standard of care for emergency vehicle operators is "appropriate regard" for the safety of all persons, and that such operator is not relieved from the consequences of his reckless disregard for the safety of others. The courts appear to use, however, the due care standard that is not now specifically stated in the current law.]

A 1996 case, *City of El Paso v. Kolster,* 931 S.W.2d 365 (Tex. App. 1996), quotes from an earlier statute that gives the standard as due regard. (The accident happened in 1991.) The court has an excellent discussion of the meaning of due regard. In many other states, courts have held that due regard is the same as ordinary negligence. The court was persuaded by a line of cases that held that due regard is a lesser standard than the ordinary negligence standard that is applied to nonemergency traffic accidents. The rationale is that if drivers on emergency runs must exercise such due regard or due care, then there is little point to the emergency vehicle statutes. The court held that the due regard standard in this context lies somewhere between ordinary care and recklessness. In my opinion, that is a very defensible position.

Code 101.055(3) set out above on its face seems to provide immunity for emergency vehicle operators, and by prohibiting suits arising "from the failure to provide or the method of providing police or fire protection." The Texas Supreme Court, in *State v. Terrell,* 588 S.W.2d 784 (Tex. 1979), held that the statute did not provide immunity to the state in an accident. A highway patrol officer was parked on the shoulder on radar duty. He detected a speeder and pulled onto the highway into the path of the plaintiff driver. He had not activated his lights or siren. The court held that this action was not a method of providing law enforcement. The court pointed out that the tort claims act must be construed as a whole. Code 101.021 waives immunity for the operation of government-owned vehicles. The language in §101.055 about the method of providing police and fire protection is to be understood as avoiding judicial review of policy decisions government must make about how much and what kind of fire and police protection to provide communities. The government must have discretion in making such decisions. (Note that some states construe the same or similar language as providing immunity.)

Texas courts recognize official or qualified immunity (public employees) and sovereign immunity (governmental entity). Texas courts give primary consideration to §101.056 and generally decide emergency vehicle cases on the basis of ministerial functions, discretionary functions, and good faith. (Note that §101.021 simply states that a governmental unit is liable for negligent operation of a motor vehicle if a private person would be liable under the same circumstances.) Only a minority of the states in this study wade into this morass in determining emergency vehicle cases, and with good reason. Whether the decision to initiate a pursuit, how to conduct it, and when and if to terminate it is ministerial or discretionary is far from clear. Some courts have held that nothing an officer does requires more discretion, while others find that it is just a matter of following rules and polices, that is, ministerial. In short, public employees, acting in the scope of their employment and exercising both good faith and discretionary functions, are in Texas entitled to official immunity. Normally, such finding ends the case because the government also is immune.

We now examine a 1994 Texas Supreme Court case that explains it all, *City of Lancaster v. Chambers,* 883 S.W.2d 650 (Tex. 1994). The case arose from a motorcycle pursuit. A DeSoto police officer, in the early morning hours, saw a motorcyclist with a passenger run a red light. The officer switched on his emergency lights and attempted to stop the motorcyclist. The motorcyclist fled instead and a high-speed pursuit ensued. The pursuit moved onto I-35, and eventually ten police vehicles from five jurisdictions engaged in the pursuit. The motorcyclist imprudently attempted to exit at high speed and hit a pole, killing himself and injuring his passenger, our plaintiff. (Note that the plaintiff here has the status of an innocent party, since the defendants did not allege that he encouraged the operator to flee. In at least one state, Illinois, the plaintiff as an occupant of the pursued vehicle would have the burden of proving that he was an innocent party. The notion that the plaintiff could do nothing is not quite accurate. Several hard punches to the kidneys might have gotten the operator's attention, for example.) The plaintiff sued the cities of DeSoto and Lancaster and their police officers who were engaged in the pursuit. The trial court granted summary judgment to the defendants on its finding that the defendants were neither negligent nor the proximate cause. The court of appeals reversed (except for the §1983 claim), and the matter came to the Texas Supreme Court.

The first issue was negligence. The supreme court agreed with the court of appeals that the negligence of the defendants was a jury question. The defendants had the statutory duty to drive with due regard for the safety of all persons, including the plaintiff. To find negligence here, the defendants would have to have been the "cause in fact", that is, to have been a substantial factor in the accident, and to have been a proximate cause. If it were foreseeable that the pursuit might reasonably lead to the accident, the pursuit could be a proximate cause. That too is a jury question.

The second issue was official immunity. Government employees who are within the scope of their authority and are exercising discretionary functions in good faith are immune. The court gives the usual definitions of ministerial and discretionary acts. A ministerial act is one where the duties of the official are prescribed and defined so as to leave nothing left for discretion, while a discretionary act requires judgment and personal deliberation, including the decision whether to undertake the act. The court of appeals took the position that the defendant officers had a ministerial duty to drive with due regard for the safety of others, and thus were not acting with discretion. The supreme court could not agree. If a public official is not negligent, he does not need immunity. There would be no need for official immunity under this view. Official immunity protects a public official even if he negligently performs a discretionary act. The court found that the decisions to apprehend and pursue if need be, and the manner of the pursuit, and whether or when to terminate it, are discretionary. The statutes are not so explicit so as to make such actions ministerial. (The court notes that courts are not unanimous. *Tice v. Cramer,* 627 A.2d 1090 (N.J. 1993) held that discretionary immunity in pursuit cases is limited to the highest levels of government and are not applicable to the conduct of pursuits. Some cases hold that pursuits are governed by department rules and policies that in effect are ministerial duties. See *City of Pinellas Park v. Brown,* 604 So. 2d 1222 (Fla. 1992). Cases that have held some decisions in the conduct of a pursuit are discretionary while others, such as a policy not join a pursuit without permission or direction, are not. In this case, the plaintiff did not make this argument.)

On the question of whether the officers were acting within the scope of their authority, the court held that they were. The court did find that whether the officers acted in good faith could not be disposed of by summary judgment, and remanded this issue to the trial court. The court noted the competing interests in pursuit accident cases: the rights of bystanders and innocent parties if an officer acts in disregard of public safety versus the injustice of subjecting officers to potential liability who are exercising discretion in performing their duties with the decisiveness and judgment required by the public good. The court set out a the following test for good faith in a police pursuit case: An officer acts in good faith if "a reasonably prudent officer, under the same or similar circumstances, could have believed that the need to immediately apprehend the suspect outweighed a clear risk of harm to the public in continuing the pursuit." The court derived the test from federal immunity law in §1983 cases. The court pointed out that the test was not a negligence or gross negligence test, but went to the issue of immunity. A plaintiff, to defeat summary judgment, must show that no reasonable officer would have acted as the defendant officer acted. Merely showing that some officers would not have so acted is not enough.

The court dismissed the plaintiff's federal claims of excessive force. The court found that the officers' conduct did not shock the conscience. A concurrence would hold that the motorcyclist's decision to flee at up to 100 mph, and attempt to exit at high speed was the only cause.

The court referred to a 1992 case, *Travis v. City of Mesquite,* 830 S.W.2d 94 (Tex. 1992). *Travis* did not involve the immunity issue, but rather proximate cause. The case arose when police officers working off-duty at a truck stop investigated occupants of a vehicle suspected of prostitution. The officers directed the driver to drive to the front of the truck stop and wait while they checked his identification. Instead, the driver fled at high speed, and there was a pursuit. The officer radioed the Mesquite police department, and two officers responded. The suspect drove the wrong way on a access road. The two off-duty officers followed, and the other two entered the access road the right way with the intention of stopping the suspect. The suspect hit an innocent driver, with one fatality and several injuries resulting. The pursuit lasted two minutes. The trial court granted summary judgment to the officers, and the court of appeals affirmed on the finding that the decision to pursue could not be a proximate cause. The supreme court reversed. As the court noted in *Lancaster* above, proximate cause involves both "cause in fact" and foreseeability. There can be concurrent causes. Whether the decision to pursue or to continue pursuit serves a law enforcement purpose that outweighs the danger to the public under the circumstances should not be a matter of summary judgment, unlike the decisions of older cases. The court could not conclude that all pursuits are reasonable no matter what the circumstances. (I would comment that when, as here, a suspect enters a one-way road driving the wrong way places the police in a difficult position. Is it better to pursue and alert others or let him go?) The court held that the defendant did not preserve the issue of immunity. That issue was discussed in the later case above.

A 1996 case in the Texas Supreme Court arose from an accident between a Waco police officer responding to a burglary call and another vehicle. The officer was investigating a burglary and was dispatched to assist another officer on a theft-in-progress call. The officer turned on his siren, air horn, and lights. He attempted a left turn onto a seven-lane road. His view of one lane was blocked by a large truck, and he collided with the plaintiff. Apparently, the plaintiff had the green signal. The defendants moved for summary judgment on the basis of immunity. The trial court denied the motion and the court of appeals affirmed. The issue before the supreme court was good faith. The plaintiffs did not contest the claims that the officer was within the scope of his employment and was performing a discretionary function. Consequently, the test for good faith as stated by the court in *Chambers* above was the issue. The officer relied on his own affidavit and an expert, James F. Smith. The statements dealt with the nature of call (thought to be purse snatching), the officer's belief that the other unit dispatched would reach the scene first, the decision to proceed on an emergency basis to assist and protect the other officer, and in general his reasons for taking the route he did.

The plaintiff did not dispute the reasons, but focused on the nature of the left turn. His expert argued that the officer did not properly weigh the risk of making a left turn when his view was obstructed, that to do so was an unnecessary risk to the public. The court held that the officer did not present conclusive evidence of good faith under the *Chambers* test. Again, the court pointed out that good faith is not a general negligence test, but is a test of what a reasonable officer would have done. An interesting three-judge dissent points out that the facts are quite unlike *Chambers*. There the officers did not know where the pursuit was going; that depends on the pursued. Here, the officer had a fixed destination, and the only evidence of bad faith is one mistake. All his other actions were discretionary and in good faith, and these should control, not one isolated traffic decision. I would comment that when cases are judged by discretionary or ministerial functions, and good faith, there is ample room for disagreement. *Wadewitz v. Montgomery,* 951 S.W.2d 464 (Tex. 1997).

Note that *Wadewitz* was not a pursuit case. A 2000 case in the Texas Supreme Court, *University of Houston v. Clark,* 38 S.W.3d 578 (Tex. 2000) consolidated two pursuit cases in one decision. One case involved a plaintiff who was injured by the driver being pursued, and the other a plaintiff who collided with an officer in pursuit of another driver. Both cases were appealed from the Fourteenth District Court of Appeals. The first case is *Clark v. University of Houston,* 979 S.W.2d 707 (Tex. App. 1998). University police arrived at the scene of a fight on the campus, and one fighter, Thomas, got into his Suzuki. Officers ordered Thomas to stay. Instead, Thomas fled, and drove around a patrol car that the officer had positioned to block Thomas. The officer radioed another officer, Williams, to stop the Suzuki. That officer turned on his overhead lights and siren, knowing, as he said, that frequently drivers will then stop. Thomas sped off, ran a red light, and injured a passenger in a vehicle he hit. Thomas had not been identified. The trial court granted summary judgment to Williams and the University. The court of appeals reversed on its finding that the defendants had not established good faith as a matter of law. The court did not rule on whether the officers were acting in a discretionary capacity. The court found that the officers' affidavits indicated their reasons for deciding to try to stop Thomas (the "need" test), but failed to meet the "risk" test. The officers did not sufficiently detail reasons why the need outweighed risk. A long dissent pointed to the officers' description of the light traffic and other factors. In additions, the officers didn't make much of a pursuit; they did try to stop Thomas by turning on the siren and lights.

The supreme court first held that the test which was set out in *Wadewitz* applies to a case where the pursued driver collided with an innocent person. True, the risk factors are different. The risk in responding to an emergency is that the official vehicle will collide with a third party, while in police pursuits the risk (usually) is that the pursued will collide with the third party. Those risks are to a great extent more speculative because the police do not know what the driver will do. The good faith test can be applied to both, taking the differences into account. The supreme court reversed the court of appeals and found that the officers acted in good faith and were entitled to official immunity. Their affidavits were sufficient. In particular, the fleeing driver sped 85–90 mph in a 35-mph-zone, and that was evidence that he driver had no regard for the safety of others and should be apprehended. However, the court remanded to the court of appeals to decide the discretionary function aspect. The court of appeals on remand found that the officers were engaged in a discretionary act and were entitled to official immunity as a matter of law, and that immunity extends to the University. Merely because an officer is ordered by his superior to pursue does not make his acts ministerial. *Clark v. University of Houston,* 60 S.W.2d 206 (Tex. App. 2001).

The second case in this consolidated case is *Ener v. Thomas,* 20 S.W.3d 712 (Tex. App. 1999). The facts are that Harris County Deputy Ener clocked a speeder on FM 2920 at 71 mph in a 55-mph-zone at around 1:00 A.M. The deputy activated his signals and pursued. The traffic was extremely light. The speeder crossed the intersection with Rhodes Road safely and the deputy followed, but "unfortunately the driver of a Ford pickup failed to yield the right of way to an emergency vehicle and a collision resulted." (I think we can presume that the pickup had the right of way.) The trial court denied summary judgment for the defendants and the court of appeals affirmed. The affidavits of the deputy failed to consider or even conceive the risk of harm to others, or consider alternatives to the pursuit.

The supreme court affirmed the court of appeals in *Ener.* The court emphasized that even if the deputy had considered alternatives, if there were any other than not to pursue, and still made the decision to pursue, that might or might not be good faith depending on the circumstances. Essentially, the deputy's affidavit did not put in the magic words "evaluating the risk and dangers of a pursuit." Therefore, he did not conclusively show that a reasonable officer would conclude that the need to immediately apprehend the speeder outweighed risks. A concurring and dissenting opinion has some interesting points. There is a danger of turning official immunity into a test of simple negligence. Good faith is supposed to protect the officer even if he is negligent. Otherwise, it's a simple negligence test. If an officer makes the needs-risk analysis (in the very short time allowed) and initiates a pursuit or begins an emergency run in good faith, are the courts to second-guess the decision at every intersection and when he encounters the very risk already considered?

I will comment that Texas has made the emergency vehicle law rather complicated with litigants having to consider scope of employment, discretionary versus ministerial acts, and good faith (if asserted as an affirmative defense) when compared to many other states. In New York, for example, it is pretty simple. Did the police act with reckless disregard for the safety of others in performing their law enforcement duties or not?

We now look at the last two supreme court cases before moving on to decisions from the courts of appeals. Note of course that new police pursuit and emergency call cases must be evaluated in light of the recent supreme court cases covered above. *Kolsteer v. City of El Paso,* 972 S.W.2d 58 (Tex. 1998) arose from a collision between an ambulance owned by the city and a third party. The ambulance was on the way to an emergency. The facts are not given, other than the fact that the ambulance operator was not wearing glasses as prescribed on her operator's license. The jury found the ambulance operator, Sloan, 55 percent at fault and the plaintiff 45 percent. The city appealed, contending that under §101.055(2) (see above) it is immune unless its employee acted with reckless disregard for the safety of others. The city also argued that the jury was wrongly instructed on simple negligence. The court held that the city did not preserve the issue of the erroneous instruction, and affirmed the trial court. A lengthy dissent thought that the court mistakenly characterized the lower court's finding. The dissent thought that holding Sloan only 55 percent at fault did not indicate reckless disregard.

One the same day, October 9, 1998, the supreme court decided a fire truck case with the same issue as in *Kolster.* The case is *City of Amarillo v. Martin,* 971 S.W.2d 426 (Tex. 1998). A firefighter was on an emergency run and was proceeding with all warning signals operating. He entered an intersection on a red signal and collided with the plaintiff driver, and the latter claimed under the Tort Claims Act for property damage. The plaintiff was awarded damages at a bench trial on the finding that the fire truck driver was negligent. The court of appeals affirmed, agreeing that the standard is negligence, not reckless disregard. Thus the supreme court faced the problem it did not have to decide in *Kolster* above. The problem is that the statutes are contradictory. Uniform Vehicle Code 11-106 was poorly drafted. It states that an emergency vehicle operator claiming the special privileges under an emergency run must exercise due care and is not protected from the consequences of reckless disregard for the safety of others. Well, which is it, due care, ordinarily held to be simple negligence, or reckless disregard? Most states have enacted the UVC provision without change and leave it up to the courts to resolve the problem. (See the Introductory Essay for a longer discussion.) Transp. Code 546.005 repeats the UVC version, with the most recent language being "appropriate regard" rather than "due regard," if that makes a difference. Also, Civ. Prac. & Rem. Code 101.021(2) states that an emergency vehicle operator responding to an emergency and observing the applicable laws and ordinances in that situation is immune unless he exhibited conscious indifference or reckless disregard.

The court begins by saying that other state courts have faced the same problem, and gives an excellent summary of the cases. One position is that mere negligence is the standard. Another position is that yes, negligence is the standard but not the negligence of civilian drivers, but negligence in emergency driving. Another version equates due regard with recklessness. The court notes that the legislature changed "due regard" to "appropriate regard," and finds that the legislature intended to hold emergency vehicle operators in emergency conditions to a reckless conduct standard. The court makes an excellent point. Civilian drivers are always held to an ordinary negligence standard. They normally never drive in an emergency. (The few cases I have found do not give such drivers the protection of the emergency vehicle statutes.) By contrast, emergency vehicle drivers are charged with protecting the public, and even a few minutes or seconds can make the difference between life and death. In addition, civilian drivers who are not driving in emergencies and have ample opportunity to see and hear sirens, horns, and flashing lights. They are expected and required by law to yield (and of course they do hundreds of times a day nationwide). Police, fire, and ambulance drivers are entitled to assume civilian drivers will respect their right of way. The court reversed on the basis of the new standard it established. A dissent thought that the previous negligence standard was fair, assuming it referred to something more than simple negligence. That is essentially the second position noted above.

We now turn to cases in the courts of appeals and will summarize them. When appropriate, they should be evaluated by the findings in the supreme court cases above. We note two cases in which the defendant officer was held to be performing a ministerial function. A 1999 case arose when a police officer rear-ended a car that had stopped because the preceding car stopped for construction. The officer was on an errand to a photography lab. The court of appeals affirmed a denial of summary judgment for the officer, thus denying immunity. The court was not impressed with the officer's argument that he was using judgment in the way he drove his car, that is, exercising discretion. Since he was not operating in an emergency, he is held to the same duty as other drivers. *Victory v. Faradineh,* 993 S.W.2d 778 (Tex. App. 1999).

A 1998 case dealt with the same issue. A motorcycle patrol officer on routine patrol was, according to a following driver, looking off to the left. He was too late in noticing a car stopped in front of him waiting to make a left turn with the appropriate signal operating. The officer skidded, fell off, and the motorcycle hit the stopped car. The officer is the defendant. He asserted immunity, arguing discretion. The court of appeals affirmed a denial. The court notes that the distinction between discretionary and ministerial acts is often one of degree. Here the officer obviously had discretion in how to drive while on patrol, but he is not immune from liability simply because he is on duty. Situations where police officers have been held immune are high-speed chases, investigations, and traffic stops, to name some activities. The plaintiff almost made a mistake initially because she granted that the officer may have been acting with discretion but did not act in good faith. Later, in a responsive pleading, after the trial court asked the parties to brief good faith, she raised the issue of ministerial function. The case was decided on that issue, not good faith. I would comment that the plaintiff had an excuse for this when one considers the murky discussions of good faith in the supreme court cases above, especially the 1994 *Lancaster* case. *City of Wichita Falls v. Norman,* 963 S.W. 2d 211 (Tex. App. 1998).

The author is not alone in calling determinations of good faith a murky area of law. The court of appeals used that word in a 1997 case, *Chapa v. Aguilar,* 962 S.W.2d 111 (Tex. App. 1997). Here the issue was good faith. An off-duty Harris County deputy constable, driving a marked police car, saw a driver in front of him run a red light. The constable turned on his overhead lights but not his siren and entered the intersection on red and collided with the plaintiff who had the green signal. The officer moved for summary judgment on the basis of official immunity as a matter of law. The motion was denied and the officer appealed. (Parties ordinarily may not appeal such denial, but a public employee can appeal if it is based on an immunity claim.) The officer argued that his evidence (his and the affidavits of two other law enforcement officers) showed that he was entitled to summary judgment as a matter of law, that the plaintiffs did not show that he did not act in good faith, and that there was no evidence of lack of good faith. The defendant officer's affidavit was short. The important part stated that the officer observed the van "blatantly and flagrantly" run a red light, that he considered the factors such as light traffic, visibility, the fact that he was on a divided street with three lanes each way, and that "I determined that I needed to attempt to apprehend the driver of the van and that the risk to the public was slight." The court of appeals noted that the plaintiff did not submit an affidavit from an expert as was done in *Wadewitz* above. This obviously was the plaintiff's mistake. Here, however briefly, the officer did consider the risks as well as the law enforcement benefits. The officer acted in good faith; the motion should have been granted. The court commented that it was not condoning reckless pursuits, "But when we subject what is essentially a split-second decision to standards we draft at our leisure, we run the risk of stifling a peace officer's ability to unhesitatingly respond to crisis." A dissent pointed out that the officer was in fact disciplined by his superiors for his action, and would hold that was evidence against the proof of good faith.

In a 1996 case, an Olmos Park police officer heard on her radio that a San Antonio police officer had been shot and needed assistance. Although the crime scene was outside her jurisdiction, she was nearby. She requested and received permission to respond to the call. She had on her emergency signals (her version) or did not (plaintiff's version). She came to a red signal, slowed to look for other vehicles, and entered the intersection. She collided with the plaintiff who was on a motorcycle. The case came to the court of appeals on the denial of the officer's motion for summary judgment. The court held that genuine issues of material fact existed regarding her good faith defense, and affirmed the denial. The officer's affidavit alone established good faith in that a reasonable officer might so conduct herself. However, the plaintiff's affidavit controverted the claim of emergency signals in operation. Since every fact in question must be construed in the plaintiff's favor on a motion for summary judgment, the motion was properly denied. *Beatty v. Charles,* 936 S.W.2d 28 (Tex. App. 1996).

A 1996 case similarly held the officer's claim of good faith to be a jury question. A San Antonio police officer on duty identified a pickup that had been reported stolen earlier. The owner also reported that the truck contained weapons and ammunition. The officer turned on his emergency lights to stop the driver, and a pursuit ensued. At one point the thief turned off his headlights, drove over a curb, and ran a stop sign. The plaintiffs contended that the officer pursued at speeds to 70 mph when the thief had his lights off, indicating recklessness. When the thief ran a second stop sign, he hit another car and injured two passengers in that car. The defendants moved for summary judgment as a matter of law. The trial court denied the motion, and the defendants appealed. There were two plaintiffs. One plaintiff, Handley, failed to respond to the officer's requests for admissions within the thirteen days required. Since he did not respond in time, the admissions are deemed admitted by operation of law. One admission stated that Officer Davis at all times acted in good faith. (Handley may still have recourse in a malpractice action if he had engaged a lawyer.) The court held that summary judgment as to this plaintiff should be granted. The court held that the second plaintiff, Bynum, raised a genuine issue of material fact. Each party submitted affidavits. The plaintiff's expert contended that no reasonable officer would conduct the pursuit without operating his siren. The city objected that the expert did not include a copy of the police department's audio tape on which he relied. However, the city failed to object to the trial judge and may not object now. The defendants also argued that use of the siren goes to the issue of negligence, not good faith. The court held that the issue does involve good faith, and construing disputed claims in favor of the plaintiff denied summary judgment as to her. *City of San Antonio v. Bynum,* 937 S.W.2d 596 (Tex. App. 1996).

A 1997 case involved the effect off a city's pursuit policy on good faith and discretion. Pharr police officer Castillo got a call from the dispatcher about a red pickup being operated without headlights. Castillo spotted the pickup speeding. He pursued with emergency signals. Finally the driver stopped, but then sped off when the officer approached on foot.

Four police units were involved in the second pursuit before the driver finally hit another vehicle when he turned into a private lot. The occupants of that vehicle were injured. The trial court denied summary judgment to the defendants. The main issue was the effect of pursuit regulations on the issue of good faith. The pursuit policy was typical. Officers are to weigh the competing benefits of law enforcement and the risks to themselves and others, notify the supervisor, join a pursuit only when authorized, and stop pursuit when the risks outweigh the benefits. A year after the general policy came into effect, Order #1 was issued. That order limited pursuit speeds to no more than 10 mph over the speed limit, permitted a pursuit only when authorized by a supervisor, required officers to stop at all traffic control devices before proceeding when safe, and prohibited pursuing outside city limits.

The plaintiff argued that both policies simply remove all discretion, and thus the officers' acts were ministerial, which removes their immunity if the orders are not followed. The court of appeals held that the two pursuit policies, when read together and considering their interpretation by the chief and the officers, made the decision to pursue a discretionary act. As to the issue of good faith, the court noted that if a reasonably prudent officer could have made the same decision, good faith is demonstrated. The officers could have believed that the suspect had tried to ram a police car, which would be a felony offense. To defeat good faith, the plaintiff would have to show that no reasonable person in the officer's position would have made the decision to pursue. The plaintiffs met this burden by the affidavit of their expert, and following the principle of resolving any doubts in favor of the nonmovant, the plaintiff, the court affirmed a denial of summary judgment. (This case shows that the defendant must prevail both on discretion and good faith, and the standard is not the same of each.) *City of Pharr v. Ruiz,* 944 S.W.2d 709 (Tex. App. 1997).

A 1994 case overruled the trial court and found that the police officers were entitled to summary judgment because they enjoyed official immunity. At 1:49 A.M. two Harris County deputy sheriffs, on routine patrol, were parked in front of a convenience store. They heard a vehicle hit a curb, and followed it, driving two patrol cars. They got the plate number, called in, and found that the vehicle was stolen. The officers saw that the driver was speeding away, and notified the dispatcher and pursued with emergency signals on. The suspect almost hit a pedestrian, ran a red light, and killed one person and injured another. Two officers' affidavits related their belief that the risks were outweighed by the necessity to pursue. The pursuit lasted about two minutes, and the officers had backed off, but the suspect continued to drive dangerously. The court of appeals held that, following *Chambers,* the decision to pursue and the manner of pursuing is discretionary. As to good faith, the plaintiffs failed to show that no reasonable officer would have pursued in the circumstances. Since the officers are immune, so must the county be immune. *Harris County v. Ochoa,* 881 S.W.2d 884 (Tex. App. 1994).

Note a 1996 case with the same result, *State v. McGeorge,* 925 S.W.2d 105 (Tex. App. 1996). There was a high-speed chase of a weaving and speeding driver that ended when the suspect crossed the center line and killed one person and seriously injured two others. The trial court denied summary judgment for the defendants, but the court of appeals reversed. The pursuing officers explained in their affidavits that it was their belief that the suspect was dangerous to the public, and that they themselves slowed as necessary for safety at intersections. The plaintiffs' expert said, in answer to a question at the deposition, that a reasonably prudent officer would conclude that it was more dangerous to chase than to let the suspect continue. With no further facts or reasons given, that alone was not enough to defeat summary judgment. A dissent thought that the majority improperly found all inferences in favor of the movant. *Texas v. George,* 925 S.W.2d 105 (Tex. App. 1996). The moral of this case is to amplify fully at the deposition, and to properly reference it on appeal.

A 1998 case is noteworthy. The plaintiff's claim was brought under Civ. Prac. & Rem. Code 101.021 (see the beginning of this section). A San Antonio police officer was dispatched to the scene of a minor accident, but when he arrived there was no accident. However, he noticed a speeding driver who ran a stop sign. The officer followed without activating his lights. The officer stayed some 400 feet behind, and eventually extended that distance to a mile behind for fear of the risks of a pursuit to the public. He explained that he only wanted to follow to see why the driver was speeding and to try to get a plate number. The driver went over a small hill on the road, and the officer then saw smoke and debris in the air, and realized that the speeder had hit something. He had in fact killed the son of the plaintiff. The court of appeals affirmed summary judgment for the defendants. The evidence showed that there was in fact no pursuit. A pursuit implies an intention to catch. The court held that the conditions required to waive immunity under §101.021 had not been met. The death did not arise from the operation of the police vehicle, nor, if the vehicle can be considered "tangible personal property," the vehicle was not "a condition or use of tangible personal or real property." *Ramos v. City of San Antonio,* 974 S.W.2d 112 (Tex. App. 1998).

We conclude the Texas section with two cases not involving law enforcement vehicles. A 2000 case arose from a fire call. The Friendswood Fire Department received a call of a trash fire that was out of control. The duty officer, Faber, responded in a Ford Explorer with its emergency signals operating. Faber approached an intersection where he had a red signal. Other traffic was stopped. Faber slowed and planned to wait for the left turn arrow, and intended to "take that route" through the intersection. The plaintiff in his vehicle responded to the siren by starting up and entering the intersection on red and colliding with another vehicle. Faber stopped 250 to 300 feet back of the accident site. Faber and the city moved for summary judgment on the grounds of official immunity and Civ. Prac. & Rem Code 101.055(2) (see above). The trial court granted the motion. The plaintiffs argued on appeal that the defendants had not sufficiently proven good faith. The court of appeals cited *Martin* above to the effect that a plaintiff, to prevail, must show that the officer knew or should have known that his actions posed a high degree of risk of serious injury to others. The court noted that Faber must act with appropriate regard for the safety of others, which *Martin* explained was intended to impose liability only for reckless conduct. The court held that there was no such conduct shown. Since the defendants are immune under the emergency response statute, there was no need to consider the allegation that the defendants did not prove good faith. *Green v. City of Friendswood,* 22 S.W.3d 588 (Tex. App. 2000). I will comment that this case is about the most implausible claim in this book. The court also held that another defendant, the individual who set the trash fire, was also entitled to summary judgment. It was not foreseeable that his actions would cause the accident.

Another 2000 case involved the judgment of a paramedic. The case was in the court of appeals three times. Joel Calfee was a paramedic for the Houston Fire Department. Calfee and another paramedic were dispatched to a report of a man who had fallen on a stick. They picked him up and decided to take him to a hospital on a nonemergency basis, that is, observing traffic laws. The patient became combative, loosened his restraints, and fell off the gurney. After stopping several times to attend to the patient, the paramedics decided to proceed as a Code II, and drive accordingly. Calfee entered an intersection on a red light and collided with the plaintiff. It was disputed whether the emergency lights were operating, but there was no evidence at trial that the siren was on. The jury returned a plaintiff's verdict, and the trial judge granted a judgment notwithstanding the verdict. The first opinion of the court of appeals, whatever it was, was withdrawn and substituted by the opinion cited at 17 S.W.2d 23. The court first discussed good faith. The court held that the paramedics discussed the pros and cons of proceeding on a Code Two, weighing the danger to the public against the benefit to the patient of getting him medical attention quickly, and decided that it was necessary. This indicated good faith. The court also found that the paramedics were exercising discretion, not performing ministerial duties. The court affirmed the granting of the JNOV.

The same court issued a supplemental opinion on the plaintiff's second motion for rehearing, primarily to respond to the dissenting opinion. The court again overruled the motion. The dissent said the majority held that a tired paramedic at the end of his shift who wanted to get a belligerent, uncooperative patient to the hospital as quickly as possible can violate laws, rules, regulations, run a red light, slam into another vehicle, and be immune as a matter of law. The dissent thought that the defendant was not acting in a discretionary manner or in good faith. The majority did not retreat. The patient was picked up at 4:00 A.M. with a stick lodged up his nose. He appeared to be under the influence of illegal drugs. The paramedics had had previous experience of such patients, who often were a danger to themselves and to the paramedics. The test for good faith is not what a reasonable person would have done, but rather what a reasonable official could have believed. The same facts indicate that the method to deal with such unruly and potentially dangerous patients must be discretionary. *Rivas v. City of Houston,* 19 S.W.3d 901 (Tex. App. 2000).

In summary, Texas takes a rather more complicated approach to emergency vehicle cases than do many other states. Texas has met the issue of what standard to apply, appropriate regard or reckless disregard, by holding that in changing "due regard" to "appropriate regard," the legislature intended to hold emergency vehicle operators to a reckless conduct standard for liability to apply. Police pursuit and emergency run cases, however, are decided by common law. Such operators are entitled to official immunity if they are in the scope of their employment, are exercising a discretionary function, and are acting in good faith. The reader will notice that several of the cases annotated above hold that the defendants were not immune as a matter of law and that the case could go to trial. The issue of what is good faith under the facts (or the parties' versions of facts) is ambiguous and may open the door for the plaintiff to go to trial or to settle.

Utah

Utah Code Ann. §63-30-10. Waiver of immunity

Immunity from suit of all governmental entities is waived for injury proximately caused by a negligent act or omission of an employee committed within the scope of employment except if the injury arises out of, or in connection with, or results from:

(15) The operation of an emergency vehicle, while being driven in accordance with the requirements of Section 41-6-14. [Note that (15) was added in 1991.]

Utah Code Ann. §41-6-14. Emergency vehicles—Policy regarding vehicle pursuits

(1) The operator of an authorized emergency vehicle, when responding to an emergency call or when in the pursuit of an actual or suspected violator of the law or when responding to but not upon returning from a fire alarm, may exercise the privileges under this section, subject to Subsections (2) through (4).

(2) The operator of an authorized emergency vehicle may:
  (a) Park or stand, irrespective of the provisions of this chapter;
  (b) Proceed past a red or stop signal or stop sign, but only after slowing down as may be necessary for safe operation;
  (c) Exceed the maximum speed limits; or
  (d) Disregard regulations governing direction of movement or turning in specified directions.

(3) Privileges granted under this section to the operator of an authorized emergency vehicle, who is not involved in a vehicle pursuit, apply only when the operator of the vehicle sounds an audible signal under Section 41-6-146, or uses a visual signal as defined under Section 41-6-132, which is visible from in front of the vehicle.

(4) Privileges granted under this section to the operator of an authorized emergency vehicle involved in any vehicle pursuit apply only when:
  (a) The operator of the vehicle sounds both an audible signal under Section 41-6-146 and used a visual signal as defined under Section 41-6-132, which is visible from in front of the vehicle;

(b) The public agency employing the operator of the vehicle has in effect a written policy which describes the manner and circumstances in which any vehicle pursuit should be conducted and terminated;
(c) The operator of the vehicle has been trained in accordance with the written policy described in Subsection (4)(b); and
(d) The pursuit policy of the public agency is in conformance with standards established by the Department of Public Safety, Division of Peace Officer Standards and Training, which shall adopt minimum standards that should be incorporated into all emergency pursuit policies adopted by public agencies authorized to operate emergency pursuit vehicles.

Utah Code Ann. §41-6-76. Emergency vehicles—Duties of respective drivers

(1) [Requires drivers to yield to emergency vehicles displaying audible and visual signals.]
(2) This section does not relieve the operator of an authorized emergency vehicle from the duty to drive with regard for the safety of all persons using the highway.

[Both statutes, the one derived from UVC 11-106 and the governmental immunity statute, are tied together. Section 63-30-10 (15) specifically grants immunity if the emergency vehicle operator is driving in accordance with §41-6-14. Unlike many governmental immunity statutes, Utah does not create an exception from immunity if the governmental employee acted with reckless disregard or similar language. Instead, the employee's conduct is to be judged by whether he observed the provisions of §41-6-14. Section 41-6-76 requires the emergency vehicle driver to drive with regard for the safety of others (not due regard). Section 41-6-14 deviates from the UVC version by permitting emergency vehicle drivers that are not involved in a police pursuit driver to use a siren or horn or visual signals, not both. (Many states follow this pattern.) If the vehicle is involved in a pursuit, the must use both audible and visual signals, his employer must have adopted pursuit regulations as described in the statute, and he himself must have been trained. Unlike some states, there is no statutory provision for officers to approach a crime scene silently or do catch-up driving and remain within the protection of the statute.]

Now we look at the cases to see how the courts have interpreted the statutes. There are few cases, and we will take the oldest ones first. A 1954 case shows that the statutory standard of care required of an emergency vehicle driver was then due regard for the safety of others. (The statute also included the reckless disregard language included in the UVC version.) A fire truck ran through a red light and collided with a driver who had the green light. The plaintiff was a passenger in that car. The fire truck driver saw the red light but proceeded when he saw other traffic yielding. The issue was whether he slowed "as may be necessary for safe operation" in running the light. The jury found for the plaintiff and the supreme court affirmed. The jury could have rationally found the car driver not negligent as a matter of law. The court agreed with the jury that the due regard provision was not satisfied merely by the use of audible and visual signals. *Jensen v. Taylor,* 271 P.2d 838 (Utah 1954).

A 1962 case reversed a jury verdict for the plaintiff and found him negligent in not hearing a siren or noticing the red lights of a police officer on an emergency run. The officer "skirted" traffic that had stopped for a red light and entered the intersection at 35–40 mph. The plaintiff had the green light. Neither driver noticed the other until it was too late to avoid a collision. Two disinterested witnesses testified that they heard the siren and in fact waited through a red, a green and then another red signal, when the accident happened. The supreme court held that the jury erred in finding the officer negligent on speed alone, while absolving the plaintiff from his duty to yield. A dissent could see no reason for invading the province of the jury. What one person hears another may not notice and still be exercising care. *Martin v. Ehlers,* 371 P.2d 851 (Utah 1962).

A 1966 case involved an ambulance. The plaintiff driver came to an intersection and saw that she had the green light. She saw a vehicle entering the intersection from the opposite direction, and noticed a "white car," the ambulance, on her left. She did not notice its emergency signals. The ambulance entered against the red light, was hit by the other vehicle, and spun into the plaintiff's vehicle. The jury found for the defendant ambulance driver. The ambulance was an authorized emergency vehicle, even though it was "designated or authorized" by another town. The supreme court held that the case was fully and fairly tried, and affirmed. *Howe v. Jackson,* 421 P.2d 159 (Utah 1966). (These first three cases show that on similar facts we can expect varying jury verdicts. Sometimes the appellate court follows the principle of affirming a verdict if there is some evidence for it, and sometimes the court reverses.)

A 1999 case in the Utah Supreme Court construed and found the governmental immunity statute unconstitutional as it related to emergency vehicle operation. The case made some important findings on duty to the public, but the offending statute was repealed so the case is not pertinent to the current law. The accident happened March 18, 1991. A highway patrol officer patrolling I-15 clocked a speeder going 10 mph over the speed limit. (In common experience, this means going about the same speed as most other drivers.) The officer tried to stop the driver to issue a ticket. The car had been stolen, a fact the officer suspected but did not know for certain. (Obviously, this was the real reason for trying to stop the driver.) The sixteen-year-old driver fled at high speed. The chase lasted for fifteen minutes, both off and on the interstate. The driver took the University Avenue exit in Provo and killed one person and injured another when he ran a red light. The plaintiff was the injured person, and she sued for her injuries and for the wrongful death of her husband. The trial court granted summary judgment to the defendants, and the court of Appeals affirmed, and the case came to the supreme court.

The plaintiff's problem was that when the accident happened Code 41-6-14(2)(a), which was in effect slightly more than one year, stated that governmental immunity was not waived for claims "resulting from the collision of a vehicle being operated by a suspected violator of the law who is being, has been, or believes he is being or has been pursued by a peace officer entity in a motor vehicle." A footnote explains that the sponsor of the amendment was concerned by a rash of "frivolous" lawsuits in California in such cases and wanted to nip in the bud any such trend in Utah. The amendment was repealed six weeks after the accident, and the legislature restored the language as currently set out in §63-30-10(15). The court declared the amendment unconstitutional as being in violation of Art. I, section 11 of the Constitution: "All courts shall be open, and every person, for an injury done to him . . . shall have remedy by due course of law . . . . " A dissent thought that the majority was too free in "constitutionalizing" existing tort law. The legislature saw its mistake a year later.

The court found that the officers owed a duty to the plaintiff as set out in §41-6-14, the duty of driving with regard for the safety of all persons using the highway. Whether the accident was foreseeable and whether there was proximate cause are jury questions. The court rejected the state's argument that the officer owed only a generalized duty to the public and not a specific duty to the plaintiff. The court cited cases from other states, since there were virtually no pursuit precedents in Utah, and remanded for trial. *Day v. State,* 980 P.2d 1171 (Utah 1999).

A 2000 case, *Lyon v. Burton,* 5 P.3d 616 (Utah 2000), arose from an intersection accident. A fire chief was monitoring a fire report in the North View Fire District, which is adjacent to the chief's own district, the Weber Fire District. The chief decided to go to the fire in a Ford Explorer. The Explorer was equipped with emergency signals, but was not otherwise marked as an emergency vehicle. The district wanted to be able to use it for other purposes such as arson stakeouts. The emergency lights were mounted behind the grill, the siren was behind the front bumper, and a light could be placed on the dash. The chief switched on the emergency lights and siren, ran a red light, and collided with a car. Two minors in the car suffered severe and permanent injuries, including brain injuries. The jury awarded $709,537 to one plaintiff and over $1 million to the other. The trial court reduced the damages to the statutory limit of $250,000 each, and denied the defendants' motion for judgment notwithstanding the verdict on the claim of immunity.

The immunity claim was based on §63-30-3(1) and §63-30-10(18)(b). The first states that governmental entities are immune from suit for any injury that results from the exercise of a governmental function, and the second for the activities of fighting fires. The supreme court first noted that it need not deal with the issue of what duty the fire chief owed to the plaintiffs. The defendant admitted that the fire chief owed a duty of care and that duty was breached. The remaining issue was whether the fire district was immune. A three-step analysis was necessary. First, was the activity a governmental function and therefore entitled to immunity under Code 63-30-3, a general grant of immunity? Second, if so, has some other statute waived that blanket immunity? Third, if blanket immunity has been waived, is there still immunity under an exception to the waiver?

As to the first question, the court found that the driving of a fire vehicle to a fire is a governmental function. As to the second question, the governmental function immunity has been generally waived by Code 63-30-10. The injuries were caused by the negligent act of an employee committed while working within the scope of employment. As to the third question, there are two possible exceptions to the waiver. Subsection (15) provides immunity if the emergency vehicle is being operated in accordance with Code 41-6-14, and subsection (18) provides immunity for the activities of fire fighting. If either apply, the defendant is immune.

The court had not previously had occasion to construe either of these two subsections. The court decided that subsection (15) applied with greater specificity, and that it should control. The fire chief was driving to a fire, and that involved "the operation of an emergency vehicle," and thus had the duty to comply with Code 41-6-14. This subsection is more specific to the activity complained of than the more general "activities of fire fighting." The fire chief had the statutory duty of proceeding past a red signal "only after slowing down as may be necessary for safe operations." The fire chief gave conflicting estimates of his speed in entering the intersection, from 10–40 mph, and the jury was entitled to pick the higher figure and find him negligent. The court also made an extensive analysis of the constitutionality of the $250,000 damage cap, and by one vote upheld its constitutionality. The court affirmed the awards, and made some findings on prejudgment interest that I will not detail here.

Vermont

Vt. Stat. Ann. tit. 23, §1015. Authorized emergency vehicles

[Vermont has enacted UVC 11-106 with only minor changes. Section 1015 states "fresh pursuit of a suspected violator of the law," rather than "pursuit of an actual or suspected violator of the law." As in many other states, Vermont permits operators on emergency runs to turn on either an audible or a visual, not both as required by the UVC. Vermont has a provision requiring emergency vehicle drivers to stop for a school bus with flashing red lights and remain stopped until the lights go off. The school bus driver has a reciprocal duty to try to get the bus off the highway, make sure no children are on the highway, and turn off his flashing lights to facilitate the movement of the emergency vehicle.]

There are two cases. The older case was brought in federal court, *Schatz v. Cutler,* 395 F. Supp. 271 (D. Vt. 1975). There were no federal issues. The pursuing officer was from New Hampshire, the accident happened in Vermont, and the pursued vehicle had Florida plates, so there was diversity of citizenship. The court determined that the law of Vermont applied. The facts were that in the early evening a Brattleboro police officer got a citizen complaint of reckless driving and suspicious persons. The officer, LaFlam, drove to Lamson Street and saw two vehicles with Florida plates. The car, which was filled with young people, sped off and LaFlam attempted unsuccessfully to stop the driver. At LaFlams's request, Officer Jenkins joined the pursuit west on Route 9. Route 9 is a two-lane road with an occasional passing lane on hills. Cutler, the driver, prevented the officers from getting in front by veering left when they tried. Eventually Jenkins pulled abreast, whereupon Cutler swerved left and forced Jenkins to lose control. Jenkins' cruiser wound up in front of Culver and LaFlam pulled in behind. When LaFlam got out, Culver managed to get away by ramming the car in front of him and reversing, and Jenkins continued the pursuit. As the pursuit passed a state police substation, two more officers joined the chase. Trooper Cronan managed to come abreast on a three-lane section, and again Cutler swerved into the cruiser. Cronan went off the road and Cutler collided head-on with the plaintiff's car.

The plaintiff sued the four officers and Cutler (who was a fugitive at the time of the trial). The defendants moved for summary judgment as not being the proximate cause. The plaintiff contended that the defendants were jointly negligent and reckless. The court noted that §1015 was a recent enactment and had not been construed, so the court was required to predict what a Vermont court would decide. The court noted without deciding between the two standards of due care and reckless disregard. The court decided that such issues of negligence, proximate cause, and concurring causes are for the jury, and denied summary judgment. The court did grant summary judgment to Trooper LaFlam. He was not able to continue the pursuit, and any initial actions that might be considered negligent were too remote to be a cause.

In a 1994 case, the Vermont Supreme Court had an opportunity but did not indicate whether a pursuing officer's conduct must be judged by due care, reckless disregard, or both. The case arose when a New Hampshire police officer, LaPointe, responded to a report of a single motorcycle accident in which the driver or the passenger was injured. The officer found glass on the road. He was not sure if the site was in New Hampshire or Vermont. The officer, driving north, met a motorcycle with a wobbly front wheel heading south. The motorcyclist was not wearing eye protection as required, so there was a reason to issue a ticket. LaPointe motioned for the motorcyclist to stop, but the latter continued into Vermont, and LaPointe pursued with emergency signals. A Vermont state trooper joined the pursuit. The chase went on for 25 miles at speeds between 40 and 60 mph (that is slow for a motorcycle pursuit). Vermont Trooper Tupper was notified, and waited for the group to come up. He got in front and straddled the center line in an attempt to stop the motorcyclist. The motorcyclist, according to the officers, tried to pass, lost control, and was killed when he crashed off the road.

The estate sued for negligence and alleged a 42 U.S.C. 1983 constitutional violation, an unlawful seizure. The supreme court affirmed summary judgment , denying the §1983 claim, and reversed summary judgment for the defense on the negligence count. The court followed a 1989 U.S. Supreme Court case, *Brower v. County of Inyo,* 489 U.S. 593. (See the California section in this book.) In *Brower,* police officers arranged for a semi to be parked across the road, and positioned a cruiser's headlights so as to shine in the eyes of the pursued driver as he came up. The U.S. Supreme Court remanded to the district court to determine whether such roadblock, under the Fourth Amendment, was an unreasonable seizure. The test is whether the officers sought to stop the suspect's freedom of movement "through means intentionally applied." (I will comment that most courts in this study do not so find if a way was left for the suspect to drive past the vehicle blocking the road, even if the suspect must slow and swerve.) The Vermont Supreme Court held that there was no unlawful seizure here because the suspect simply lost control. (I will comment again that a motorcyclist needs little room to get by. Apparently here he was going too fast for a curve.) A dissent thought that this "rolling roadblock" was an unlawful seizure.

The court remanded the negligence claims for trial. The court notes that under New Hampshire law, the officer may pursue for a felony, and the officer may in fresh pursuit follow into Vermont. However, the pursuit must be conducted under Vermont statute T.23, §1015, which requires due care and prohibits reckless disregard. Considering the length of the chase, the allegations that some of the officers knew the identity of the motorcyclist, the closeness of the pursuing cruisers, and the attempt to slow the pursued right before a curve, the case may not be disposed of by summary judgment. *Morais v. Yee,* 648 A.2d 405 (Vt. 1994).

Virginia

Va. Code Ann. §46.2-920. Emergency vehicles exempt from regulations in certain situations

A. The driver of any emergency vehicle, when such vehicle is being used in the performance of public services, and when such vehicle is operated under emergency conditions, may, without subjecting himself to criminal prosecution;

1. Disregard speed limits, while having due regard for safety of persons and property;
2. Proceed past any steady or flashing red signal, traffic light, stop sign, or device indicating moving traffic shall stop, if the speed of the vehicle is sufficiently reduced to enable it to pass a signal, traffic light, or device with due regard for the safety of persons and property;
3. Park or stop notwithstanding the other provisions of this chapter;
4. Disregard regulations governing a direction of movement of vehicles turning in specified directions so long as the operator does not endanger life or property;
5. Pass or overtake, with due regard to the safety of persons and property, another vehicle at any intersection;
6. Pass or overtake with due regard to the safety of persons and property, while en route to an emergency, stopped or slow-moving vehicles, by going to the left of the stopped or slow-moving vehicle either in a no-passing zone or by crossing the highway centerline;
7. Pass or overtake with due regard to the safety of persons and property, while en route to an emergency, stopped or slow-moving vehicles, by going off the paved or main traveled portion of the roadway on the right. Notwithstanding other provisions of this section, vehicles exempted in this instance will not be required to sound a siren or any device to give automatically intermittent signals.

B. The exemptions granted to emergency vehicles by subsection A of this section shall apply only when the operator of such vehicle displays a flashing, blinking, or alternating emergency light or lights as provided in §§46.2-1022 and 46.2-1023 and sounds a siren, exhaust whistle, or air horn designed to give automatically intermittent signals, as may be reasonably necessary, and, only when there is in force and effect for such vehicle either (i) standard motor vehicle liability insurance covering injury or death to any person in the sum of at least $100,000 because of bodily injury to or death of one person in any one accident, and, subject to the limit for one person, to a limit of $300,000 because of bodily injury to or death of two or more persons in any one accident, and to a limit of $20,000 because of injury or destruction of property of others in any one accident or (ii) a certificate of self-insurance issued pursuant to §46.2-368. Such exemptions shall not, however, protect the operator of any such vehicle from criminal conduct constituting reckless disregard of the safety of persons and property. Nothing in this section shall release the operator of any such vehicle from civil liability for failure to use reasonable care in such operation.

C. [This section defines emergency vehicles as police vehicles used in pursuit or in responses to an emergency call, prison or detention vehicles operated in an emergency, fire vehicle responding to an alarm, ambulances, and so on.]

[Section 46.2-920 is similar to UVC 11-106, but has differences. Part B requires the operator to display both visual and audible signals when on emergency runs. Part A does not require an audible signal when passing stopped or slow-moving traffic on the right or left, and part (A)(5) specifically permits emergency vehicles to pass in intersections. Part B requires liability insurance in force before the emergency vehicle operator is permitted to enjoy the privileges of emergency vehicle operation. The standard of car required is given as reasonable care and due regard for the safety of persons and property. Reckless disregard can subject the emergency vehicle operator to criminal prosecution.]

Now for the rest of the story. A 2000 case, *Nationwide Mut. Ins. Co. v. Hylton,* 530 S.E.2d 421 (Va. 2000) shows that sovereign immunity, a common law doctrine in Virginia, applies to emergency vehicle cases. A state trooper, Dehart, was on patrol in Henry County. He came to an intersection and stopped at a red light behind a truck. When it turned green, he saw a pickup driver run the red light without stopping and make an improper turn. DeHart intended to stop the erring driver and glanced in his mirror before moving into the right lane. The truck in front suddenly stopped, and DeHart rear-ended it. That driver was the plaintiff. The plaintiff sued DeHart, and the trial court granted the officer summary judgment on the finding that he had sovereign immunity. The supreme court affirmed, referring to its decision in *Colby,* 1991 (see below). Note that the officer was considered to be in an emergency in that he was preparing to stop the red-light runner. This is one of the feeblest emergencies covered in the cases in this book, but the point is that the officer was beginning to exercise his discretionary, not ministerial, law enforcement duties and was protected by sovereign immunity. (Nationwide Insurance Company is in the caption of the case because the jury returned a judgment of $100,000 against Nationwide as the plaintiff's uninsured/underinsured carrier. The court set aside the award because Nationwide was not a party in the case before the trial judge.)

Since the supreme court in *Nationwide* indicated that *Colby v. Boyden,* 400 S.E.2d 184 (Va. 1991) is still the law in Virginia, we look at it next. Paula Colby entered an intersection on a green light in the city of Virginia Beach and was hit on the left side by a city police officer pursuing a red-light runner. There were serious and permanent injuries. Shortly before the accident, Officer Boyden saw a driver run a red light. He turned on his blue lights and pursued. The driver sped, swerved from lane to lane, and ran a second red light. Boyden sounded his siren for a short burst and also entered the intersection and hit the plaintiff. The plaintiff alleged negligence and gross negligence against Boyden, and Boyden asserted sovereign immunity. (Note that in many states the asserted immunity would be termed official, not sovereign immunity, but the court here refers to sovereign.) The trial judge granted Boyden summary judgment, and the case came to the supreme court.

The plaintiff alleged that the officer's acts in a pursuit case are ministerial, not discretionary, relying on a D.C. case that involved Virginia police officers, *Bricoe v. Arlington County,* 738 F.2d 1352 (D.C. Cir. 1984). The supreme court here declined to follow *Briscoe* (because the D.C. Circuit applied the law of the District, not Virginia). Both jurisdictions protect government employees performing discretionary acts who are not negligent. But in the District the standard for liability is proof of negligence, while in Virginia it is gross negligence. The court commented that *Colby* is the first instance in which sovereign immunity has been asserted in a Virginia pursuit case. Virginia applies four factors in determining if the act complained of is discretionary or ministerial. They are: the nature of the function of the government employee, the extent of the government's interest and involvement in the function, the degree of control and direction exercised over the employee by the government, and whether the act involved discretion and judgment. (Note that there are many such formulations in deciding whether an act is discretionary or ministerial. This is Virginia's version.)

There is no question that the first two factors apply. Enforcement of traffic laws is a primary governmental function, and the city is heavily involved through the maintenance of law enforcement staffs. As to the third and fourth factors, the city does exercise supervision over police officers through policies and guidelines dealing with emergency situations, but the actual conduct of a pursuit (and the decision to start) requires discretion and judgment. The officer was exercising a discretionary function.

The second issue is which standard to apply, negligence or gross negligence. The court states that precedent cases (they are not pursuit cases) require gross negligence. The plaintiff pointed out that the final sentence in subsection (B) says: "Nothing in this section shall release the operator of any such vehicle from civil liability for failure to use reasonable care in such operation." The four cases cited by the plaintiff were not quite on point. The court decided that the standard is gross negligence. After pointing out some logical problems if the statute were interpreted as the plaintiff argued (not very , in the author's opinion), the court held that the legislature had not explicitly abrogated sovereign immunity, and the court would not so find by implication. In other words, the legislature will have to do more than add the reasonable care standard in the statute. The court found that the defendant's acts were not grossly negligent. Some degree of care negates gross negligence. Here the officer activated his emergency signals, entered the intersection only slightly over the speed limit, saw the plaintiff and took some action (unsuccessfully) to avoid the collision. Three dissenting judges thought the standard of reasonable care set out in the statute could not be more clear, and should control. I will comment that Virginia is not the only state in which governmental immunity trumps the emergency vehicle statute.

A 1997 case arose from a car-ambulance collision. The ambulance was being driven with its emergency signals on. The operator ran a red light and hit the plaintiffs' vehicle. Smith, the ambulance driver, pleaded guilty to reckless driving. The several claims were consolidated for trial. The defendants pleaded sovereign immunity, and the trial judge so granted. The judge released the rescue squad as party defendant and ruled that Smith could be liable only if his acts amounted to gross negligence. During trial the insurer of Smith and the squad, according to the plaintiffs, agreed that if the jury found for Smith, the insurer would settle for $350,000, and the same amount if the jury awarded a lesser amount. The jury found for Smith and the insurer refused to pay. The trial judge meanwhile granted a new trial because of an erroneous instruction, to the effect that if the car driver was contributorily negligent, a passenger could not recover damages. A second trial could not agree on its verdict, and the third trial again found for Smith. The supreme court held that this "high-low" agreement was not enforceable because the plaintiffs refused an offer of $350,000 following the first verdict and thus repudiated the offer. Big mistake.

The court also held that Smith was entitled to sovereign immunity and was not grossly negligent. The plaintiffs argued that Smith was not yet on an emergency call and thus was not exercising discretion, but had ordinary driver duties. The facts are sketchy, but it appears that Smith was proceeding to a location where he could receive a radio signal to precisely identify the location of the emergency. The court held that Smith had been dispatched to the scene and was thus exercising a discretionary governmental function. The jury correctly found that Smith was not grossly negligent, notwithstanding some language in §46.2-920. *Smith v. Settle,* 492 S.E.2d 427 (Va. 1997).

A 1992 fatal accident case was brought to federal court. A state trooper on patrol met a speeder. The officer turned around to "pace" the car and check its speed without operating his lights or siren. The approximate speed of the officer is not given. The road took a dip, and when the officer came out of the dip he struck a car making a left turn into a high school. Both occupants in the car were killed. The defendants moved to dismiss on immunity and abstention grounds, among others. The court ruled on the federal constitutional claims only. (The court commented in a footnote that the defendants might successfully make a state immunity claim.) The first claim was unreasonable seizure under the Fourth Amendment. The court held that there was no evidence that the officer intended to seize the deceased persons, or in fact any persons, in pacing the speeder. The plaintiff was essentially arguing negligence. The plaintiff failed to allege an intent to seize in his complaint. As to the second claim, that of substantive due process, the court compared the allegation with a Fourth Circuit case, *Temkin v. Frederick County,* 945 F.2d 716 (4th Cir. 1991). (See the Maryland section.) The court there found that *Temkin* did not meet the shocks-the-conscience test even though the pursuit was longer and in general more egregious than the facts indicated here. The court dismissed the case for failure to state a constitutional claim. *Hicks v. Leake,* 821 F. Supp. 410 (W.D. Va. 1992).

Note an older case in which a Norfolk police officer pursued a driver who had run a red light and thereafter sped and drove recklessly in trying to escape. The fleeing driver skidded on a turn, lost control, and was hit by the officer. The officer sued for injuries, and the jury awarded damages. The supreme court affirmed. The trial judge correctly refused to instruct on speeding and following too closely. All traffic infractions are not specifically enumerated as exempt in the emergency vehicle statute, but they are inherently waived for law enforcement officers pursuing suspects. *Smith v. Lamar,* 188 S.E.2d 72 (Va. 1972).

In summary, Virginia requires a plaintiff to prove gross negligence on the part of emergency vehicle drivers to defeat immunity, as long as the driver is performing a discretionary function. Virginia takes a wide view of what is discretionary in emergency runs. The gross negligence standard is a judicial standard, and controls over the standard of reasonable care set out in §46.2-920. The last decision so holding was a 4-3 split, so stay tuned.

Washington

Wash. Rev. Code Ann. §46.61.035. Authorized emergency vehicles

[UVC 11-106 is adopted with one change. See the Introductory Essay for the UVC text. The change made is to permit emergency vehicle drivers to sound audible signals "when necessary to warn others," rather than to make such sounding necessary required at all times.]

Washington, as all states do, retains sovereign immunity for discretionary acts or omissions, but Washington by case law restricts such immunity to high-level planning decisions. Such discretion does not extend to typical law enforcement acts or omissions. A 1975 case makes this clear. A state patrol officer identified a speeding Corvette on I-5 at 65 mph *over* the speed limit. The speeder was clocked at one time at 140 mph. There was a pursuit and an attempted rolling roadblock to no avail. Several of the seven officers involved thought that the pursuit was futile and dangerous to others, but that belief was not forwarded to the dispatcher or to other pursuing officers. (Of course, other officers might not have agreed that the risk to the public outweighed necessity of stopping the speeder, so it is not clear why the court thought this was significant.) The speeder lost control, crossed the median, and collided with an oncoming driver and both drivers were killed. The estate of the innocent driver sued the estate of the speeder, the state, and Seattle officers. The trial court granted the government defendants summary judgment, and the Washington Supreme Court reversed. There were genuine issues of material fact regarding whether the pursuit should have been called off, whether the statutory duty of due regard was observed, whether departmental pursuit policies were followed, and whether the officers' actions were a proximate cause. The court did not agree with the defendants that the duty of due regard applied only when the police vehicle itself is involved in the accident. The statute refers due regard for the safety of all persons. The court reversed the trial court's finding that the officers were engaged in discretionary acts. The court held that these acts were operational or ministerial, not basic policy decisions made at the highest executive level. *Mason v. Bitton,* 534 P.2d 1360 (Wash. 1975).

A principle of traffic law is that a driver, observing the rules of the road and maintaining proper lookout, is entitled to assume that other drivers are doing the same until he knows or should have known otherwise. More precisely, a driver is entitled to expect others to yield to his right of way until he sees that they are not so yielding. The principle applies equally to emergency vehicle drivers. A 1983 case involved a fire truck on a fire run. There was a collision at an intersection, and an occupant of the car involved was killed. There was conflicting evidence on speed and who had the green light. The fire department counterclaimed for property damage. At trial the judge denied plaintiff's motion for a directed verdict on liability and the color of the traffic light. The jury found for the defendant and awarded $24,000 on its property damage claim. The supreme court held that the instruction stating that every person using the street or highway and exercising ordinary care has a right to expect the same from others correctly stated the law. The emergency vehicle driver is to use due care, a traditional negligence standard, in driving within the privileges granted by §46.61.035. The instruction does not imply that an emergency vehicle driver may blindly presume that all others will yield, but must use due care to see that they do. The court also found the refusal to instruct on the sudden emergency doctrine was correctly rejected. The plaintiff was not without fault in creating the emergency, and thus may not claim the benefit of sudden emergency. The court affirmed the verdict. *Brown v. Spokane County Fire Protection Dist.,* 668 P.2d 571 (Wash. 1983).

A 1965 case involved whether a fire truck was validly on an emergency run. A Spokane fire station received a call requesting help. A two-year-old child had locked himself in the bathroom, and the grandmother needed help in getting him out. The station responded with a Kenworth triple combination hose wagon (apparently a very large vehicle), and the truck was dispatched with lights flashing and siren screaming. According to the fire truck driver and passengers, the driver saw the traffic light ahead change to red and coasted to the intersection at 25 to 30 mph. The fire truck operator thought that all traffic had stopped, and entered the intersection. The fire truck hit a farm truck that had the green light broadside and killed all three occupants. Numerous witnesses testified that they heard the siren (even those inside buildings) and saw the flashing lights. The weather conditions, typical for the Northwest in February, were drizzly. The two trucks locked together and skidded for 196 feet, so this was gruesome accident. The jury returned defense verdicts, and the trial court granted a new trial. The supreme court reversed and ordered a dismissal.

The primary issue was whether the fire truck was on an emergency run. The controlling statute at the time referred to an emergency vehicle actually responding to an emergency call *within the purpose for which such emergency vehicle has been authorized* (italics supplied). The plaintiff argued that the call was not a fire department purpose, so proceeding under emergency signals was not allowed. The court held that the fire officer who took the call and ordered the No. 10 wagon to respond did so in good faith and did not act arbitrarily. The burden to prove that he did not was on the plaintiff. The court further held that the farm truck operator was negligent as a matter of law. The court emphasizes that fourteen disinterested witnesses were aware of the emergency signals. *Lakoduk v. Cruger,* 296 P.2d 690 (Wash. 1956).

A 1954 case similarly held that the plaintiff driver failed to yield and was negligent. An air force ambulance was dispatched from an air base to a Tacoma suburb in response to a report of an airplane crash. The ambulance operator sounded both audible and visual signals, and reduced his speed from 65 mph to 35 mph when he entered an intersection that had a red blinking light for him. The plaintiff station wagon driver hit the side of the ambulance. The case was brought in federal court, and the judge found the plaintiff to be the sole and proximate cause and dismissed the complaint. *Gilpin v. United States,* 120 F. Supp. 812 (W.D. Wash. 1954).

A 1952 case also involved a fire truck owned by the United States. For some years the city of Vancouver and a nearby army post, Vancouver Barracks, had an agreement that when one dispatched a fire truck on an emergency run, the other would send a fire truck to the station to be on standby duty. The city fire department sent a truck out on a fire call and an army fire truck set out, with emergency signals operating, to proceed to the city station. The army truck entered an arterial highway at 25 mph and hit a state patrol car, whose driver was on routine duty. The State of Washington paid benefits to the trooper and sued the army for reimbursement. The issue was whether the army unit was responding to an emergency call. It was conceded that if it was, the trooper had the duty to yield. The case was brought in federal court, and the trial judge found for the defendant. The closest case on point cited to the Ninth Circuit was an old Ohio case, *Staudenheimer v. City of Newark,* 23 N.E.2d 845 (Ohio App. 1939). There one fire truck was sent out, and a truck from another station was sent to replace it at the station and had an accident on the way. The court held that the both trucks were answering the alarm, and held that the second truck was on an emergency run. The court followed that precedent and held that the army unit was on an emergency and moreover, that the operator acted with due regard for the safety of others. The court affirmed. *State of Washington v. United States,* 194 F.2d 38 (9th Cir. 1952).

A 1971 case involved a Seattle ordinance that required emergency vehicle drivers to come to a full stop and to allow traffic to clear before continuing. The state statute permits such vehicles on runs to "proceed past a red or stop sign, but only after slowing down as necessary for safe operation." The accident arose when an ambulance driver, with emergency signals operating, slowed to about 5–10 mph at a red light, saw no traffic, and proceeded. The plaintiff driver hit the right side of the ambulance. The court notes that the plaintiff had her windows up and the radio on. The court held that the state statute must control, and affirmed the trial court in finding the plaintiff negligent. *Seaberg v. United States,* 448 F.2d 393 (9th Cir. 1971).

A 1977 case found a private motorcyclist doing funeral escort duty to be a de facto emergency vehicle and was entitled to have the privileges of emergency vehicles. Note that this is not necessarily the law in other jurisdictions, and it is a little shaky here. The motorcyclist is permitted to use his flashing red light, but may not use a siren. The court held that the state had approved the type of emergency light the motorcyclist was using, although the escort service had not registered or obtained written permission. (I have been on bicycle events where a following vehicle displays a portable flashing yellow light, usually roof-mounted, and I doubt that such vehicle would be considered an emergency vehicle.) The court held that the motorcyclist, who was injured when he collided with a car, could claim the right of way but could not claim the full benefits of bona fide emergency vehicles. The jury properly considered his contributory negligence and reduced his award accordingly. *Boyle v. Emerson,* 561 P.2d 1110 (Wash. App. 1977).

In summary, Washington deals with the emergency vehicle issue by applying the standard of due regard, not reckless disregard. Both are in the statute, because the state adopted the Uniform Vehicle Code almost without change, but the courts apply due regard. This should make Washington a more plaintiff-friendly jurisdiction than many other states. No immunity is allowed. Perhaps this accounts for the fact that there are few recent cases. My guess is that the settlement factor is higher than in most states in these emergency vehicle cases.

West Virginia

W. Va. Code §17C-2-5. Authorized emergency vehicles

[West Virginia has enacted UVC 11-106 without change. See the Introductory Essay for text. As in many states, West Virginia has enacted a provision, §61-5-17, that makes it a crime to flee from a law enforcement officer. Penalties include both fine and imprisonment, with enhanced penalties if the flight results in property damage, injury, or death. Flight while under the influence of alcohol or illegal drugs also results in enhanced penalties.]

There are few cases, and we begin with a pursuit case, *Peak v. Ratliff,* 408 S.E.2d 300 (W. Va. 1991). The plaintiff was an innocent bystander. The case was one of first impression. Two state troopers in one vehicle were in pursuit of a burglary suspect who had previously eluded arrest in a chase the previous day by escaping on foot after he went off the road in his stolen car. The pursuit came about when a tipster phoned the state police and said that the suspected burglar, Akers, was driving towards Princeton on Route 20 in a stolen pickup. Akers was a suspect in a series of burglaries, and several counties had issued arrest warrants, so the officers had a duty to attempt an arrest. The officers saw Akers coming from the opposite direction. They turned around, Akers sped up, and the officers turned on siren and lights and followed. They radioed for a roadblock to be attempted at an intersection 2.5 miles ahead. The site was a two-lane road with blind curves. The speed limit on the road varied between 35 and 45 mph, depending on whether there was a curve or a straight stretch. The character of the area is somewhat residential, having side streets, a school and a golf course. Akers' speed was estimated to be between 60 and 100 mph. Akers passed slower traffic on blind curves and forced oncoming traffic off the road. The officers slowed on those occasions to wait for oncoming traffic to get out of the way. Akers testified later that the officers were never more than 100 yards behind, although the officers lost sight of Akers on curves. When Akers came to the intersection with Route 460, he encountered a line of stopped traffic. He was at that time out of sight of the pursuing officers. He intended to cut though a gas station and enter the oncoming traffic's lane. He hit the plaintiff's car and the plaintiff was injured.

The jury awarded damages against the officer and the state. The trial judge then granted a judgment notwithstanding the verdict, and the case came before the supreme court. The issue was which standard applies, due regard or reckless disregard? The trial judge instructed that for the plaintiff to prevail, the officers' conduct must exhibit reckless disregard for the safety of others, and the judge found that the evidence did not meet this standard. The problem arises, and I have commented on this where it has come up in other jurisdictions, because the Uniform Vehicle Code gives two standards. West Virginia Code §17C-2-5 states: "The foregoing provisions shall not relieve the driver of an authorized emergency vehicle from the duty to drive with due regard for the safety of all persons, nor shall such provisions protect the driver from the consequences of his reckless disregard for the safety of others."

The supreme court looked to others states, and found that most have not focused on this dichotomy. Many have focused on the due regard standard. (At least one has suggested that an emergency vehicle operator must use due regard not to drive with reckless disregard.) The court noted that its own closest precedent, *McClanahan* (see below), is an ambulance case that involved the ambulance itself. Here the pursed driver hit the plaintiff. The court suggests that the apparent disparity between the standards can be understood for the due regard standard to apply to the operation of the emergency vehicle itself, as where the emergency vehicle itself directly causes an accident. The statute that requires drivers to yield to emergency vehicles, §17C-9-5(b) states that nevertheless the emergency vehicle driver is not relieved from the duty of due regard for the safety of others. It does not include reckless disregard; that could indicate that the legislature did not intend for the latter to apply to the driving of the emergency vehicle.

The court decided that common sense dictates that reckless disregard applies to a pursuit case. Chasing a suspected or actual law violator is not the same as proceeding to a fire. The pursued driver himself dictates in large part the course of events. Law enforcement officers should not be held to a due regard standard in this difficult and dangerous duty, but rather to reckless conduct or gross negligence. If officers have exhibited reckless conduct, they may not escape liability even if not involved in the accident but were a substantial cause. The court then considered the facts and held that the defendants were not grossly negligent, and affirmed. There is no specific formula for making this judgment; all the circumstances must be considered.

The ambulance case noted above is *McClanahan v. Putnam County Commission,* 327 S.E.2d 458 (W. Va. 1985). The accident scene was a no-passing zone on a two-lane road. The plaintiff car driver was preparing to make a left turn from the northbound lane when she noticed an ambulance overtaking with its lights flashing. She thought it was sufficient distance behind to make the turn, but the ambulance driver tried to pass and struck the plaintiff in the southbound lane. The jury found that the plaintiff was 51 percent at fault. In West Virginia, the plaintiff gets nothing in this situation. On appeal the plaintiff argued that the emergency vehicle statute, 17C-2-5, does not include a specific exemption for an emergency vehicle to pass left of center in a no-passing zone nor within 100 feet of an intersection. The specific exemption that is closest to these facts is 17C-2-5(a)(4): "The driver of an authorized emergency vehicle may (4) disregard regulations governing direction of movement of turning in specified directions." (The legislature made a mistake and enacted "movement of turning" rather than "movement or turning.") The plaintiff argued that the trial court erred in not giving her requested instructions to the effect that the defendant improperly attempted to pass in a no-passing zone and drove left of center within 100 feet of an intersection.

The court held that it would correct the legislative mistake of "of" for "or," and that an emergency vehicle driver on a run may disregard regulations governing "turning in specified directions." The court noted that the defendant still must exercise due care in so disregarding regulations, and the jury's finding that he did so exercise would not be disturbed.

A 1968 case involved a motorcycle and a fire truck (obviously a mismatch). The accident happened in Elkins at First Street and Kerens Avenue. The fire truck was responding to an alarm and did not stop at the stop sign. The motorcyclist was on the favored street and hit the side of the truck, and the motorcyclist was severely injured. The jury awarded $25,315, but the trial judge set aside the verdict, and the Supreme Court of Appeals heard the case. The court affirmed judgment for the defendant. The evidence showed that the fire truck entered the intersection slowly, about 5 mph, which indicated that the operator slowed as necessary for safety. (I would comment that motorcycles, and this bike probably was a small one, are much harder for drivers to see or notice than are cars and trucks. Experienced motorcyclists know this and drive accordingly.) There was conflicting evidence about whether the truck's siren was sounding. The court found that the two witnesses who did not hear the siren indicted negative evidence only, that is, the testimony does not prove the siren was not working. Also, there was evidence that the plaintiff was speeding. *Davis v. Cross,* 164 S.E.2d 899 (W. Va. App. 1968).

West Virginia is noteworthy in making a clear distinction between due regard and reckless disregard. The latter applies only to pursuit cases, and only then in ones in which the police vehicle is not involved in the crash or is a substantial factor in causation.

Wisconsin

Wis. Stat. Ann. §346.03. Authorized emergency vehicles

(1)–(3) [These subsections set out UVC 11-106(a)–(c). See Introductory Essay for text. One minor change in the Wisconsin version is that 346.03(2)(c) states that emergency vehicles on runs may exceed the speed limit, while the UVC version permits exceeding the speed limit so long as life and property are not thereby endangered.]

(3) Except as provided in subsection (4M), a law enforcement officer operating a police vehicle shall otherwise comply with the requirements of sub. (3) relative to the giving of audible and visual signals but may exceed the speed limit without giving audible and visual signals under the following circumstances:
  (a) If the officer is obtaining evidence of a speed violation.
  (b) If the officer is responding to a call which the officer reasonably believes involves a felony in progress and the officer reasonably believes any of the following:
    1. Knowledge of the officer's presence may endanger the safety of a victim or other person.
    2. Knowledge of the officer's presence may cause the suspected violator to evade apprehension.
    3. Knowledge of the officer's presence may cause the suspected violator to destroy evidence of a suspected felony or may otherwise result in the loss of evidence of a suspected felony.
    4. Knowledge of the officer's presence may cause the suspected violator to cease the commission of a suspected felony before the officer obtains sufficient evidence to establish grounds for arrest.
  (4m) A law enforcement officer operating a police vehicle that is a bicycle is not required to comply with the requirements of sub. (3) relative to the giving of audible and visual signals.
    5. The exemptions granted to the operator of an authorized emergency vehicle by this section do not relieve such operator from the duty to drive or ride with due regard under the circumstances for the safety of all persons nor do they protect such operator from the consequences of his or her reckless disregard for the safety of others.
    6. Every law enforcement agency that uses authorized emergency vehicles shall provide written guidelines for its officers and employees regarding exceeding speed limits under the circumstances specified in sub. (4) and when otherwise in pursuit of actual or suspected violators. The guidelines shall consider, among other factors, road conditions, density of population, severity of crime and necessity of pursuit by vehicle. Each law enforcement agency shall review its written guidelines by June 30 of each even-numbered year and, if considered appropriate by the law enforcement agency, shall revise the guidelines.

Wis. Stat. Ann. §893.80. Claims against governmental bodies

(4). No suit may be brought against any . . . political corporation . . . or any agency thereof for the intentional torts of its officers, officials, agents or employees . . . nor may any suit be brought for acts done in the exercise of legislative, quasi-judicial or quasi-legislative functions.

[Wisconsin has adopted UVC 11-106, and has further spelled out when police officers may speed without giving emergency signals. The statute further requires law enforcement agencies to have written speeding and pursuit guidelines and gives several factors they must cover. It also requires that they be reviewed every two years.]

The most important case is a recent pursuit case, *Estate of Cavanaugh v. Andrade,* 550 N.W.2d 103 (Wis. 1996). It shows that both the statute governing emergency operation and governmental immunity may apply. The facts were that a Milwaukee police officer, Andrade, was on patrol in the city in the late evening. He saw a driver pass several cars stopped at a red light and speed through the intersection. The officer followed and turned on his lights and siren two blocks later. The officer suspected that the driver was intoxicated. The suspect, Zergoski, started to pull over but again sped away. Six to seven blocks later the officer found Zergoski stalled in an intersection, and got the plate number. Zergoski restarted and sped on an arterial street in a residential neighborhood. Witnesses estimated his speed at 60 to 80 mph, with the officer one block or less behind. Some seventeen blocks after the initial event, the suspect ran a red light and hit Cavanaugh, who died after five months in a coma.

The parties eventually stipulated to damages of $250,000, the statutory cap for negligent operation of a motor vehicle owned by a municipality. (Apparently the plaintiff misread the statute and first stipulated to $100,000, but the trial court held that agreement void as not being a meeting of the minds.) The jury found for the plaintiff and apportioned damages as 75 percent Zergoski, 23 percent city for negligent implementation of its pursuit policy, and 2 percent Andrade, the officer involved. The court of appeals reversed as to the city on its finding that the city was immune from liability, and affirmed the 2 percent finding as to the officer because he was negligent in the operation of a vehicle.

The case was appealed to the Wisconsin Supreme Court. It reinstated the finding of 23 percent fault against the city. The city is immune for performance of a discretionary function, §893.80(4). The immunity issue is whether the city's duty of carrying out §346.03(6) is ministerial or discretionary. That section requires written guidelines that cover, among other items, four specific factors, including "severity of crime and necessity of pursuit." In a very close reading of the relevant parts of the city's pursuit policy, the court found that the overall policy did not cover "severity of crime." Strictly speaking, severity of crime is identified as a factor to consider in making a decision of whether to ram a pursued vehicle or use a police car to set up a barricade. Such action should be taken only for a serious felony or when the suspect poses a substantial risk of serious injury or death to the public. In this case, the jury could find that neither factor was present, and the rest of the guidelines did not include the magic words in covering pursuits for traffic infractions. The court also found that the implementation of subsection (6) was a ministerial act, which does not trigger immunity.

The court found that the jury could reasonably find that the officer's actions were a substantial factor in producing the injury, as it related to the city's liability. Neither the officer nor his supervisor specifically stated in their testimony at trial that they considered the severity of the crime as a reason to consider abandoning the pursuit. Zergoski himself testified that he would flee only as long as he was chased.

As to Officer Andrade, the court decided that his decision to pursue was a discretionary act for which he was immune. However, he must conduct the pursuit within the confines of §346.03, and use due care in the operation of the vehicle. He is entitled to speed; he must speed with "due care." The court followed an Oklahoma case (see that section), *Kelly v. City of Tulsa,* 791 P.2d 826 (Okla. App. 1990). The court reversed as to Andrade, finding insufficient evidence that his negligence, if any, was a causative factor. The issue was whether the officer's actions were causative. The plaintiff may not rely on allegations of excessive speed or speed too fast for conditions, because those decisions are discretionary. However, "management and control" of the vehicle, such as following too closely or not keeping a proper lookout, are ministerial acts and the officer can be found negligent in these. (Is everyone thoroughly confused by this distinction? I am. "I'm speeding, but that's okay because it's discretionary. Oops! I'm following too closely. Better back off, that's ministerial.") The court reversed the jury finding and decided that because the officer remained a block or so behind, his acts were not causative.

The final issue was whether the stipulation was $100,000 or $250,000. The court held that it was the former because of the finding that Andrade was not negligent, and the applicable statute refers to the negligent operation of a motor vehicle. There were two long dissents. One noted that many other state courts would make a different finding. (It will be no secret to the readers of this book that virtually every variant imaginable is the position of some jurisdiction.) One dissent thought the finding on the city's liability was wrongly decided and reasoned.

We now look at some nonpursuit cases. In a 1988 ambulance case, an ambulance operator approached a red signal in a 30-mph-zone. He slowed from about 55 mph to about 40 mph. He noticed a white car coming from the left and took his foot off the gas. He saw the car's front dip as though the driver were braking, and did not slow. However, the car accelerated into the intersection and the ambulance hit it broadside. The ambulance operator was convicted of violating a city ordinance requiring stopping at a red light. The defendant argued on appeal that he was entitled to "proceed past a red or stop signal or stop sign, but only after slowing down as may be necessary for safe operation." The court of appeals affirmed. Emergency vehicle operators have limited privileges not available to others, including running red lights. Such operators still must drive with due regard for the safety of others, and here the defendant did not. There was a rather eccentric dissent. *City of Madison v. Polenska*, 421 N.W.2d 862 (Wis. App. 1988).

A 1976 case also involved an ambulance. Again, the ambulance driver entered an intersection on red and hit two minor pedestrians. The ambulance was preceded by a police motorcycle escort, also with its emergency signals operating. The jury found the ambulance operator 80 percent at fault and the children 20 percent. The supreme court affirmed. The defendant wanted the court to follow an old California case that held that by using his emergency signals, an emergency vehicle driver is exercising due care and may not be held negligent. He can be held liable only if he is acting with reckless disregard. The court declined to follow this reasoning, and held that the defendant may be liable for lack of due care. Here he did not notice the children until he was ten feet from them, which indicated lack of proper lookout. There was no particular reason why the children, ages eight and ten, should expect an ambulance to be following the police motorcycle. *Montaldo v. Fond Du Lac County,* 76 N.W.2d 279 (Wis. 1976).

A 1973 case arose when two volunteer firemen, responding to the same alarm, collided at an intersection that had four-way stop signs. Neither stopped. Both were driving private cars, one had a flashing light and a siren, the other a flashing light only. The jury found the driver without the siren 85 percent at fault, and he appealed. The supreme court affirmed on its finding that the evidence was sufficient to show that the defendant's flashing light met visibility requirements. The court does not explain why the plaintiff was found 85 percent at fault, or indicate whether he also was operating his flashing light. This case isn't worth much. *Pittman v. Lieffring,* 207 N.W.2d 610 (Wis. 1973).

A 1963 case is unusual. The plaintiff was being chased by a deputy sheriff around midnight on Highway 41 in Washington County when he ran into a roadblock and was injured. The plaintiff's allegation was that he had been stopped by Deputy Brown for a traffic violation and "immediately thereafter Brown chased the plaintiff and attempted to arrest him at high rates of speed for another traffic violation." (It sounds like we are missing some facts here.) Brown called ahead for a roadblock, and Deputy Becker ordered a semi-truck driver to block the road, and the plaintiff hit the semi. The plaintiff alleged an illegal roadblock for the purpose of catching a traffic violator, creating a trap, and not using the proper lighting at the roadblock. The case involves the plaintiff's claim against the truck driver. The supreme court found that the officer had the right, arising from the common law of hue and cry, and the power to call up a posse, to commandeer a private vehicle. Drivers of those vehicles have a duty to comply with the order. Such drivers, as the defendant here, are not then responsible for the roadblock, or how it is conducted. Thus the defendant is not liable for failure to set out a flare, and so on. In other words, the defendant cannot do right and wrong at the same time. *Kagel v. Brugger,* 119 N.W.2d 394 (Wis. 1963).

A case involving a motorcycle pursuit that ended when the motorcyclist crashed was filed in federal district court. Colfax police officers in the early morning hours noticed a seventeen-year-old motorcyclist, whom they had previously chased, drive within the speed limit in town. In other words, they knew who he was. One of the officers also noticed a car go by without it lights on. He pulled out to stop and talk to the driver. As he came up behind the motorcyclist, the officer switched on his flashing lights and turn signal to pass the bike. However, the motorcyclist thought that he himself was thereby signaled to stop, and fled at high speed instead. He did not want a ticket for not having a motorcycle endorsement on the license. Two officers pursued the motorcyclist for several miles on Highway 40 at speeds up to 100 mph. The motorcyclist came to a curve, lost control and crashed. At that time the officer was fifteen to thirty feet behind. As the motorcyclist lay semicomatose on the ground, one officer remarked to the other, "We finally got the little bastard." (He was not wearing a helmet and had suffered severe brain injury.)

The trial judge denied summary judgment to the defendants. He found a factual issue as to whether there was a seizure of the plaintiff. The plaintiff may attempt to show that the close pursuit made it dangerous for him to stop, and thus constitute an unreasonable seizure. The court notes that the claim of unlawful force is also a Fourth Amendment claim, and if there was no seizure, there can be no unlawful force. The court also declined to rule on state law negligence claim. *Moyer v. Dunn County,* 691 F. Supp. 164 (W.D. Wis. 1988).

In summary, Wisconsin has but three pursuit cases, and one with a case in Wisconsin will need to thoroughly review *Cavanaugh*. Other than that, there are not many cases. I have omitted ones over seventy-five years old as not useful.

Wyoming

Wyo. Stat. Ann. §31-5-106. Authorized emergency vehicles

[Wyoming has enacted UVC 11-106 without change (see the Introductory Essay for text), except that UVC 11-106(c), the part dealing with emergency signals, is omitted. Section 31-5-952 regulates horns and other warning devices. It states that the operator of an emergency vehicle on an emergency run shall sound the warning device "when reasonably necessary to warn pedestrians and others of the approach thereof." Section 31-5-928 specifies the type of emergency lights permitted but does not specifically state when they shall be used.]

Wyo. Stat. Ann. §1-39-112. Liability of peace officers

A governmental entity is liable for damages resulting from tortious conduct of peace officers while acting within the scope of their duties.

Wyo. Stat. Ann. §1-39-105. Liability; operation of motor vehicles

A governmental entity is liable for damages resulting from bodily injury, wrongful death or property damage caused by the negligence of public employees while acting within the scope of their duties in the operation of any motor vehicle, aircraft or watercraft.

Wyo. Stat. Ann. §1-39-102(a). Purpose [of Governmental Claims Act]

[T]his act is intended to retain any common law defenses which a defendant may have by virtue of decisions from this or any other jurisdictions.

[The Wyoming Governmental Claims Act of 1979, §§1-39-101 to 121, abolished all judicially created categories such as governmental or proprietary functions and discretionary or ministerial acts previously used by the courts to determine immunity or liability, according to section 1-39-102. However, the law does not impose strict liability. Note also that the act does not abolish common law defenses. The act limits claims to $250,000 for one claimant and $500,000 for all claimants arising out of one "transaction or occurrence." There is no special exemption for law enforcement or for providing fire protection.]

Only three cases have been identified. A 1926 case arose from an accident between the plaintiff's car and a fire truck. The fire truck was responding to a fire alarm sounded in the city. The plaintiff heard the alarm and stopped along a row of parked vehicles, and the fire truck skidded into the plaintiff's car. The plaintiff claimed that the city had negligently made the street slippery by watering, that the operator negligently drove too fast on the wet street, and that the operator was not taking the shortest route to the fire, but drove down the most congested street in Casper, Center Street, to show-off. The supreme court followed its precedents, holding that the city was not liable for exercising a governmental function, that of fire protection. *White v. City of Casper,* 249 P. 562 (Wyo. 1926).

There are two police pursuit cases. The first is a 1986 case, *DeWald v. State,* 719 P.2d 643 (Wyo. 1986). This was a fatal accident case in which a fleeing suspected drunk driver hit and killed a driver stopped at a red light. A highway patrolman was on duty in Laramie. A citizen flagged him down and reported that a man who appeared extremely intoxicated just left town driving a large yellow car and heading north on U.S. 30. The officer headed north and soon saw a yellow Oldsmobile making a U-turn. The officer turned and followed, and the driver weaved across the yellow line several times which indicated possible intoxication. The officer turned on his flashing lights and touched his siren a couple of times, but the driver did not respond. Another officer who had been following the action on his radio came out, made a U-turn, and pulled alongside the driver and signaled him to stop. The driver looked at the officer and accelerated rapidly. The officers pursued up to 55 mph as the driver sped toward downtown Laramie, but then backed off. The driver then hit the innocent victim.

The widow settled with the driver and pursued her case against the state, the patrol, and the officers. She proceeded under the Governmental Claims Act, alleging several acts of negligence. The allegations primarily were that in the circumstances the manner of the pursuit did not take into account the danger to the public, and various allegations of improper training and failure to establish guidelines to deal with this kind of situation. The defendants moved for summary judgment. The trial judge so granted, finding that the officers had qualified immunity in that they acted in good faith and reasonably. The court found that the state and the highway commission were immune as a matter of law, and that there was no dispute of material facts that would alter the decision, and thus no issue for a jury.

The first issue for the supreme court was whether the retention of common law defenses (§1-39-102(a)) was also intended to retain immunity defenses. If so, would that immunity trump the statutory exceptions to immunity, §§1-39-105 and 1-39-112? It was not clear just what the legislature intended. The court followed its precedents, primarily *Kimberly v. City of Green River*, 663 P.2d 871 (Wyo. 1983) (false arrest claim), in holding that common law immunity survived the passage of the Governmental Claims Act. Second, the court held that, following the doctrine of the state of Washington (see *Mason v. Britton* in the Washington section of this book), the decision to pursue a suspected law violator does not involve the high level of discretion required to trigger such immunity. However, the officers here were not negligent. Their duty to catch and stop drunk drivers must be upheld. The officer is responsible only for the negligent operation of his own vehicle. In the facts of this case, because the officer did not collide with the plaintiff's car, the officer can be liable only if his conduct amounts to gross or wanton conduct amounting to bad faith. The officer should not be liable either for his decision to pursue or not to pursue. The defendant officers here were not the proximate cause. The causation is too strained and remote. If the officers were not negligent, neither can their employers be liable. A dissent thought that the issue of good faith was properly a jury question.

The second case is *Teton County Sheriff's Dept. v. Bassett,* 8 P.3d 1079 (Wyo. 2000). The case involved a pursuit and a roadblock. Highway patrol officers were in pursuit of Ortega, who was wanted in two jurisdictions and was considered armed and dangerous. The pursuit was on Highway 89 north of Jackson, and it was high-speed. Ortega repeatedly swerved toward oncoming traffic in apparent attempts to cause a crash that would impede the pursuit. The officers called for a roadblock to be set up at Moran Junction, but sheriff's deputies, who had been called upon to assist, decided to set it up farther south at the junction near Antelope Flats Road. They hoped that Ortega would turn off the highway to avoid the roadblock and run over spikes that would flatten his tires. Meanwhile, the plaintiffs, en route home from fishing, turned onto Highway 89 south at Moran Junction. There were several police vehicles parked along the road, and none of the officers warned the plaintiffs of the dangerous situation rapidly developing. The plaintiffs came to the roadblock, and "surprised officers began frantically gesturing for them to go through as a deputy sheriff moved his car for their passage." The plaintiffs understandably slowed to 30 mph and went through the roadblock. Ortega came up a 100 mph or more, went through the same opening, and crashed into them. They were injured and Ortega was caught and arrested.

The jury at trial found the plaintiffs not at fault, and allocated fault as 40 percent highway patrol, 20 percent sheriff's officers, and 40 percent National Park Service, not a party. (The case does not indicate why the jury found the National Park Service at fault, but it must have been because Park Service officers were parked along the highway being briefed by a highway patrol officer when the plaintiffs pulled onto Highway 89, and failed to warn or stop the plaintiffs.) Ortega's name was not placed on the verdict form.

There were three issues on appeal to the supreme court. Should Ortega's name have been placed on the verdict form as an "actor," in accordance with the comparative fault act, §1-1-109(c)? Should the jury have been instructed that the plaintiffs can recover damages only if the officers' conduct was extreme or outrageous in a case in which the fleeing suspect hit the plaintiff's car? And was Patrolman Wilson (who did not attempt to prevent plaintiffs from turning onto the highway) entitled to qualified immunity?

Section 1-1-109 defines "actor" as a person or entity whose fault is determined to be a proximate cause whether or not the actor is a party to the litigation. The trial court did not put Ortega's name on the verdict form because his acts were willful and wanton or intentional, and could not be compared to the merely alleged negligent conduct of the defendants. The court noted that in 1994 the legislature changed the word "negligence" to "fault" in the relevant part of the statute. The plaintiff argued that legislative history shows that the words, "reckless," "wanton," culpable," and "intentional" were struck from Senate File No. 35, which became the language of the amendment. Thus, the plaintiff argued, the legislature did not intend to apply the stricken words to the comparative fault act. The court did not agree, and held that "fault" was intended to include all culpable conduct. Ortega's name must be put on the verdict form in the new trial, and his fault assessed.

The court also reversed because the trial judge refused to instruct that the officers could be at fault only if their conduct was extreme and outrageous. The court said that *DeWald* above must be followed. *DeWald* held that when the plaintiff is hit by the pursued driver, the officers involved in the chase are negligent only if their actions were extreme and outrageous and were reasonably a proximate cause.

As to qualified immunity, the court held, as above, that only high-level discretionary actions at the executive level were immune, and the trial court correctly refused to grant the state's motion for judgment as a matter of law.

In summary, in pursuit cases in which the police do not collide with the plaintiff, they are not liable unless their actions were a proximate, not remote, cause, and their conduct was outrageous and extreme. We can presume this is similar to the standards of gross negligence, reckless disregard, or willful and wanton conduct. While there are no appellate cases so far, perhaps we can presume that when a police vehicle collides with the plaintiff, the officer is will be held to a due care standard. Then perhaps the emergency vehicle statute, §31-5-106 would apply. That statute apparently has never been construed.

Part IV

Appendices

Appendix I

Fire Apparatus Specifications: Pumper/Rescue

Cab/Chassis Specifications

F-1: One (1) <u>**CAB AND CHASSIS**</u>

The cab and chassis shall be a Spartan Gladiator, model GA42L, long four door, 20" raised roof, aluminum tilt cab, built specifically for the fire service by a publicly held U.S. parent company, specializing in chassis design for all fire service applications.

BIDDER COMPLIES WITH PREVIOUS PARAGRAPH: YES/NO

F-2: One (1) <u>**FRAME AND LINER**</u>

Side rails shall be channel type 10 1/4" × 3 1/2" × 3/8" with an inner channel 9 7/16" × 3 1/8" × 3/8" of 110,000 psi heat treated steel, an RBM of 3,437,500 in. lbs. and a section modulus of 31.25 cu.in.. A minimum of seven (7) fully gusseted bolted assembly crossmembers shall be installed using grade "8" flanged head bolts and flanged lock nuts.

The area between the axle suspension hangers shall be free of any holes or fasteners in the flanges. No welding shall be incorporated in attachment of components. All frame dimensional cutting shall be by a plasma cutter. All relief areas shall be cut in at a 30 degree angle with a two inch radius at intersection points with edges ground smooth to prevent a stress focal point.

The frame shall carry a lifetime warranty to the original purchaser.

BIDDER COMPLIES WITH PREVIOUS PARAGRAPH: YES/NO

F-3: One (1) <u>**FUEL TANK**</u>

The fuel tank shall have a capacity of 68 gallons, made of 12 gauge phosphate coated steel with chromate epoxy exterior finish. Fuel tank must be I.C.C. approved with certification plate attached. Tank shall be mounted under the frame, behind the rear axle on strap hangers with a "U" strap bolted front and rear so the tank can be easily dropped and removed. Tank shall have vent port to facilitate rapid filling without "blow-back." I.C.C. roll over ball check vent shall be installed. Dual draw tubes and dual sender ports shall be installed. 2" NPT fill ports shall be available for right or left hand fill. A 1/2" NPT drain plug shall be centered in the bottom of the tank. Rubber reinforced supply and return hoses with reusable fittings shall be installed tank to engine.

BIDDER COMPLIES WITH PREVIOUS PARAGRAPH: YES/NO

F-4: One (1) <u>BUMPER EXTENSION</u>

The front bumper shall be one piece, polished stainless steel, 12" high, two rib wrap-around, extended 24" ahead of the cab.

BIDDER COMPLIES WITH PREVIOUS PARAGRAPH: YES/NO

F-5: One (1) <u>TOW HOOKS</u>

Two (2) heavy duty chrome plated tow hooks shall be installed under the bumper and bolted directly to the chassis frame with grade "8" bolts.

BIDDER COMPLIES WITH PREVIOUS PARAGRAPH: YES/NO

F-6: One (1) <u>AIR HORNS</u>

Dual Grover Stuttertone 24" air horns shall be installed recessed in the front bumper 1 right and 1 left of the center trash line well. A 3/8" air line "teed" equal distance from each horn shall be installed. An additional 1200 cu. in. reservoir with a 90 psi protection valve shall be installed to prevent depletion of brake air system. Actuation shall be accomplished by horn button on steering wheel and a right hand officer's foot switch.

BIDDER COMPLIES WITH PREVIOUS PARAGRAPH: YES/NO

F-7: One (1) <u>FRONT AXLE</u>

The front axle shall be a Rockwell FL-943 with a 5" drop and shall have a GAWR of 21,500 lbs.

BIDDER COMPLIES WITH PREVIOUS PARAGRAPH: YES/NO

F-8: One (1) <u>FRONT TIRES AND WHEELS</u>

Front tires shall be Goodyear 425/65R 22.5 18 ply "J" tubeless radial G-286 highway tread with 22.5 × 12.25, 10 stud disc wheels. Tires and wheels shall be rated at 20,000 lbs.

BIDDER COMPLIES WITH PREVIOUS PARAGRAPH: YES/NO

F-9: One (1) <u>OIL LUBRICATED FRONT WHEEL BEARINGS</u>

Front axle wheel bearings shall be oil lubricated with an oil level visual inspection window.

BIDDER COMPLIES WITH PREVIOUS PARAGRAPH: YES/NO

F-10: One (1) <u>FRONT SHOCK ABSORBERS</u>

Monroe # 70150 double acting 2-3/4" diameter shock absorbers shall be installed for the front axle.

BIDDER COMPLIES WITH PREVIOUS PARAGRAPH: YES/NO

F-11: One (1) <u>STEERING</u>

A Ross TAS-65 hydraulic power assist steering gear shall be provided. The hydraulic power steering pump shall be gear driven from the engine. The steering ratio shall be 20.4:1 and have 5.4 turns stop-to-stop.

BIDDER COMPLIES WITH PREVIOUS PARAGRAPH: YES/NO

F-12: One (1) <u>STEERING COLUMN AND WHEEL</u>

The steering column shall be a 7 position tilt and 2.25" telescopic type with a 20" steering wheel. The steering wheel shall be covered with black absorbite padding. The steering column shall contain a horn button, self-canceling turn signal switch, four-way hazard switch and head lamp dimmer switch on turn lever.

BIDDER COMPLIES WITH PREVIOUS PARAGRAPH: YES/NO

F-13: One (1) DISC BRAKES

The front axle shall have Rockwell ADB-1560 disc type brakes with 15" vented rotors providing a total of 120 sq.in. of braking area and automatic slack adjusters.

BIDDER COMPLIES WITH PREVIOUS PARAGRAPH: YES/NO

F-14: One (1) REAR AXLE

The rear axle shall be a Rockwell RS-26-180 with single reduction gearing and shall have a special fire service rated capacity of 27,000 Lbs. GAWR. The gear ratio shall produce a maximum speed of **60 mph** at engine governed rpm.

The bidder to supply the axle ratio at time of bid, the purchaser is estimating a 5.86:1 ratio

BIDDER COMPLIES WITH PREVIOUS PARAGRAPH: YES/NO

F-15: One (1) ABS BRAKES

A Rockwell "WABCO" anti-lock braking system shall be installed on Rockwell axles. A dash mounted anti-lock lamp shall be provided to notify driver of a system malfunction. A momentary test switch shall be installed to test the system for diagnostic codes.

BIDDER COMPLIES WITH PREVIOUS PARAGRAPH: YES/NO

F-16: One (1) DIFFERENTIAL LOCK

The rear axle shall have a driver controlled differential lock. This feature shall allow the main differential to be locked and unlocked when the vehicle is stationary or during low speed operation to provide maximum wheel end traction.

BIDDER COMPLIES WITH PREVIOUS PARAGRAPH: YES/NO

F-17: One (1) REAR TIRES AND WHEELS

Rear tires shall be Goodyear Unisteel 12R-22.5 16 ply "H" tubeless radial G124 mud and snow tread with 22.5 × 8.25, 10 stud disc wheels. Tires and wheels shall be rated at 27,000 lbs.

BIDDER COMPLIES WITH PREVIOUS PARAGRAPH: YES/NO

F-18: One (1) OIL LUBRICATED REAR WHEEL BEARINGS

The rear axle shall have oil lubricated rear wheel bearings.

BIDDER COMPLIES WITH PREVIOUS PARAGRAPH: YES/NO

F-19: One (1) REAR SUSPENSION

The rear suspension shall be vari-rate self-leveling captive slipper type, 57.5" × 3" spring with one adjustable and one fixed torque rod and without shackles. The spring capacity must meet or exceed the capacity of the rear axle.

BIDDER COMPLIES WITH PREVIOUS PARAGRAPH: YES/NO

F-20: One (1) REAR BRAKES

Rear brakes shall be 16 1/2" × 7" "S" cam type with automatic slack adjuster provided by the axle manufacturer.

BIDDER COMPLIES WITH PREVIOUS PARAGRAPH: YES/NO

F-21: One (1) BRIGHT WHEEL TRIM

The front and rear axle wheels shall have bright nut covers. Bright front hub caps with hole to view oil seal window and bright rear axle top hat shall be provided.

BIDDER COMPLIES WITH PREVIOUS PARAGRAPH: YES/NO

F-22: One (1) AIR BRAKE SYSTEM FOR SINGLE REAR AXLE

In compliance with NFPA standards a rapid air build-up air brake FMVSS-121 system shall be provided. It shall include air reservoirs with a total of 4136 cu.in. air capacity. A parking brake on the spring actuated chambers on the rear axle brakes with a push-pull valve on the instrument panel shall be installed. The rear axle spring brakes are to automatically apply in case of air pressure drop below 60 psi with a mechanical means for releasing the spring brake chambers.

BIDDER COMPLIES WITH PREVIOUS PARAGRAPH: YES/NO

F-23: One (1) AIR DRYER

A Midland spin-on desiccant model DA33100 air dryer with an automatic heated moisture ejector shall be installed in the air brake system.

BIDDER COMPLIES WITH PREVIOUS PARAGRAPH: YES/NO

F-24: One (1) MANUAL DRAINS ON AIR TANKS

Manual drains shall be installed on all three reservoirs of the two tank system with lines extended to edge of body.

BIDDER COMPLIES WITH PREVIOUS PARAGRAPH: YES/NO

F-25: One (1) NYLON AIR LINE TUBING

A dual air system plumbed with reinforced nylon tubing air lines shall be installed.

BIDDER COMPLIES WITH PREVIOUS PARAGRAPH: YES/NO

F-26: One (1) ADDITIONAL AIR RESERVOIR

An additional 1200 cu.in. air reservoir shall be installed. it shall be isolated and be plumbed with a pressure protection valve on the reservoir supply side.

BIDDER COMPLIES WITH PREVIOUS PARAGRAPH: YES/NO

F-27: One (1) AIR COMPRESSOR

The air compressor on the engine shall be a Holset model QE-296 capable of producing 16.5 CFM at 120 psi tank pressure and 2100 engine rpm. It shall be gear driven, engine oil pressure lubricated and cooled by the engine cooling system. The air compressor shall have a 5-year warranty.

BIDDER COMPLIES WITH PREVIOUS PARAGRAPH: YES/NO

F-28: One (1) ENGINE

The engine shall be a Cummins Celect Plus electronic M11-350E developing 350 hp at 1800 rpm, Gov. at 2100 rpm, with 1350 lb.ft. of torque at 1200 rpm. It shall be an in-line,

6 cylinder, 4 cycle, turbocharged, charge air cooled engine with a displacement of 661 cu.in.

A wiring harness with connectors extending to the pump panel area with circuits provided for a tachometer, oil pressure and engine temperature gauges shall be provided.

A spin-on engine coolant filter with shut-off valve shall be provided.

A secondary spin-on by-pass type oil filter shall be part of the engine's lubrication system.

BIDDER COMPLIES WITH PREVIOUS PARAGRAPH: YES/NO

F-29: One (1) EXHAUST SYSTEM

The exhaust system shall be installed under the frame with the discharge to the **left side**
forward of the rear tires. A muffler and .065 wall aluminized steel exhaust tubing supported by bolted on frame brackets shall be installed. Stainless steel flex tubing is to be installed between exhaust pipe and muffler. System joints shall be connected with lapping band clamps.

BIDDER COMPLIES WITH PREVIOUS PARAGRAPH: YES/NO

F-30: One (1) TAIL PIPE CHROME EXTENSION

A radiused chrome exhaust tail pipe extension, to direct exhaust gases downward, which terminates on the left hand side of the chassis, forward of the rear tires, shall be installed with a galvaneal band clamp.

BIDDER COMPLIES WITH PREVIOUS PARAGRAPH: YES/NO

F-31: One (1) AIR CLEANER

The air cleaner shall be dry type with a replaceable element, it shall have an outside air intake with an ember separator filter and an instrument panel air restriction gauge to show when the air cleaner element requires replacement.

BIDDER COMPLIES WITH PREVIOUS PARAGRAPH: YES/NO

F-32: One (1) COOLING SYSTEM

The cooling system shall have sufficient capacity to keep the engine properly cooled under all conditions of road and pumping operations. The radiator shall have bolted top and bottom tanks, the top tank shall be full deaeration with sight glass and the bottom shall have an integral transmission oil cooler with a drain cock. The cooling system shall contain a 50/50 ethylene glycol/water mix suitable for operation to -34 degrees F and shall be shutterless with rapid warm-up features. All radiator hoses shall be silicone. All radiator hose clamps shall be stainless steel.

The system shall incorporate a heavy duty fan, installed on the engine and includes a shroud. Recirculation shields shall be installed to ensure that air which has passed through the radiator is not drawn through it again.

BIDDER COMPLIES WITH PREVIOUS PARAGRAPH: YES/NO

F-33: One (1) ENGINE-PUMP HEAT EXCHANGER

A single bundle type heat exchanger, shall be provided so that water from the pump does not come in contact with the engine coolant to allow the use of water, plumbed without a shut-off valve, from the discharge side of the pump for cooling the engine coolant. The heat exchanger is to be installed between the engine and the radiator.

BIDDER COMPLIES WITH PREVIOUS PARAGRAPH: YES/NO

F-34: One (1) **<u>HIGH IDLE SWITCH FOR ELECTRONIC ENGINE</u>**

A momentary rocker switch shall be installed in the rocker switch panel to provide a preset elevated engine speed. The high idle shall operate only with the park brake on and transmission selector in neutral.

BIDDER COMPLIES WITH PREVIOUS PARAGRAPH: YES/NO

F-35: One (1) **<u>FUEL WATER SEPARATOR</u>**

A Fleetguard FS1242

BIDDER COMPLIES WITH PREVIOUS PARAGRAPH: YES/NO

F-36: One (1) **<u>TRANSMISSION</u>**

The transmission shall be an Allison MD 4060PR 4 speed/5 speed automatic with electronic controls and an output retarder modulated via release of the throttle pedal and an additional 1/3 via modulation of the service brake pedal. The transmission shall be plumbed to an external cooler in the engine-to-radiator coolant hose in a full flow coolant system application to insure adequate cooling. The transmission shall have two internal oil filters. The pressure sensitive range selector touch pad is to be to the right of the driver with mode legends back lighted for night use. Fourth gear hold-in range may be accomplished by wiring for a pumping application. The transmission gear ratios shall be: 1st 3.51:1, 2nd 1.91:1, 3rd 1.43:1, 4th 1.00:1, 5th .74:1, rev 4.80:1.

BIDDER COMPLIES WITH PREVIOUS PARAGRAPH: YES/NO

F-37: One (1) **<u>DRIVE LINES</u>**

All drive lines shall be 1710 heavy duty series with "glide coat" spines on all slip shafts.

BIDDER COMPLIES WITH PREVIOUS PARAGRAPH: YES/NO

F-38: Two (2) **<u>DRIVELINE GUARD</u>**

A strap shall be installed, hanging from a crossmember, to capture a failed driveline shaft.

BIDDER COMPLIES WITH PREVIOUS PARAGRAPH: YES/NO

F-39: One (1) **<u>ELECTRICAL SYSTEM</u>**

A single starting system shall be installed per NFPA 1901. The electrical system shall be 12 volt, suppressed per SAE J551 with (6) Delco BCl-31 625 CCA batteries, 3/0 welding type dual path starter cables per SAE J541, appropriate gauge cross link 311 deg. F. insulation chassis wiring circuit numbered with legends and 250 deg. F. minimum high temperature flame retardant loom.

The following shall be supplied:

- heavy duty battery disconnect switch
- keyless ignition switch
- starter button
- push/pull headlight switch with turn actuated instrument lamp rheostat
- windshield wiper switch, 2 speed intermittent with integral washer switch
- self-canceling steering column mounted turn signal switch with integral hazard and head lamp dimmer switch

- twelve backlit rocker switches and legends for accessories
- master rocker switch, installed below eye level to the right of the instrument panel
- thermal reset breakers and relays behind dash panel

Instrument panel includes:
- electronic tachometer
- electronic speedometer with trip odometer (non-registering in pump mode)
- air filter restriction gauge
- air pressure 1 gauge
- air pressure 2 gauge
- voltmeter gauge
- engine hourmeter gauge with oil pressure galley switch
- transmission oil temp. gauge
- engine oil pressure gauge
- engine coolant temp. gauge
- fuel level gauge

Red warning lamps for the following:
- low air 1
- low air 2
- low oil pressure
- high coolant temp.
- high trans. temp.
- park brake

Green lamps for:
- battery on
- ignition on
- directional indicator lamps

A blue high beam indicator lamp.
An audible tone warning alarm with lamp test switch for:
- low air system one
- low oil pressure
- high coolant temp.
- high trans. temp.

The interior lighting includes:
- dome lamp with spotlight over each step (door activated and individually switched)
- 6.5"L × 3"H clear light in the door panel dual swivel map light module with center dome light mounted in headliner over engine tunnel
- 4.5"L × 2.5'W × 1.75"H red flashing open door lamp overhead center

The exterior lighting includes:
- four rectangular halogen headlamps with "Daytime Running" lamp feature which will illuminate the headlamps to 80% brilliance when the ignition switch is in the on position

- round side turn signals and marker lamps five I.C.C. cab marker lights shall be installed on the cab face above the windshield.

An apparatus wiring panel shall be installed on the officer side bulkhead below dash.

A loose gauge package shall be supplied including:
- tachometer
- oil pressure gauge
- coolant temp gauge with senders installed and wiring to pump panel

BIDDER COMPLIES WITH PREVIOUS PARAGRAPH: YES/NO

F-40: One (1) 290 AMP NIEHOFF ALTERNATOR

A 290 amp 12 volt Niehoff alternator shall be provided. The alternator shall have positive threaded type belt tensioning adjustment.

BIDDER COMPLIES WITH PREVIOUS PARAGRAPH: YES/NO

F-41: One (1) CLASS I LOAD MANAGER – 12 VOLT

One (1) CLASS 1 voltage load manager shall be provided on the apparatus to automatically monitor the 12 volt electrical system and shall sequence the 12 volt lighting loads on and off at exact intervals. This unit shall be configured to control seven (7) electrical loads and shall be programmable to set the priority of each load.

The purchaser shall approve the load shedding sequence.

BIDDER COMPLIES WITH PREVIOUS PARAGRAPH: YES/NO

F-42: One (1) HALOGEN WARNING LIGHTS

Two halogen alternate flashing warning lights with red polycarbonate lens shall be installed on the front of the cab above the head lamps with an outer amber polycarbonate lens arrow directional signal.

BIDDER COMPLIES WITH PREVIOUS PARAGRAPH: YES/NO

F-43: One (1) CODE 3 – OSCILASER (RED)

One (1) CODE 3 Model #OL150 Oscilaser warning light(RED) shall be provided. The unit will have a 50 watt halogen bulb with an oscillating aluminum reflector. The oscilaser will be controlled from a switch located on the emergency light switch panel in the cab.

The location shall be in the center of the grille centered 44" from the ground.

BIDDER COMPLIES WITH PREVIOUS PARAGRAPH: YES/NO

F-44: One (1) HAND HELD SPOTLIGHT

An Optronics #KB-4003 hand-held spotlight shall be hung on a bracket mounted on the side of the engine tunnel next to the officer. It shall have a coil-cord, a momentary switch and a 400,000 candle power lamp.

BIDDER COMPLIES WITH PREVIOUS PARAGRAPH: YES/NO

F-45: One (1) MAP LIGHT

A Roxter goose neck style instrument panel map light with switch at base shall be installed on the right hand side of the dash panel.

BIDDER COMPLIES WITH PREVIOUS PARAGRAPH: YES/NO

F-46: One (1) BATTERY CONDITIONER

A Kussmaul battery conditioner with built-in isolator regulating equal voltage to all batteries shall be installed in the cab on the bulkhead in front of the officer. It is to be powered by a 20 amp 1 20V inlet receptacle located in the left hand step area.

BIDDER COMPLIES WITH PREVIOUS PARAGRAPH: YES/NO

F-47: One (1) KUSSMAUL AUTO-EJECT RECEPTACLE

A Kussmaul 20 amp auto eject electrical receptacle with cover and weather box shall be installed on the left side ahead of the front door. It shall automatically eject the plug when the starter button is depressed.

BIDDER COMPLIES WITH PREVIOUS PARAGRAPH: YES/NO

F-48: One (1) CLEAR DOME LAMP

A 7" clear dome lamp with switch shall be mounted in the headliner of the cab front to the left of the officer seat.

BIDDER COMPLIES WITH PREVIOUS PARAGRAPH: YES/NO

F-49: TWO (2) RED DOME LAMPS IN CREW AREA

Two 7" red dome lamps with switches shall be installed in the crew area headliner over the outer rear facing seats.

BIDDER COMPLIES WITH PREVIOUS PARAGRAPH: YES/NO

F-50: FOUR (4) CLEAR DOME LAMPS IN CREW AREA

Four 7" clear dome lamps with switches shall be installed in the crew area headliner , three to be installed in the walk area and one to be installed over the command console area.

BIDDER COMPLIES WITH PREVIOUS PARAGRAPH: YES/NO

F-51: One (1) AUXILIARY 12 VOLT RECEPTACLE

A auxiliary 12 volt receptacle(cigarette lighter) shall be provided in the cab dash on the officer side.

BIDDER COMPLIES WITH PREVIOUS PARAGRAPH: YES/NO

F-52: One (1) BACKUP ALARM

A backup alarm shall be installed at the rear of the chassis and must have a background noise adjustable output level not less than 97 dbs. It will automatically activate when the transmission is placed in reverse.

BIDDER COMPLIES WITH PREVIOUS PARAGRAPH: YES/NO

F-53: One (1)GLADIATOR LFD 20" RAISED ROOF TILT CAB

The cab shall be a Spartan Motors Gladiator LFD (long four door), 20" raised roof, aluminum tilt cab, capable of seating 10 firefighters.

The roof shall be raised 20" with look-out windows and a solid center area in the rear upper portion.

The cab shall have four side doors. The rear side doors shall be flush, .19" thick skin, 34.00" × 97.00" with an opening of 31.00" × 94.00" and an enclosed two step entrance with a lower step 22.00"W × 8.00"D and an inner step 23.75"W × 8.00"D.

The side rear door windows shall be roll-down type 27.50" × 26.00" (715 sq. in.) each.

The cab front doors shall be flush 43.00" × 77.00" full length .19" thick aluminum skin with hidden stainless steel door hinges with .375" pin and bright aluminum diamond plate covered inside steps. The step area shall be 10.5" × 32.5" with an 18" inner step height.

The front doors shall have a full roll down window 27.00" × 26.00" with a total glass area of 702 sq.in. each.

Push button exterior door latches, suitable for use with firefighter mittens, shall be provided on all cab side doors. Interior latches shall be flush paddle-type incorporated into a vacuum formed upper door panel.

The lower half of the inner door panel shall be covered with aluminum diamond plate.

The cab shall be of the Eurospace interior design allowing for easy communication inside the cab. The cab overall length shall be 141.50" with 67.50" from the centerline of the front axle to the back of the cab.

The rear cab wall shall be .125" thick aluminum and shall be covered with padded vinyl to match the headliner. The rear floor to the headliner shall be 75.75" high.

The seat riser for the rear facing seats shall be a minimum of 52.50" from the back wall.

The cab front skin and floor shall be .19" thick aluminum. The inside width shall be 88.00" and the front floor to the headliner above the driver and officer shall be 78.00" high.

All glass used in the cab shall be automotive tint. The windshield shall have a maximum of 2808 sq.in. area and be of the wrap-around design 52.00" wide and 27.00" height for maximum visibility. Left and right windshield shall use the same interchangeable glass. All cab windows shall be the same height as the windshield to provide full panoramic visibility.

Interior of the cab shall have a vacuum formed dash and insulated engine cover and headliner. Rubber covered grab handles shall be installed for entering and exiting the front of the cab with an 18" handle on the officer side and an 11" handle on the driver side. The cab floor shall be covered with a gray pebble grain, non-slip vinyl covered, foam backed, sound deadening, multi-layer insulating floor mat.

The engine cover shall be hinged to permit routine engine maintenance without the need of tilting the cab. The engine cover shall utilize a gas operated cylinder to lift and keep the cover open. The cover shall be of fiberglass sound deadening construction with a heavily insulated underside of multi-layer foam with a foil cover and retention pins. The top of the cover shall be vacuum formed.

A 57,600 btu front overhead heater and defroster, located between two padded sunvisors, shall be provided. A locking glove box, 4"H × 11 .75"W × 6.75"D shall be installed on the right hand side of the cab dash.

Intermittent parallel arm-type electric wipers with separate motors and electric powered "wet arm" type windshield washers shall be provided. Access to the wiper motors shall be through an access panel located on the front cab face.

A stainless steel front grille, 40.00"W × 30.00"H, with a minimum free air intake area of 488.8 sq.in. shall be installed on the cab front. Polished stainless steel side air vents, 7.00" × 23.00" with 87.6 sq.in. area of free air flow each, shall be installed on the cab sides.

Stainless steel trim, 1 2" high with upper and lower trim band affixed without holes and fasteners (equal in height of the front bumper), shall be installed on the bottom sides of the cab.

BIDDER COMPLIES WITH PREVIOUS PARAGRAPH: YES/NO

F-54: One (1) CAB TILT ACTUATION

The entire cab shall tilt 45 degrees (LO-PRO 30 degrees) to allow for easy maintenance of the engine and transmission.

The cab tilt actuation shall be with an electric over hydraulic lift pump with a control box on a pennant for safe visual operation. The lift system shall have an ignition interlock and red lock down indicator lamp which shall illuminate when holding "down" switch to indicate safe road operation. Two cab tilt cylinders shall be provided with velocity fuses in each cylinder port. The cab pivots shall be 1.90" ball and be anchored to frame brackets with 1.25" dia. studs.

Two spring loaded hydraulic hold down hooks (one on E/R cab) outboard of the frame shall be installed for holding the cab securely to the frame.

BIDDER COMPLIES WITH PREVIOUS PARAGRAPH: YES/NO

F-55: ONE (1) CREW CAB COMPARTMENTS

Two (2) exterior compartments shall be supplied, one each side, behind the crew doors. The compartments shall be located under the cab floor with access only from the exterior of the cab. Aluminum overlap compartment doors shall be supplied, vertically hinged on the rear side, with a D-ring slam spring loaded type latch and a compartment door retention chain.

BIDDER COMPLIES WITH PREVIOUS PARAGRAPH: YES/NO

F-56: One (1) AIR SUSPENDED DRIVERS SEAT

Drivers seat shall be high back SEATS INC. "911" 4-way air suspended type with air control valve located at center lower front of seat. The suspension mechanism shall be enclosed by a rubber bellows.

A lap belt and 3 point shoulder harness with automatic retractors shall be installed.

BIDDER COMPLIES WITH PREVIOUS PARAGRAPH: YES/NO

F-57: One (1) SCBA OFFICER SEAT

Officers seat shall be a SEATS INC. "911" SCBA style for the quick donning of an air pack. Seat back shall include a spring loaded hinged headrest and tank retention clips.

A lap belt and 3 point shoulder harness with automatic retractors shall be installed.

BIDDER COMPLIES WITH PREVIOUS PARAGRAPH: YES/NO

F-58: Two (2) SCBA OUTER FORWARD FACING FLIP-UP SEATS

A quantity of two, outer forward facing horseshoe style SEATS INC. "911" seat backs shall be installed on the back wall of the cab. A "911" seat bottom shall be installed on a flip-up seat base. The tank brackets and straps shall be installed by the body builder.

BIDDER COMPLIES WITH PREVIOUS PARAGRAPH: YES/NO

F-59: One (1) SCBA SEATS

The two outer rear facing crew area seat positions shall be SEATS INC. "911" SCBA type seats for the quick donning of air packs. Seat backs shall include spring loaded hinged headrest and tank retention clips.

A lap belt with an automatic retractor shall be installed on each seat.

BIDDER COMPLIES WITH PREVIOUS PARAGRAPH: YES/NO

F-60: Four (4) CREW AREA SHOULDER HARNESSES

The forward facing crew area seats shall be equipped with a black 3-point shoulder harness and lap belt assembly.

BIDDER COMPLIES WITH PREVIOUS PARAGRAPH: YES/NO

F-61: One (1) GRAY INTERIOR CAB TRIM

All cab interior vinyl surfaces, including seats, shall be gray. The pebble grain, non-slip vinyl covered, foam backed, sound deadening, multi-layer insulating floor mat shall be gray.

BIDDER COMPLIES WITH PREVIOUS PARAGRAPH: YES/NO

F-62: One (1) WINDSHIELD DEFOGGER FANS

Two individually switched all metal construction 6" windshield defogger fans shall be installed on overhead heater in center of cab directly in front of the windshield.

BIDDER COMPLIES WITH PREVIOUS PARAGRAPH: YES/NO

F-63: One (1) AIR CONDITIONING

Cab shall have a 40,000 BTU combination heater/air conditioning unit with the condenser installed ahead of the radiator. The evaporator with adjustable diffusers shall be mounted on the top-rear portion of the engine tunnel. The air conditioner shall have a temperature control and a three speed blower capable of circulating 750 cubic feet of air per minute. The heater shall be plumbed with a shut-off valve at the engine.

BIDDER COMPLIES WITH PREVIOUS PARAGRAPH: YES/NO

F-64: One (1) CAB SIDE WINDOWS

Fixed cab side windows, 1 6.00" × 26.00" (416 sq.in.) shall be installed behind the front cab doors. Each window shall be the same height as the windshield so as to provide maximum visibility.

BIDDER COMPLIES WITH PREVIOUS PARAGRAPH: YES/NO

F-65: One (1) ASSIST HANDLES

Four 18" knurled anti-slip one piece stainless steel exterior assist handles shall be installed, one next to each cab door.

BIDDER COMPLIES WITH PREVIOUS PARAGRAPH: YES/NO

F-66: One (1) REAR VIEW MIRRORS

Combination rear view mirrors, with flat 16" × 6" head and 7 1/2" lower convex mirrors, mounted on stainless steel bow swing away-type arms, shall be installed. All mirror heads shall be aluminum center mounted to produce less vibration.

BIDDER COMPLIES WITH PREVIOUS PARAGRAPH: YES/NO

F-67: One (1) WHEEL-WELL LINERS

Full width aluminum wheel well liners shall be installed, liners shall be 18" wide, with the outer edge polished.

BIDDER COMPLIES WITH PREVIOUS PARAGRAPH: YES/NO

F-68: One (1) CAB PAINTING

The cab shall be painted two tone with a break line below the cab windows through the mirror brackets and door handles.

All cab painting must be completed prior to the installation of glass, accessories or any other cab trim to assure complete paint coverage and maximum corrosion protection.

The entire cab must be disc ground to remove any surface oxidation or surface debris that may hinder the paint adhesion.

Upon the application of required body fillers and their preparation, the cab shall be primed with a coating designed for corrosion resistance and surfacer-paint adhesion. The entire cab then shall be coated with an intermediate surfacer that is designed to fill minor surface defects, provide an adhesive bond between the primer and the paint, and improve the color and gloss retention of the color coats. The cab shall then be finish sanded and painted with 2 to 4 coats of an acrylic urethane base-coat clear-coat type system designed not only for color retention but to resist acid rain and most atmospheric chemicals found on the fire ground or emergency scene. The cab interior shall be painted the same color as the exterior.

Since excessive chipping is not desired, the maximum overall film thickness shall not exceed 5 mils.

The paint system shall be warranted for 7 years against cracking, checking, or peeling and loss of gloss caused by chalking or fading.

Cab underside and doors shall be rustproofed with a 10 year or 100,000 mile warranty certificate against perforation issued in the Fire Department's name.

UPPER CAB SHALL BE DuPONT # 817

LOWER CAB SHALL BE DuPONT # 7744

BIDDER COMPLIES WITH PREVIOUS PARAGRAPH: YES/NO

F-69: Two (2) SERVICE MANUALS AND PARTS LIST

Two copies of the chassis service manual and parts list with wiring and air plumbing diagrams shall be provided with an engine and transmission service manual.

BIDDER COMPLIES WITH PREVIOUS PARAGRAPH: YES/NO

F-70: One (1) FIRE EXTINGUISHER

A 2.5 lb. BC DOT approved fire extinguisher shall be installed at drivers area in the cab.

BIDDER COMPLIES WITH PREVIOUS PARAGRAPH: YES/NO

F-71: One (1) DOT EMERGENCY FOLDING TRIANGLES

One set of DOT approved emergency triangle are to be supplied with the chassis.

BIDDER COMPLIES WITH PREVIOUS PARAGRAPH: YES/NO

END OF CAB/ CHASSIS SPECIFICATIONS

START FIRE BODY SPECIFICATIONS

F-72: One (1) <u>FRONT BUMPER OPEN TRASH LINE COMPARTMENT</u>

One (1) open trash line compartment constructed of .188" smooth aluminum shall be provided in the front bumper extension. Plastic Mateflex grating and drain holes shall be supplied in the floor of compartment.

The compartment shall hold 200' of 1.75" double jacket fire hose and a fog nozzle. Two (2) ZICO CLPR tool clips shall be installed to hold the nozzle. The hose and nozzle shall be supplied by the purchaser.

The location shall be in the center of the extended bumper. The purchaser shall approve the exact location of the mounting clips at a factory visit.

BIDDER COMPLIES WITH PREVIOUS PARAGRAPH: YES/NO

F-73: One (1) <u>FRONT MUD FLAPS</u>

The front tires shall have a set of black mud flaps mounted behind the front chassis wheels.

BIDDER COMPLIES WITH PREVIOUS PARAGRAPH: YES/NO

F-74: One (1) <u>REAR AXLE MUD FLAPS</u>

The rear tires shall have a set of black mud flaps mounted behind the rear chassis wheels.

BIDDER COMPLIES WITH PREVIOUS PARAGRAPH: YES/NO

F-75: Five (5) <u>SCBA BOTTLE BRACKET FOR SEAT BACK</u>

Five (5) ZIAMATIC SCBA air bottle brackets model UN-6-3O-2SF shall be mounted in the seat back with a pull release restraint Zico #CRS strap.

The Fire Department supplied SCBA equipment shall be SCOTT 2.2's.

BIDDER COMPLIES WITH PREVIOUS PARAGRAPH: YES/NO

F-76: One (1) <u>FIRE PUMP SPECIFICATIONS</u>

PUMP MODEL HALE PSD 1250 GPM

A HALE model PSD 1250 GPM single stage centrifugal pump shall be designed to be mounted on the chassis frame rails and shall be driven by a split shaft HALE "MG" gear box. The pump body and related parts shall be of fine grain alloy cast iron. All moving parts in contact with water shall be of high quality bronze or stainless steel.

The pump impeller shall be fine grain bronze, hand ground and individually balanced. Wrap-around labyrinth impeller clearance rings shall be bronze, easily renewable without replacing impellers or pump volute body. The pump shaft shall be heat treated stainless steel and shall be supported by deep groove ball bearings. Pump drive unit shall withstand up to 16,000 lbs. ft. of engine torque.

Both the input and output gearbox drive shafts shall be of heat treated chrome nickel steel and at least 2.75" in diameter. All drive and pump gears shall be cut spur design, made from chrome nickel steel. The gear box shall be provided with an oil cooler.

PUMP MOUNTING

The pump shall be bolted to steel angles that are bolted to the chassis side rails, using Grade 8 bolts. The fire pump shall be easily removable at rear of apparatus.

DRIVE LINE

Hollow-tube drivelines and universals shall be of a torque rating equal to or greater than the maximum net transmission output torque ratings. An "0' ring bolted drop guard shall be provided around drive shaft.

NO EXCEPTIONS ALLOWED TO PREVIOUS PARAGRAPH YES/NO

F-77: One (1) PRESSURE RELIEF VALVE

The HALE fire pump shall be equipped with a Model P-30 automatic pressure control device. The device shall be a bronze, variable pressure setting valve with ample capacity to prevent undue pressure rise as outlined in applicable NFPA standards. The relief valve shall be normally closed and shall open against pump pressure with a control light to signal the opening. In the event of relief valve control failure, the pump is to remain operational for the complete range of the pump's rated capacity without requiring the closing of any emergency or "IN CASE OF FAILURE" control valves.

The relief valve control wheel and indicator light shall be mounted on the operator's pump panel in the right rear compartment.

BIDDER COMPLIES WITH PREVIOUS PARAGRAPH: YES/NO

F-78: One (1) HALE PRIMING SYSTEM

A HALE model #ESP oil-less electrically driven priming pump shall be provided and will be a positive displacement vane type. One (1) priming control, located at the pump operator's panel in the right rear compartment, shall open the priming valve and start the priming motor.

BIDDER COMPLIES WITH PREVIOUS PARAGRAPH: YES/NO

F-79: One (1)HALE MIDSHIP MECHANICAL WATER SEAL

The HALE pump shaft shall have a high quality, self-adjusting, maintenance free mechanical seal.

BIDDER COMPLIES WITH PREVIOUS PARAGRAPH: YES/NO

F-80: One (1)THERMAL BYPASS COOLER SYSTEM

The pump shall be equipped with a HALE thermal bypass cooler valve that will automatically dump pump water through a .375" discharge line to the water tank when pump water temperature exceeds 120 degrees. The valve shall be equipped with an integral strainer and shall reset automatically.

BIDDER COMPLIES WITH PREVIOUS PARAGRAPH: YES/NO

F-81: One (1) PUMP SHIFT—AIR OPERATED

A HALE air operated pump shift shall be pneumatically controlled using a power shifting cylinder. The power shift control valve shall be mounted in the cab, and be labeled "PUMP SHIFT." The apparatus transmission shift control shall be furnished with a positive locking shift lever, preventing accidental shifting of the chassis transmission.

A GREEN indicator light shall be located in the cab, and be labeled "PUMP ENGAGED." The light shall not activate until the pump shift has completed its full travel into pump engagement position. second GREEN indicator light shall be located in the cab and be labeled OK TO PUMP." This light shall be energized when both the pump shift has been completed and the chassis automatic transmission has obtained converter lockup (4th gear lockup).

BIDDER COMPLIES WITH PREVIOUS PARAGRAPH: YES/NO

F-82: One (1) <u>INDICATOR LIGHT—PUMP SHIFT AND LOCKUP COMPLETED</u>

One (1) pump panel mounted GREEN indicator light shall be positioned above the throttle control and labeled "WARNING: DO NOT OPEN THROTTLE UNLESS LIGHT IS ON".

The light shall be energized when both the pump shift has been completed, chassis automatic transmission has obtained fourth gear lockup, and the chassis parking brake is set.

BIDDER COMPLIES WITH PREVIOUS PARAGRAPH: YES/NO

F-83: One (1) <u>SUPPLEMENTARY HEAT EXCHANGER COOLING SYSTEM</u>

The chassis manufacturer shall furnish a heat exchanger for supplementary engine cooling during fire pump operations. A valve mounted at the operator's panel shall open a fire pump water line allowing cooling water to the heat exchanger that is mounted in the engine radiator hose. The exchanger will allow cooling water from the fire pump to circulate around the engine radiator coolant without mixing or coming in direct contact with the engine coolant. An engraved metal label shall identify the valve and indicate the direction of opening and closing.

BIDDER COMPLIES WITH PREVIOUS PARAGRAPH: YES/NO

F-84: One (1) <u>PUMP CERTIFICATION TEST – 1250 GPM</u>

The pump shall be tested and certified by UL, as defined in applicable NFPA standards. The testing shall be conducted at the factory of the contractor and prior to delivery of the completed unit. A copy of the test results shall be supplied at delivery of the apparatus along with a stamped test plate attached to the pump operator's position.

The pump shall deliver the following rated capacities:

1250 GPM at 150 PSI net pump pressure
1250 GPM at 165 PSI net pump pressure
875 GPM at 200 PSI net pump pressure
625 GPM at 250 PSI net pump pressure

BIDDER COMPLIES WITH PREVIOUS PARAGRAPH: YES/NO

F-85: One (1) <u>PUMP PAINT</u>

The pump and attached fixed piping shall be painted.

BIDDER COMPLIES WITH PREVIOUS PARAGRAPH: YES/NO

F-86: One (1) <u>DISCHARGE MANIFOLD</u>

The discharge manifold shall be constructed with black iron square piping, sized to provide sufficient waterflow to required outlets. The manifold shall be fixture built and attached to the pump system with bolt flanges or Victaulic couplings.

The piping system shall utilize galvanized piping and Class I high pressure flexible piping with a burst pressure of 1000 PSI and 30 HG vacuum rating. All flexible high pressure hose couplings shall be stainless steel.

The fire pump, all piping, and associated equipment shall be hydrostatically tested to 250 PSI in conformance to applicable NFPA standards.

BIDDER COMPLIES WITH PREVIOUS PARAGRAPH: YES/NO

F-87: One (1) ELKHART INTAKE DUMP VALVE

One (1) ELKHART intake dump valve shall be provided and mounted on the suction side of the pump. The discharge of the dump valve shall be a minimum of 2.5" diameter and be plumbed to the curb side of the apparatus.

The discharge pipe shall terminate with a chrome 2.5" NST male adapter and will be labeled "INTAKE RELIEF VALVE DISCHARGE, DO NOT CAP."

The discharge valve pressure setting to be set at 110 p.s.i..

BIDDER COMPLIES WITH PREVIOUS PARAGRAPH: YES/NO

F-88: One (1) HALE MASTER DRAIN VALVE AND CLASS I BLEEDER VALVES

The HALE master pump drain assembly shall consist of a bronze master drain with a rubber disc seal, a universal joint and a handwheel control. The master drain shall provide independent ports for low point drainage of the fire pump and auxiliary devices. All gated inlets and discharges (1.5" and larger in size) shall have Class 1 quarter turn bleeder valves or automatic Class 1 drain valves.

BIDDER COMPLIES WITH PREVIOUS PARAGRAPH: YES/NO

F-89: One (1) ELKHART ELECTRIC OPERATED VALVES

All in-line and pump valves shall be ELKHART,ELECTRIC OPERATED, swing-out, full flow, ball valves ALL VALVES TO HAVE A 3/4" RATCHET OVER-RIDE.

BIDDER COMPLIES WITH PREVIOUS PARAGRAPH: YES/NO

F-90: Eight (8) ELKHART ELECTRIC VALVES WITH INDICATOR

Eight (8) ELKHART ELECTRIC VALVES WITH SWITCH INDICATORS LIGHTS shall be provided for noted valves.

2.0" front bumper trash line
3.0" tank suction line (tank to pump)
2.5" tank fill discharge (pump to tank)
2.5" discharge left rear
2.5" discharge left rear
2.5" discharge right rear
2.5" discharge right rear
3.0" discharge left rear

BIDDER COMPLIES WITH PREVIOUS PARAGRAPH: YES/NO

F-91: One (1) ELKHART ELECTRIC OPERATED VALVE CONTROL

One (1) ELKHART electric operated valve control shall be provided with the following:

A. 12 volt DC motor
B. A toggle switch and indicator lamp assembly for each valve, mounted on pump operator's panel
C. RED, YELLOW, GREEN valve position indicator lights
D. Valve open and close "AUTO TRAVEL"
E. Manual valve actuation

Controls provided for the following valves:

1. The 4.0" rear suction.

BIDDER COMPLIES WITH PREVIOUS PARAGRAPH: YES/NO

F-92: One (1) <u>ELKHART ELECTRIC OPERATED VALVE CONTROL</u>

One (1) ELKHART electric operated valve control shall be provided with the following:

A. 12 volt DC motor

B. A toggle switch and indicator lamp assembly for each valve, mounted on pump operator's panel

C. RED, YELLOW, GREEN valve position indicator lights

D. Valve open and close "AUTO TRAVEL"

E. Manual valve actuation

Controls provided for the following valves:

1. The 6.0" center rear suction

BIDDER COMPLIES WITH PREVIOUS PARAGRAPH: YES/NO

F-93: One (1) <u>4.0" PUMP INTAKE – LEFT REAR OF APPARATUS</u>

One (1) 4.0" diameter inlet shall be piped to the rear of the apparatus body. to the left of the rear 6" intake. An inline full flow ball valve shall be provided for the inlet with an ELKHART electric operator and shall be controlled at the pump operators panel.

The inlet shall be furnished with a 4.0" nst male × 4.0" Storz adapter with a removable screen.

A class I bleeder valve shall be installed

BIDDER COMPLIES WITH PREVIOUS PARAGRAPH: YES/NO

F-94: One (1)<u>6.0" CENTER REAR INLET</u>

One (1) 6.0" diameter inlet shall be piped to the center rear of the apparatus body. The inlet shall terminate with a 6.0" NST male chrome plated adapter with a removable screen. The valve shall be operated by a control on the pump operators panel.

BIDDER COMPLIES WITH PREVIOUS PARAGRAPH: YES/NO

F-95: One (1)<u>6.0" DIAMETER ELKHART BUTTERFLY VALVE</u>

A 6.0" diameter ELKHART butterfly valve shall be provided on the apparatus. The valve shall be operated by a control on the pump operators panel.

The valve controller shall be ELKHART electric.

The valve shall be located on the 6.0" rear suction.

BIDDER COMPLIES WITH PREVIOUS PARAGRAPH: YES/NO

F-96: One (1) <u>6.0" SUCTION CAP</u>

One (1) Chrome long handle 6.0" diameter pressure rated cap shall be provided with NST threads.

BIDDER COMPLIES WITH PREVIOUS PARAGRAPH: YES/NO

F-97: One (1) <u>3.0" TANK SUCTION CONNECTION</u>

The tank suction line shall be provided with a 3.0" diameter ball valve with 3.0" piping and flex hose with stainless steel hose clamps. To prevent pressurization of the water tank there shall be a swing check valve located in the plumbing. A tank to pump water flow test shall be completed at time of pump testing. The flow rate shall be per applicable NFPA standards.

The valve shall be controlled with a ELKHART electric valve.

BIDDER COMPLIES WITH PREVIOUS PARAGRAPH: YES/NO

F-98: One (1) TANK FILL DISCHARGE (2.5")

A swing-out, 2.5" full flow ball valve shall be furnished in the pump-to-tank line. A 2.5" diameter Class 1 flexible hose shall connect the pump to the tank. The tank fill discharge shall be controlled from the operators pump panel with a ELKHART electric valve and will have an engraved metal label located next to the control.

BIDDER COMPLIES WITH PREVIOUS PARAGRAPH: YES/NO

F-99: One (1)2.0" DISCHARGE, FRONT OF APPARATUS

One (1) 2.0" diameter discharge shall be piped to the front of the apparatus, terminating with a 1.5" NST male thread × 2" NPT female thread chicksan swivel. An incline full flow ball valve shall be provided for the discharge and shall be controlled at the pump operator's panel.

A class I automatic bleeder valve shall be furnished for the discharge.

The final location of the front trash line shall be approved by the purchaser at a preconstruction conference.

BIDDER COMPLIES WITH PREVIOUS PARAGRAPH: YES/NO

F-100: Two (2) 2.5" LEFT REAR BODY DISCHARGE

Two (2) 2.5" diameter discharges shall be piped to the left rear of the apparatus, terminating with a chrome plated 30 degree 2.5" NST male adapter. An in line, full flow ball valve shall be provided and shall be controlled with a ELKHART electric valve controller at the pump operators panel. A .75" bleeder valve shall be furnished for the discharge.

BIDDER COMPLIES WITH PREVIOUS PARAGRAPH: YES/NO

F-101: Two (2) 2.5" RIGHT REAR BODY DISCHARGE

Two (2) 2.5" diameter discharges shall be piped to the right rear of the apparatus, terminating with a chrome plated 30 degree 2.5" NST male adapter. An inline, full flow ball valve shall be provided and shall be controlled with ELKHART electric valve controller at the pump operators panel. A .75" bleeder valve shall be furnished for the discharges.

BIDDER COMPLIES WITH PREVIOUS PARAGRAPH: YES/NO

F-102: One (1) 3.0" RIGHT REAR BODY DISCHARGE

One (1) 3.0" diameter discharge shall be piped to the right rear of the terminating with a chrome plated 30 degree 3.0" NST male elbow adapter. An inline, full flow ball valve shall be provided and shall be controlled with a ELKHART electric valve controller at the pump operator's panel. A .75" bleeder valve shall be furnished for each discharge.

BIDDER COMPLIES WITH PREVIOUS PARAGRAPH: YES/NO

F-103: One (1) STAINLESS STEEL PUMP PANEL

The pump panel shall be brushed 14 gauge #304 stainless steel and shall be installed in the rear compartment on the curb side of the apparatus with 3/8" stainless~steel bolts and locking nuts with teflon inserts. The upper instrument panel shall have a full width stainless steel continuous hinge that shall swing down for easy access to the interior of the pump enclosure.

Engraved, color coded, Vision Mark metal labels shall be furnished for the discharges, intakes, and the controls and indicators.

BIDDER COMPLIES WITH PREVIOUS PARAGRAPH: YES/NO

F-104: One (1) PUMP PANEL LIGHTS

Four (4) Weldon #2025 lights with CLEAR lenses shall be installed under the right side panel light hood. The right side pump panel lights shall be switched with the operators pump panel light switch.

BIDDER COMPLIES WITH PREVIOUS PARAGRAPH: YES/NO

F-105: One (1) PUMP ENCLOSURE LIGHT

Two (2) KD #5283344 lights, (with switch in enclosure) shall be provided in the pump enclosure.

BIDDER COMPLIES WITH PREVIOUS PARAGRAPH: YES/NO

F-106: One (1) MASTER PUMP DISCHARGE GAUGE

One (1) 4.5" diameter Class 1 liquid filled (30-0-600 PSI) pressure gauge with an engraved, color coded, metal label, "PUMP PRESSURE," shall be provided on the pump instrument panel. The face of the gauge shall be white, with black letters.

BIDDER COMPLIES WITH PREVIOUS PARAGRAPH: YES/NO

F-107: One (1) MASTER PUMP INLET GAUGE

One (1) 4.5" diameter Class 1 liquid filled (30-0-600# PSI) compound gauge with an engraved, color coded, metal label, "PUMP INTAKE" shall be provided on the pump instrument panel. The face of the gauge shall be white, with black letters.

BIDDER COMPLIES WITH PREVIOUS PARAGRAPH: YES/NO

F-108: Seven (7) 3.5" LINE DISCHARGE GAUGES

Seven (7) 3.5" diameter Class 1 liquid filled (0-600 PSI) pressure gauges with an engraved, color coded, metal label shall be provided on the pump instrument panel. One (1) gauge shall be supplied for all discharges 1.5" or larger. The face of the gauges shall be white, with black letters.

BIDDER COMPLIES WITH PREVIOUS PARAGRAPH: YES/NO

F-109: Two (2) FLOW AND PRESSURE INDICATOR

Two(2) Class 1 "Flowminder Plus" , pressure and GPM indicators that are designed for 2.5" diameter plumbing and reads between 60 and 600 gallons per minute shall be provided. The "flowminder Plus" shall show GPM or water pressure on .625" incandescent digits located on the units display. The indicator will also have a memory that shall display the total gallons of water that has flowed through the monitored discharge.

The flow meters shall be installed on the left rear and right rear outside 2.5" discharges.

BIDDER COMPLIES WITH PREVIOUS PARAGRAPH: YES/NO

F-110: One (1) I.C. ENGINE MONITOR SYSTEM WITH ALARM

The apparatus shall be equipped with an Innovative Controls engine monitor system with audible alarm and RED warning lights. The system shall be mounted on the pump operators panel and will include:

1. One (1) 2" engine coolant temp. gauge (100–230 degrees F)
2. One (1) 2" engine oil pressure gauge (0–100 PSI)
3. One (1) 3" engine tachometer
4. One (1) audible alarm (Engine monitor shall be wired through pump shift circuit)

5. Two (2) RED warning lights

BIDDER COMPLIES WITH PREVIOUS PARAGRAPH: YES/NO

F-111: One (1) CLASS 1 ELECTRIC VERNIER THROTTLE –PUMP PANEL

The apparatus shall be equipped with a Class 1 electric throttle system. This system, specifically designed for fire apparatus, shall monitor and control the engine providing power to the fire pump.

The system shall control the engine speed when the pump system has been placed into gear. The system shall monitor engine RPM's and shall maintain the engine's selected speed.

An interlock system shall be provided to prevent the advancement of the engine speed until the apparatus parking brake is applied, the chassis transmission is in the proper gear, and the fire pump gearbox is properly engaged. When the above conditions have been met, the "OK TO PUMP" light shall be illuminated.

BIDDER COMPLIES WITH PREVIOUS PARAGRAPH: YES/NO

F-112: One (1) TEST TAPS FOR GAUGES

Test taps for pump intake and pump pressure shall be provided on the pump panel and be properly labeled.

BIDDER COMPLIES WITH PREVIOUS PARAGRAPH: YES/NO

F-113: One (1) BOOSTER TANK LEVEL GAUGE

One (1) MC Products "Fluidmeter" four (4) light indicator shall be located on the pump operators panel and shall monitor the booster tank level.

BIDDER COMPLIES WITH PREVIOUS PARAGRAPH: YES/NO

F-114: One (1) AIR HORN PUSH-BUTTON

An air horn control button with metal label shall be provided on the pump operators panel.

BIDDER COMPLIES WITH PREVIOUS PARAGRAPH: YES/NO

F-115: One (1) MIC BOX

A mic box that is approximately 12" high × 18" wide × 18" deep with a spring close door shall be provided. The door shall have a brush finish. The mic box shall be mounted in the pump operator's area. One (1) 110 VAC duplex outlet wired to the shore power system shall be supplied in the box to power a radio charger. The purchaser shall approve the final design and exact location at a factory visit.

BIDDER COMPLIES WITH PREVIOUS PARAGRAPH: YES/NO

F-116: One (1) 1000 GALLON FIBERGLASS BOOSTER RECTANGLE TANK (6" OVERFLOW)

FIBERGLASS BOOSTER WATER TANK SPECIFICATIONS

The apparatus shall be equipped with a Custom Fiberglass Products, rectangular shaped fiberglass water tank. The tank shall carry a CFP lifetime warranty.

The tank shall be constructed using laminated fiberglass, with .375" thick floor and .3125" to .375" thick baffles, bulkheads and sides. The exterior coating shall be of WHITE jelcoat, while the final interior coating shall be polyester resin. There shall be a 1" radius on all longitudinal corners. The longitudinal and transverse tank baffles shall be designed to provide maximum water and air movement between compartments.

The .375" thick tank cover shall be recessed type, solid laminate, and fully removable. The tank cover gasket shall be closed-cell neoprene and run around the full perimeter. The tank cover shall be bolted into place with stainless steel bolts on 4" centers around the perimeter of the lid and 8"centers on all transverse baffles.

The bottom of the tank shall be secured within vertically mounted angles. Cross members shall be placed across the tank top and bolted to a longitudinal angle which shall be welded to the inside of the body. The water tank fill tower shall be a minimum of 10" square, as high as hose bed sides, and equipped with a spring loaded lid with a removable nonferrous screen.

The water tank overflow pipe shall be constructed from 4" Schedule 40 PVC pipe. The overflow shall be piped to exit behind rear axle. The water tank sump shall be a minimum of 10" square × 3" deep and located near the front of the tank.

The water tank to pump connection shall be fitted with a flexible hose connection to tank suction line. The water tank drain connection shall be 3" NPT Schedule 80 female nipple with plug and located at the bottom of the sump. The pump to tank refill connection shall be fitted with a deflector shield inside tank.

Tank shall be completely removable from the apparatus body without disturbing the body. The tank shall be secured in place with bolted in crossbars above the tank. The tank will rest on cushions made of nitrile impregnated non-rotting anti-chaffing insulator on all surfaces where tank is in contact with mounts and supports.

F-117: One (1) STAINLESS STEEL BODY CONSTRUCTION

BODY MOUNTING

Prior to the construction of this vehicle, various chassis attachments such as battery boxes, air reservoirs, mufflers, tail pipes, filters, and other bolt-on frame attachments may be relocated to permit full utilization of the chassis for equipment compartments.

The attachment of the apparatus body shall be at a minimum of four (4) points along the chassis frame rails. The rear mounting shall be easily accessible bolts through the body substructure caps and the chassis frame rails. Forward mounting points shall be grade 8 U-bolts around the chassis frame and provided with coil springs to permit movement between the chassis and body.

BODY SUBSTRUCTURE–STAINLESS STEEL

A chassis frame rail cap shall be constructed from a formed 10 gauge stainless steel inverted channel. The body rail channels shall encapsulate .5" × 3" U.H.M.W. plastic isolators along the full length of each chassis frame rail. The isolators shall permit the necessary movement between the body and chassis.

Body cross members of 2" × 3" × 11" gauge stainless steel tubing shall be welded to the body frame rail channels. The tubes shall extend full width of the body for compartment wall structural support. Additional 2" × 3" stainless steel cross members shall be installed on not more than 18" centers for support of the booster tank.

Two (2) 10 gauge stainless steel 4" × 6" support members shall be installed under compartment floors. The supports shall be formed "U" sections that will extend the full width beneath the compartment floors.

BODY AND COMPARTMENT FABRICATION

The compartment and body construction shall be fabricated from not less than 12 gauge Type 304 stainless steel material. This shall include compartment front panel, vertical side, and rear walls, door frames, compartment dividing walls, and compartment ceilings. The 12 gauge Type 304 stainless steel compartment floor construction shall permit easy cleaning with a "sweep out" type design. A 4" channel shaped double break for rub rail attachment shall be provided for the reinforcement of the floor.

The compartment door frame shall be of channel shaped double break design on the sides. The wheel well panels and door frames shall be set back 1" from outside edge of rub rails for protection of doors and body panels.

All compartments shall be provided with double louvered ventilation units in the lower back wall of the compartments. These louvered vents shall be approximately 6" square.

The sheet metal fabrication of the module shall be constructed in a welding fixture. All sheet metal welding shall be with inert gas continuous feed welders only. Only stainless steel bolts, nuts, and sheet metal screws shall be used in mounting exterior trim, hardware equipment. The use of pop rivets in any portion of the body and compartment construction shall not be acceptable, including attachment of trim or moldings. All seams in the compartments shall be sealed with continuous silicone base waterproof caulking sealants.

BIDDER COMPLIES WITH PREVIOUS PARAGRAPH: YES/NO

F-118: One (1) ADDITIONAL FIRE BODY CONFIGURATION

A. The Fire body shall be designed and constructed so the top of the fire body is flush with the top of the chassis cab.

BIDDER COMPLIES WITH PREVIOUS PARAGRAPH: YES/NO

B. There shall be a access door to the pump on the rear of the apparatus. This door shall allow access to the pump compartment . The access is to be as large as possible.

BIDDER COMPLIES WITH PREVIOUS PARAGRAPH: YES/NO

F-119: One (1) ALUMINUM TREAD PLATE

Bright .188" aluminum tread plate shall be provided at the rear of the apparatus body, directly below the rear opening of the hose bed and on the left and right side compartment tops.

BIDDER COMPLIES WITH PREVIOUS PARAGRAPH: YES/NO

F-120: One (1) WHEEL WELL EXTERIOR ALUMINUM TREADPLATE PANEL

The exterior panel of the wheel well enclosure shall be constructed of .188" aluminum diamond plate bolted in place.

BIDDER COMPLIES WITH PREVIOUS PARAGRAPH: YES/NO

F-121: One (1) WHEEL WELL LINER

The inner wheel well liner shall be constructed of 16 gauge stainless steel, bolted in place. Prior to installation, the complete wheel well area shall be undercoated.

BIDDER COMPLIES WITH PREVIOUS PARAGRAPH: YES/NO

F-122: One (1) WHEEL WELL FENDERETTE

The body wheel well openings shall be equipped with round radius, polished stainless steel fenderettes bolted in place with black rubber welting.

BIDDER COMPLIES WITH PREVIOUS PARAGRAPH: YES/NO

F-123: One (1) FUEL FILL DOOR

The apparatus fuel system shall be equipped with a Cast Products fuel fill door, located in the rearward area of the driver side wheel well. The door shall be cast aluminum with a brushed finish, label provided, "DIESEL FUEL ONLY."

BIDDER COMPLIES WITH PREVIOUS PARAGRAPH: YES/NO

F-124: One (1) BRUSHED STAINLESS STEEL RUB RAIL ASSEMBLY

The rub rail assembly shall be constructed of 3" × 1.5" × 11" gauge stainless steel tubing with a BRUSHED finish. This unit shall be bolted to the lower exterior edge of the apparatus body, with .25" nylon spacers and shall extend 1" from the compartment doors, with a 3" vertical scuff surface.

BIDDER COMPLIES WITH PREVIOUS PARAGRAPH: YES/NO

F-125: One (1) HOSEBODY – STAINLESS STEEL

The hosebody sides shall be constructed of 12 gauge stainless steel. The hosebed panels shall be reinforced and covered with an unpainted, brushed stainless steel inner hosebed panel. The inner panel shall provide a smooth surface for the hosebed sides allowing easy removal and loading of fire hose.

Uni-Strut type channel tracks shall be provided for the mounting of hose bed partitions. There shall be two (2) located at the front and one (1) at the rear of the hosebed.

The hosebed shall provide the following Fire Department supplied hose storage:

1: 200' 1.75" double jacket hose (one left, one right) single stack
2: 400' 1.75" double jacket hose (left) single stack
3: 300' 2.50" double jacket hose (right)single stack
4: 1500' 4.0" hose w/storz fittings (center)

BIDDER COMPLIES WITH PREVIOUS PARAGRAPH: YES/NO

F-126: One (1) HOSEBED DECKING

The hosebed decking shall be constructed with aluminum extrusions.

BIDDER COMPLIES WITH PREVIOUS PARAGRAPH: YES/NO

F-127: Five (5) ALUMINUM HOSEBED DIVIDER ADJUSTABLE

Five (5) adjustable hosebed partitions constructed of .375" mill finished aluminum shall be installed.

BIDDER COMPLIES WITH PREVIOUS PARAGRAPH: YES/NO

F-128: Five (5) HOSEBED DIVIDER HANDHOLE

Each hosebed partition shall be provided with a handhole cutout approximately 3" wide × 8" long.

BIDDER COMPLIES WITH PREVIOUS PARAGRAPH: YES/NO

F-129: Two (2) HYPALON HOSE BED COVERS

The apparatus shall be equipped with two (2) black hypalon hosebed covers without rear flap and a velcro fastening system front and sides. The covers are to be half covers and are to allow for a left and right cover with a walkway in the hose bed. The final design to be approved at a factory visit.

BIDDER COMPLIES WITH PREVIOUS PARAGRAPH: YES/NO

F-130: One (1) <u>BODY SPECIFICATIONS</u>

EXTERIOR COMPARTMENT DIMENSIONS

ROAD SIDE (Driver)

No. 1: Front corner compartment with the approximate interior dimensions of:
48" wide, 72" high and 28" deep in the lower area and 13" deep in the upper area.
Approximate door opening of: 45" wide and 66" high.

No. 2: Ahead of wheel well compartment with the approximate interior dimensions of:
48" wide, 72" high, and 28" deep in the lower area and 13" deep in the upper area.
Approximate door opening of: 45" wide and 66" high.

No. 3: Over wheel well compartment with the approximate interior dimensions of:
60" wide, 42" high, and 13" deep.
Approximate door opening of: 60" wide and 39" high.

No. 4: Rear of wheel well compartment with the approximate interior dimension of:
57" wide, 72" high, and 28" deep in the lower area and 13" deep in the upper area.
Approximate door opening of: 52" wide and 66" high.

CURB SIDE (Right)

No. 5: Front corner compartment with the approximate interior dimensions of:
48" wide, 72" high, and 28" deep.
Approximate door opening of: 45" wide and 66" high.

No. 6: Ahead of wheel well compartment with the approximate interior dimensions of:
48" wide, 72" high, and 28" deep.
Approximate door opening of: 45" wide and 66" high.

No. 7: Over wheel well compartment with the approximate interior dimensions of:
60" wide, 42" high, and 28" deep.
Approximate door opening of: 60" wide and 39"high.

No. 8: Rear of wheel well compartment with the approximate interior dimensions of:
57" wide, 72" high, and 28" deep in lower area and 13" deep in upper area. The pump operators panel shall be located in this compartment.
Approximate door opening of: 52" wide and 66" high.

BIDDER COMPLIES WITH PREVIOUS PARAGRAPH: YES/NO

F-131: One (1) <u>LADDER COMPARTMENT</u>

A vertically mounted ladder storage compartment shall be provided on the apparatus. The compartment shall be an integral part of the body construction. Compartment access shall be through a hinged door at the rear of the apparatus.

Fiberglass type slide guides shall be provided within the compartment to separate the ladders and allow ease in ladder removal.

The ladder compartment to hold the 35" ladder, 14" roof ladder, the attic ladder, and the pikepole.

The location shall be on the drivers side of the apparatus with final approval at the preconstruction conference.

BIDDER COMPLIES WITH PREVIOUS PARAGRAPH: YES/NO

F-132: ONE (1) <u>UPPER LEFT COMPARTMENT W/ TREADPLATE LIFT-UP DOOR</u>

The top of the body on the street (driver) side shall be equipped with integral equipment storage compartments with lift-up doors. All seams shall be continuously welded to prevent the entry of moisture within the body.

The estimated dimensions of these compartment are 21" wide × 20" deep × 187" long

The lift up doors shall be .125" aluminum treadplate with .125 aluminum box pans. 12 gauge stainless steel hinges and dual gas openers shall be installed. The compartment door openings shall be flanged upward 2" with a 1" reverse lip, to prevent water from running into compartments when doors are closed. Two (2) T-handle flexible draw latches shall be furnished to hold down each lid with a rod type handle mounted on the front face to lift the lid.

The floor of each compartment shall have mateflex grating and One (1) 1" drain piped to the underside of the body (in front compartment corner).

BIDDER COMPLIES WITH PREVIOUS PARAGRAPH: YES/NO

F-133: ONE (1) <u>UPPER RIGHT COMPARTMENT W/ TREADPLATE LIFT-UP DOOR</u>

The top of the body on the curb (passenger) side shall be equipped with integral equipment storage compartments with lift-up door. All seams shall be continuously welded to prevent the entry of moisture within the body.

The estimated dimensions of these compartments are 11" wide × 20" deep × 187" long. The lift up doors shall be .125" aluminum treadplate with .125 aluminum box pans. 12 gauge stainless steel hinges and dual gas openers shall be installed. The compartment door openings shall be flanged upward 2" with a 1" reverse lip, to prevent water from running into compartments when doors are closed. Two (2) T-handle flexible draw latches shall be furnished to hold down each lid with a rod type handle mounted on the front face to lift the lid.

The floor of each compartment shall have mateflex grating and One (1) 1" drain piped to the underside of the body (in front compartment corner).

BIDDER COMPLIES WITH PREVIOUS PARAGRAPH: YES/NO

F-134: One (1) <u>ROLLUP DOOR CONSTRUCTION</u>

The apparatus shall be equipped with Rollomatic Robinson Series #34 compartment shutter doors. The doors shall be constructed of box frame type anodized aluminum slats. The slats exterior surface shall be flat, with the rear surface concave to prevent loose equipment from jamming door.

The aluminum extrusions shall be equipped with nylon end shoes that slide in extruded aluminum tracks for easy operation.

The shutter door shall have a pretension operator sealed in an alloy drum, and positioned in the compartment for maximum clearance and head room.

The shutter door shall be a completely weather-resistant assembly. Neoprene seals shall be provided on the side, bottom, and upper portions of the door, and in between each slat.

The roll-up door shall decrease the compartment door opening approximately 4" in width, 9-1/2" in height, and 2" in compartment depth.

BIDDER COMPLIES WITH PREVIOUS PARAGRAPH: YES/NO

F-135: Eight (8) LIFT BAR LATCH WITH LOCK

Latching of the Rollomatic exterior compartment shutter doors shall be with an aluminum, spring loaded lift bar. The lift bar shall latch itself under two (2) cam-shaped strike blocks mounted on the outer door frame of the compartment. The base of the door shall have a heavy duty 4" × 3" extruded angle reinforcement.

BIDDER COMPLIES WITH PREVIOUS PARAGRAPH: YES/NO

F-136: Eight (8) ROLL-UP DOORS

Eight (8) Rollomatic roll-up doors shall be provided for compartments as follows:

1. Four (4) side compartments on drivers side of body.
2. Four (4) side compartments on curb side of body.

BIDDER COMPLIES WITH PREVIOUS PARAGRAPH: YES/NO

F-137: Two (2) UPPER HORIZONTAL DOOR HEADER – ROLL-UP DOORS

The upper horizontal door header above the roll-up doors shall be constructed of .188" aluminum painted to match the Dupont 817U.

BIDDER COMPLIES WITH PREVIOUS PARAGRAPH: YES/NO

F-138: One (1) COMMAND CENTER AND PLANNING MAP TABLE

The center rear wall of the cab shall be provided with a command center and map/planning table. This unit shall be constructed of 3/16" (.188") brushed finish aluminum.

The table shall be approximately 36" high to the top of the front from the floor and 39" high at the rear. It shall be 36" wide × 24" deep with a slanted drafting board type table top. The table top shall be hinged at the rear to lift-up for access to a small storage area. The storage area shall be approximately 8" high at rear × 36" wide × 24" front to rear.

The table top shall be 3/16" (.188") aluminum covered with a sheet of clear Plexiglas. A "Tactical Command" sheet shall be provided under the Plexiglas.

The bottom of the unit under the table area shall consist of smooth .188" aluminum with access doors and adjustable shelves.

Over the desk shall be a bookcase approximately 36" wide × 30" high. it shall have adjustable shelves and access doors.

The purchaser shall approve the blueprint design at the preconstruction conference and the final approval at a factory visit.

BIDDER COMPLIES WITH PREVIOUS PARAGRAPH: YES/NO

F-139: One (1) REAR BODY STEP

The rear step of the apparatus shall be bolted to the lower edge of the rear compartments with a 1" space for water drainage.

The assembly shall be of one piece construction, 12" deep minimum. The unit shall be constructed of bright .188" aluminum tread plate and have a double channel break on all edges with angled corners for marker light mounting.

BIDDER COMPLIES WITH PREVIOUS PARAGRAPH: YES/NO

F-140: Four (4) FOLDING STEPS

Four (4) folding steps, 8" square, die cast aluminum with stainless steel springs shall be installed.

The location shall be approved by the purchaser at pre-completion inspection.

BIDDER COMPLIES WITH PREVIOUS PARAGRAPH: YES/NO

F-141: One (1) <u>REAR BODY HANDRAILS – PAIR</u>

Two (2) vertical handrails shall be provided on the apparatus. Each handrail shall be constructed of extruded aluminum with groves for maximum gripping ability, and chrome plated zinc die cast end stanchions.

The locations shall be approved by the purchaser at preconstruction conference print sign off.

BIDDER COMPLIES WITH PREVIOUS PARAGRAPH: YES/NO

F-142: One (1) <u>REAR TOW PLATES – PAIR</u>

A pair of rear painted tow plates with 3" I.D. holes, constructed with 3/4" steel plate shall be provided on the apparatus. These shall be bolted directly to the rear chassis frame rails.

BIDDER COMPLIES WITH PREVIOUS PARAGRAPH: YES/NO

F-143: Seven (7) <u>UNI-STRUT ADJUSTABLE SHELVING TRACKS</u>

Seven (7) compartment side walls shall be equipped with four (4) stainless steel vertically mounted adjustable tracks that are bolted in place for adjustable shelving and equipment mounting.

The location shall be compartments No. 1 through No. 7.

BIDDER COMPLIES WITH PREVIOUS PARAGRAPH: YES/NO

F-144: Seven (7) <u>SHELVING</u>

Seven (7) shelves constructed of 3/16" (.188") thick #3003 smooth aluminum plate shall be supplied and be mounted with cast aluminum shelf brackets.

Each shelf shall have a 3" broken front edge (2" up and 1" back), and a 3" broken rear edge (2" down and 1" forward), for added strength and reinforcement.

The locations shall be in approved by the purchaser at preconstruction conference with final approval at the pre-completion inspection.

Tentative locations : compartments 2(2), 4(3),6(2)

BIDDER COMPLIES WITH PREVIOUS PARAGRAPH: YES/NO

F-145: Two (2) <u>250# SLIDING EQUIPMENT TRAY</u>

Two (2) sliding trays constructed of 3/16" (.188") Type #3003 aluminum, and equipped with 250# capacity slide tracks shall be supplied. The tracks shall be Jonathan zinc plated steel with stainless steel ball bearings mounted in triple track rails and will slide approximately 18 inches.

The tray edges shall be broken with a minimum 3" flange on each side and double channel lip on front pull out side, with welded corners to form a box type tray surface. The trays shall slide fully out of the compartment with gas operators to hold the trays in both "in" and "out" positions.

The tentative locations to be in compartments 2 & 6 (lowest position) with final approval by the purchaser at the preconstruction conference.

BIDDER COMPLIES WITH PREVIOUS PARAGRAPH: YES/NO

F-146: One (1) <u>SLIDE-OUT/DROP-DOWN EQUIPMENT TRAY</u>

One (1) slide-out/drop-down equipment tray shall be provided on the vehicle. The tray shall be constructed of 3/16" (.188") brushed aluminum, and shall have broken edges on each side and double channel lip on front pull out side, with welded corners to form a box type surface. The tray shall be designed to hold 250 lbs.

The tray shall be equipped with four (4) roller bearing wheels in a continuous "C" shaped galvanized track assembly allowing the tray to roll out of compartment while dropping downward at approximately 30 degree angle. The tray shall be equipped with spring-type locking devices, for locking the tray in the "IN" position.

The tentative location to be in compartment 3 with final approval by the purchaser at pre-completion inspection.

BIDDER COMPLIES WITH PREVIOUS PARAGRAPH: YES/NO

F-147: Two (2) ADJUSTABLE ALUMINUM PEGBOARD TOOL PANEL

Two (2) adjustable and vertical tool board assembly shall be constructed of .188" aluminum pegboard, and shall be mounted on Uni-strut tracks, secured with 3/8" bolts and acorn nuts with Teflon inserts.

The tentative location to be compartment 4 with approved by the purchaser at pre completion inspection.

BIDDER COMPLIES WITH PREVIOUS PARAGRAPH: YES/NO

F-148: Twelve (12) PVC AIR BOTTLE MODULE 12 BOTTLES AND Three (3) THREE FIRE EXTINGUISHERS

A storage rack for self contained breathing apparatus bottles shall be provided. The rack shall be constructed of a .188" brushed aluminum frame with PVC plastic tubes to hold the department supplied SCBA bottles.

The storage rack shall be constructed with a 2" slant from the compartment opening to the rear of the compartment, designed to hold the bottles in place when the compartment door is open. MATEFLEX shall be supplied on the vertical wall behind the rack to prevent the bottles from hitting the wall.

The fire extinguishers to be held are 2.5 gallon pressure water, 20 lb ABC and 15 lb CO_2

The number of SCBA tubes shall be twelve (12) and three (3) fire extinguishers ,and their location shall be in compartment #7.

The final design to be at a factory visit.

BIDDER COMPLIES WITH PREVIOUS PARAGRAPH: YES/NO

F-149: Six (6) PLASTIC TOOL BOX

Six (6) 12" wide × 12" deep × 22" long plastic tool boxes shall be provided on the vehicle. The tool box shall be constructed of 1/2" plastic with hand hole cutouts on each end.

The location shall be on plastic angle slides to be determined at a factory visit.

BIDDER COMPLIES WITH PREVIOUS PARAGRAPH: YES/NO

F-150: One (1) STOKES BASKET STRETCHER AND BACKBOARD MOUNTING

One (1) 3/16" (.188") aluminum Stokes basket/stretcher horizontal shelf shall be provided.

Below the basket/stretcher area shall be a "U" shaped slide-in shelf for backboards constructed of 188" smooth aluminum.

The Stokes basket/stretcher area shall be approximately 10" high, 32" wide, and 86" across. The backboard slide-in unit shall be 4" high, 24" wide and 86" across. Both units shall be bolted in place. (Equipment shall be department supplied.)

The location shall be in the front full through compartment, allowing the stretcher to be removed from either side of the apparatus. The location shall be in the front are of compartments #1 and 5

The final approval of the design to be at the preconstruction/print sign-off conference.

BIDDER COMPLIES WITH PREVIOUS PARAGRAPH: YES/NO

F-151: One (1) 12 VOLT WIRING SYSTEM

The apparatus shall be equipped with a heavy duty 12 volt wiring system. The system shall include all parts, components, switches, relays, and wiring to insure complete operation. All body module electrical equipment shall be served by circuits, separate and distinct from the vehicle chassis circuits.

System wiring shall be function, number, and color coded, and will be mounted in protective loom harnessed to body members with bolted on clips and nylon wire ties.

The edges of all metal members which wiring passes through shall be deburred and provided with edge protection or large rubber grommets. All electrical connections shall be with mechanical type fasteners.

All 12 volt wiring from the cab to the body shall be joined at a main terminal box in the body module. This panel shall allow the split of the cab and body wiring so the body module could be removed. This main body panel shall house the automatic reset breakers and relays.

The panel box shall be a Class 1 power distribution relay board with selectable input polarity and load management input for the use of any load manager. This relay board shall have visual diagnostic for negative and positive input as well as relay output. All connection terminals shall have protective covers where necessary. Additional terminal boxes shall be installed in the chassis cab or body as required.

BIDDER COMPLIES WITH PREVIOUS PARAGRAPH: YES/NO

F-152: One (1) 12 VOLT RADIO Wiring SYSTEM

One red fused #10 wire directly to the positive side of the battery and terminated at the command console area shall be provided.

One black #10 wire directly to the positive side of the battery and terminated at the command console area shall be provided.

One red fused #14 wire that is connected to the 12 volt positive voltage of the vehicle and switched by the master switch, and terminated in the command center area shall be provided.

The wires are to terminate in the command console area with the final approval for location and routing to be at the preconstruction conference.

BIDDER COMPLIES WITH PREVIOUS PARAGRAPH: YES/NO

F-153: One (1) BODY CLEARANCE LIGHTS

The apparatus shall be equipped with body clearance lights with chrome plated die-cast zinc guards. The lights shall be the proper color and type as required by Federal Regulations.

BIDDER COMPLIES WITH PREVIOUS PARAGRAPH: YES/NO

F-154: One (1) <u>REAR LICENSE PLATE MOUNT WITH LIGHT</u>

There shall be a license plate mount with light supplied at the rear of the apparatus. The purchaser shall approve the exact location at a preconstruction conference.

BIDDER COMPLIES WITH PREVIOUS PARAGRAPH: YES/NO

F-155: One (1) <u>WHELEN STOP, TAIL, TURN, AND BACK-UP LIGHTS</u>

The rear of the apparatus body shall be equipped with individual WHELEN Model #64HC series lights. The rear "tail" & stop lights shall be red, the turn lights shall be amber with a black arrow, and the back-up lights will be clear. The lights shall be mounted in a diamond tread step/bezel.

BIDDER COMPLIES WITH PREVIOUS PARAGRAPH: YES/NO

F-156: Eight (8) <u>UNDERBODY/GROUND LIGHTS</u>

Eight (6) WELDON Model #2025 underbody/ground lights shall be angle mounted under the apparatus body. The lights shall have a clear lens and shall be <u>wired to the battery disconnect switch.</u>

The exact location of these lights to be under the cab (4) (driver and officer, and each crew door under compartment 2 and 6, under compartment 4 and 8.

BIDDER COMPLIES WITH PREVIOUS PARAGRAPH: YES/NO

F-157: Eight (8) <u>COMPARTMENT LIGHTS – WALL MOUNTED</u>

The exterior compartments of the apparatus shall be equipped with two (2) Weldon model #2025 light fixtures mounted on the interior side walls or the ceiling of the compartment. Each light shall have a clear lens.

The purchaser shall approve the location of each light depending on the layout of the compartment.

BIDDER COMPLIES WITH PREVIOUS PARAGRAPH: YES/NO

F-158: Eight (8) <u>COMPARTMENT LIGHTS – AUTOMATIC DOOR SWITCH</u>

Each exterior compartment shall have the lights controlled by an automatic door switch.

BIDDER COMPLIES WITH PREVIOUS PARAGRAPH: YES/NO

F-159: Eight (8) <u>WHELEN OPTI-SCENE LIGHT</u>

Eight (8) WHELEN Model #97 series Opti-Scene lights with flanges and relampable halogen bulbs shall be provided. The light fixtures shall have a built-in lens 13 degree reflector. The lights shall be controlled by switches located in the cab. The switching shall be left, right, rear. The lens color shall be clear.

The locations will be at the following:

Two (2) on cab one left one right just behind the front doors

Two (2) on left fire body top one front, one rear

Two (2) on right fire body one front, one rear

Two (2) on rear of fire body below the emergency flashing red model 97's

BIDDER COMPLIES WITH PREVIOUS PARAGRAPH: YES/NO

F-160: One (1) <u>WHELEN 4300 NFPA LIGHTBAR</u>

One (1) WHELEN 4300 NFPA AL light bar shall be provided on the apparatus. The lightbar shall measure 56" long. ALL LIGHTS TO BE FORWARD AND SIDE FACING

The lightbar will be controlled by a labeled rocker switch located on the emergency light switch panel in the cab.

The lens color shall be red.

The location shall be forward facing on the cab roof .

BIDDER COMPLIES WITH PREVIOUS PARAGRAPH: YES/NO

F-161: Six (6) <u>WHELEN HALOGEN FLASHING LIGHT</u>

Six (6) Series #97 halogen lights with chrome plated flanges shall be provided. The light fixtures shall measure 7" high × 9" wide and will have the relampable feature. The lights shall be controlled by a switch located on the emergency light switch panel in the cab.

Lens color shall be red.

Location shall be at the top of the fire body sides,front and rear and on the rear of the apparatus as the top set of lights, with an alternating sequence.

BIDDER COMPLIES WITH PREVIOUS PARAGRAPH: YES/NO

F-162: Two (2) <u>WHELEN HALOGEN FLASHING LIGHT (amber lens)</u>

Two(2) Series #97 halogen lights with chrome plated flanges shall be provided. The light fixtures shall measure 7" high × 9" wide and will have the relampable feature.

Lens color shall be amber lens.

Location shall be mounted in an bezel One (1) each side in the rear under the red emergency lights and the under the scene lights.

These lights are to be wired to the ignition of the apparatus and be operating whenever the apparatus is running.

BIDDER COMPLIES WITH PREVIOUS PARAGRAPH: YES/NO

F-163: Two (2) <u>WHELEN HALOGEN FLASHING LIGHT (red/clear split lens)</u>

Two (2) Series #64 halogen lights with chrome plated flanges shall be provided. The light fixtures shall measure 4" high × 6" wide and will have the relampable feature. The lights shall be controlled by a switch located on the emergency light switch panel in the cab.

Lens color shall be red/clear split lens.

Location shall be mounted in an angled bezel. One (1) each side in the front bumper.

BIDDER COMPLIES WITH PREVIOUS PARAGRAPH: YES/NO

F-164: Two (2) <u>WHELEN HALOGEN FLASHING LIGHT (red lens)</u>

Two (2) Series #64 halogen lights with chrome plated flanges shall be provided. The light fixtures shall measure 4" high × 6" wide and will have the relampable feature. The lights shall be controlled by a switch located on the emergency light switch panel in the cab.

Lens color shall be red lens.

Location shall be mounted in a bezel One (1) each side over the rear wheel.

BIDDER COMPLIES WITH PREVIOUS PARAGRAPH: YES/NO

F-165: Three (3) <u>WHELEN HALOGEN FLASHER – FOUR (4) OUTLET</u>

Three (3) WHELEN Model #UHF-64, four (4) outlet universal halogen flashers shall be provided.

BIDDER COMPLIES WITH PREVIOUS PARAGRAPH: YES/NO

F-166: One (1) ELECTRONIC SIREN

One (1) FEDERAL SIGNAL Model #PA-300M electronic siren shall be mounted in the cab. This unit shall feature electronic air horn, wail, yelp, hi-lo, and shall have a hard wired microphone.

The exact location shall be approved by the purchaser at a factory visit.

BIDDER COMPLIES WITH PREVIOUS PARAGRAPH: YES/NO

F-167: One (1) FEDERAL Q2B SIREN

One (1) FEDERAL SIGNAL Model #Q2B mechanical siren (pedestal type) shall be provided on the apparatus. The siren shall be activated with a foot switch installed on the driver side cab floor. A siren/brake switch shall be installed on the chassis dash.

The location shall be on the left side of the front bumper.

BIDDER COMPLIES WITH PREVIOUS PARAGRAPH: YES/NO

F-168: One (1) OFFICER SIREN FOOT SWITCH

There shall be a foot switch provided on the officer side of the cab floor to activate the Q2B siren.

BIDDER COMPLIES WITH PREVIOUS PARAGRAPH: YES/NO

F-169: One (1) CAST PRODUCTS ALUMINUM SPEAKER – 100 WATT

One (1) CAST PRODUCTS Model #SH-3015 100 Watt speaker shall be mounted on the front of the chassis. The speaker shall be wired to the electronic siren located in the cab.

The location shall be in the front bumper.

BIDDER COMPLIES WITH PREVIOUS PARAGRAPH: YES/NO

F-170: One (1) CHROME FIRE BELL

There shall be a 12" chrome fire bell mounted on the apparatus. The bell shall have a lanyard for operation routed inside the chassis cab.

The location shall be on the right side of the front bumper.

BIDDER COMPLIES WITH PREVIOUS PARAGRAPH: YES/NO

F-171: Three (3) RADIO ANTENNA INSTALLATION

Three (3) radio antennas supplied by the purchaser shall be mounted on the cab roof.

The purchaser shall approve the exact location on a factory visit.

BIDDER COMPLIES WITH PREVIOUS PARAGRAPH: YES/NO

F-172: One (1) ONAN 10.0 KW DIESEL GENERATOR

The apparatus shall be equipped with a ONAN Model #10.0 HDKAQ/1 1475A single phase, radiator cooled, diesel driven generator with 10000 watt capacity at 120/240 volt, amps. The generator shall be radiator cooled with a pusher fan.

The generator muffler and flexible exhaust pipe shall be securely supported and shall be shielded or insulated to prevent heating of the body, electrical components or equipment mounting.

The generator oil drain shall be fitted with a quarter turn drain valve and a hose for remote draining.

The generator fuel system shall be connected to the chassis main fuel tank by a separate fuel line installed directly from the tank with a shut-off valve provided close to the generator. An electric fuel pump shall be installed.

The generator shall be mounted on shock and anti-vibration rubber mountings and be equipped with a removable lifting bracket.

The generator shall weigh approximately 500 lbs., and shall have the following approximate dimensions: 24" high × 21" wide × 36" long.

The generator location shall be over the booster tank in the front of the hose bed and under the light tower if possible.

STARTING SYSTEM

The generator starting system shall be provided by a battery powered starter motor with the 12 volt power supplied from the chassis battery system. The ignition switches and preheat switches shall be located at the generator circuit breaker panel area and the emergency console in the cab.

COOLING

Proper ventilation for the generator unit shall be provided on the apparatus. The installation shall permit the operation of the generator with the vehicle in the stationary condition.

The incoming air flow shall provide adequate air flow for operation of the generator.

SAFETY CONTROLS AND GAUGES

The generator unit shall include the following safety controls and gauges that are installed on the generator or in the generator circuitry:

One (1) automatic high engine temperature shutdown

One (1) automatic low oil pressure shutdown

One (1) oil pressure gauge on generator

One (1) generator hour meter

One (1) water temperature gauge on generator

120 VOLT WIRING DEVICES AND FIXTURES

The complete wiring and generator installation shall conform to National Fire Protection Association Standard #1901.

DISTRIBUTION AND CIRCUIT BREAKER PROTECTION

To properly control the electrical generator output a circuit breaker panel shall be provided on the apparatus. This panel shall incorporate one (1) master circuit breaker and twelve (12) individual load circuit breakers.

The panel location shall be approved by the purchaser at a factory visit.

All electrical wiring shall be fine stranded copper Type THHN. The wire shall be sized to load and circuit breaker rating: 8 gauge on 50 amp circuits, 10 gauge on 30 amp circuits, 12 gauge on 20 amp circuits.

CONDUIT

All 120/240 volt wiring in the apparatus body shall be through CARFLEX flexible moisture resistant reinforced conduit, with proper seal-tight connectors and hardware.

LABELING OF EQUIPMENT

All circuit breakers shall be labeled. Metal engraved labels shall be provided for all outlets indicating output amperage, voltage, and phase.

BIDDER COMPLIES WITH PREVIOUS PARAGRAPH: YES/NO

F-173: One (1) AUXILIARY GENERATOR CONTROLS

The generator shall have additional generator start-stop controls provided. The generator control station shall include the following:

One (1) generator preheat switch.

One (1) generator stop-start momentary switch.

One (1) generator "run" green pilot light.

The controls shall be located in the cab at a location approved by the purchaser at a factory inspection trip.

BIDDER COMPLIES WITH PREVIOUS PARAGRAPH: YES/NO

F-174: One (1) GENERATOR METERING PANEL

To properly monitor the electrical generator performance and load demands during operation, a metering panel shall be provided on the apparatus. This panel shall be separate from the circuit breaker panel.

The installation shall be equipped with the following LCD digital panel instruments:

One (1) line frequency monitor with backlit display

One (1) AC voltage, with backlit display

Two (2) AC current (amps) with backlit display

The panel location shall be approved by the purchaser at a factory visit.

BIDDER COMPLIES WITH PREVIOUS PARAGRAPH: YES/NO

F-175: Four (4) 120 VOLT RECEPTACLE

Four (4) 120 volt 20 amp. twist-lock outlets with weatherproof covers shall be provided with wiring in flexible conduit to circuit breaker panel.

The receptacle locations, cover color, and NEMA configuration shall be approved by the purchaser at a factory visit.

BIDDER COMPLIES WITH PREVIOUS PARAGRAPH: YES/NO

F-176: One (1) ELECTRIC CABLE REEL – 120 VOLT – ELECTRIC REWIND

Hannay series #ECR-l616 electric cable reel, (16" spool width), with electric rewind, shall be provided on the apparatus. The reel shall have 30 amp, four conductor wiring and four (4) fully enclosed collector rings.

The reel's rewind motor shall be wired to the chassis battery system with heavy duty stranded copper cable. The rewind button shall be a Hannay model #90030 sealed push button momentary switch with rubber cap located on the cord reel.

The wiring shall terminate in a sealed conduit box at the reel with mechanical type connectors for quick removal of reel. The reel shall be wired with #8 gauge, four wire, cable to the breaker box and include a 30 amp double pole, 240 volt, breaker.

The reel location shall be in compartment 4 with final approval by the purchaser at a preconstruction conference.

BIDDER COMPLIES WITH PREVIOUS PARAGRAPH: YES/NO

F-177: One (1) ELECTRICAL CABLE

One (1) two hundred foot (200') length of 8/4 yellow electrical cable, type SEO W-A, 20 amp, 120 volt wire, shall be provided and installed on the cable reel. A female outlet shall be provided on the end of the cable with a NEMA configuration approved by the purchaser.

BIDDER COMPLIES WITH PREVIOUS PARAGRAPH: YES/NO

F-178: One (1) CIRCLE-D OUTLET BOX

One (1) circle-D Model PF-51G, four (4) outlet electrical junction box with 12" pigtail and plug shall be provided. The unit shall have an integral pilot light to note electrical current, and shall be equipped with four (4) 12O volt twist-lock receptacles with hinged weather proof covers.

There shall be an aluminum holder for the outlet box provided.

NEMA configuration for receptacles and holder location shall be approved by the purchaser at a factory visit.

BIDDER COMPLIES WITH PREVIOUS PARAGRAPH: YES/NO

F-179: One (I) LIGHT TOWER

The apparatus shall be equipped with a WILL-BURT Model #NS 15-6000 Night Scan pneumatically-raised mast. The mast shall have an extended height of approximately 24 feet.

The light tower location shall be on front of the hose bed with final approval at the preconstruction conference.

The tower shall be capable of full extension in 45 seconds and sustain a 65 mph wind. The pneumatic tower shall be in four sections, with a lowered unit 68" long × 40" wide × 10" high above body. The tower shall be supplied by auxiliary air storage off the air brake system.

The light tower shall be equipped with a rotator motor and gearbox unit. The assembly shall turn the mast from 0 degrees to 360 degrees.

A hand-held remote control pendant, connected to 30 ft. cable shall be provided to control the mast. A four-way switch on the pendant shall control the vertical movement as well as the left and right rotation of the mast. Any upward tower movement from the nested position shall energize "red" warning lights at the tower control area, and in the cab.

The remote control location shall be approved by the purchaser at a factory visit.

The apparatus shall be equipped with a special nesting bracket that contains cushioned rests to secure the lighting equipment in the down position.

Four (4) 1500 watt quartz floodlights shall be mounted on tower.

BIDDER COMPLIES WITH PREVIOUS PARAGRAPH: YES/NO

F-180: One (1) AIR CASCADE SYSTEM (FOUR (4) BOTTLE 6000#)

SCOPE

This specification defines the requirements for a compressed air system used to provide air suitable for human respiration with self-contained breathing apparatus (SCBA). The air system is to be designed for mobile operation.

PIPING AND INSTALLATION REQUIREMENTS

The air system shall be installed to the highest of standards as they apply to this vehicle. The elbows, ties, couplings, and other fittings shall all have a working pressure of 6000 PSI, with a safety factor of 3 to 1.

FLEXIBLE AIR HOSE PIPING REQUIREMENTS

The breathing air system shall be Synflex high pressure breathing flexible air hose.

AIR CONTROL LINE VALVES AND PRESSURE GAUGES

The air system shall be equipped with line valves designed exclusively for pressurized breathing air systems. The panel control valves shall be DACOR, PARKER or WHITEY.

The pressure gauges shall be Span liquid filled (2-1/2" diameter)

DOT AIR CYLINDER SPECIFICATIONS

The cylinders shall be provided with approved brass valves and approved safety relief burst disc.

The cylinders shall be painted "Bright Yellow." "Breathing Air" shall be painted on cylinder in 1" high black lettering.

Number of Cylinders: Four (4)

Pressure: 6000#

Diameter: 9-1/2"

Height: 60"

Weight: 188# (empty)

Mounting Location: In compartment # 5 with final approval by the purchaser at preconstruction conference.

The air cylinders shall be securely mounted. The cylinders shall be accessible for removal, testing, and maintenance. The cylinders shall not be subject to damage, wear, or vibration in the mounting and support system.

EXTERIOR CASCADE PANEL CONSTRUCTION

The exterior cascade and air system control panel shall be removable or hinged and constructed of 14 gauge brushed stainless steel. Location of panels shall be approved by the purchaser at a preconstruction conference.

AIR CONTROL PANEL

The following equipment shall be provided on the air control panel:

1. Four (4) air storage cylinder quarter turn control valves
2. Four (4) 2-1/2" air cylinder pressure gauges
3. One (1) adjustable pressure regulator
4. One (1) 2-1/2" master air pressure gauge upstream of regulator
5. One (1) 2-1/2" air pressure gauge downstream of regulator
6. One (1) Dacor type SCBA fill valve.
7. One (1) No. 4 JIC male thread air outlet connection for SCBA fill hose
8. One (1) 2-1/2" air gauge to indicate SCBA cylinder pressure
9. One (1) CGA #677 air inlet/unregulated air outlet with dust cap to main manifold
10. One (1) DACOR air valve on inlet

SCBA FILLING HOSE AND FILL STATION

The air system shall be equipped with an SCBA filling hose assembly. The units shall be suited for heavy duty use and shall be suited for flexibility and easy on/off connections to SCBA bottles.

1. Number: One (1)
2. Length of Hose: 10' long
3. Connector: CGA #346/347 hand wheel
4. Bleed Valve: One (1) on fill hose
5. Hose Data: 1/4" I.D. 6000# PSI

One (1) portable SCBA cylinder holding device shall be provided as follows:

Fill Tubes: Two (2)

Inner Tube: PVC liner (8" I.D.)

Bottom of Tube: Plastic Mateflex

Exterior Box: .125" aluminum tread plate with two (2) handles

Two (2) loops with carabiner clips to retain SCBA fill hose, connected to portable holding box.

BIDDER COMPLIES WITH PREVIOUS PARAGRAPH: YES/NO

F-181: One (1) ALCO-LITE MODEL ALPF-415-14. 14 FT. ROOF LADDER

One (1) ALCO-LITE 14 foot fiberglass roof ladder with folding steel roof hooks on one end and steel spikes on the other end shall be provided on the apparatus. Ladder shall meet or exceed all latest NFPA Standards.

BIDDER COMPLIES WITH PREVIOUS PARAGRAPH: YES/NO

F-182: One (1) ALCO-LITE MODEL ALPF-430-35 35 FT. (3) SECTION LADDER

One (1) ALCO-LITE 35 foot, three (3) section fiberglass extension ladder shall be provided on the apparatus. The ladder shall meet or exceed all the latest NFPA standards.

BIDDER COMPLIES WITH PREVIOUS PARAGRAPH: YES/NO

F-183: One (1) ALCO-LITE MODEL ALPF-480-10. 10 FT. FOLDING ATTIC LADDER

WITH MOUNTING

One (1) ALCO-LITE 10 foot folding fiberglass attic ladder with mounts shall be provided on the apparatus. The ladder shall meet or exceed all the latest NFPA Standards.

BIDDER COMPLIES WITH PREVIOUS PARAGRAPH: YES/NO

F-184: One (1) HARD SUCTION RACK – HORIZONTAL MOUNTED

The apparatus shall be equipped with a hose bed mounted hard suction hose rack. The rack is to be constructed of .188" aluminum in One (1) square tubes designed to be located in the hose bed.

The purchaser shall approve final design and location at preconstruction conference.

BIDDER COMPLIES WITH PREVIOUS PARAGRAPH: YES/NO

F-185: One (1) 10' FIBERGLASS PIKE POLE

One (1) 10' NUPLA Model SPD-l0fiberglass pike pole shall be provided on the apparatus.

BIDDER COMPLIES WITH PREVIOUS PARAGRAPH: YES/NO

F-186: One (1) PIKE POLE TUBE MOUNTING BRACKET

A 2-1/4" l.D. aluminum tube shall be provided on the apparatus for pike pole mounting. The location shall be determined by the purchaser at the preconstruction conference.

BIDDER COMPLIES WITH PREVIOUS PARAGRAPH: YES/NO

F-187: One (1) PAINTING AND FINISHING SPECIFICATIONS

The final finishing of the apparatus body shall be to fire apparatus standards; exhibiting excellent gloss, durability, and color retention properties. The entire exterior of the apparatus body shall be painted with DuPont Series #6000 paint for good chemical and solvent resistance together with exceptional toughness that resists marring and stone chipping plus lower maintenance costs.

The apparatus body shall be sanded smooth on all exterior surfaces to assure removal of imperfections in metal surface. All bare metal shall be chemically cleaned and conditioned. The body shall receive a corrosion resistant epoxy primer coat. This undercoating shall be lightly sanded to assure a smooth surface for the finish coat. All flush-mounted lights, drip moldings, and other equipment shall also be removed prior to final finishing. This shall assure finish paint under all equipment.

The roll up compartment doors are to remain natural (aluminum) finish.

The paint shall be DuPont #6000 Series:

Color: lime-green

Number: 7744

BIDDER COMPLIES WITH PREVIOUS PARAGRAPH: YES/NO

F-188: One (1) DETAILING OF APPARATUS CAB

The apparatus cab shall be waxed and buffed prior to delivery of completed apparatus.

F-189: One (1) CAB REFLECTIVE LETTERING WITH BLUE OUTLINE

The apparatus cab shall be provided with reflective letters using the highest quality materials and craftsmanship available to insure an outstanding visual effect, and a durable finish. 3-M adhesive reflective letters, outlined with a fine blue line.

Total number of letters: Forty (40)

Size: 4"

Total number of letters: Forty (40)

Size: 6"

Final lettering will be given at a factory visit.

BIDDER COMPLIES WITH PREVIOUS PARAGRAPH: YES/NO

F-190: One (1) CAB AND BODY SCOTCHLITE STRIPING

Reflective stripes shall be affixed to the perimeter of the apparatus. The stripe or combination of stripes shall conform to ASTM D4956, Standard Specifications for Retroreflective Sheeting for Traffic Control, Type III, Class 1 or 3. At least 50 percent of the cab and body length on each side, at least 50 percent of the width of the rear and at least 25 percent of the width of the front of the apparatus shall have reflective material affixed to it.

Total length of striping: feet, six (6") width.

Number of stripes: 1

BIDDER COMPLIES WITH PREVIOUS PARAGRAPH: YES/NO

END OF REQUIRED FIRE BODY SPECIFICATIONS
START OF OPTIONAL BODY SPECIFICATIONS

REQUIRED OPTIONS

The following are required option bids to be bid separately on page two of the bid sheet.

F-191: One (1) PAINTED ROLL-UP DOORS

The roll-up doors to be painted with the same Dupont 7744 paint as the lower body.

The paint finish to include a seven (7) year warranty against marking and staining from the rubber inserts.

F-192: One (1) REAR LADDER

The top of the apparatus shall be accessible from the ground by a railing type ladder. The ladder shall be constructed of 1-1/4" diameter stainless steel tubing mounted to chrome plated brass holders.

The ladder shall be approximately 16" wide, with rungs on an 18" maximum spacing. The location shall be the rear of the apparatus.

START REQUIRED EQUIPMENT SPECIFICATIONS

F-193: REQUIRED EQUIPMENT

The equipment in the following specifications are to be supplied and mounted (if required) with delivered apparatus. The specifications are listed with a description and are to be listed on the "**EQUIPMENT BID**" form.

F-194: One (1) DISCHARGE CAP

One (1)1.5" chrome plated cap and chain shall be provided with NST threads.
BIDDER COMPLIES WITH PREVIOUS PARAGRAPH: YES/NO

F-195: Three (3) ADAPTER

Three (3) 2.5" NST female × 1.5" NST male chrome plated reducers shall be provided.
BIDDER COMPLIES WITH PREVIOUS PARAGRAPH: YES/NO

F-196: One (1) ADAPTER

One (1) 3.0" NST female × 2.5" NST male chrome plated reducer shall be provided.
BIDDER COMPLIES WITH PREVIOUS PARAGRAPH: YES/NO

F-197: One (1) ADAPTER

One (1) 4.0" Storz × 3.0" NST female swivel, lightweight adapter shall be furnished.
BIDDER COMPLIES WITH PREVIOUS PARAGRAPH: YES/NO

F-198: Two (2) STORZ CAP

Two (2) 4.0" Storz lightweight blind cap and chain shall be provided on the apparatus.
BIDDER COMPLIES WITH PREVIOUS PARAGRAPH: YES/NO

F-199: One (1) 20 lb. ABC EXTINGUISHER

One (1) ABC pressure operated dry chemical extinguisher shall be provided with the apparatus. The unit shall have a pressure gauge and come filled with a dry chemical extinguishing agent.
BIDDER COMPLIES WITH PREVIOUS PARAGRAPH: YES/NO

F-200: One (1) 2.5 GALLON PRESSURE WATER EXTINGUISHER

2.5 gallon pressure water extinguisher shall be provided with the apparatus.
BIDDER COMPLIES WITH PREVIOUS PARAGRAPH: YES/NO

F-201: One (1) CO_2 EXTINGUISHER

15 lb. CO_2 fire extinguisher shall be provided with the apparatus. The extinguisher shall have heavy wall construction and shall be pressurized.
BIDDER COMPLIES WITH PREVIOUS PARAGRAPH: YES/NO

F-202: One (1) FIRE EXTINGUISHER MOUNTING

Storage for the three (3) fire extinguishers shall be in the tubes in PVC air bottle module.

SEE PVC AIR BOTTLE STORAGE DESCRIPTION
The location shall be compartment #7.
BIDDER COMPLIES WITH PREVIOUS PARAGRAPH: YES/NO

F-203: One (1) HARD SUCTION HOSE 6.0" × 13'

One (1) section of 6.0" × 13' of KOCHAK pvc hard suction hose shall be provided on the apparatus. The hose shall be light-weight type with pyrolite, nst threads, long handle couplings.

BIDDER COMPLIES WITH PREVIOUS PARAGRAPH: YES/NO

F-204: One (1) KOCHAK LOW LEVEL STRAINER

One (1) KOCHAK model LL60 low level strainer with 6.0" NST threads shall be supplied on the apparatus.

F-205: Twelve (12) 4" ANGUS HI-VOL OR EQUIVALENT HOSE 100'

Twelve (12) lengths of 4" Angus Hi-Vol 100 foot lengths with Storz connectors are to be supplied at time of delivery with the completed apparatus.

BIDDER COMPLIES WITH PREVIOUS PARAGRAPH: YES/NO

F-206: One (1) 4" ANGUS HI-VOL OR EQUIVALENT HOSE 50'

One (1) length of Angus Hi-Vol 50 foot lengths with Storz connectors are to be supplied at time of delivery with the completed apparatus

BIDDER COMPLIES WITH PREVIOUS PARAGRAPH: YES/NO

F-207: Two (2) 4" ANGUS HI-VOL OR EQUIVALENT HOSE 25'

Two (2) lengths of Angus Hi-Vol 25 foot length with storz connectors are to be supplied at time of delivery with the completed apparatus

BIDDER COMPLIES WITH PREVIOUS PARAGRAPH: YES/NO

F-208: Nine (9) 1.75" PONN-LITE

Nine lengths of Ponn-Lite 1.75" attack hoses are to be supplied at delivery. The colors and quantities are as follows:

1.75" × 100' color red	200 feet (2 lengths)
1.75" × 100' color blue	200 feet (2 lengths)
1.75" × 100 color orange	200 feet (2 lengths)
1.75" × 100 color green	300 feet (3 lengths)

BIDDER COMPLIES WITH PREVIOUS PARAGRAPH: YES/NO

F-209: Four (4) STREAMLIGHT LITEBOX

Four Streamlight Liteboxs with AC chargers are to be supplied and mounted, and wired to the shoreline power. the location will be determined at the preconstruction conference.

BIDDER COMPLIES WITH PREVIOUS PARAGRAPH: YES/NO

F-210: Two (2) ANGUS STORZ WRENCH SETS

Two angus Storz wrench sets are to be supplied and mounted with the apparatus. The mounting location and mounting to be determined at the delivery inspection.

BIDDER COMPLIES WITH PREVIOUS PARAGRAPH: YES/NO

F-211: One (1) <u>HYDRANT,SPANNER WRENCHES AND BRACKET</u>

One hydrant,spanner wrenches and bracket -470 to be supplied and mounted with the apparatus. The mounting location and mounting to be determined at the delivery inspection.

BIDDER COMPLIES WITH PREVIOUS PARAGRAPH: YES/NO

F-212: One (1) <u>HALLIGAN TOOL WITH MOUNTS</u>

One HALLIGAN tool (36") with mounts are to be supplied and mounted with the apparatus. The location and mounting to be determined at the delivery inspection.

BIDDER COMPLIES WITH PREVIOUS PARAGRAPH: YES/NO

F-213: One (1) <u>KWIK-KONES</u>

One KWIK-KONE set of four traffic cones with storage bag to be supplied with the completed apparatus.

BIDDER COMPLIES WITH PREVIOUS PARAGRAPH: YES/NO

F-214: Two (2) <u>WHEEL CHOCK, FOLDING, ZICO #SAC-44 , WITH HOLDER</u>

Two (2) ZICO Model #SAC-44 folding wheel chocks with holders shall be provided and mounted with the apparatus.

BIDDER COMPLIES WITH PREVIOUS PARAGRAPH: YES/NO

F-215: One (1) <u>FLAT HEAD AX</u>

One (1) NUPLA Model FPC 6E 6 lb. flat head ax with fiberglass handle shall be provided on the apparatus.

BIDDER COMPLIES WITH PREVIOUS PARAGRAPH: YES/NO

F-216: One (1) <u>PICK HEAD AXE</u>

One (1) NUPLA Model APC 6E 6 lb. pick head ax with fiberglass handle shall be provided on the apparatus.

BIDDER COMPLIES WITH PREVIOUS PARAGRAPH: YES/NO

F-217: Two (2) <u>AX BRACKET</u>

Two (2) chrome plated ax brackets with tulip clips shall be mounted on the apparatus. The location shall be determined by the purchaser at a factory visit.

BIDDER COMPLIES WITH PREVIOUS PARAGRAPH: YES/NO

END OF ALL SPECIFICATIONS

Appendix II

Revised Federal Specifications for Ambulances

KKK-A-1 822D dated November 1, 1994
Supersedes KKK-A-1 822C dated January 1, 1990

The changes in this specification are based on input from emergency medical technicians, paramedics, vehicle maintenance personnel, ambulance manufacturers, and equipment/component suppliers to the ambulance industry.

The specification has been coordinated with the Ambulance Manufacturers Division (AMD) of the National Truck Equipment Association, the American Ambulance Association (AAA), and the Federal Interagency on Emergency Medical Services (FICEMS). The content is based on performance, function, and design requirements necessary to provide for a safe, reliable, highly functional ambulance while allowing flexibility for purchasers to customize the ambulance for individual needs.

This specification should be carefully read prior to ordering an ambulance, to help separate "real from perceived" needs, and to avoid requiring the ambulance manufacturer to supply equipment or designs contrary to good engineering practices, and the requirements and intent of this specification.

When using this specification, particular attention should be given the NOTES, NOTICES, and WARNINGS placed in critical areas throughout the document. This will help avoid serious errors that could have a deleterious effect on the ambulance safety, reliability, and the ability to "complete the mission."

THIS SPECIFICATION IS APPROVED BY THE COMMISSIONER,
FEDERAL SUPPLY SERVICE, GENERAL SERVICES ADMINISTRATION,

FOR THE USE BY ALL FEDERAL AGENCIES.
Copies of this specification are available from the following sources:

Federal civilian agencies can obtain copies by writing to:

General Services Administration
Centralized Mailing List Service (7CAFL)
P.O. Box 6477
Fort Worth, TX 76115

The mailing code for this standard is AMBU-0001.

State, County, Local Governments, and private sector activities:
GSA, Federal Supply Service Bureau
Specifications Section, Suite 8100
470 East L'Enfant Plaza, SW.
Washington, DC 20407
Telephone: (202) 755-0325 or (202) 755-0326

Department of Defense:
Defense Printing Service
700 Robbins Avenue
Building 4D
Philadelphia, PA 19111-5094
Attn: Customer Service
Telephone: (215) 697-2667/2179
AUTOVON: (215) 442-2667/2179

Bidders and government contractors:
Copies may be requested from the Business Service Centers at GSA regional offices In Boston, New York, Washington (DC), Philadelphia, Atlanta, Chicago, Kansas City (MO), Fort Worth, Houston, Denver, San Francisco, Los Angeles, and Seattle.

Comments, information, and questions may be sent to:
Chief, Engineering Branch, FCAE-E
Automotive Commodity Center
General Services Administration
Washington, DC 20406
Telephone: (703) 306-4672

1 SCOPE, PURPOSE, AND CLASSIFICATION

1.1 SCOPE

This document covers certified, tested, commercial type, Emergency Medical Services (EMS) ambulances (except military field ambulances) built on chassis that are suitable for the intended application and meet the requirements herein. The ambulances are front or rear wheel driven (4 × 2) or four wheel driven (4 × 4) and warranted as specified in Section 6. This document may be used to procure an ambulance and the applicable additional systems and equipment. Purchasers should follow the ordering data in Paragraph 6.2 to aid them with the preparation of their specification, requisition, and contract.

1.1.1 DEFINITION OF AMBULANCE

The ambulance is defined as a vehicle for emergency medical care which provides; a driver's compartment; a patient compartment to accommodate an emergency medical technician (EMT)/paramedic and two litter patients (one patient located on the primary cot and a secondary patient on a folding litter located on the squad bench) so positioned that the primary patient can be given intensive life support during transit; equipment and supplies for emergency care at the scene as well as during transport; two-way radio communication; and, when necessary, equipment for light rescue/extrication procedures. The ambulance shall be designed and constructed to afford safety, comfort, and avoid aggravation of the patient's injury or illness.

1.1.2 PURPOSE

The purpose of this document is to describe ambulances which are authorized to display the "Star of Life" symbol. It establishes minimum specifications, test parameters, and essential criteria for ambulance design, performance, equipment, appearance, and to provide a practical degree of standardization. The object is to provide ambulances that are nationally recognized, properly constructed, easily maintained, and, when professionally staffed and provisioned, will function reliably in pre-hospital or other mobile emergency medical service.

1.1.3 CERTIFIED "STAR OF LIFE"

The ambulance manufacturer/contractor shall furnish the purchaser(s) citing this specification an authenticated certification and label (see 3.19) that certifies a "Star of Life" ambulance and equipment complying with this specification and applicable amendments (if any) in effect on the date of manufacture (see 4.3). Ambulance vehicles so certified may display the registered "Star of Life" symbol, as defined by the U.S. Department of Transportation (DOT) and the National Highway Traffic Safety Administration (NHTSA), see Figure 1.

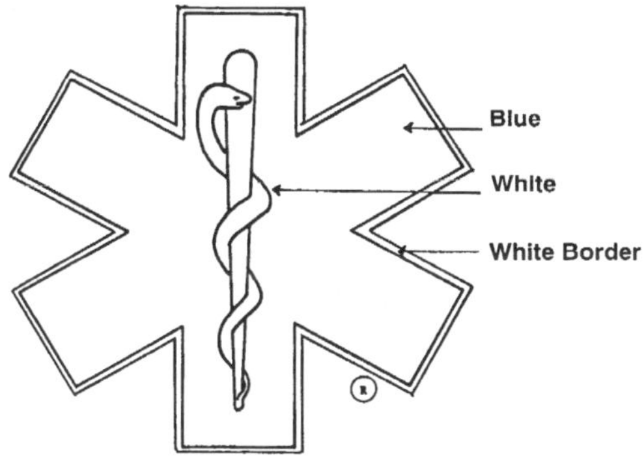

Figure 1

1.2 CLASSIFICATION

1.2.1 AMBULANCE TYPES,CLASSES, AND FLOOR CONFIGURATIONS

The authorized "Star of Life" ambulances shall be of the following (to specify, see 6.2-b & c):

NOTE:FOR OPTIONAL ADVANCED LIFE SUPPORT (ALS) APPLICATIONS, USERS SHOULD CONSIDER SPECIFYING A MODULAR (TYPE I OR III) AMBULANCE. MODULAR AMBULANCES PROVIDE ADDITIONAL SPACE AND COMPARTMENTATION FOR CARDIAC MONITORS, DRUG CASES, ETC. THE BASIC LIFE SUPPORT (BLS) CONFIGURATION IS STANDARD ON ALL TYPES.

Type I—Conventional truck, cab-chassis with modular ambulance body (3.1.2)

CLASS	CONFIGURATION
1—Two rear wheel driven (4 × 2).	A or B
2—Four wheel driven (4 × 4).	A or B

Configuration A—Elevating cot and squad bench for ALS (3.1.5.1 and 3.11.4)
Configuration B—Elevating cot and a squad bench for BLS (3.1.5.2)

Type II —Standard van, integral cab-body ambulance (3.1.3)

CLASS	CONFIGURATION
1—Two rear wheel driven (4 × 2)	A** or B
2—Four wheel driven (4 × 4)*	A** or B

* Requires conversion of chassis to four wheel drive (4 × 4) (see 3.1.6)
** On a Type II, ALS features are limited (see 6.2-c)

Type III —Cutaway van, cab-chassis with integral or containerized modular body ambulance (3.1.4)

CLASS	CONFIGURATION
1—Two rear wheel driven (4 × 2)	A or B
2 - Four wheel driven (4 × 4)*	A or B

* Requires conversion of chassis to four wheel drive (4 × 4) (see 3.1.6)
Configuration A—Elevating cot and squad bench for ALS (3.1.5.1 and 3.11.4)
Configuration B—Elevating cot and squad bench for BLS (3.1.5.2)

2 APPLICABLE DOCUMENTS

2.1 The following documents form a part of this specification, to the extent specified. Unless a specific issue is identified, the issue in effect, on date of invitation for bids or request for proposal, shall apply.

FEDERAL SPECIFICATIONS:
L-S-3OO—Sheeting and Tape, Reflective Nonexposed Lens, Adhesive Backing
RR-C-901/3—Cylinders, Compressed Gas: with Valve or Plug and Cap; ICC3aa

FEDERAL STANDARDS:
Federal Standard No.297—Rustproofing of Automotive Vehicles

Activities outside the Federal Government may obtain copies of Federal Specifications, Standards, and Handbooks as outlined under General Information in the Index of Federal Specifications and Standards and at the prices indicated in the Index. The Index, which includes cumulative monthly supplements as issued, is for sale on a subscription basis by the Superintendent of Documents, U.S. Government Printing Office, Washington, DC 20402.

Single copies of this document and other Federal Specifications required by activities outside the Federal Government for bidding purposes are available from Business Service Centers at the General Services Administration Regional Offices in Boston, New York,

Philadelphia, Washington (DC), Atlanta, Chicago, Kansas City (MO), Fort Worth, Denver, San Francisco, Los Angeles, and Seattle (Auburn, WA).

Federal Government activities may obtain copies of Federal Specifications, Standards, and Handbooks, and the Index of Federal Specifications and Standards from established distribution points in their agencies.

MILITARY STANDARDS:
MIL-STD-1223—Nontactical Wheeled Vehicles, Painting, Identification Marking and Data Plate Standards

Copies of Military Specifications and Standards required by suppliers in connection with specific procurement functions should be obtained from the procuring activity or as directed by the contracting officer.

LAWS AND REGULATIONS
40 CFR 86: Control of Air Pollution from New Motor Vehicles and New Motor Vehicle Engines
47 CFR, PART 89: Public Safety Radio Services (FCC)
49 CFR 393: Federal Motor Carrier Safety Regulations (FMCSR)
49 CFR 571: Federal Motor Vehicle Safety Standards (FMVSS)

The Code of Federal Regulations (CFR) and the Federal Register (FR) are for sale on a subscription basis by the Superintendent of Documents, U.£ Government Printing Office, Washington, DC 20402. When indicated, reprints of certain regulations may be obtained from the Federal agency responsible for issuance thereof.

2.2 OTHER PUBLICATIONS
The following documents form a part of this specification to the extent specified. Unless a specific issue is identified, the issue in effect on date of invitation for bids or request for proposal shall apply.

STATE OF CALIFORNIA MOTOR VEHICLE CODE

Application for copies should be addressed to Department of Motor Vehicles, 2570 24th Street, Sacramento, California 95809.

THE TIRE AND RIM ASSOCIATION, INC., YEARBOOK

Application for copies should be addressed to The Tire and Rim Association, Inc., 175 Montrose W Ave., Suite 150, Copley, Ohio 44321. Phone (216) 666-8121.

SOCIETY OF AUTOMOTIVE ENGINEERS (SAE), INC., STANDARDS AND RECOMMENDED PRACTICES
J163—Low Tension Wiring and Cable Terminals and Splice Clips
J537—Storage Batteries
J541—Voltage Drop for Starting Motor Circuits

J551—Measurement of Electromagnetic Radiation from Motor Vehicles and Devices
J553—Circuit Breakers
J561—Electrical Terminals, Eyelet and Spade Type
J575—Tests for Motor Vehicle Lighting Devices and Components
J576—Plastic Materials, for Use in Optical Parts Such as Lenses and Reflectors of Motor Vehicle Lighting Devices
J578—Color Specification for Electric Signal Lighting Devices
J595—Flashing Warning Lamps for Authorized Emergency, Maintenance, and Service Vehicles
J638—Test Procedure and Ratings for Hot Water Heaters for Motor Vehicles
J639—Safety Practices for Mechanical Vapor Compression Refrigeration Equipment or Systems Used to Cool Passenger Compartment of Motor Vehicles
J689—Approach, Departure, and Ramp Breakover Angles
J682—Rear Wheel Splash and Stone Throw Protection
J683—Tire Chain Clearance
J771—Automotive Printed Circuits
J858—Electrical Terminals, Blade Type
J928—Electrical Terminals, Pin and Receptacle Type
J994—Backup Alarms, Performance Test and Application
J1054—Warning Lamp, Alternating Flashers
J1127—Battery Cable
J1128—Low Tension Primary Cable
J1292—Automobile, Truck, Truck-Tractor, Trailer, and Motor Coach Wiring
J1349—Engine Power Test Code, Spark Ignition and Diesel
J1318—Strobe Warning Lights

Application for copies should be addressed to the Technical Division, SAE, 400 Commonwealth Dr., Warrendale, PA 15096

NATIONAL TRUCK EQUIPMENT ASSOCIATION / AMD:
AMD Standard 001—Static Load for Ambulance Body Structure Test
AMD Standard 002—Body Door Retention Components Test
AMD Standard 003—Oxygen Tank Retention System Test (Main and Portable Bottles)
AMD Standard 004—Litter Retention System Test
AMD Standard 005—Ambulance 12-Volt DC Electrical Systems Test
AMD Standard 006—Sound Level Test Code
AMD Standard 007—Carbon Monoxide Levels for Patient Compartment Interiors
AMD Standard 008—Ambulance Patient Compartment Grab Rail (Load Test)
AMD Standard 009—Alternating Current (AC) Electrical Systems

Application for copies should be addressed to the Ambulance Manufacturer's Division (AMD) of the National Truck Equipment Association, 37400 Hills Tech Drive, Farmington Hills, MI 48331, (810) 489-7090.

ILLUMINATION ENGINEERING SOCIETY:
Guide for Calculating the Effective Intensity of Flashing Signal Lights

Application for copies should be addressed to the Illumination Engineering Society, 345 East 47th Street, New York, New York 10017.

AUTOMOTIVE MANUFACTURERS EQUIPMENT COMPLIANCE AGENCY (AMECA):
Approval of Motor Vehicle Safety Equipment (emergency lights and sirens)

Application for copies should be addressed to the American Association of Motor Vehicle Administrators, 4200 Wilson Blvd. Suite 600, Arlington, VA 22203. Technical society and technical association specifications and standards are generally available for reference from libraries. They are also distributed among technical groups and using Federal agencies.

2.3 ORDER OF PRECEDENCE
In the event of a conflict between the text of this specification and the references cited herein, the text of this specification shall take precedence.

3 (A 1.1) REQUIREMENTS

3.1 GENERAL VEHICULAR DESIGN, TYPES, AND FLOOR PLAN

3.1.1 DESIGN
The ambulance and the allied equipment furnished under this specification shall be the manufacturer's current commercial vehicle of the Type, Class, and Configuration specified. The ambulance shall be complete with the operating accessories, as specified herein, and furnished with such modifications and attachments as may be necessary to enable the vehicle to function reliably and efficiently in sustained operation. The design of the vehicle and the specified equipment shall permit accessibility for servicing, replacement, and adjustment of component parts and accessories with minimum disturbance to other components and systems. The term "heavy-duty", as used to describe an item, shall mean in excess of the standard quantity, quality, or capacity and represents the best, most durable, strongest, etc., part, component, system, etc., that is commercially available on the OEM chassis.

3.1.2 TYPE I AMBULANCE. (A 2.1)
Type I vehicle, Class 1 or 2, Configuration A or B, shall be a chassis furnished with a 2-door enclosed cab. The chassis-cab shall be suitable for subsequent mounting of a transferable, modular, ambulance body conforming to the requirements specified herein (see Type I).

[See Chapter 5, Figure 5.5 for diagrams of a Type I Ambulance]

TYPE 1
For vehicles of 203 cm (80 in.) or more overall width.
Recommended lamp and reflector locations in accordance with the Federal Motor Vehicle Safety Standard No.108 and Federal Ambulance Specification KKK-A-1 822.

NOTE:
Lamps and reflectors may be mounted at other practicable locations provided location and visibility requirements of Federal Motor Vehicle Safety Standard No 108 and Federal Ambulance Specification KKK-A-1 822 are met. (Drawings are not to scale.)

(A 2.2) TYPE I
CONVENTIONAL, CAB-CHASSIS, WITH MODULAR AMBULANCE BODY
AMBULANCE TYPE I—FMVSS NO.108

1. Headlamp, (2) white [(4) white optional]
*2. Front side marker lamp, (2) amber
3. Front side reflector, (2) ember
4. Front turn signal, (2) amber, (includes vehicular hazard warning signal flasher)
5. Front identification lamps, (3) amber
6. Front clearance lamp, (2) amber
7. Rear side marker lamp, (2) red
8. Rear side reflector, (2) red
9. Rear identification lamps, (3) red
10. Rear clearance lamp, (2) red
11. Rear reflector, (2) red
12. Rear, stop, tail, lamp, (2) red
12a. Rear turn signal, (2) amber. Includes vehicular hazard warning signal flasher. (optional location)
13. Rear backup lamp, (1) white
14. Rear license plate lamp, (1) white
15. Front warning light, (2) red
15a. Front warning light, (1) white
16. Rear warning light, (2) red
16a. Rear warning light, (1) amber
17. Side warning light, (2) red per side
18. Grille light, (2) red
19. Intersection lights, (1) red per side
20. Side floodlight, (2) white
20a. Rear floodlight, (1) white
21. Spotlight, (white) hand held
21a. Spotlight, (white), (optional location)

*Flashes with front/rear turn signal and vehicular hazard warning signal flasher

3.1.3 TYPE II AMBULANCE. (A 2.1, A 2.2)
Type II vehicle, Class I or 2, Configuration A or B, shall be OEM chassis manufacturer's commercial, long wheelbase, integral van. This vehicle shall be suitable for subsequent ambulance conversion/modification in compliance with the requirements herein (see TYPE II).

[See Chapter 5, Figure 5.6 for diagrams of a Type II Ambulance]

TYPE 2
For vehicles of 203 cm (80 in.) or more overall width
Recommended lamp and reflector locations in accordance with the Federal Motor Vehicle Safety Standard No.108 and Federal Ambulance Specification KKK-A-1 822.

NOTE:
Lamps and reflectors may be mounted at other practicable locations provided location and visibility requirements of Federal Motor Vehicle Safety Standard No 108 and Federal Ambulance Specification KKK-A-1 822 are met. (Drawings are not to scale)

AMBULANCE TYPE II—FMVSS NO.108

1. Headlamp, (2) white [(4) white optional]
*2. Front side marker lamp, (2) amber
3. Front side reflector, (2) amber
4. Front turn signal, (2) amber (includes vehicular hazard warning signal flasher)
5. Rear side marker lamp, (2) red
6. Rear side reflector, (2) red
7. Rear reflector, (2) red
8. Rear stop, tall & turn signal lamp, (2) red, (turn signal section may be amber, includes vehicular hazard warning signal flasher)
9. Rear backup lamp, (1) white
10. Rear license plate lamp, (1) white
11. Front warning light, (2) red
11a. Front warning light, (1) white
12. Rear warning light, (2) red
12a. Rear warning light, (1) amber
13. Side warning light, (2) red per side
14. Grille light, (2) red
15. Intersection lights, (1) red per side
16. Side floodlight, (2) white
16a. Rear floodlight, (1)
17. Spotlight, hand held, white
17a. Spotlight (white, optional location)

*Flashes with front/rear turn signal & vehicular hazard warning signal flasher.

3.1.4 TYPE III AMBULANCE. (A 2.1, A 2.2)
Type III, Class I or 2, Configuration A or B shall be a "cutaway" van with a transferable, modular, ambulance body or unitized cab-body mounted on a chassis. The chassis with unitized cab-body or front section cab-chassis shall be suitable for the subsequent fabrication, conversion, or modification into an ambulance incorporating the requirements and the equipment specified herein (see TYPE III).

[See Chapter 5, Figure 5.7 for diagrams of a Type III Ambulance]

TYPE III
For vehicles of 203 cm (80 in.) or more overall width.
Recommended lamp and reflector locations in accordance with the Federal Motor Vehicle Safety Standard No.108 and Federal Ambulance Specification KKK-A-1 822.

NOTE:
Lamps and reflectors may be mounted at other practicable locations provided location and visibility requirements of Federal Motor Vehicle Safety Standard No 108 and Federal Ambulance Specification KKK-A-1 822 are met. (Drawings are not to scale)

AMBULANCE TYPE III—FMVSS NO.108

1. Headlamp, (2) white [(4) white optional]
*2. Front side marker lamp, (2) amber
3. Front side reflector, (2) amber
4. Front turn signal, (2) amber, (includes vehicular hazard warning signal flasher)
5. Front identification lamps, (3) amber
6. Front clearance lamp, (2) amber
7. Rear side marker lamp, (2) red
8. Rear side reflector, (2) red
9. Rear identification lamps, (3) red
10. Rear clearance lamp, (2) red
11. Rear reflector, (2) red
12. Rear, stop, tall, lamp, (2) red
12a. Rear turn signal, (2) amber (Includes vehicular hazard warning signal flasher) (optional location)
13. Rear backup lamp, (1) white
14. Rear license plate lamp, (1) white
15. Front warning light, (2) red
15a. Front warning light, (1) white
16. Rear warning light, (2) red
16a. Rear warning light, (1) amber
17. Side warning light, (2) red per side
18. Grille light, (2) red
19. Intersection lights, (1) per side
20. Side floodlight, (2)
20a. Rear floodlight, (1,)
21. Spotlight (white) (hand held)
21a. Spotlight (white) (optional location)

*Flashes with front/rear turn signal and vehicular hazard warning signal flasher

3.1.5 CONFIGURATION OF PATIENT COMPARTMENT
Unless otherwise specified (see 6.2-c), Configuration "B", Basic Life Support (BLS) shall be provided in the patient compartment. All litters shall be loaded to position the patient's head forward in the vehicle.

3.1.5.1 CONFIGURATION "A" (ALS)

When specified for Types I or III (see 6.2-b), Advanced Life Support (ALS) applications, one primary patient shall be on a wheeled elevating cot and a secondary patient lying on a folding stretcher or combination stretcher chair on the squad bench, or one primary patient and three secondary seated patients on the squad bench (see 3.11.4) and one seated EMT/ paramedic. The primary cot shall be center mounted or, when specified (3.15.4 M-21), a dual position mount shall be furnished. Unless otherwise specified (see 3.15.3-28), a CPR seat conforming to all specification requirements shall be furnished on the left (street) side of the body facing the patient's thoracic region and include a seat belt and padding at the EMT's head region. Space shall also be furnished in the action wall/counter area for placement of a cardiac monitor/defibrillator. When specified (see 3.15.4 M-25), a crash stable mounting bracket for securing the cardiac monitor/defibrillator shall be furnished. A locking drug kit compartment per 3.15.3-14 and high intensity light per 3.15.3-8 shall also be furnished. Two I-V hangers (per 3.11.9) shall be provided for the primary patient, one at the patient's head and one at the patient's lower extremities.

3.1.5.2 CONFIGURATION "B" (BLS)

Unless otherwise specified (see 6.2-c), Configuration "B" shall provide for one primary patient on a wheeled elevating cot and one secondary patient on a folding stretcher or combination stretcher/chair on a squad bench, which is also capable of accommodating three seated patients (see cots 3.11.5) and one seated EMT (see 3.10.3).

NOTE: IT IS PREFERABLE THAT THE PRIMARY PATIENT'S COT BE POSITIONED ON THE LEFT (STREET SIDE OR CENTERED IN THE AMBULANCE PATIENT COMPARTMENT. THE ANATOMY OF THE TRACHEAL BRONCHIAL TREE IS SUCH THAT AN UNCONSCIOUS PATIENT WILL BE LESS LIKELY TO ASPIRATE FLUIDS INTO THE TRACHEA IF PLACED ON THEIR LEFT SIDE WITH THE RIGHT SIDE UP.

3.1.6 FOUR WHEEL DRIVE, CLASS 2, 4 × 4

When a 4 × 4 chassis is specified (see 6.2-b), the additional curb weight of the 4 × 4 chassis above the 4 × 2 chassis will reduce the payload proportionally. When available, a Class 2 ambulance shall be an original chassis manufacturer's (OEM) 4 × 4 chassis for Type 1 ambulances, or an OEM 4 × 2 model with a professionally engineered conversion to a four wheel drive (4 × 4) conforming to all applicable requirements herein. All workmanship, welding, mechanical fit, grade, and quality of components and materials used in conversions shall be equal to or greater than OEM manufacturer's production 4 × 4 units. Conversion components shall not interfere with other body, chassis, or mechanical parts through the complete range of suspension and wheel angle travel and allow proper alignment of axles. The tracking of the front/rear axles shall be identical on both sides of the vehicle and wheelbase. When available, the chassis manufacturer's OEM components shall be furnished, including, but not limited to: spring hangers, shackles, drive axle, integral transmission/transfer case, universal joints, steering linkage, stabilizer bars, radius and torque rods, transfer case shaft linkage, brake calipers, pads, rotors, shock absorbers, and springs. When available, the chassis manufacturer's guidelines/ requirements for 4 × 4 conversions shall be followed. The design of the 4 × 4 conversion shall minimize the height of the vehicle's chassis. The 4 × 4 converter shall include a complete chassis modifier FMVSS certification and sticker.

The 4 × 4 converter shall provide to the purchaser a full parts and labor warranty covering all added 4 × 4 parts and materials, including workmanship and design. The warranty shall also cover all OEM components affected or modified by the conversion process. This warranty shall be at least equivalent, in mileage and time, to the OEM chassis manufacturer's warranty, including any extended warranties required or furnished. Accompanying each conversion shall be complete manuals showing operation, maintenance, and repair procedures, chassis manufacturer's part numbers, drawings for components used in the conversion, and dimensioned drawings for manufactured parts, alignment procedures, and specifications.

3.2 VEHICLE, AMBULANCE COMPONENTS, EQUIPMENT, AND ACCESSORIES

The emergency medical care vehicles; including chassis, ambulance body, equipment, devices, medical accessories, and electronic equipment shall be standard commercial products, tested and certified to meet or exceed the requirements of this specification. The ambulance shall comply with all Federal Motor Vehicle Safety Standards (FMVSS) and other Federal and state regulations applicable or specified for the year of manufacture. The chassis, components, and optional items shall be as represented in the chassis manufacturer's current technical data. The ambulance body, equipment, and accessories shall be as represented in their respective manufacturer's current technical data. Data shall be limited to specifications and technical materials identical to that furnished to the authorized company representatives and shall be furnished to the engineering/technical offices of the procuring activity and the supplier's (contractor's) offices prior to acceptance of the ambulance. The supplier shall provide total standardization and interchangeability between similar vehicles, equipment, items, and accessories specified for all ambulance units under each contract.

3.3 RECOVERED MATERIALS

All equipment, material, and articles required under this specification are to be new or fabricated from new materials produced from recovered materials. The term "recovered materials" means materials which have been collected or recovered from solid waste and reprocessed to become a source of raw materials, as opposed to virgin raw materials. None of the above shall be interpreted to mean that the use of used or rebuilt products is allowed under this document.

3.4 VEHICLE OPERATION, PERFORMANCE, AND PHYSICAL CHARACTERISTICS

3.4.1 OPERATION AND PERFORMANCE

Unless otherwise specified by the purchaser, all requirements in Section 3.4 shall be met with the ambulance loaded to meet or exceed the minimum specified payload including all specified equipment and devices installed and operating at the maximum power-consuming condition, i.e., air conditioning, lights, radio(s), etc., with the chassis performing in accordance with the specified chassis manufacturer's technical data. The vehicle shall be capable of operating safely and efficiently under environmental conditions outlined herein or as specified in the invitation for bid, contract, or order. When specified by the purchaser for ambulances requiring higher GVWR's to carry additional specialty equipment, e.g.

medical; rescue; neonatal; (see 6.2-d), lower performance levels are acceptable in Paragraphs 3.4.6 through 3.4.8.2.

3.4.2 TEMPERATURE CONDITIONS

The ambulance, including all required systems, equipment, and medical devices (except when medical equipment is regulated by another agency) furnished inside the ambulance, when serviced and maintained in accordance with the OEM manufacturer's recommendations (4.4), shall be capable of being stored without damage or deterioration in ambient temperatures of –34° C to 52° C (–30° F to 125° F).

The ambulance and ambulance equipment shall be cold soaked for 6 hours at –34°C (–30° F) followed by a 1 hour cold soak at –18° C (0° F). All of the vehicles non-engine-powered equipment systems shall then be tested and operated at the –18° C (0° F) temperature. The engine shall then be started and all remaining chassis and ambulance systems shall be tested.

The ambulance and ambulance equipment shall be heat soaked at 53° C (125°F) for 6 hours followed by 1 hour heat soak at 43° C (110° F). All the vehicles non-engine-powered systems shall then be tested and operated at the 43° C (100° F) temperature. The engine shall then be started and all remaining chassis systems and equipment shall be tested.

Medical devices, such as resuscitation and suctioning apparatus, shall be cold tested for the required flow performance on 12 volt DC vehicle power and 115 volt AC power modes (see 3.12.1.1). Type certification from medical and other equipment manufacturers is acceptable.

3.4.3 NOISE AND SOUND LEVEL LIMITS, EXTERIOR

Unless more stringent sound levels are regulated by the states and municipalities where the ambulance will be based, the exterior noise level produced by the vehicle, except siren, shall not exceed Federal regulations.

3.4.4 VEHICLE PERFORMANCE

The ambulance shall provide a smooth, stable ride, with minimum noise and vibration. Test under 4.4.4.

3.4.5 BRAKES

The ambulance braking system as delivered to the user shall comply to performance values required by Federal Motor Vehicle Safety Standards (FMVSS).

3.4.6 sSPEED

The vehicles shall be capable of a sustained speed of not less than 105 km/hr (65 mph) over dry, hard surfaced, level roads, at sea level, and passing speeds of 113 km/hr (70 mph) when tested under normal ambient conditions.

3.4.7 ACCELERATION

Vehicle shall have a minimum average acceleration, at sea level, of 0-88 km/hr (0-55 mph) within 25 seconds. Test shall be performed under normal ambient conditions. Test under 4.4.4.

3.4.8 GRADEABILITY

At full payload, the vehicle shall be capable of meeting the following requirements. The determination shall be made by actual test or chassis manufacturer's certified computer prediction or chassis manufacturer's certification.

3.4.8.1 GRADEABILITY AT SPEED
89 km/hr (55 mph) on a 3% (1.72 degrees) grade.

3.4.8.2 MINIMUM LOW SPEED GRADEABILITY
The minimum low speed gradeability of 8 km/hr (5 mph) on a 35% (19.3 degrees) grade is required for Class I (4 × 2) vehicles and 45% (24.2 degrees) grade for Class 2 (4 × 4) vehicles in the low 4 × 4 range.

3.4.9 FUEL RANGE
Unless otherwise specified (see 6.2-e), the ambulance shall be capable of being driven for at least 402 km (250 miles) without refueling under the conditions in 4.4.4.

3.4.10 FORDING
The vehicle shall be capable of three fordings, without water entering patient and equipment compartments while being driven through a minimum of 20 cm (8 in.) of water, at speeds of 8 km/hr (5 mph), for a distance of at least 305 m (100 feet). Test under 4.4.4.

3.4.11 VEHICLE PHYSICAL DIMENSIONAL REQUIREMENTS

3.4.11.1 LENGTH
Overall length of the ambulance shall not exceed 671 cm (22 ft), including bumpers, but excluding rear step and bumper guards. Purchaser may specify (see 6.2-f) additional length if necessary to accommodate special equipment or larger bodies but should consult with the manufacturer to determine that other safety/handling characteristics are not compromised.

3.4.11.2 WIDTH
Unless otherwise specified (see 6.2-g), the overall (OA) width of ambulance bodies having single rear wheels shall be between 200 and 213 cm (79 and 84 in.), excluding mirrors and lights. The ambulance body sides, on a chassis with dual rear wheels, shall be within +/- 5 cm (+/- 2 in.) of the overall width of the tires (outside sidewalls) (see 3.5.6, 3.6.5.6, and 3.9.7). Tires shall riot extend beyond the fenders. The maximum ambulance body width shall not exceed 244 cm (96 in.), excluding mirrors and lights. The wide track axle shall be furnished on dual rear wheeled vehicles, unless a narrow ambulance body is specified by the purchaser (see 6.2-g).

3.4.11.3 HEIGHT
Unless otherwise specified (see 6.2-h), overall height of the ambulance at curb weight (see 3.5.1) shall not exceed 279 cm (110 in.), including roof mounted equipment, but excluding two-way radio antenna(s).

3.4.11.4 GROUND CLEARANCE
The lowest part of the vehicle, when loaded to the GVWR, shall have a minimum of 15 cm (6 in.) of ground clearance. The vehicle's body (skirts) components shall provide not less than 20 cm (8 in.) of running clearance.

3.4.11.5 ANGLE OF APPROACH, RAMP BREAKOVER, AND DEPARTURE
The ambulance, loaded to the GVWR (including payload 3.5.2) with bumpers and rear step (down if folding style), shall provide not less than the following clearance, measured in accordance with SAP J689

Approach angle	20 degrees
Ramp breakover	10 degrees
Departure angle	10 degrees

3.4.11.6 TURNING RADIUS
Shall not be greater than chassis OEM standard.

3.4.11.7 FLOOR HEIGHT
The finished floor (loading) heights shall not exceed 84 cm (33 in.) on Class 1 (4 × 2) vehicles and 97 cm (38 in.) for Class 2 (4 × 4) vehicles. The height shall be measured at curb weight plus full payload but minus patients and personnel.

3.5 VEHICLE WEIGHT RATINGS AND PAYLOAD

3.5.1 CURB WEIGHT
Curb weight is the total weight of the complete ambulance and is defined as: chassis (including batteries, spare tire, jack tire changing tools), cab, body, minimum required equipment per this specification, and a full complement of fuel, lubricants, and coolant, excluding optional equipment in 3.15.3 and 3.15.4.

3.5.2 PAYLOAD ALLOWANCE
Each vehicle's payload allowance shall be determined and displayed in accordance with the weight/payload certification form (sticker) depicted on page 14 **[below]**. Sticker shall be located in a conspicuous location in the vehicle. Payload shall be over and above the specified curb weight (see 3.5.1) of the vehicle and include specified options, miscellaneous medical equipment, and communications equipment that is appropriately distributed within the vehicle.

WARNING: DUE TO POTENTIAL HAZARDS AND DAMAGE TO VEHICLE'S CHASSIS, THE AMBULANCE SHALL NOT BE OVERLOADED. THE PURCHASER SHOULD CONSULT WITH THE AMBULANCE AND/OR CHASSIS MANUFACTURER(S) TO DETERMINE ACTUAL RESERVE CAPACITY ABOVE THE SPECIFIED MINIMUM REQUIRED HEREIN.

Unless additional payload is specified (see 6.2-l) by the purchaser, the required minimum payload per vehicle shall be as follows:

1. Single, rear wheeled, modular ambulances (Type I or III)—680 kg (1500 lbs.)
2. Single, rear wheeled, van ambulances (Type II)—771.11 kg (1700 lbs.)
3. Dual, rear wheeled, modular ambulances (Type I or III)—793.79 kg (1750 lbs.)

The payload shall be evenly distributed throughout the vehicle and shall consist of the following:

1. Driver and EMT at 79 kg (175 lbs.) each
2. Two patients at 79 kg (175 lbs.) each
3. Main and portable oxygen bottles
4. Stretchers, cots, and patient handling equipment
5. Portable, removable, medical devices
6. Durable and disposable medical items
7. Optional vehicle equipment such as battery charger, inverter, or any other optional items or equipment not supplied as a base vehicle requirement
8. Communications equipment
9. Extrication and rescue equipment

NOTICE: THIS VEHICLE, AS MANUFACTURED, CONFORMS TO THE PAYLOAD REQUIREMENTS OF THE FEDERAL AMBULANCE SPECIFICATION KKK-A-1822. USERS SHALL NOT LOAD VEHICLE ABOVE THE GVWR. USERS SHOULD DETERMINE THAT THE ACTUAL LOAD, TO BE PLACED ON THE VEHICLE, DOES NOT EXCEED THE USABLE PAYLOAD.

Ambulance Manufacturer's Name
Chassis Model, Year of Manufacture

1.	Vehicle Type and Class	
2.	Chassis Manufacturer GAWR-Front	kg(lbs.)
3.	Chassis Manufacturer GAWR-Rear	kg(lbs.)
4.	Chassis Manufacturer GVWR	kg(lbs.)
5.	Minimum Payload per KKK-A-1 822 or as Specified. (Para. 3.5.2)	kg(lbs.)
6.	Curb Weight-Front Base Vehicle	kg(lbs.)
7.	Curb Weight-Rear Base Vehicle	kg(lbs.)
8.	Curb Weight Base Vehicle (Item 6 Plus Item 7) (Para. 3.5.1)	kg(lbs.)
9.	Payload of Basic Vehicle (Item 4 Minus Item 8) Must Meet or Exceed Item 5	
10.	Total Weight of Options as Specified on Vehicle	kg(lbs.)
11.	REMAINING USABLE PAYLOAD (ACTUAL WEIGHT USER MAY ADD) (Item 9 Minus Item 10)	kg(lbs.)

CAUTION: THIS STICKER/PLATE SHALL BE MOUNTED ON THE BODY (MODULE) IN A CONSPICUOUS LOCATION.

3.5.3 GROSS VEHICLE WEIGHT RATING (GVWR)

The ambulance gross vehicle weight rating shall equal or exceed the combination of the vehicle's curb weight and payload weight. The chassis manufacturer's rating label shall show the actual gross vehicle weight rating (GVWR) and the gross axle weight rating (GAWR) of the vehicle (also see 3.19).

3.5.4 WEIGHT DISTRIBUTION

To provide for maximum safety, purchasers and manufacturers shall locate vehicle mounted components, equipment, and supplies to provide a vehicle that is laterally balanced and has front/rear loading that is proportional to axle ratings. The weight distribution of a properly loaded ambulance on a level surface shall permit conformance to the braking requirements in FMVSS 105. The right and left wheel of each axle (at curb weight see 3.5.1) of a completed ambulance, shall be weighed to determine weight distribution. The weight between The right and left side of a given axle, when on a level surface, shall be within 5 percent when calculated in accordance with 4.4.3.1. The ambulance manufacturer shall locate the center of gravity (CG) of The ambulance/ ambulance body to determine and assure the purchaser that the CG of the completed ambulance does not exceed the maximum limits set by the chassis manufacturer. Consideration shall be given by the purchaser and manufacturer to locate equipment and components to permit inherently proper lateral balance, front/rear axle loading, and center of gravity. Ballasting (dead weight) shall not be used for proper weight distribution, counter balancing, leveling, or lowering the center of gravity.

3.5.5 RATINGS

Vehicle and component ratings shall be The chassis manufacturer's published ratings and shall not be raised above The chassis manufacturer's rating. When ratings are not published or chassis has been modified, The verification data of The component manufacturer's rating shall be made available to The appropriate offices of The procuring activity.

3.5.6 CAB TO AXLE (CA), TYPE I AND III VEHICLES

Cab to axle (CA) dimension of the vehicle chassis shall permit a minimum of 50 percent of the outside body length forward of the rear axle centerline, plus cab to body clearance. Bodies designed with wheel openings shall have the rear wheels centered, +/- 5 cm (2 in.) longitudinally, within the opening, and +/- 13 mm (0.5 in.) laterally.

3.6 CHASSIS, POWER UNIT, AND COMPONENTS

3.6.1 CHASSIS-FRAME

The chassis shall include the chassis manufacturer's ambulance component package or, if not available, chassis manufacturer's heavy duty components that will enhance vehicles reliability, durability, safety, and comfort. The chassis-frame and components shall be constructed to withstand the strains of on-off road service and any special service and equipment requirements specified herein. All chassis (including cab) components shall be as represented in the chassis manufacturer's technical data.

3.6.2 VEHICLE LUBRICATION

The chassis components, devices, accessories, and added equipment requiring lubrication shall be fully equipped with lubrication fittings, as provided by the chassis or equipment manufacturer.

3.6.3 POWER UNIT, ENGINES

3.6.3.1 POWER UNIT
The power unit shall meet or exceed the required vehicle performance specified herein (see 3.4), at not more than the engine manufacturer's recommended operating engine speed. It shall be of such design and construction that it will provide a smooth flow of power, at all RPM's, without undue vibration, strain, or overheating of engine parts.

3.6.3.2 ENGINE LOW TEMPERATURE STARTING
The engine shall start satisfactorily without the aid of preheating devices (except glow plugs) at -12°C (10°F), for the diesel engine, and -28.9°C (-20°F) when a gasoline engine is specified (see 6.2-j). When specified by the purchaser, auxiliary engine/battery heating and/or starting devices (see 3.6.3.2.1) shall be provided to meet other starting requirements. Verification test to be performed during environmental test.

3.6.3.2.1 POWER PLANT HEATERS, MINIMUM -34° C (-30° F) STARTABILITY
In addition to the chassis manufacturers standard block heater on diesel engines, when specified (see 3.15.3-16)), the manufacturer's standard power plant heater(s) (coolant or immersion type crankcase heaters and battery heaters) shall be furnished for the gasoline engine, which permits starting an engine at -34°C (-30°F).

3.6.3.3 DIESEL ENGINE
The chassis manufacturer's highest horsepower/displacement diesel engine and power train shall be provided, complying with requirements of 3.4. Net horsepower rating shall conform to SAE J1349. Additionally, the chassis manufacturer's block heater shall also be furnished.

3.6.3.4 GASOLINE ENGINE
When specified (see 6.2-j), a gasoline engine shall be furnished which permits warranted operation on unleaded gasoline, when operated in accordance with OEM's manual. When specified (see 6.2-yy) or required for export, the vehicle shall be capable of warranted operation on either leaded or unleaded gasoline. Engine horsepower and torque requirements shall be sufficient to comply with the requirements specified in 3.4. Net horsepower ratings shall conform to SAE J1349. Vehicle(s) being offered/supplied with catalytic converters shall be identified to all purchasers.

NOTICE: PURCHASERS SHOULD BE AWARE THAT, BECAUSE OF THE SEVERE OPERATING CONDITIONS TO WHICH AN AMBULANCE IS EXPOSED, THE USE OF A GASOLINE ENGINE RESULTS IN SUBSTANTIALLY HIGHER UNDERHOOD AND UNDER VEHICLE TEMPERATURES WHICH COULD SERIOUSLY IMPACT VEHICLE RELIABILITY, DEPENDABILITY, SAFETY, AND MAY RESULT IN SOME MANUFACTURER'S CHASSIS WARRANTY BEING VOIDED.

3.6.4 POWER UNIT COMPONENTS

3.6.4.1 OIL FILTER
The oil filter shall be the chassis manufacturer's standard for the engine offered.

3.6.4.2 AIR FILTER
The air filter shall be the chassis manufacturer's standard for the engine offered.

3.6.4.3 AIR POLLUTION CONTROL (see 6.2-k)
Vehicles destined for the 50 states, the District of Columbia, Puerto Rico, the Virgin Islands, Guam and American Samoa shall comply with the Environmental Protection Agency (EPA) regulations governing Control of Air Pollution from New Motor Vehicles and New Motor Vehicle Engines in effect on date of manufacture of the chassis/engine. In addition, vehicles destined for California shall comply with State of California regulations governing air pollution control in effect the date of manufacture. When specified (3.15.3-17), the vehicle shall comply with the EPA high altitude regulations. For export vehicles, the chassis/engine manufacturer's export vehicle emission package shall be furnished that complies with destination requirements.

3.6.4.4 FUEL SYSTEM
The fuel system shall conform to FMCSR, Subpart E, 393.65 and 3.4.9 herein. Unless otherwise specified (see 6.2-e), fuel tank(s) capacity shall be the maximum available from the chassis manufacturer. When more than one tank is furnished, tanks shall be interconnected and, when available from the chassis manufacturer, separately controlled and shall have a dash-mounted fuel gauge controlled by a selector switch which will permit separate fuel level readings for each tank. When available from the chassis manufacturer, the Class 2 ambulance shall have the fuel tank(s) protected by a metal shield (skid-plate) and shall be located in an area which will he free from hazards encountered in off-road, cross-country operation. A permanent plate at the fuel filler opening shall be furnished specifying type of fuel required.

3.6.4.5 COOLING SYSTEM
The engine cooling system shall be a closed, air free, liquid state type with a coolant overflow recovery tank. The supplier shall provide the heaviest duty components and maximum size cooling system available from the chassis manufacturer. The cooling system design shall maintain the engine at safe operating temperatures at all drivable altitudes and grades encountered during on and off road vehicle use. Verification test: the cooling system shall be capable of maintaining a safe stabilized engine operating temperature for a minimum of one hour, at a minimum ambient temperature of 38˚C (100˚F), at any altitude up to 3048 m (10,000 ft). For gasoline engines, the cooling system shall be capable of maintaining a safe operating temperature for not less than 40 minutes, at sea level, in a minimum ambient temperature of 35˚C (95˚F). Testing shall be at the engine speed required to maintain the generating systems as specified in 3.7.6.

3.6.4.5.1 ANTIFREEZE
The cooling system shall be protected with a 50/50 solution of permanent type antifreeze.

3.6.4.6 EXHAUST SYSTEM
Vehicle shall be equipped with an exhaust system in accordance with Federal Motor Carrier Safety Regulations, Part 393.83. The exhaust system shall be suspended using not less than three hangers, excluding the manifold attachment. The exhaust shall discharge at the vertical side(s) of the ambulance at a maximum distance of 25 mm (1 in.) beyond the side of the module away from the door(s) to minimize fumes and contaminants entering the

interior. On modular vehicles, the tailpipe outlet shall not terminate within 15 cm (6 in.) of the vertical axis of the fuel tank filler opening(s). Modifications or extensions made to the OEM exhaust system shall meet or exceed chassis manufacturer's requirements in terms of back pressure, components, design, and workmanship.

3.6.5 DRIVE TRAIN

3.6.5.1 DRIVE TRAIN COMPONENTS

The drive train and component's torque capacity shall meet or exceed the maximum torque developed in the lowest gear ratio by the engine. Transmission and controls shall comply with FMVSS 101 and 102.

3.6.5.2 AUTOMATIC TRANSMISSION

Unless otherwise specified (see 3.6.5.2.2), the chassis manufacturer's automatic transmission shall be provided. The transmission shall provide not less than three speeds forward and one reverse and shall be equipped with the chassis manufacturer's heaviest duty oil cooler.

3.6.5.2.1 AUTOMATIC TRANSMISSION AUXILIARY HEAVY-DUTY OIL COOLER

When specified (see 3.15.3-20). A heavy-duty fin type auxiliary transmission oil cooler in addition to the OEM manufacturer's basic cooler (3.6.5.2) shall be provided and installed, in series, in a suitable location except behind the radiator.

3.6.5.3 TRANSFER CASE FOR CLASS 2 AMBULANCE

Unless otherwise specified (see 6.2-1), a part-time four wheel drive system shall be provided for Class 2 vehicles. These vehicles shall be furnished with 2-speed transfer case that is integral with the main transmission. A shift pattern located on the shift knob or decal, within the cab, shall be provided. Front-wheel drive hubs (see 3.6.5.9) and special traction rear-end differential (see 3.6.5.8), shall also be furnished on 4 × 4 vehicles. When furnished as part of a Class 2 conversion, the differential shall be identical to or heavier duty than the unit furnished on the OEM's original 4 × 4 Type I, Class 2 chassis.

3.6.5.3.1 HEAVY-DUTY SKID PLATE (CLASS 2, (4x4))

Unless otherwise specified (see 6.2-I) and when available from the chassis manufacturer, a heavy-duty skid plate shall be installed protecting engine, transmission, and transfer case from ground contact if there is less than 38 cm (15 in.) of clearance to the ground. Skid plate shall be demountable for servicing the engine and transmission. Openings shall be provided to enable draining of transmission and servicing the underside of the engine. For fuel tank protection, see 3.6.4.4.

3.6.5.4 CLUTCH

Vehicles with manual transmissions shall be furnished with the manufacturer's largest, heavy-duty clutch available for the engine and model offered.

3.6.5.5 DRIVE-LINE

The drive-line (drive shaft, U-joints, etc.) shall be balanced and supported to perform throughout the design speed range without whipping or vibrating. Modifications resulting from lengthening the wheelbase or from 4 × 4 conversion shall be of chassis manufacturer's approved design.

3.6.5.6 AXLE, RATINGS, RATIOS

Axle ratings shall exceed the load imposed on each axle, measured at the ground, when vehicle is loaded. When specified (see 6.2-I) that the vehicle is used in mountainous terrain. manufacturer shall provide the highest numerical axle ratio. The chassis OEM "wide dual" rear axle shall be furnished on dual rear wheel chassis, unless otherwise specified (see 6.2-I) for a narrow modular body.

3.6.5.7 BRAKE SYSTEM SERVICE AND PARKING

Chassis manufacturer's heaviest duty, power assisted brakes, linings, and parking brake shall be furnished on the chassis model offered.

3.6.5.8 SPECIAL TRACTION (REAR END) DIFFERENTIAL

Unless otherwise specified (see 6.2-1) or if not available from chassis manufacturer, all vehicles shall have a positive traction, limited slip differential or automatic, locking type differential.

3.6.5.9 DRIVE HUBS, FRONT WHEEL, (4 × 4, CLASS 2)

Unless otherwise specified (see 6.2-1), and when available from the chassis manufacturer, automatic type front wheel locking hubs shall be provided. Additionally, when a 4 × 4 conversion is furnished, the heaviest duty front hubs available from the axle manufacturer shall be furnished.

3.6.5.10 SUSPENSION

Vehicle shall be equipped with matched sets (front and rear) of spring, torsion, or air suspension system components. Components shall have a rated capacity in excess of the load imposed on each member. For better riding quality, vehicle springs shall have chassis manufacturer's lowest clamped deflection rating (smoothest riding suspension) for the model offered. Suspension stabilizer bar(s) shall be furnished for front and rear, when available from chassis manufacturer for the model chassis furnished. Only corrections permitted by the chassis manufacturer to compensate for lean due to normal spring tolerance variations are permitted. Correction of lean due to imbalance is not permitted. VEHICLE SHALL BE BALANCED.

3.6.5.11 SPRING STOPS

The chassis manufacturer's standard spring bumpers and axle stops shall be furnished. The stops/bumpers shall prevent the wheel and axles from striking the engine, oil pan, fenders, and body under all conditions of operation. Additionally, when a 4 × 4 conversion is furnished, spring stops/bumpers shall be provided and, at a minimum, be of OEM quality.

3.6.5.12 SHOCK ABSORBERS

Shock absorbers, double-acting type, heaviest duty available from chassis manufacturer for model offered, shall be furnished on the front and rear axles, Class 2 (4 × 4) vehicles shall additionally be furnished with quad front shocks when available from chassis manufacturer.

3.6.6 STEERING

The OE chassis manufacturer's standard, power assisted steering shall be furnished.

3.6.7 WHEELS

Unless otherwise specified (see 6.2-m), Types land III vehicles shall be equipped with dual rear wheels. Type II vehicles shall be equipped with single, front and rear wheels. Wheels shall conform to the recommendations of the Tire and Rim Association, Inc,, and shall be identical in type, size, and load rating for all wheels on the vehicle, including the spare.

3.6.8 TIRES

Unless otherwise specified or permitted by purchaser (see 6.2-n), tires shall be OEM tubeless, steel belted radials with "All Season Type Tread" for the GVWR of the vehicle. All tires furnished shall be identical and comply to FMVSS 120. To provide a softer ride for a loaded ambulance, the tires shall be inflated only to the tire manufacturers cold inflation air pressure absolutely necessary to support the load on each tire measured at the ground. The manufacturer shall provide instruction for the correct front and rear tire pressures required for the ambulance GVWR.

3.6.9 INNER TUBES

When tube type tires are furnished, inner tubes shall conform with the recommendations of the Tire and Rim Association, Inc.

3.6.10 SPARE TIRE AND STORAGE

Unless otherwise specified by the purchaser (see 6.2-o), one inflated spare wheel/tire assembly identical to those on the vehicle shall be furnished. The spare assembly shall be stored in an accessible weather protected compartment or area. When available from the chassis manufacturer, the spare tire shall be located under the rear of the vehicle. The spare tire assembly shall be accessed by the chassis manufacturer's winch type carrier. An equivalent winch carrier assembly may be retrofitted on other chassis. When locating spare tire assembly under the vehicle, all other specification requirements (e.g. departure angle, ground clearance, weight distribution) shall be in compliance. Access door shall be fitted with suitable latch and lock when accessible from outside the vehicle. Loose or removable panel(s)/door(s) shall not be acceptable. The spare tire assembly shall be accessible without removal of the patients from the compartment. The spare assembly and tools shall have restraining device(s) to eliminate rattling and dislocation in case of an accident. The carrier design shall enable removal and mounting of the spare assembly using only the tools specified in 3.6.13 (see 3.11.3).

3.6.11 TIRE CHAIN CLEARANCE

Tire chain clearance shall be provided for all driving wheels per SAE J683. Additionally, sufficient chain clearance shall be provided to permit off road operation with the ambulance loaded to the maximum payload.

3.6.12 WHEEL TIRE BALANCING

Wheel/tire, hubs, and brake drum assemblies of the vehicle shall be in balance to a minimum of 112 km/hr (70 mph).

3.6.13 TOOLS (TIRE CHANGING)
Vehicle shall be furnished with tools required for exchanging mounted tire assembly with the spare assembly and shall include at least a jack, jack handle, and wheel nut wrench. The jack height, when closed, shall permit its location under the axle or other satisfactory lift point at any wheel with the tire flat. The jack, without having to block the jack, shall be capable of raising any wheel of the loaded vehicle to a height adequate to permit removal and replacement of the wheel and tire assembly. Tools shall be stored in accordance with 3.6.10.

3.6.14 HUBCAPS
Manufacturer's (OEM) standard hubcap or wheel cover shall be fitted on each wheel except the spare. Vehicles with dual wheels and Class 2 modular vehicles (see 3.16.2.1) do not require hubcaps.

3.7 ELECTRICAL SYSTEM AND COMPONENTS

3.7.1 ELECTRICAL SYSTEM
The ambulance electrical system shall be equipped with, but not limited to, the following: dual, chassis manufacturer's 12 volt batteries; generating, starting, lighting, visual and audible warning systems; specified electronics equipment and devices (including master consoles located in the cab and patient compartment); and other specified accessory wiring. The electrical systems and equipment shall comply with all applicable FMVSS, Federal Motor Carrier Safety Regulations (FMCSR), and shall also conform to all the applicable SAE recommended standards and practices, whether or not specifically referenced in this document, while complying with the subparagraphs herein. ALL ELECTRICAL AND ELECTRONIC COMPONENTS SHALL BE SELECTED TO MINIMIZE ELECTRICAL LOADS THEREBY NOT EXCEEDING THE VEHICLE'S GENERATING SYSTEM CAPACITY. All electrical system components and wiring shall be readily accessible through access panels for checking and maintenance. All switches, indicators, and controls shall be located and installed in a manner that facilitates easy removal and servicing. All exterior housings of lamps, switches, electronic devices, and fixtures shall be corrosion resistant and weatherproofed.

Electrical fixtures attached to the sides of the ambulance below the 191 cm (75 in.) level shall be near flush mounted and not protrude more than 51 mm (2 in.), except for such items as spotlights and ventilators. All electrical devices and equipment installed, which produce RFI, shall include filters, suppressors, or shielding to prevent electromagnetic radiation and the resultant interference to radios and other electronic equipment (see 3.7.12). Vehicles equipped with electronic engine controls shall be immune from interference caused by radio transmissions.

WARNING: ELECTRICAL LOADS SHALL BE MINIMIZED. ADDITIONAL ELECTRICAL LOADS ABOVE THOSE REQUIRED BY THIS SPECIFICATION SHOULD BE AVOIDED BECAUSE HEAVY LOADS CAUSE ALTERNATOR AND BATTERY FAILURES. FAILURE OF THESE COMPONENTS WILL LIKELY RESULT IN THE AMBULANCE NOT BEING ABLE TO COMPLETE ITS MISSION.

The electrical system shall include patient compartment outlets for 12 volt power (see 3.7.7.3) for medical equipment. A driver compartment console-mounted "module disconnect switch or device" (see 3.7.7.4) shall be provided which controls the equipment as defined in Figure 5 [omitted].

3.7.1.1 WARNING INDICATORS

The electrical system shall incorporate a warning light panel located in the driver's compartment. It shall provide indicator lights for showing: open patient compartment entry door(s) (see 3.10.8); open equipment compartment door(s); and when batteries are turned on by the battery disconnect switch (see 3.7.7). The "Door Open" warning lights shall be red, flash, and approximately 13 mm (1/2 in.) in diameter, or equal, in area. The battery indicator light shall be green with a lens approximately 13 mm (1/2 in.) in diameter, or equal, in area. Warning indicators shall be identified and marked per 3.7.11.

3.7.2 WIRING INSTALLATION

The ambulance body and accessory electrical equipment shall be served by circuit(s) separate and distinct from vehicle chassis circuits. All wiring provided by the ambulance manufacturer shall be copper and conform to all the SAE J1292 requirements and shall have type SXL or GXL high temperature crosslinked polyethylene, or better, insulation conforming to SAE 31127 and ii 128. The use of multiconductor or ribbon cables are permitted provided they are not exposed to underhood or under vehicle temperatures/ conditions. The wiring shall be permanently color coded or marked the entire length of the wire for identification with easily read numbers and/or letters and routed in conduit or high temperature looms with a rating of 149°C (300°F). When cables are supplied by a component manufacturer to interconnect system components, these cables need not be continuously color coded/identified. They shall be coded/identified at the termination or interconnection points. All added wiring shall be located in accessible, enclosed, protected locations and kept at least 15 cm (6 in.) away from exhaust system components. Electrical wiring and components shall not terminate in the oxygen storage compartment except for the oxygen controlled solenoid, compartment light, and switch plunger or trigger device. Wiring necessarily passing through an oxygen compartment shall be routed in a metallic conduit (see 3.11.3). All conduits, looms, and wiring shall be secured to the body or frame with insulated metal cable straps in order to prevent sagging and movement which results in chafing, pinching, snagging, or any other damage. All apertures on the vehicle shall be properly grommeted for passing wiring and conform to SAE 1292. All items used for protecting or securing the wiring shall be appropriate for the specific application and be standard automotive, aircraft, marine, or electronic hardware.

Electrical panels that are accessible to accidental contact shall have a protective cover, shield, etc. to prevent shorts that can result in injury, fire, or damage to the electrical system.

3.7.2.1 WIRING CRITERIA

All wiring (including grounds), devices, switches, outlets, etc., except circuit breakers, shall be rated to carry at least 125 percent of the maximum ampere load for which the circuit is protected.

A service loop of wire or harness shall be provided at all electrical components, terminals, and connection points that permits at least two replacements of the wire(s) termination. All splices and terminals provided shall comply with SAE J163, J561, or J928 as applicable. All wiring between the cab/ chassis and module in Type I and III ambulances shall be connected to a terminal strip(s), block(s), or multi-pin connector(s) near the point of entry to the patient module. All terminals shall be permanently numbered or coded. Terminal strip(s), block(s), or multi-pin connector(s) shall be readily accessible for checking and service.

The ambulance electrical system shall incorporate a master circuit breaker panel with circuit breakers or other electronic, non-disposable, current protection devices, in each circuit, which comply with SAE 3553 (Type I) or SAE 3553 (Type II), if circuit breaker is readily accessible for resetting by the driver or EMT. When multiconductor cables/ ribbon cables are used for low current (self limiting) circuits, additional fuses/circuit breakers are not required. Additionally, one 15 ampere circuit breaker shall be provided for future use. For high current circuits, where SAP Type I breakers are not commercially produced, protection for these circuits may be provided with other types of circuit breakers. All circuit breakers shall be securely mounted, easily removable, and readily accessible for inspection and service. All electrical and electronic components, switches, connectors, circuit breakers, lamps, and indicators, including the vehicle batteries, shall be marked with an easily read identification code number and/or letter. Complete, highly legible, wiring diagrams and schematics, including identification codes and parts list for the ambulance's standard and optional equipment furnished, shall be included in the service manual and be supplied with each ambulance in accordance with paragraph 6.8.

To provide RF grounding, the module shall be connected to the chassis with a minimum 19 mm (3/4"), braided, ground strap with soldered ends that are secured to clean metal surfaces on the body and frame with star washers, etc. (Note: Regular stranded copper wire while providing a DC ground, does not provide RF grounding.)

3.7.2.2 PRINTED CIRCUITS

When printed circuits are utilized, they shall conform to SAE J771. Printed circuit boards shall be securely mounted and protected from physical damage and accidental shorts. Printed circuit board connections and components shall conform to all other specification requirements.

3.7.3 SPACE RESERVED

3.7.4 WINDSHIELD WIPERS AND WASHERS

Vehicle shall be equipped with dual, electric, multi-speed, windshield wipers and washer complying with FMVSS 104. When specified (see 6.2-p), the OEM intermittent wipers shall be furnished.

3.7.5 HORNS

The chassis manufacturer's dual electric horns shall be furnished (see siren section for switching arrangements, (3.14.6)).

3.7.6 ELECTRICAL GENERATING SYSTEM

Unless otherwise specified (see 6.2-q), the ambulance shall be equipped with a generating system rated at not less than 135 amperes, at 14 volts, with an underhood temperature of 93° C (200°F). As a minimum, the generating system shall be capable of supplying at its regulated voltage, at 93°C (200°F), the continuous electrical load which consists of the following electrical equipment and systems: engine/transmission control system; headlights (low beam); all FMVSS 108 lights; windshield wipers (low speed); cab air conditioning (at coldest setting with highest blower speed); radio in receiving mode (or equal load, if not equipped); patient module dome lighting (in the high intensity setting); patient module air conditioning (at coldest setting with highest blower speed); emergency warning lighting system (in the daytime "primary" mode) (3.8.2); and 20 amp medical load or equal.

WARNING:
1. ADDITIONAL (OPTIONAL), CONTINUOUS LOAD, ELECTRICAL DEVICES OR SYSTEMS, (ADDITIONAL WARNING LIGHTS ON BOARD 115 VOLT AC SUPPLY, ETC.), ARE PERMITTED, PROVIDING RESERVE OR ADDITIONAL GENERATING CAPACITY IS AVAILABLE OR ADDED.

2. MANUFACTURERS/PURCHASERS SHOULD NOTE THAT THE "COLD", MAXIMUM RPM, OUTPUT RATING OF THE ALTERNATOR MAY BE SUBSTANTIALLY HIGHER THAN THE HOT 930 C (2000 F) OPERATIONAL SPEED RATINGS. FOR EXAMPLE, THE COLD, HIGH SPEED RATING MAY BE 165 AMPS WHILE THE HOT OPERATIONAL SPEED RATING MAY BE APPROXIMATELY 135 AMPS.

The generating system shall supply the maximum electrical load, at the regulated voltage, at 930 C (2000 F) underhood temperature, and with an engine speed not exceeding 40 percent of the furnished engine's SAE net HP, RPM rating. The throttle control device, specified in 3.7.6.1, shall control the engine RPM necessary to maintain the heating and air conditioning systems, at full operating capacity, and to maintain the generating system's required output when the vehicle is stationary. The 12 volt electrical system shall incorporate an ammeter (see 3.7.6.2) and a voltmeter or voltage warning device (sec 3.7.6.3) which are functionally connected as shown in Figure 5 [omitted]. The final stage manufacturer/supplier shall test each ambulance prior to delivery and provide, to the purchaser, a written certification (tag) indicating the amount of generating capacity remaining, at the regulated voltage, at 93°C (200°F), after supplying the total electrical load as manufactured (including the purchaser options). Testing and tagging shall be in accordance with "AMD Standard 005, Ambulance Electrical System."

3.7.6.1 ENGINE HIGH-IDLE SPEED CONTROL, AUTOMATIC

An engine high-idle speed control shall be furnished on all vehicles which automatically increases the engine speed (RPM) to sustain the ambulance's total continuous electrical load and maximum heating/air conditioning output. The device shall be preset so that, when activated, it will operate the engine at the appropriate RPM (see 3.7.6). The device shall operate only when switched to the "ON" position and the transmission is in "NEUTRAL," "PARK." For transmissions without a "PARK" position, the device shall only function with the transmission in "NEUTRAL" and with the parking brake applied. The device shall disengage when the operator depresses the service brake pedal or the transmission is placed in gear, and automatically reengages when the service brake is released or when the transmission is placed in neutral or park. The device shall be furnished by the chassis manufacturer when available.

NOTE: DIESEL ENGINE MANUFACTURERS STATE THAT THEIR ENGINES SHOULD BE SET AT AN ELEVATED SPEED DURING EXTENDED IDLE PERIODS. ADDITIONALLY, THE ELEVATED IDLE SPEED REDUCES TORSIONAL VIBRATION WHICH CAUSES ALTERNATOR BELT SLIPPAGE.

3.7.6.2 AMMETER

The electrical system shall incorporate a center scale ammeter or equivalent electronic digital display which is capable of indicating a current of +/- 150 amperes to or from (charging or discharging) the vehicle's batteries. The ammeter, or equivalent electronic, digital display shall incorporate an external shunt or current monitoring device which does not exceed 150 millivolts at maximum current. The ammeter and shunt, or equivalent device, shall have a combined accuracy of +/- 10 percent of the full scale reading. The meter shall be mounted in a location highly visible to the vehicle operator and shall be illuminated for night operation. The shunt or monitoring device shall be protected against physical damage, weather road spray, and shall be mounted in an easily accessible location which shall minimize the length of the power cables.

3.7.6.3 VOLTMETER OR VOLTAGE MONITOR

A voltmeter, or equivalent electronic device, illuminated for nighttime operation, which constantly monitors the 12 volt electrical system or a warning device and indicates abnormally high or low electrical system voltages, shall be furnished. The device furnished must be mounted so it is clearly visible to the driver at all times.

3.7.7 BATTERY SYSTEM

Two 12 volt batteries (or additional batteries as required by the chassis manufacturer), with a labeled "Battery Disconnect Switch," shall be furnished. A switch operated solenoid may be employed to provide the battery disconnect function. Unless otherwise specified (see 3.15.3-5), the batteries shall be equivalent to the chassis OEM batteries. When high cycle batteries are specified (Delco 1150 or 1151, or equal), performance ratings for each battery shall not be less than 625 cold cranking amps and 180 minutes reserve capacity.

Battery ratings shall conform to SAE J537. Batteries shall be located in a ventilated area, sealed off from occupant compartments, and shall be readily accessible for servicing and removal. When batteries are mounted in the engine compartment, they shall be provided with a heat shield as a safeguard against high underhood temperatures.

Unless the "Battery Disconnect Switch" is clearly visible to the driver, in the seated position, a green indicator light, indicating the batteries are "ON" (see 3.7.1.1), shall be furnished. Battery selector device wiring and added/modified starting motor circuits shall meet or exceed the SAE J 1127 for high temperature SGX wire and SAE J541 for maximum voltage drop requirements for 12 volt heavy-duty applications. The "Battery Disconnect Switch" shall be wired per Figure 5 [omitted], page 87, and shall select both batteries simultaneously

When specified (see 6.2-R), the batteries, on Types I & Ill vehicles, shall be located in the module, sealed from the interior, or on Type II vehicles, in a compartment or lower skirt. Batteries shall be on a slide out tray or be readily available for service. When relocating batteries, the method of relocation shall be approved by the chassis manufacturer.

NOTE: MODULE BATTERY LOCATIONS CAN PROVIDE FOR IMPROVED FRONT/ REAR AND LATERAL WEIGHT DISTRIBUTION AND FOR A LOWER CENTER OF GRAVITY.

3.7.7.1 BATTERY CHARGER OR CONDITIONER

When specified (see 3.15.3-7), either a 12 volt DC taper type battery charger or automatic conditioner shall be provided. The charger/conditioner shall be listed by an organization meeting the requirements in Paragraph 3.7.8 and connected to the dual 12 volt DC battery system as shown in Figure 5 [omitted]. The charger shall be capable of supplying a minimum of 10 amperes charging current. The charger/conditioner shall be permanently mounted, in the vehicle, in a properly ventilated, accessible location and wired to the 115 volt AC utility power as specified in 3.7.8 and Figure 6 [omitted]. Battery charger controls shall include an "ON"/"OFF" switch and a charging indicator. When a battery conditioner is provided, it shall monitor the battery state of charge and, as necessary, automatically charge or maintain the batteries without gassing, depleting fluid level, overheating, or overcharging.

3.7.7.2 SPACE RESERVED

3.7.7.3 INTERNAL 12 VOLT DC POWER

Unless otherwise specified by the purchaser (see 6.2-s), the patient compartment shall be furnished with a 12 volt DC, 20 ampere capacity, separately protected circuit, with two outlet receptacles. This circuit shall also include a (low voltage drop) "Schottky" diode to isolate medical equipment batteries from any electrical loads that the remainder of the ambulance electrical system may impose. The "Schottky" diode shall be heat-sink mounted, have an inverse voltage rating of at least 45 volts and also be rated to carry the maximum short circuit current, until the circuit breaker opens. The diode shall be physically located in an accessible location and be electrically connected between the circuit breaker and the "action wall" mounted receptacle. Unless otherwise specified (see 6.2-s), the receptacles shall be a military type connector of the following generic designation, MS3112E12-3S or its interchangeable commercial equivalent. The polarity of the connector shall be as follows: Pin A (+ 12V), Pin B (Ground), Pin C (not used). The receptacles shall be located on a vertical surface of the "action wall" or other location specified by the purchaser. The mating plug attached to the medical equipment shall be an MS3116F12-3P or its interchangeable commercial equivalent. The polarity for the plug shall be the same as above. Two of these unwired plugs shall be furnished and tagged with polarity requirements and shall be connected to the receptacles. (NOTE: These connectors are widely available directly from most major industrial electronics distributors.)

When required or specified (see 6.2-s) for constant power source for medical equipment battery charging, a 12V protected circuit shall be furnished that connects directly to the battery side of the load disconnect switch.

3.7.7.4 MASTER MODULE DISCONNECT SWITCH OR DEVICE

This switch shall be located in the driver's compartment, be legibly marked, illuminated when "ON", and rated to carry at least 125 percent of the circuit's maximum current.

3.7.8 115 VOLT OR 115/230 AC UTILITY POWER

Unless otherwise specified, the ambulance shall be furnished with a 2-wire plus ground

115 volt AC wiring system that is separate and distinct from the vehicle's 12 volt DC wiring system(s). The AC electrical system, including wiring and associated equipment, shall comply with AMD Standard 009. Listing shall be by a nationally recognized testing laboratory, recognized by OSHA under Appendix A to 29 CFR 1910.7. The AC system is to be utilized while the vehicle is stationary for powering maintenance devices, medical equipment battery chargers, and when specified (see 3.15.3 & 6.2), vehicle battery conditioner/charger, and any other device(s) deemed necessary by the purchaser. The AC system shall incorporate a ground fault circuit interrupter (GFCI) device and a minimum 15 ampere circuit breaker which can be used as a master AC disconnect switch. The GFCI and circuit breaker may be an integral unit. The AC wiring shall utilize stranded wire in armored cable, nonmetallic sheathed cable, "Type SO" or better flexible cable rated at 600V and 900° C, covered with a minimum 1490° C (3000° F) flame retardant wire loom, or approved wire in conduit.

When an onboard AC supply is specified (see 3.7.8.3), an automatic transfer switch shall be furnished which turns off this onboard AC supply (interlock) and disconnects its output, when the AC utility power is applied. When the 12V DC battery charger (see 3.15.3-7) and any power plant heater(s) (see 3.15.3-16) are furnished and connected to this system, they shall be wired so that they can be energized only from the utility power, and not the onboard AC supply. The onboard AC system shall not be utilized for operational ambulance interior lighting, such as dome and cotlights.

3.7.8.1 UTILITY POWER CONNECTOR

Unless additional capacity is required, a 115 volt AC (male) plug (NEMA 5-15P), rated at 15 amps (similar to a Levitan 4937, Arrow-Hart 527WP or equal), with spring loaded cover assembly suitable for wet locations, shall he installed on the driver's side of the ambulance body in close proximity to driver's door. The connection shall be permanently labeled with the following:

THIS CONNECTION IS FOR [115] VOLT AC, 60 Hz, 15 AMPERE SUPPLY.

NOTE: THE ABOVE LABEL SHALL REFLECT ACTUAL VOLTAGE AND CURRENT CONNECTIONS IF OTHER THAN SHOWN.

This receptacle shall energize the vehicle's internal AC circuit(s) from an exterior power source (utility power). The purchaser's stationary utility power circuit supplying the ambulance's 115 volt AC power should incorporate ground fault protection. A proper mating, weatherproof, 15 ampere or more when required, female receptacle (NEMA 5-15R) shall also be furnished without cable and tagged specifying the size, type of wire necessary, and the polarity of the future hookup.

3.7.8.2 ELECTRICAL 115 VOLT AC RECEPTACLES

The patient compartment shall be furnished with a pair of 2-wire plus ground duplex 115 volt AC receptacles. Receptacles shall be near flush mounted with one outlet located on the vertical, primary patient, action wall and the other vertically located at the torso area of the secondary patient (squad bench). Both outlets shall be at least 31 cm (12 in.) from any oxygen outlet. An indicator shall be located within each 115 volt AC receptacle as a line monitor indicating a live (hot) circuit. The receptacles shall be labeled with the following: "115 VOLT AC."

3.7.8.3 SOLID STATE INVERTER FOR ON BOARD 115 VOLT AC POWER

When specified under 3.15.3-6, a DC to AC inverter shall be provided in conjunction with the onboard 115 volt AC wiring system (3.7.8.) The device furnished shall be capable of continuously delivering at least 750 watts of sine wave type regulated AC power to safely power all types of electrical or electronic loads and maintain a frequency of 60 +/-4 Hz. The single phase RMS output sine wave type regulated voltage shall not rise to more than 135 volts nor drop to less than 105 volts from no load to full load, at an 80 percent power factor, over the 12.8 to 14.8 range of the DC input voltage. The device shall be capable of operating all types of AC loads, including resistive and reactive (either inductive or capacitive) over the temperature range in 3.4.2. The installation shall include an "ON-OFF" switch to activate the device. This switch shall be located on the EMT's control panel or cab console, have a red "ON" indicator, and be labeled with the following: "115 VOLT AC Inverter." A decal shall be provided near this switch which states that "WHEN OPERATING THE 115 VOLT AC SYSTEM, ALL UNNECESSARY 12 VOLT DC ELECTRICAL LOADS SHOULD BE TURNED OFF." The inverter shall have integral GFCI protection and shall be wired per Figures 5 and 6 [omitted].

3.7.9 DRIVER COMPARTMENT CONTROLS

In addition to the left-hand drive controls and switches, the final stage ambulance manufacturer shall provide and locate, within easy normal reach of the driver, the specified controls and instruments. The battery and module disconnect switches or devices (see 3.7.7.4) shall be different in feel from each other and be supplied with a handle or knob which feels different to the touch than the other switches or be physically isolated from them.

3.7.10 PATIENT COMPARTMENT CONTROLS

Unless otherwise specified by the purchaser (see 6.2-t), the patient compartment controls, switches, and instruments shall be panel mounted and located within normal reach of the seated EMT (3.10.3).

3.7.11 MARKING OF SWITCHES, INDICATORS, AND CONTROL DEVICES

All switches, indicators, and control devices supplied by the end product manufacturer of the ambulance shall be clearly visible to the ambulance personnel. They shall be perceptively and permanently identified with at least 12 point letters for the noun or function and 8 point for the remainder of the legend. The identifications shall be contrasting colors etched or engraved in plastic or metal, grouped according to function, mounted in illuminated or backlit panel(s) or console.

3.7.12 ELECTROMAGNETIC RADIATION AND SUPPRESSION

In addition to OEM chassis, all added electrically operated or electrical generating devices, including alternators, air conditioning, warning light systems, and medical equipment, shall be electromagnetic radiation suppressed, filtered, or shielded to prevent interference to radios and telemetry equipment aboard the vehicle and the surrounding area and shall not exceed SAE J551 limits. Type certification for these devices is acceptable. When specified by the purchaser (see 6.2-u), the completed ambulance vehicle shall be tested and certified to demonstrate that the RFI does not exceed the maximum limits of SAE J551. When specified, electrically operated medical equipment furnished shall comply to MIL-

STD-461, Electromagnetic Interference Characteristics Requirements for Aeromedical Equipment.

3.8 LIGHTING AMBULANCE EXTERIOR AND INTERIOR

3.8.1 AMBULANCE EXTERIOR LIGHTING

The basic exterior ambulance lighting shall comply to FMVSS Std. No.108 and the requirements herein and include amber colored front and rear directional signals and hazard warning lights (except on Type II ambulances, if amber lenses are not available from OEM). The lower front and rear side marker lights shall flash in conjunction with the directional signals. Backup light(s)/loading light(s), clearance lamps (when applicable), ambulance emergency lights (3.8.2), floodlights (3.8.3), and spotlight(s) (3.8.4), shall be furnished as specified. The ambulance manufacturer shall furnish light assemblies that are stainless steel, plastic, or other weather resistant materials, that are installed in a manner that will not cause electrolysis of light housings or vehicle body.

3.8.2 AMBULANCE EMERGENCY LIGHTING

(A 6.1) As specified (see 6.2-v), the strobe or halogen incandescent emergency lighting system shall provide the ambulance with 360 degrees of conspicuity for safety during its missions. The system must display highly perceptible and attention-getting signals that function in a modal system and convey the message in the PRIMARY MODE—"Clear the Right-of-Way" and in the SECONDARY MODE—"Hazard, Vehicle Stopped on Right-of-Way." The ambulance standard warning light system shall not impose a continuous average electrical load exceeding 35 amperes.

WARNING:PURCHASERS SHOULD NOT SPECIFY WARNING LIGHTS BEYOND THOSE REQUIRED HEREIN. ADDITIONAL LIGHTING WILL UTILIZE RESERVE ALTERNATOR CAPACITY AND COULD RESULT IN OVER LOADED ELECTRICAL SYSTEMS. (SEE WARNINGS IN PARAGRAPHS 3.7.1. AND 3.7.6)

Additional warning lights are not required but, if specified (see 6.2-v), shall not obscure the light output of the standard warning light system. Additional warning lights furnished shall be separately switched. Any warning devices furnished, in addition to the specified system, shall be compensated for with reserve or additional generating capacity as required in 3.7.6.

3.8.2.1 EMERGENCY LIGHTING SYSTEM CONFIGURATION

The ambulance standard emergency warning light system shall contain twelve fixed red lights, one fixed clear light and one fixed amber light. These lights shall function in a dual mode system as shown in the Table I below and meet the physical and photometric requirements of 3.8.2.2. The upper body warning lights shall be mounted at the extreme upper corner areas of the ambulance body, below the horizontal roofline. The single clear light shall be centered between the two front facing, red, upper corner lights. The standard warning lights shall not be obstructed by doors or other ancillary equipment. The single amber light shall be centered between the two rear facing red lights. The red "grille" lights shall be located at least 76 cm (30 in.) above the ground and below the bottom edge of the windshield and be laterally separated by at least 46 cm (18 in.), measured from centerline to centerline of each lamp. The lateral facing intersection lights shall be mounted as close as possible to the front upper edge of each front fender. All warning lights furnished shall be mounted to project their highest effective intensity beams on the horizontal axis, (see 3.8.2.4).

3.8.2.2 PHOTOMETRIC AND PHYSICAL REQUIREMENTS

Each emergency light shall flash on/off 75 to 80 times per minute, with each light having a minimum of 129 sq cm (20 sq. in.) of illuminated viewing area. All warning lights shall project a beam spread of at least 5 degrees up and 5 degrees down and at least 45 degrees left and right of H-V. Each light shall produce a gradual gradient of effective intensities from the H-V to all the extreme test point coordinates as shown below when tested at 13.6 volts. The effective intensities shall be determined in accordance with the Illumination Engineering Society's (IFS) Guide for Calculating the Effective Intensity of Flashing Lights.

3.8.2.3 SWITCHING ARRANGEMENTS

The emergency light switches shall be wired and arranged to provide the warning light signal modes and combinations as specified. All emergency light switches shall be labeled (see 3.7.11) and each primary/secondary mode switch shall have an amber or red indicator light to show the driver which mode is activated. When strobe lights are furnished or, when specified (see 6.2-w) for incandescent lights, a day-night switch shall be provided. When specified (see 6.2-w) from the lighting system manufacturer, an automatic switch to secondary mode in "Park" with manual override to "Primary Mode" shall be furnished. Additionally, when specified (see 6.2-w) from the lighting system manufacturer, the lighting system shall provide a ambient light the secondary mode. A sensing circuit to automatically switch to the night position while in manual override to the daytime (bright) mode shall be provided. Operators manual (3.20) shall include suggestive management instructions for the warning systems.

3.8.2.4 HARDWARE CONSTRUCTION AND INSTALLATION

The emergency lighting system shall be comprised of components and devices that comply to the general requirements and tests of SAE J575g, J576d, J578, and J551, as applicable for the unit. Warning lights shall be firmly fastened to reinforced body surfaces and include aiming wedges to compensate for sloped body surfaces or mold release angles on roof caps. The lights shall be aimed either mechanically or optically on the horizontal axis with a tolerance of +0 degrees to -3 degrees. All switches, connectors, and wiring shall he rated to carry a minimum of 125 percent of their maximum ampere load. When halogen lamps are used, the duty cycle of any lamp shall not exceed 50 percent. When strobe lights are furnished, all high voltage leads and connections shall be insulated and enclosed.

3.8.2.5 TESTS, WARNING, LIGHT SYSTEM

The lighting manufacturers shall furnish and certify or the ambulance manufacturer shall measure and record the total average current toad of the standard emergency warning light system on the vehicle as manufactured, when operated in the mode which draws maximum current. This load current test shall be conducted during the "ambulance's electrical system test" (3.7.6 and AMD Std. 005). The warning light system and related components and devices shall conform to temperature conditions in 3.4.2 and be tested and approved by an Automotive Manufacturers Equipment Compliance Agency (AMECA) accredited laboratory and listed with the AMECA for compliance with the requirements in this specification.

3.8.3 FLOOD AND LOADING LIGHT (EXTERIOR)

Flood and loading lights shall be not less than 191 cm (75 in.) above the ground and unobstructed by open doors. Floodlights shall be located on the left and right sides of the ambulance and firmly fastened to reinforced body surfaces, below the roofline. The lamp(s) H-V shall be projected downward, either mechanically or optically, at an angle of 12 to 18 degrees from the horizontal plane and the lighting on each side of the ambulance shall provide a minimum of 800 beam candle power and produce a flood light pattern similar to a 4419 sealed beam lamp. Floodlight switches shall be located on the cab console and control each side independently. Loading light(s) shall provide a minimum of 500 candle power and shall illuminate the area surrounding the back loading doors. The light(s) shall produce a light pattern equivalent to a 4406 sealed beam. Loading light(s) shall automatically be activated when rear doors are opened and may be incorporated with the FMVSS backup lighting system.

3.8.4 SPOTLIGHT

A hand held spotlight shall be provided with a minimum 100,000 CF lamp, corrosion-proof housing with momentary switch, and minimum 244 cm (8 ft.) heavy-duty coiled cord. It shall be hard wired to the vehicle 12 volt DC system (for anti-theft reasons) and stowed in an applicable storage holder in a compartment/area, accessible to the driver and passenger. When specified (see 3.15.3-25), a remote controlled spotlight shall be furnished, having an "ON"/"OFF" panel switch and variable speed fingertip control. Searchlight shall be a minimum 13 cm (5 in.) diameter, 100,000 CP sealed beam. Lights shall be operable over 360 degrees horizontal sweep and approximately 90 degrees vertical range. Remote control, exterior, light housings shall be chrome plated, bronze, or brass construction.

3.8.5 AMBULANCE INTERIOR LIGHTING

The basic interior ambulance lighting configuration shall be designed to minimize electrical loads and include: A driver's compartment dome light; instrument panel lights; master switch panel; and console light(s). When specified (see 3.15.3-33), a map light for the front seat passenger shall be furnished. Lighting shall be designed and located so that no glare is reflected into the driver's eyes or his line of vision, from switch control panels or other areas that are illuminated while the vehicle is in motion. The patient compartment dome lighting (3.8.5.1) shall be sufficient to illuminate the step-well (3.10.12). The EMT's control panel shall be separately illuminated. All lights shall have lamp shells and housings grounded.

3.8.5.1 PATIENT COMPARTMENT ILLUMINATION

Normal white illumination (dome and EMT's switch panel lighting) in the patient compartment shall not be less than 15 foot candle intensity, measured along the centerline of the clear floor without any outside ambient light. The primary cot shall be provided with a minimum of 35 foot candles of illumination measured on at least 90 percent of the cot's surface area. Blue light(s) or lenses shall not be used. Patient compartment lights shall not be powered by the vehicle's 115 volt AC system if so equipped. The patient compartment dome lighting (in the dim setting) and loading lamp(s) shall be automatically activated when the patient compartment doors are opened. All interior dome lighting, including "checkout" lights, shall be near flush mounted and not protrude more than 3.8 cm (1.5 in.). The use of fluorescent lighting which operates on 12 volts DC, meets the above performance and interference requirements of paragraph 3.7.12, can be used in lieu of incandescent lighting. Fluorescent fixture(s) shall have a removable cover that positively locks in place. The fluorescent tube shall be positively locked in place to preclude loosening due to vehicle movement or vibration.

Dome lighting shall not consume more than 15 amps in the bright setting and shall have two separately protected and controlled circuits. Switches, electronic controls, or fire-proofed rheostats may be used to control lighting.

3.8.5.2 PATIENT COMPARTMENT "CHECKOUT LIGHTS"

When specified (see 3.15.3-9), two "patient compartment checkout lights" shall be furnished with 6 candle power lamps, or equal, and 5 minute timer switch that is wired directly to the ammeter shunt (battery side of load disconnect switch). One of the light fixtures shall be located towards the front of the patient compartment and one towards the rear. The checkout lights may be integrated into the standard patient compartment lighting by activating two of the lamps in the "dim" circuit.

NOTE: USE OF CHECKOUT LIGHTS MINIMIZES BATTERY CYCLING, PRECLUDES THE NECESSITY FOR TURNING ON BATTERIES, AND USE OF HIGHER POWER COMPARTMENT LIGHTING.

3.9 CAB-BODY DRIVER COMPARTMENT AND EQUIPMENT.

3.9.1 DRIVER'S COMPARTMENT, CAB-BODY STRUCTURE

All cab compartments shall be of sufficient size to accommodate a driver and assistant, with space to perform driving and control activities. The cab (Type 1) or the cab-body Types II and III) shall be organized and designed with the specified and required equipment and accessories for ease of operation and safety.

The partition and door, when provided between cab and patient compartment on Type II and Type III vehicles, shall not interfere with or restrict the OEM's full range of front seat adjustment.

The cab and cab-body shall comply with Federal Motor Vehicle Safety Standards 101,102, 103, 104, 105, 107, 108, 111, 113, 115, 120, 124,205,206, 207,208, 209, 210, 211, 212, 219, 301, and 302. Types II and Ill vehicles, of the integral cab-body design, shall be equipped with front doors and window openings equal to the Type I conventional truck cab and conform to 3.9.2 through 3.9.8. All ambulance Types shall be provided with a partition between the cab or driver's compartment and the patient's compartment (see 3.10.2 and 3.10.15).

3.9.2 CAB-BODY PROVISIONS

Cab-body section shall provide a right and left side, weatherproofed, forward hinged door with crank operated side windows and, when available from the OEM: crank or push open vent windows; door stops; external key operated door locks with two sets of keys; trim or closed panels and headliner (washable vinyl upholstery, or flooring type materials); floor covering (OEM's heat, noise and appearance trim packages); panel mounted instruments; and seat(s). All exposed interior surfaces shall be painted. Hardware and other exterior exposed metal trim shall be chrome plated, stainless steel, or anodized aluminum. When specified by the purchaser (see 3.10.15.2 and 3.15.3-29), a console shall be furnished that will contain a map book, clip board, radio control head(s) speaker(s), etc. The console shall be furnished per purchaser's specifications. Additionally, the vehicle shall be furnished with at least the following equipment:

a. Dual sunvisors (padded)
b. Armrests, mounted on each side door
c. Compartment ventilation, other than windows
d. Key operated ignition/starter switch
c. Ammeter and voltmeter (see 3.7.6.2 and 3.7.6.3)
f. Fuel gauge(s)
g. Oil pressure gauge
h. Engine temperature gauge
i. Speedometer with odometer
j. Environmental controls (heater-defroster/air conditioner, etc.)
k. Seatbelts and shoulder harness for driver and passenger, when available from chassis manufacturer
1. Dual outside mirrors (3.9.5)
m. Cab lighting and controls
n. Tinted windshield
o. Dual electric horn(s)
p. Throttle control (see 3.7.6.1)

3.9.3 CAB COMPARTMENT DRIVER AND ASSISTANT SEAT

Unless otherwise specified (see 6.2-x), the driver's compartment shall be furnished with at least two individual bucket-type seats (driver and assistant) or, when specified (see 6.2-x), bench style seating for three. The seats shall be frame constructed with cushioned springs or foam rubber, padded and upholstered to provide maximum riding comfort. Unless otherwise specified (see 6.2-x), the seats shall be covered & with fire-retardant, washable, nonabsorbent material. Non OEM seats shall have 40 oz. reinforced vinyl upholstery. Driver's seat shall have the OEM's full, unobstructed range of adjustment.

3.9.4 CONTROLS AND OPERATING MECHANISM

All controls and operating mechanisms shall be located for left-hand drive. Lever controls, equipment, items, and devices shall be installed, located, and stowed for the convenience of the purpose intended and shall not interfere with the ambulance personnel or patient's ingress or egress of respective compartments.

3.9.5 OUTSIDE REARVIEW MIRRORS

Unless otherwise specified (see 6.2-y), dual, firmly secured, vibrationless, rearview mirrors, having a combination flat/convex mirror system, shall be furnished. Convex shall be a minimum 13 cm (5 in.) diameter. Mirrors shall have the largest usable reflective area practical for the designed ambulance, totaling at least 806 square centimeters (125 square in.) per vehicle. Unless the mirror assembly is manufactured as an integral assembly, all four mirror head faces shall be independently adjustable. Hardware and mirror heads shall be chromed, polished stainless steel, or polished aluminum.

3.9.6 BUMPERS AND STEPS

Chassis manufacturer's standard chrome bumper shall be furnished in the front of the chassis. The rear of the ambulance shall be furnished with a sturdy, full-width, rear bumper, with step secured to the vehicle's chassis-frame. The rear bumper and step shall be adequate to support a test weight of 227 kg (500 pounds) without flexing. A step shall be furnished at the back (loading) door opening, if the patient compartment floor is more than 46 cm (18 in.) above the ground. Unless otherwise specified by the purchaser for Class 2 vehicles (see 6.2-z), the step installation shall be equal distance +/-A 5 cm (2 in.) between the ground and the floor. The bumper-step shall be designed to prevent the accumulation of mud, ice, or snow and made of anti-skid open grating metal (a folding style step if specified in 3.15.3-21). These steps shall not be located or exposed to the interior of the ambulance when the door(s) are closed. All necessary steps shall be at least the width of the door opening for which they are provided. The step's tread shall have a minimum depth of 13 cm (5 in.) and a maximum depth of 25 cm (10 in.). If the step protrudes more than 18 cm (7 in.) from the rear of the vehicle, a foldup step shall be furnished.

3.9.6.1 OVERRIDE FRONT GRILLE GUARD

When specified (see 3.15.3-22), an override protection (high-rise heavy-duty grille guard) shall be furnished to protect the vehicle grille, siren speakers, grill lights, radiator, and air conditioner condenser. The heavy-duty grille guard shall be custom fit to the front bumper and bolted to the vehicle frame. The grille guard shall be of all steel construction with not less than 9.5 mm–10 mm (.318") thick side plates and two horizontal 5.0 cm (2 in.) diameter tube crossmembers. The grille guard shall be of a height to extend from the bottom edge of the front bumper to not less than 20.3 cm (8 in.) above the top edge. Unless otherwise stated in ordering data, the grille guard shall be chrome plated or painted to match the vehicle exterior color.

WARNING:PURCHASERS SHOULD BE AWARE THAT GRILLE GUARDS ADD ADDITIONAL FRONT END WEIGHT AND SHOULD NOT BE ORDERED UNLESS ABSOLUTELY NECESSARY.

3.9.6.2 BUMPER GUARDS

When specified (3.15.3-22) or required for protection of grille mount siren speakers which extend beyond the face of the bumper, a set of bumper guards shall be furnished. Bumper guards shall be rubber faced, 30 cm (12 in.) to 36 cm (14 in.) high, polished aluminum, chromed steel, or polished stainless steel. They shall bolt or clamp to the bumper or chassis, and shall protect the bumper, siren speakers, grill lights, etc.

3.9.7 BODY PROTECTION

3.9.7.1 FENDERS

Fenders and wheel housings (see 3.10.13) shall be provided to cover all tires. Fender extension(s) over dual rear tires are permitted on ambulance bodies to cover tires, providing they are less than 20 cm (8 in.) wide, reinforced, secured, and may be painted black, unless polished metal is specified (see 3.4.11.2 and 6.2-aa).

3.9.7.2 RUNNING BOARDS

When specified (see 3.15.3-30), bright finish, aluminum running boards shall be provided from the rear of the front wheel opening to the rear of the cab (Type I); to the front of the rear wheel opening (Type II); and to the front of the modular body (Type Ill). All running boards shall be securely attached to the cab and/or body, as required, and includc nonmetallic spacers to prevent contact of dissimilar metals and prevent trapping of dirt and water between the running board and the cab body. All running boards shall be gusseted and provide brackets to prevent flexing, sagging, and damage. Design of the running boards may incorporate mud flaps. Running board installation shall conform to chassis manufacturer's recommendation.

3.9.7.3 MUD FLAPS

When specified (see 3.15.3-31), mud flaps, at least as wide as the tire(s), shall be provided per SAE J682 behind the front and rear wheels and shall be reinforced at the point of attachment to the vehicle. Mud flaps may be incorporated into the running boards.

3.9.7.4 MUD/STONE GUARDS

When specified (see 3.15.3-15), bright finish, aluminum diamond plate shall be provided to protect the lower part of Type I and Type III ambulances.

3.9.7.5 RUB RAIL

When specified on Type I and III bodies (see 3.15.3-19), a body rub rail, minimum 25.4 mm (.119") thick × 50.8 mm (2") wide, shall be provided that extends the full length of the right and left sides of the lower third section of the module. The rub rail shall be firmly attached to the body.

3.9.7.6 FUEL FILL SPLASH PLATES

The painted surface of the ambulance body shall be protected from discoloration due to spilled fuel during refueling. Protection shall be provided by a drain in the fuel fill housing(s) or by splash plate(s) under the fuel fill opening.

3.9.8 ENGINE HOOD

Engine hood and cowl shall be fitted to prevent precipitation, heat, odors, and noise from entering the interior of the cab and body. Cab compartment engine covers, on Types II and Ill ambulances, shall be removable for easy access to engine and components.

3.9.9 CAB CONNECTING BELLOWS FOR TYPE I VEHICLE

A flexible, weather-tight bellows, fabricated from EPDM, Hypalon, sheet or molded rubber, or other durable materials that meet the temperature requirements herein and resist ozone, sunlight, oil, fungus, and will not crack, rot or deteriorate, shall be provided between the Type I cab and the modular body. Bellows shall be designed for proper fit and finish and be able to absorb lateral, vertical, and torsional displacement due to body/ cab movement. Window in the cab or body shall be of the sliding type, shall be aligned, and connect with the modular body window opening (see 3.10.2,3.10.14 and 3.10.15) and shall conform to requirements of the partition.

3.10 AMBULANCE BODY AND PATIENT AREA

3.10.1 BODY ACCOMMODATIONS

The ambulance body and patient compartment shall be sufficient in size to transport occupants, as specified in Configuration A or B (see 3.1.5), and accommodate and store all the stretchers, cots, and litters through the range of dimensions as specified in Table II. There shall be space around the patients to permit a technician to administer life support treatment to at least one patient during transit (see dimensional parameters 3.10.4).

3.10.2 CAB/PATIENT COMPARTMENT ACCESS WINDOW

The ambulance and body bulkheads shall have an aligned window opening of at least 968 square centimeters (150 square in.), for visual checking and voice communications between the cab and the patient's compartment for Type I vehicles and, when specified (see 3.10.15.2), in lieu of walk-through door on Types II & III. The window shall be latchable from the cab side and shall be an adjustable, transparent, shatterproof panel (see 3.9.9 and 3.10.14).

3.10.3 EMERGENCY MEDICAL TECHNICIAN (E) SEATING

The EMT shall be provided with a seat equipped with a safety halt and a padded back and a padded headrest. Unless otherwise specified (see 6.2-bb), the seat shall be not less than 46 cm (18 in.) deep by 46 cm (18 in.) wide and 38 to 46 cm (15 to 18 in.) high, measured to the top of the seat cushion (see 3.11.1.1) and adjustable a minimum of 10 cm (4 in.) (front to rear) when a bucket type seat is furnished. The EMT shall be seated at the head of the primary patient, near the bulkhead or partition behind cab compartment, facing rearward (see 3.1.5.) The space under the seat may be designed as a storage compartment or utilized for other equipment. The EMT seat shall be upholstered per 3.11.4. When an ambulance manufacturer's fabricated seat is furnished, the padded backrest shall be the maximum practical width and height. When specified and available (see 6.2-BB), an EMT seat with integral infant/child safety seat shall be furnished and conform with all requirements including FMVSS.

3.10.4 PATIENT COMPARTMENT INTERIOR DIMENSIONAL PARAMETERS

The patient compartment shall provide, but not be limited to, a minimum of 9.2 cubic meters (325 cubic feet) of space, less 1 cubic meter (35 cubic feet) for cabinets, while complying with the following:

Length: Unless otherwise specified or permitted by the purchaser (see 6.2-cc), length, measured from the partition to the inside edge of the rear loading doors at the floor, shall be at least 310 cm (122 in.). This length in the compartment shall provide at least 64 cm (25 in.) and not more than 76 cm (30 in.) of unobstructed space at the head of the primary patient, measured from the face of the backrest of the EMT seat to the forward edge of the Style I cot. A minimum of 25 cm (10 in.) shall be provided, from the end of the cot to the rear loading doors, to permit clearance for traction or long board splints.

Width: The width of the compartment, after installation of the cabinets, shall provide 46 cm (18 in.) +/-A 16 cm (6 in.) of clear aisle walkway between cot and base of squad bench, with the cot located in the street side (non-centered) position.

Height: Unless otherwise specified by the purchaser (see 6.2-dd), the patient compartment shall provide at least 152 cm (60 in.) height, over the primary patient area, measured from floor to ceiling panels.

3.10.5 BODY, GENERAL CONSTRUCTION

For modular construction, the body shall be all welded aluminum or, when specified or approved by the purchaser (see 6.2-ee), other lightweight, inherently corrosion resistant materials of equal, or greater, strength. The exterior of the body shall be finished smooth with symmetrically radiused corners and edges, including rub-rails when specified (see 3.15.3-19), and shall include doors and windows specified herein. Ambulance body, as a unit, shall be designed and built to provide impact and patient compartment penetration resistance and shall be of sufficient strength to support the entire weight of the fully loaded vehicle on its top or side, if overturned, without separation of joints or permanently deforming roof bow or reinforcements, body posts, doors, stringers, floor, inner linings, outer panels, rub-rails, and other reinforcements. Wood, or wood products, shall not be used for structural framing. As evidence that the ambulance body meets the above criteria, the manufacturers body (fabricated, modified, or converted), excluding the conventional cab, shall furnish for each body model (Type) a certification that the ambulance body meets Static Load Test for Ambulance Body Structure, AMD Standard No.001.

Additionally, the roof structure, liner, and outer skin or cap shall be designed and constructed to prevent separation and the possible ejection of patients or crew in the event of an accident and shall include welded and, if necessary, reinforced attachments to the body.

WARNING: ANY ABSORBENT MATERIAL SUCH AS CARPETING, FABRIC, OR INSIDE/OUTSIDE PLASTIC TYPE CARPETING, ETC. THAT RESISTS CLEANING AND DECONTAMINATION SHALL NOT BE USED IN ANY STORAGE OR PATIENT COMPARTMENT.

3.10.6 AMBULANCE BODY STRUCTURE

All parts of the ambulance body and attachments shall be fastened with rust-resistant fasteners in a manner which will preclude loosening. Cabinets, benches, partitions, oxygen cylinder holders, guide rails, and cot holders shall be attached to metal tapping plates and/or framing welded to the body structure. These components shall be fastened by welding, bolting, or self-tapping machine screws, on at least 46 cm (18 in.) centers or less, as applicable to the component being installed. Sheet metal, self-tapping wood/metal screws, nails, staples, etc. shall not be used in assembling the ambulance structure, except for self-threading sheet metal screws used for light trim panels and for retention of wood flooring.

Vehicles furnished with fiberglass/plastic, exterior roof panel shall have the center section embedded with metal wire screening (see 3.14.3 for a radio antenna ground plane). Ambulance bodies with an extended roof shall have the roof structural members permanently fastened to structural members of the body (welded, bolted, and sealed) to prevent separation in an accident. Body structure shall, as necessary, include gusseting to provide diagonal strength. Drip rail(s) shall be provided around the entire modular body and have drain points at each corner. Drip rails shall also be furnished over each entry and compartment door. When structure permits, drip rail shall extend 2.5 cm (tin.) beyond the edge of the doors.

Body skirt(s) shall not extend more than 8 cm (3 in.) below the vehicle cab/body. The body, roof, and panel joints shall be watertight. All openings between the chassis-body and occupant carrying compartments, including the bulkhead between cab and body of Type I and III (see 3.9.9), shall be sealed to prevent intrusion of water, dust, and exhaust gases.

3.10.7 BODY MOUNTING

On Types I and III ambulance bodies, to reduce stress on body and frame, minimize height above the frame, and isolate the patient compartment from noise and vibration, full floating, automotive style, rubber body mounts or other chassis manufacturer's approved body mounting systems shall be furnished with a minimum four per side on bodies up to 371 cm (146 in.) and five per side on bodies over 371 cm (146 in.). Reinforcements or filler blocks shall be used where mounting device(s) may deform frame flanges. Mounting devices shall be locked units which will minimize loosening, but which may be re-tightened, if necessary. Modular bodies shall be interchangeable with other chassis of the same manufacturer having the same cab to axle (CA) dimension (reference 3.5.6).

3.10.8 DOORS

Two patient compartment door openings shall be provided. There shall be a door opening on the right forward side and at the rear of the body for loading a patient on a cot. The side opening shall have a single forward hinged door for Types I and III, double hinged doors for Type II, or if specified (see 6.2-ff) a sliding type door for Type II vehicles shall be furnished. Door(s) shall provide a minimum right-side clear opening of 76 cm (30 in.) wide and of 160 cm (63 in.) high for Types land III and the chassis manufacturer's standard opening for Type II vehicles. Unless otherwise specified (see 6.2-ff), eg: wide offset doors, rear loading door(s) shall cover a clear opening of not less than 117 cm (46 in.) in height for Type II and a minimum height of 117 cm (46 in.) for Types I & Ill and 112 cm (44 in.) in width for Types I and III and the chassis manufacturer's standard opening for Type II vehicles. All ambulance body doors shall be equipped with not less than 1613 sq. cm (250 square in.) of safety glass area per door. Doors shall be designed for easy release and meet the requirements of 3.10.9. A "Door-Open" warning device shall signal (indicate in the cab) when doors are not closed (see 3.7.1.1). Each door shall have effective compression or overlapping seals to prevent leakage of exhaust fumes, dust, water, and air. Doors may contain and be equipped with recessed compartments as applicable to the interior for storage of supplies and devices.

Patient compartment doors, on Types I & III, shall be flush or near flush style, full box type construction with removable inner panel. Inner panel shall be finished with a durable, washable type material and include trim moldings around all unfinished, exposed edges.

A red light or reflector, minimum 7.6 cm (3 in.) diameter, shall be installed, one on the interior surface of each side or rear door. The reflectors shall be so positioned as to provide maximum visibility when the doors are in the fully open position. At the manufacturer's option, red reflective tape, conforming to L-S-300, Type I, Class 3, reflectively 1, may be furnished. The tape shall be 10 cm (4 in.) wide and shall extend the full width of each rear door panel.

3.10.8.1 PROTECTION OF PATIENTS AND CREW

Head cushions shall be provided at the upper interior areas of the door frames to prevent head injury to patients or personnel. Similar cushions shall also be furnished at other areas that may be capable of causing injury.

3.10.9 DOOR LATCHES, HINGES, AND HARDWARE

Door latches, hinges, and hardware furnished by chassis and ambulance manufacturers shall comply with FMVSS 206. When doors are open, the hinges, latches, and door-checks shall not protrude into the access area. All doors shall have hardware or devices to prevent inadvertent opening and closing. To facilitate entry and exit from the vehicle, a minimum 15 cm (6 in.), tubular or semi-oval, minimum 19 mm (3/4") wide (diameter), grab handle shall be provided on the inside of each door or the adjacent body structure (in addition to a door operating handle). Door stops, to prevent damage to body sides and a handle with latches, operable from inside (even if key locked externally) and from outside of the body after unlocking, shall be furnished. One external operated lock, with key per door opening, shall be provided. All patient compartment door locks shall be identically keyed. Hardware shall be chrome plated, bright finish, stainless steel, or anodized aluminum. Inside door latch handles shall be designed and placed so they cannot be operated when accidentally hit or used as a grab handle. Ambulance body, side and rear door hardware, installed by the ambulance body manufacturer, shall be tested to ensure installation meets or exceeds the requirements of AMD Standard 002—Body and Door Retention Components Tests.

3.10.10 FLOOR

Floor shall be at the lowest level permitted by the chassis/body, but not exceed the height permitted in 3.4.11.7. It shall be flat, except when the area near the rear entrance door is sloped for a lower entering height, and with the exception of cot related hardware, shall be unencumbered in the door(s) access and work area. All floor areas shall withstand a distributed load of 73 kg/1000 sq. cm (150 lbs./sq. ft.). Metal floors shall be reinforced to eliminate "oil canning" and insulated against outside heat and cold. The subfloor of the Types I and Ill patient compartment shall be water resistant. When plywood is utilized, it shall be water resistant, not less than 13 mm (1/2 in.) thick, 5 ply minimum, supported by body framework. Under the subfloor of the Type I and III vehicles shall be a aluminum heat shield/splash pan, minimum 1.3 mm (0.050"), with silicone or other non-hardening sealant evenly distributed around it's perimeter. The subfloor of the Type II patient compartment shall be not less than 13 mm (1/2 in.) thick density, marine or exterior grade plywood. Fiberglass, aluminum, or other non-hydroscopic composites, with at least the equivalent strength of plywood may be used as the subfloor. Particle board or equivalent type materials are not acceptable. Additionally, noise/thermal insulation shall be provided under the floor as necessary.

Voids or pockets, where water or moisture can become trapped to cause rotting and unsanitary conditions, are not acceptable. Voids and pockets shall be filled with sealer or caulking compound. Flooring shall extend the full length and width of the patient compartment or body (including space under the cabinets, unless otherwise insulated).

3.10.11 FLOOR COVERINGS AND COLOR

Floor covering shall be easily cleaned, sanitized, and harmonize with the interior color and decor of the patient compartment. The floor covering shall be seamless, one piece, no wax type, solid linoleum, vinyl, or poured epoxy not less than 1.6 mm (1/16 in.) thick and permanently applied to the subfloor. The floor material shall cover the entire length and width of the compartment's working area. The covering of joints (corners, etc.), where the sidewalls and covering meet, shall be sealed and bordered with corrosion resistant cove molding or the covering shall extent at least 7.6 cm (3 in.) up the sidewalls. Cabinets at floor level shall be finished with floor covering material. Floor covering shall comply with FMVSS 302.

3.10.12 STEPWELL (SIDE DOOR)

Steps shall be provided in the door openings, if the floor is more than 46 cm (18 in.) above the ground. Stepwell, when applicable, shall be the enclosed two-step type. Height of the bottom step shall not exceed 46 cm +/- 5 cm (18 +A 2 in.) (Types I and III vehicles), and 56 cm (22 in.) (Type II) with the exception of Class 2 vehicles. Stepwells shall be lighted by interior lights and all step surfaces shall be fabricated from reinforced, bright finish, aluminum diamond plate or covered with anti-slip material.

3.10.13 WHEELHOUSINGS

Wheelhousings of Type I and III vehicles shall include metal or plastic splash shields between the body wheelhousing and the wheel(s) (see 3.9.7) extending over the top of the tires to the bottom of the body side skirting. Wheelhouse openings shall allow for tire chain usage and easy tire removal and service. Chassis manufacturer's standard wheelhousings will be acceptable. Wheelhousings shall be undercoated or rustproofed.

3.10.14 WINDOWS

Glazing shall comply to FMVSS 205. Unless otherwise specified (see 3.15.3-12), the patient's compartment shall not have windows, except the viewing panel in the partition or bulkhead(s) (see 3.10.15) and in the side and rear doors (see 3.10.8). All windows shall be tinted and, unless otherwise specified (see 6.2-gg), upper windows shall open and, when vertical windows are furnished, shall have provisions to prevent accidental closure.

3.10.15 BULKHEAD/PARTITION FOR TYPE II AND III VEHICLES

A full height and Width partition or bulkhead (with or without compartments), having rollbar characteristics and a walk-through opening with a door shall be placed between the driver and patient's compartment. This partition shall be located directly behind the driver and companion seats when in the rearmost position. The partition shall be secured on the sides, ceiling, and floor by welding or bolting to tapping plates.

3.10.15.1 DOOR I WALK-THROUGH

Unless otherwise specified (see 6.2-hh), the door opening shall be at least 43 cm (17 in.) wide and 117 cm (46 in.) high and shall provide an aisle between the compartments. The door, shall have at least a 968 sq cm (150 square in.), transparent, shatterproof viewing panel in the center section at the driver's eye level. The door shall be secured with a driver's side self-latching device in the open and closed positions. (see 3.10.2).

3.10.15.2 WALK-THROUGH DELETION

When specified (see 6.2-hh), the partition door shall be deleted and a window, as specified in 3.10.2, shall be furnished. When specified (see 6.2-hh), additional compartments shall also be furnished in the space normally occupied by the door opening, including the area above the required pass through window. Further, a console (see 3.15-3-29) may be specified between the driver and passenger seat.

NOTE: PURCHASER SHOULD CONSIDER DELETING THE WALK-THROUGH BECAUSE, IN ADDITION TO PROVIDING INCREASED SPACE FOR COMPARTMENTS AND THE CAPABILITY OF ADDING A RADIO CONTROL HEAD/MAP BOOK CONSOLE (SEE 3.9.2 AND 3.15.3-29) BETWEEN THE FRONT SEATS, THE

DELETION OF THE DOOR ALSO PROVIDES FOR ISOLATION OF THE DRIVER FROM VIOLENT PATIENTS AND THE SEPARATION OF INDIVIDUALS THAT MAY BE ACCOMPANYING THE PATIENT WHO COULD INTERFERE WITH PATIENT CARE.

3.10.16 INSULATION

The entire body, sides, ends, and roof of the patient's compartment shall be completely insulated to enhance the performance of the environmental systems, specified in 3.4.2 and 3.13, and prevent external noise from entering the vehicle interior. The insulation shall be a non-settling type, vermin-proof, mildew-proof, fire retardant, nontoxic, and nonhygroscopic. If fiberglass insulation is used, it shall not be exposed to water, e.g. door panels.

3.10.17 INTERIOR SURFACES

The interior of the body shall be free of all sharp projections. All hangers or supports for equipment and devices shall be mounted as flush as possible with the surrounding surface. The finish of the entire patient compartment, including interiors of storage cabinets, shall be: impervious to soap and water, disinfectants, mildew, fire resistant, comply with FMVSS 302; and be easily cleaned/disinfected (carpeting, cloth, and fabrics are not acceptable). Interior body lining and cabinetry materials, excluding the cab compartment (see 3.9), shall be selected to minimize dead weight and shall be plastic laminated or painted aluminum panels or plastic laminated wood or of other lightweight, equivalent strength materials such as fiberglass, composites, and fiberglass reinforced plastics. Panels shall be installed in a manner that prevents sagging, deflection, warpage, or vibration. The ceiling/headliner shall be of similar lightweight materials. For interior colors, see 3.16.2.

3.11 STORAGE COMPARTMENTS

Storage compartments shall be furnished for all items required by this specification and or specified by the purchaser and include storage for, but not be limited to, backboards, portable cots/litters, stair chairs, and any other specified patient handling devices. Any absorbent material such as carpeting, fabric, or inside/outside plastic type carpeting, etc. that resists cleaning and decontamination shall not be used in any storage or patient compartment.

3.11.1 INTERIOR STOWAGE ACCOMMODATIONS

Unless otherwise specified by the purchaser (see 6.2-u), the interior of the patient compartment shall provide a minimum volume of 1 cubic meter (35 cubic feet) of enclosed stowage cabinetry, compartment space, and shelf space which shall be conveniently located for medical supplies, devices, and installed systems as applicable for the service intended. Enclosed compartments and spaces shall be located at, in, or on the partition, sidewalls, overhead, squad bench, technician seat, and doors. Compartment(s) under the floor, with opening panel(s) inside the patient compartment, shall not be acceptable.

At a minimum stowage shall be required for the following items as shown below:

ITEM	MINIMUM VOLUME CUBIC METERS	CUBIC FEET
Medicine dispensary cabinet(s)	0.17	6
Medical supplies cabinet	0.25	9
Linen supplies	0.11	4
Trash receptacle compartment and sharps disposal container	0.06	2
Oxygen installation (main) (see 3.12)	0.17	6
Oxygen unit (portable)	0.06	2
Telemetry equipment	0.06	2
Radio equipment & antenna	0.03	1
Storage miscellaneous	0.06	2
Vacuum aspirator unit	0.03	1
Air conditioning unit	As required	
Heating system	As required	

3.11.1.1 LOCATION OF MEDICAL EQUIPMENT AND SUPPLIES

Location of medical supplies and equipment shall be dictated by their relative importance and shall be readily accessible to the Emergency Medical Technician (EMT). Priority shall be given to items necessary to cope with life threatening conditions at the scene and in transit. The equipment and supplies necessary for airway maintenance, ventilation, oxygenation, and suction shall be at the head of the primary litter (action area). When items for cardiac monitoring/defibrillation, mechanized CPR, administration of intravenous fluids/ medications, and the monitoring of blood pressure are furnished, they shall also be readily available to the EMT at the action area. Supplies, devices, tools, etc., shall be stored in enclosed compartments and drawers designed to accommodate the respective items. All medical devices and equipment shall be stowed or properly fastened in/on the action area or in cabinets or to the rail system(s) brackets, when specified (3.15.4 Ml), to prevent items from becoming projectiles in the patient compartment that can cause injury while the vehicle is in motion or involved in an accident.

3.11.1.2 INTERIOR COMPARTMENT WITH EXTERIOR ACCESS

When specified by the purchaser, on Types I and III vehicles (sec 3.15.3-32), an interior/ exterior compartment with 2 shelves shall be provided in the front, right corner of the body or other location as specified. This compartment shall also have an exterior door and, unless otherwise specified (see 3.15.3-32) or permitted by the purchaser, a roll-up, inside door with an enclosed, roller assembly. When viewed from the vehicle's interior, this compartment shall have the following minimum dimensions: W 76 cm (30"), D 46 cm (18"), and H 102 cm (40"). The door shall be equipped with the necessary gaskets, seals or other vibration and noise silencing devices, a positive latch and a grab handle to facilitate opening and closing. The shelves shall be adjustable and have an approximate 3/4" lip covered with automotive edge molding.

3.11.1.3 WASTE AND SHARPS DISPOSAL

A.A trash receptacle compartment, with closure over opening, for general waste shall be furnished with a full size plastic/rubber trash can and disposable plastic liners, with 12 spare liners. The trash compartment shall be accessible to the EMT seat and to the head of the squad bench. The consignee shall furnish bags/liners that meet OSHA requirements for disposing regulated medical waste as defined by OSHA.

B.A crash stable sharps receptacle compartment/storage area shall be furnished for retention of a sharps container that meets OSHA requirements.

3.11.2 EXTERIOR STORAGE ACCOMMODATIONS

On Types I & III and when specified (see 6.2-u) on Type II vehicles, outside entered, weatherproof, storage compartments shall be provided in the ambulance body for extrication, spare tire, oxygen, and miscellaneous equipment. When specified, (see 6.2-li), additional exterior storage compartments shall be furnished. Exterior compartment doors and hardware shall be flush or near flush style construction. All doors shall have spring or gas tube type, hold open devices that permit one hand closure. Hardware (hinges, locks, latches, etc.) shall be rust resistant. All exterior compartments shall have latches with locks and shall be keyed alike. All exterior compartments, except the long narrow backboard storage compartment, shall be automatically lighted when opened. When top and vertically hinged doors are furnished, they shall employ ratchet or spring loaded or gas tube hold open devices.

3.11.2.1 EXTRICATION EQUIPMENT AND STORAGE

The extrication equipment compartments, maximum of two, shall provide at least .28 cu meters (10 cu feet) of storage accommodations. Small hand-tools shall be contained in a single pouch type carrying device. An example of items and other miscellaneous extrication equipment which shall be stored in the extrication storage compartment(s) follows. The contractor shall furnish specific items as specified under 3.15.3-18:

a. One wrench; 30 cm (12"), adjustable, open end
b. One screwdriver; 30 cm (12"), regular blade (slot)
c. One screwdriver; 30 cm (12"), Phillips type
d. One hacksaw; with 12, wire blades (carbide)
e. One pliers; 25 cm (10"), vise grip and one pair, channel lock pliers
f. One hammer; 1.4 kg (3 lb.), 38 cm (15") handle (engineer style)
g. One fire axe; flat head
h. One crowbar; 130 cm (51"), pinch point
i. One wrecking bar; 61 cm (24"), combination tool
j. One bolt cutter; 91 cm (36"), jaw opening of 32 mm (11/4") or greater.
k. One power jack; portable, hydraulic and spreader tool kit, hand powered, minimum 4,064 kg (4 ton) capacity
l. One shovel; pointed blade (folding type)
m. One tin snip; double action, minimum 20 cm (8")
n. Two ropes; synthetic, kernmantle, 15 m (50') × 19 mm (3/4") diameter each
o. Three pairs gloves; leather (gauntlets)
p. Three pairs goggles; clear eye protective
q. One cold chisel; and one center punch; 13 mm (1/2") × 305 mm (12")
r. One cold chisel; 25 mm (1") × 305 mm (12")
s. One seatbelt cutter
t. Two utility knives; curved blade
u. One rope; weighted, polypropylene, 30 m (100 ft.), 13 cm (1/2") diameter,

v. One pair lineman's rubber gloves with leather shells
w. Two lights; portable, battery operated
x. One fire blanket and case (mm. 152 cm (5 ft.) × 183 cm (6 ft.))
y. Two baling hooks
z. Three hard-hats with face and eye protection device
aa. One spring loaded, window punch
bb. Twelve (12) hardwood, shoring blocks, 51 mm (2") × 102 mm (4") × 254 mm (10") with rope handles
cc. Four blocks; hardwood, cribbing, 102 mm (4") × 102 mm (4") × 304 mm (12") with rope handles
dd. Four blocks; hardwood, cribbing, wedge shaped, with rope handles
ee. One come-along, 1814 kg (2 ton), chain type with hooks
ff. Two alloy steel, rescue, pull chains, 3 m (10 ft.) minimum, with grab hooks and rings
gg. One gun kit; air cutting, 1722 kPa (250 psi), with cylinder and chisels.
hh. One truck jack, hydraulic, 2722 kg (3 ton)
ii. Two extrication straps, synthetic fabric, 2.7 m (9 feet), with quick release buckles
jj. One extrication, loop sling, 2.5 cm (1") wide nylon, or equal, 183 cm (6 ft.) circumference with closure ring

3.11.3 STORAGE COMPARTMENTS AND CABINETS DESIGN

Storage cabinets, drawers, and kits shall be easily opened but shall not come open in transit. For rapid identification of contents, medical supply cabinets above The litter patients, shall have shatterproof, transparent or lightly tinted, sliding doors provided with two recessed metal finger cups, near flush grip, or low profile handles. Storage compartments shall be divided into sections, shelves shall be adjustable, drawers shall be marine style slide or tilt, and all shall be removable. Sliding doors for cabinets designed to carry light weight items such as dressings, bandages, etc. shall automatically latch or be fitted with friction holding devices when in a closed position. Other compartments designed to carry heavier items such as first aid and drug kits, portable and spare oxygen bottles, suction units, etc. shall have positively locked latches that are bolted to The door and The door frame structure and are designed to remain closed in an accident, including rollovers. Use of sheet metal or wood screws is not acceptable. Side cabinet shelves shall be no more than 51 cm (20 in.) in depth, when located above The vehicle belt level. Storage compartments, cabinets, and support equipment area interior surfaces shall be finished in accordance with 3.10.17. To preclude injury, in The event of an accident, all cabinets shall be firmly anchored (bolted or welded) to tapping plates of The body structure (3.10.6). Tops of The cabinets and shelves shall be surrounded by a lip of not less than 13 mm (1/2") in height covered in a soft, pliable molding. All shelves shall be continuously adjustable. Storage for The main oxygen cylinder (see 3.12.1) shall be accessible for replacement from an outside position. The oxygen compartment shall be provided with at least a 58 sq. cm (9 sq. in.) louvered device to dissipate/vent leaking oxygen to The outside of The ambulance. Oxygen cylinder compartment shall not be utilized for storage of any other equipment. Any wiring and electrical devices within this compartment shall comply with 3.7.2. Oxygen cylinder(s) shall be mounted with a restraining device(s), as required for The crash worthiness tests of AMD Standard 003, Oxygen Tank Retention System.

3.11.4 SQUAD BENCH, SEATS, AND BACKRESTS

The squad bench shall have post cups and/or wheel cups to prevent lateral and longitudinal movement and support The Style 3 folding stretcher, combination stretcher chair specified (see 3.10.1 and 3.11.5), or other purchaser specified stretcher (3.15.4 Ml 6, or Style S). The squad bench top shall be between 48 cm (19") and 56 cm (22") wide, a minimum of 183 cm (72") long, and a height of between 43 cm (17") and 51 cm (20") measured from The floor to The top of The padded seat. Top opening squad bench platform(s) shall be fitted with an automatic hold open and closed latching device for safety. All seats in The patient compartment shall be padded, at least 46 cm (18") wide, and have The largest practical padded backrest. Padding furnished shall be rubber or polyester urethane foam of a medium to firm density, in covered, finish thickness of at least 6.3 cm (2.5 in.) for seats, and approximately 5.1 cm (2 in.) for head and backrests. All padding and upholstery shall be fire retardant and meet FMVSS 302. Additionally, The upholstery shall be nonabsorbent, washable and impervious to disinfectants.

Squad bench shall be furnished with at least three sets of lap (Type I) safety belts for seated occupants and for retention of a cot/stretcher when positioned on The squad bench. The belts shall be positioned at The patients chest, hips, and legs. Seatbelts shall comply to FMVSS 209. The anchorages for The side facing seatbelt assembly shall withstand a minimum of 1134 kg (2,500 lbs.) force when tested in accordance with FMVSS 210-SS.1. Other side facing seats shall have a Type I lap safety belt.

Sealed storage shall be provided for The specified oxygen bottle, when located under The rear of The squad bench, with separate compartment under The front of The bench, and The maximum possible cubic foot volume. The squad bench cover latching device(s) shall be a quick release, slam type latch similar to a Cleveland Model 3300 TL with 3400-15 striker pin. For configuration A (ALS) or, when specified (see 3.15.3-28), a side facing "CPR" seat located on The street side of The vehicle at The patients thoracic region shall be furnished. This seat shall also be furnished with a scat belt meeting The above requirements and shall have rear and lateral padding at The head areas.

3.11.5 STRETCHERS, COTS, AND LITTERS

When specified (see 3.15.4 M1O and 6.2-jj), The ambulance supplier shall provide The following: a Style I (Ferno Washington Model 30 or equal) or Style 2 (Ferno Washington Model 92 ES or equal, with roll-in feature) cot for The primary patient and a Style 3 folding stretcher or a combination stretcher chair, each with legs and two wheels designed to permit a patient to be carried (or wheeled) on stairways and through other narrow areas (see 3.11.5.1), or equivalent. Stretchers, cots, and litters shall conform to The dimensions of Table II (see 3.10.1). Length and width measurements shall be taken at The metal framing, excluding joint fittings. Wheeled cots shall include foot and head pulls, when available from The cot manufacturer, and a polyester foam mattress at least 8 cm (3 in.) thick or an equivalent mattress covered with vinyl coated, nylon fabric or other nonporous fabric conforming to FMVSS 302, or equivalent, and restraint straps per 3.11.8.1.

3.11.5.1 COMBINATION STRETCHER CHAIR

When specified in 3.15.4 M16, a Ferno Washington Model 107B Combination Stretcher Chair, or equivalent, shall be furnished, in lieu of The stored stairway stretcher specified in 3.11.5. The stretcher shall conform to The dimensions of Style 3 in Table II (see 3.10.3). Stretcher shall have posts, wheels, folding handles at foot-end, straps, and convert to a stair chair or stretcher.

3.11.6 SEAT SAFETY BELTS AND ANCHORAGES

All seats shall comply with FMVSS 207. Safety belts and anchorages shall comply with FMVSS 208, 209, and 210, except as otherwise specified (3.11.4). Seatbelts shall have retractor devices for all seat positions in The vehicle, including The squad bench.

3.11.7 LITTER FASTENERS AND ANCHORAGES

A crash stable side or center mounting Ferno-Washington Model 175 type cot fastener assembly with quick release latch shall be furnished. It shall secure The Style I or Style 2 wheeled cot or infant transporter to The ambulance body. Style 4 and Style 5 litters need not be provided with fastening devices unless otherwise specified (see 6.2-jj). The installed cot fastener device(s) for wheeled cots shall be tested to comply with AMD Standard 004, Litter Retention System. Additional cot related hardware is permitted, provided The patient compartment exit/entry is not encumbered with The cot in place. The furnished devices shall have a bright colored finish, if The device presents a tripping hazard in The entry/exit area when The cot is removed.

3.11.8 PATIENT RESTRAINTS

3.11.8.1 COT STRAPS

At least three strap type restraining devices (chest, hip, and knee) shall be provided per stretcher, cot, and litter to prevent longitudinal or transverse dislodgement of The patient during transit. Additionally, The head of The cot shall be furnished with upper torso (over shoulder) restraints that mitigate forward motion of The patient during severe braking or in a frontal impact accident. Restraining straps shall incorporate metal-to-metal quick release buckles, be not less than 51 mm (2 in.) wide, and fabricated from nylon or other materials easily cleaned and disinfected.

3.11.8.2 INFANT/CHILD CARRIER

When specified (see 3.15.4 M26) an infant/child spinal immobilization/carrier device shall be furnished that safely secures those patients to The cot.

3.11.9 I.V. HOLDERS FOR INTRAVENOUS FLUID CONTAINERS

Two near flush style, I.V. ceiling holders specifically designed for holding I.V.'s, with a velcro type strap to hold and control I.V. bags/bottles, shall be provided. The ceiling holders shall not protrude more than 2.5 cm (1 in.), be located on or adjacent to The side wall, one at The head of The primary patient and one at The head of The secondary patient's cot (squad bench). When an ALS configuration (see 3.1.5.1) is specified, one additional I.V. hanger shall be furnished for each patient, at The lower extremities. When specified (3.15.4 M2), a fold-down, rigid, telescoping I.V. pole and holder, with a 107 cm (42 in.) minimum height (when extended), shall be provided on The left side of The Style I cot.

3.12 OXYGEN AND SUCTION SYSTEMS AND EQUIPMENT

The ambulance manufacturer, as required, shall furnish devices/equipment under 3.12 through 3.12.4 complying to this specification or to other Medical Associations approved standard(s) (e.g. ASTM). The specified device shall be tested and certified by an independent, medical testing laboratory for EMS use.

3.12.1 OXYGEN, MAIN SUPPLY AND INSTALLATION

The ambulance shall have a hospital type piped medical oxygen system capable of storing and supplying a minimum of 3,000 liters (see cylinder data 6.7) of medical oxygen. Unless otherwise specified (see 6.2-kk), The main oxygen supply shall be from a single "M" size cylinder which consignee will provide and install at The time The vehicle is placed in service. Cylinder(s), when furnished, shall conform to Federal Specification RR-C-901, DOT Specification 3AA2015 and MS-39226-8. Unless otherwise specified by The purchaser (see 6.2-kk), The oxygen cylinder shall be located in an area that provides for proper balance and weight distribution (see Paragraph 3.5.4).

The cylinder controls shall be accessible from The inside and outside of The vehicle. The bottle pressure gauge or equivalent device shall be visible from The EMT's seat and/or the squad bench. the oxygen bottle shall be accessible for changing from the exterior of the body and, on Type II vehicles, through a sealed door. Access to the bottle, when located under the squad bench, shall also be through the rear half of a two section, squad bench cover.

If necessary, or when specified (see 3.15.4 M8), a remote bottle pressure indicator, such as a multi-light indicator showing "Full," "3/4", "1/2" and "1/4," and "Empty" or a digital indicator or other analog type that displays bottle capacity, in at least 25 percent increments, shall be furnished and mounted on the EMT panel. Systems with audible alarms shall have a resetable, audible alarm that sounds when the capacity is down to 20-25 percent, A shutoff valve or solenoid, when specified, shall be furnished in the 344 kPa (50 psi) line, and shall be located on the EMT panel. If a solenoid is utilized, a readily accessible, emergency bypass valve shall be furnished and identified. the purchaser shall specify the type of quick disconnect, action wall device (see 6.2-II) to be used to interface the flow meter and oxygen powered, manually triggered (resuscitator) valve (see 3.15.4 M7 and M20). E.g. DISS, NCG, or OHIO DIAMOND.

A cylinder changing wrench shall be furnished, chained, and clipped within the oxygen cylinder compartment. THE contractor shall install all other components and accessories required for the piped oxygen system which shall include, but not be limited to: a pressure regulator preset to 344 +/- 34 kPa (50 +/- 5 psi) line pressure, nonferrous piping and/or low pressure, electrically conductive, hose approved for medical oxygen at the flow rate specified in 3.12.1.1. INDUSTRIAL OR WELDING TYPE OXYGEN HOSE SHALL NOT BE USED. Oxygen piping shall be concealed and not exposed to the elements, securely supported to prevent damage, and be readily accessible for inspection and replacement. Oxygen shall be piped to a self-sealing duplex oxygen outlet station for the primary patient and, when specified (see 3.15.4 M9), for the secondary patient. Oxygen outlet shall be located within 89 cm (35") from the center of the patient's head when in the supine position with the cot located in the position closest to the action wall. One of the outlets shall be for a flow meter/humidifier, and the second oxygen outlet shall be used for quick disconnect plug-in devices not requiring humidification. Outlets shall be adequately marked and identified (see 3.7.11) and not interfere with the suction outlet. the entire system shall be leak tested (see 4.4.6).

3.12.1.1 OXYGEN PRESSURE REGULATOR AND FLOWMETER

The medical, oxygen pressure reducing and regulating valve with inlet filter at the cylinder shall have line relief valve set at 1378 kPa (200 psi) maximum, and a gauge range of 0 to 17225 kPa (0 to 2,500 psi) [27,560 kPa (4,000 psi) tested] with the gauge scale graduated in not more than 100 PSI increments. The regulator shall be easy to connect and preset, with a locking adjustment, at 344 +1-34 KPa (50 +/- 5 psi) line pressure, permitting a minimum 100 LPM flow rate at a bottle pressure of 689 kPa (150 psi) (empty). In order to show actual oxygen flow rate, the flow meter, when specified (3.15.4 M7), shall be a pressure compensated type (with floating ball indicator-Thorpe type). It shall have calibrations and be continuously adjustable over a minimum range of 0 to 15 liters per minute (LPM). The flowmeter shall withstand a minimum 1378 KPa (200 psi) inlet pressure without failure or damage. Flowmeter(s) shall be installed vertically and shall be readable from the EMT seat and squad bench. Both the oxygen regulator and flowmeter, when specified (see 3.15.4 M7), shall have standardized connections, be electrically conductive from inlet to outlets (rated hose), and incorporate an inlet filter. These devices shall maintain accurate readings and calibrations under ambulance operation and shall be unaffected by temperature conditions of 3.4.2. All settings, calibrations, safety valves, limiters for pressure, and flowmeters shall maintain accuracy to within +/- 10 percent. Servicing, parts, and instructional manuals shall be provided and included in 3.20. Each device shall be color coded green and also be approved for use by recognized national medical associations and societies. THE oxygen regulator and flowmeter shall each be permanently identified with the manufacturer's name, model number, calibrated conditions, and specific markings including warning/caution information.

3.12.2 PORTABLE OXYGEN UNIT

When specified (3.15.4 M17), a portable oxygen unit of at least 300 liters ('D" size cylinder, see 6.7) shall be provided having a yoke regulator with pressure gauge, dual 344 kPa (50 psi) ports, regulator/dial type flowmeter, delivery tube, and oxygen mask(s). The unit shall be capable of delivering an oxygen flow of at least 15 liters per minute through the flowmeter and 40 liters per minute to the manually triggered, oxygen (resuscitator) valve in 3.12.2.2. A full spare cylinder of oxygen for this unit shall be furnished and stored. Portable cylinder(s) and kit shall be secured and, when located in patient compartment, shall include a crash stable, quick release bracket that meets the test requirements in the AMD Oxygen Tank Retention Standard 003. A quick release mounting bracket(s) shall also be provided for the portable oxygen unit and spare bottle(s) when located in a secure compartment (see 3.11.1.2) or other purchaser specified (see 6.2-mm) location not requiring a crash stable mounting system..

3.12.2.1 SQUEEZE BAG-VALVE-MASK SYSTEM

When specified (3.15.4 M6), a portable squeeze bag-valve-mask kit shall be furnished. The bag-valve-mask system shall comply with the following.

a. Two self-refilling bags without sponge rubber inside. One bag of 1,000 cc +/- 10% volume for adults and one bag of about 700 cc, but not more than 750 cc volume, for children and infants
b. Three see through inflatable face-masks, one each in adult, child, and infant sizes. Two masks covering the full range of adult through pediatric sizes may be furnished in lieu of three masks
c. Standard 15 mm/22 mm fittings
d. A true non-breathing valve

e. An oxygen reservoir system for delivery of 80-100 percent oxygen through an ancillary oxygeinlet. The system shall be capable of accepting at least 15 liters/mm. of oxygen without valve jam.

3.12.2.2 OXYGEN POWERED, MANUALLY TRIGGERED (RESUSCITATOR) VALVE

When specified (3.15.4 M20), an oxygen powered manually triggered (resuscitator) valve shall be provided. This device shall have a see through inflatable face mask, interface with the action wall outlet, via the specified quick disconnect device, and/or the dual port regulator/flowmeter of the portable oxygen system, and meet the following requirements.

a. Provide oxygen for breathing and non-breathing patients
b. Provide manually triggered flow rates in accordance with the American Heart Association (AMA) or other recognized medical/standards organization(s)
c. Provide not less than 85 percent oxygen
d. Be equipped with a safety valve having a release pressure setting of 60 cm of water for adults

3.12.2.3 AUTOMATIC RESUSCITATOR

When specified (3.15.4 M29), an automatic resuscitator/volume ventilator shall be furnished. The device's features and performance shall conform to AHA or other medical/standards organization requirements. The device shall be furnished with a connection hose and ambulance and/or portable oxygen system compatible quick disconnect fittings.

3.12.3 SUCTION ASPIRATOR, PRIMARY PATIENT

Unless otherwise specified (see 6.2-nn)ı an electrically powered suction aspirator system shall be furnished with an illuminated switch and a panel mounted, labeled, quick disconnect inlet device on the EMT panel. The suction pump shall be located in an area that is accessible but sound and vibration insulated from the patient compartment. The pump shall be vented to the vehicle's exterior. When specified by the purchasers (see 6.2) for gasoline powered vehicles, an engine powered suction system may be furnished and shall include a hospital type self-sealing plug-in outlet. The system shall include a corrosion resistant reservoir having a volume of not less than 10.5L (640 cu. in.) which is connected to the engine manifold through a check valve. All components, electrical, pressure and/or vacuum, and other lines and accessories, shall be securely mounted yet readily accessible. The electric type aspirator system shall be connected per Figure 5 [omitted]. The aspirator system shall provide a free air flow of at least 20 LPM and achieve a minimum of 300 mm (11.81 in.) Hg vacuum within four seconds after the suction tube is closed. The rate of flow and pressure of the engine vacuum aspirator system shall be attainable continuously during the entire normal range of vehicle operation (i.e., accelerating, decelerating, at rest, climbing hills, etc.). A vacuum control and a shutoff valve, or combination thereof, shall be provided to adjust vacuum levels and to discontinue aspiration instantly. A vacuum indicator gauge of 76 mm +/- 13 mm (3 +/- 0.5") in diameter, with numerical markers at least every 100 mm Hg and a total range of 0 to 760 mm Mg, shall be provided. The collection bottle or bag shall be nonbreakable and transparent with a minimum 1,000 ml capacity. The following accessories shall be furnished: one suction rinsing water bottle; one semi-rigid pharyngeal suction tip, with thumb suction control port, (nonmetallic) of sufficient diameter to allow aspiration of semi-solid gastric contents; a 3 meter (10 foot) length of transparent or translucent, non-kinking suction tubing which shall not collapse under high suction; and a minimum often spare collection bags when the bag type system is furnished.. To assure high air flows and free passage of aspirate, minimum inside diameter for the suction tubing and tubing connectors shall be at least 6.4 mm (1/4 in.). The other suctioning catheters shall be provided by the user. An operator's manual containing clearly illustrated instructions for operation, cleaning, assembly/disassembly, decontamination/sterilization, troubleshooting, parts list, etc., shall be provided. The apparatus shall be clearly marked with manufacturer's name, address, and any applicable standards ratings (see 3.20).

3.12.4 PORTABLE SUCTION ASPIRATOR

When specified (3.15.4 M18), a portable suction aspirator having the following characteristics shall be provided. The unit shall be capable of operation from the vehicle's 12 volt DC electrical system. It shall also operate from an integral battery supply which is rechargeable and which will allow the unit to meet the air flow and suction requirements of this paragraph for at least 20 minutes of continuous operation. Gas powered portable suction aspirators are acceptable if they meet the performance requirements specified herein. The portable suction aspirator shall provide an air flow of at least 20 liters per minute at the end of the suction tube and a vacuum level of at least 300 mm Hg to he reached within 4 seconds after the tube is clamped. The portable aspirator shall be provided with the following accessories: one suction rinsing water bottle, one nonmetallic and pharyngeal suction tip with thumb suction control port of sufficient diameter to allow aspiration of semi-solid gastric contents. The suction tubing shall be translucent and shall not kink or collapse under high suction. The inside diameter of the tubing shall be at least .64 mm (1/4"). The other suctioning catheters shall be provided by the user. The internal battery supply shall be rechargeable from the vehicle's 12 volt DC electrical system and from 115 volts AC. The 115 volt AC charging system need not be integral to the portable suction unit. The unit shall be provided with applicable power cords and plugs (MS 3116F12-3P). The polarity of the DC plug shall be as follows: Pin A (+12 volts), Pin B (ground), Pin C (not used). An operator's manual containing clearly illustrated instructions for operation, cleaning, assembly/disassembly, decontamination/sterilization and troubleshooting, parts list, etc., shall be provided. The apparatus shall be clearly marked with manufacturer's name, address, and any applicable standards ratings (see 3.20).

3.13 ENVIRONMENTAL: CLIMATIC AND NOISE PARAMETERS

3.13.1 ENVIRONMENTAL SYSTEMS

All ambulances shall be equipped with a complete climate environmental system(s) to supply and maintain clean air conditions and specified level of inside temperature in both driver and patient compartments. The various systems for heating, ventilating, and air conditioning may be separate or combination systems which shall permit independent control of the environment within each compartment. All ambulances shall be equipped with heating, ventilating, and air conditioning systems that can be made to collectively operate using recirculated air and outside ambient air and shall be capable of maintaining interior temperature of 200° C to 26° C (68° F to 780° F) when operated between -18° C to **350°** C (0° F to **950°** F) outside ambient temperature. The air systems shall be high volume capacity with low velocity delivery for minimum draft circulation while providing a positive pressure within each closed compartment (positive pressure may be attained through the fresh air ventilation system either apart from or as an integral part of the heating/air conditioning system). Environmental system components shall be readily accessible for servicing at the installed location(s). Connecting hoses for heating and the air conditioning system shall be supported by rubber insulated metal clamping devices at least every 30 cm (12 in.). The driver and patient compartment environments shall not in any way be dependent upon each other.

3.13.2 DRIVER'S COMPARTMENT ENVIRONMENTAL EQUIPMENT

The driver's compartment shall be furnished with the chassis manufacturer's hot water, fresh air, high capacity, heater, defroster, and air conditioning system. The patient compartment heater shall be installed in a manner that does not adversely affect the chassis manufacturer's FMVSS 103 performance certification for windshield defrosting.

3.13.3 PATIENT COMPARTMENT ENVIRONMENTAL EQUIPMENT

The patient compartment shall be heated, ventilated, and air conditioned in accordance with the criteria specified herein. During the performance testing (see 4.4.2) of the heating and air conditioning systems, the use of fans or blowers, within the test chamber, is only permitted to provide air circulation to help maintain the required ambient test temperature. The use of these fans or blowers to direct air through the vehicle's radiator and/or condenser(s) is not permitted.

3.13.4 HEATING CRITERIA

The heating system(s) shall have sufficient capacity to simultaneously raise the temperature at the midpoints, in each compartment, to a minimum dry bulb temperature of 20°C (68°F), at all twelve test points (9 patient compartment and 3 cab) within 30 minutes. Test conditions: the ambulance (with doors open) shall be cold soaked for three hours in an ambient temperature of (00 F), then the engine started and allowed to run at the high idle setting (see 3.7.6) while the transmission is in park or neutral. Three verification readings shall be made (start, middle, and final) by nine equally spaced test thermocouples in the patient compartment, and three in the cab compartment. Heating equipment may be in the recirculating mode and all compartment openings, including partition doors/windows and exhaust vents, shall be closed. Heater(s) furnished shall comply to applicable Motor Carrier Safety Regulation 393.77. Heater(s) shall, to the maximum extent possible, be connected to the chassis manufacturers furnished interconnection points. When interconnection points are not furnished, chassis manufacturers unused engine water pump supply ports shall be utilized. All added heater hose shall meet or exceed chassis manufacturers requirements and when specified (see 6.2-fff), silicone hoses shall be furnished.

3.13.5 AIR CONDITIONING CRITERIA

The air conditioning system(s) shall have sufficient capacity to simultaneously lower the temperature at the midpoints in each compartment to a maximum dry bulb temperature of 26°C (78°F), at all twelve (9 patient compartment plus 3 cab) test points within 30 minutes. Test conditions: the ambulance (with doors open) shall be heat soaked for three hours in an ambient temperature of 35°C (95°F), then the engine started and allowed to run at high idle setting (see 3.7.6), while the transmission is in park or neutral, with test verification and thermocouples placement as specified in 3.13.4. Air conditioning equipment may be in the recirculating mode and all compartment openings, including partition doors/windows, shall be closed. When available, chassis manufacturers interconnection points shall be utilized.

3.13.5.1 AUXILIARY AIR CONDITIONING CONDENSER

When specified by the purchaser (see 6.2-oo) or required for high temperature geographic areas or furnished as standard equipment, an auxiliary condenser shall be provided which will allow for maximum system performance, based on the air conditioning and ambulance manufacturer's recommendations. If the condenser is located above the cab, it shall not block the emergency lights. All added refrigeration lines and fittings shall be mechanical fittings compatible with OEM components furnished by the chassis manufacturer. An auxiliary condenser may be furnished or specified to function in lieu of the chassis manufacturer's condenser provided it meets or exceeds all other system performance requirements.

3.13.6 VENTILATION CRITERIA

Ventilation system(s) of the driver and patient compartments shall provide a complete change of ambient air within each compartment at least every two minutes with the vehicle stationary. Ventilation shall be separately controlled within each compartment. Fresh air intakes shall be located towards the front of the vehicle and exhaust vents shall be located on the upper rear of the vehicle. The patient compartment shall be ventilated by the air delivery system of the environmental equipment (heater-air conditioner) or by separate system(s), such as power intake, exhaust ventilator(s).

3.13.7 ENVIRONMENTAL CONTROLS

Adjustable, manual or thermostatically operative controls shall permit heating and/or air conditioning and ventilation in either compartment without affecting the other compartment. Switches and controls shall be located in EMT's panel and/or remote panel and identified for function and operating position (see 3.7.9 through 3.7.11). Switches, wiring and control components shall be rated at a minimum of 125 percent of the maximum electrical load of the blower(s) and other components (see 3.7.2). Blower or fan system shall have at least three speeds (excluding "OFF"). Separate noncorroding brass, bronze, stainless steel, plastic or other inherently corrosion proof shutoff valves, for the patient compartment hot water heating system, shall be provided to allow servicing and shutoff. The use of vacuum operated shutoff valves is acceptable provided the above criteria is met and the valve provides inherent sealing when vacuum is removed. This sealing shall prevent engine cooling system pressure and water pump pressure from causing any leakage when vacuum is removed. Air systems shall have adjustable louvers to direct the flow of air.

3.13.8 PATIENT COMPARTMENT SOUND LEVEL CRITERIA

Unless otherwise specified (see 6.2-pp) by Federal or State regulations, the patient compartment decibel level shall not exceed 80db and shall be tested in accordance with AMD STANDARD 006.

3.14 COMMUNICATIONS

3.14.1 COMMUNICATION EQUIPMENT

The ambulance and console(s) shall be configured for two-way radio(s) (mobile and portable), intercom, public address, electronic siren, and telemetry system(s). These systems, when specified (see 6.2-qq), herein or purchaser furnished/specified shall be installed and ready to use. Switches and controls shall be panel mounted and conform to 3.7.9 through 3.7.11. Communications equipment shall meet the applicable FCC rules and required state and local area EMS radio communications.

3.14.2 RADIO (MOBILE) PROVISIONS

All ambulances shall be provided with sufficient ventilated space for a two-way radio (including convenience features), antenna openings, ground plane, terminal wiring for 12V power and ground, and mounting panels (see 3.14.3) required for the installation and servicing of mobile radios. Unless otherwise specified (see 6.2-qq), the consignee will locally install mobile radio unit(s) or shall provide any additional data, equipment and a coordinator to insure compatibility of specified systems.

3.14.3 ANTENNA, CABLE, AND MOUNTING

THE contractor shall provide each ambulance with a ground plane, antenna opening, and coaxial lead-in wire from the ventilated radio storage area/compartment to a temporarily sealed antenna port. The antenna and/or mounting shall be on the centerline of the patient compartment roof. An antenna port shall be provided in the patient's compartment directly under the mounting area. The port shall provide at least a 10 cm × 10 cm (4" × 4") opening, and be marked as specified in 3.7.11. All nonmetallic roofs shall be equipped with at least a 102 cm 102 cm (40" × 40") metal ground plane molded into the roof (see 3.10.6). A conduit and fish lead shall be provided when necessary to fish the cable from the antenna to the radio. Unless otherwise specified (see 6.2-qq), the antenna cable (lead-in) shall be a labeled, RG/58U or equal. Approximately 46 cm (18") of extra cable shall he provided at the antenna base (port) and at least 91 cm (3 feet) at/in the radio compartment provided with a PL-259 connector soldered to the end of the cable. Purchaser shall specify any changes or additional requirements.

3.14.4 TWO-WAY (MOBILE) RADIO EQUIPMENT

When specified (see 6.2-qq), a solid state, mobile, two-way radio, meeting all purchaser requirements, shall be furnished. The transmitter power output shall be as specified.

3.14.5 INTERCOM SYSTEM

When specified (3.15.3-4), solid state intercom equipment shall be provided between the driver and patient compartments. The driver shall have the capability to place the patient compartment in standby "talk" mode and to control any necessary talk/listen switching. The intercom system shall be independent of the two-way radio equipment. Cab compartment intercom shall be mounted in accordance with 3.7.9 and patient compartment intercom in accordance with 3.7.10, and comply to 3.7.11.

3.14.6 SIREN—PUBLIC ADDRESS SYSTEM.

(A7.1) A combination electronic siren with integral public address system and other upgrading features, including radio interface capability, shall be provided. Dual speakers shall be installed, outside the vehicle, in the bumper/hood area. Speakers shall not protrude beyond the face of the bumper or bumper guards. The microphone shall be a noise canceling type. The siren's control shall permit the following sounds, Manual, Wail, and Yelp. Other applicable sounds such as "Rapid Yelp"; "Air Horn" or composite type sounds, etc. may be specified (see 6.2-rr) or furnished. The system shall provide plug-in connections and connecting cables; with control capability from the passenger seat and driver's horn ring by means of a siren/horn switch or, when specified (see 6.2-rr), by a foot switch; and all other necessary hardware for the siren/public address operation. The public address amplifier shall be independent of the two-way radio, except that a common microphone and control housing group may be employed. The illuminated (in siren mode) "Horn/ Siren" switch shall be provided on the driver's console.

The siren, with the exception of cancellation effects due to dual speakers, when tested in a full anechoic chamber that conforms to ANSI Standard 81.13-1971, with test equipment and methods conforming to California Administrative Code, Title 13, Article 8, shall be capable of producing a continuous warning sound at a minimum level of 123 db, A-weighted, at 3 meters (10 feet) on axis in the "Wail mode" with "Yelp" falling within 1 dba with 13.6 volts +/- 1% input, at a fundamental frequency in the range of 500 to 2,000 Hz maximum.

The output over the sweep range shall not drop to less than 116 dba. The speakers shall be located in the configuration that is representative of the vehicle on which they will be mounted.

In the "Wail" mode the siren shall have a sweep rate of 10–18 cycles per minute and in the "Yelp" mode, a sweep rate of 150 to 250 cycles per minute. All sweep modes shall cover a range of at least one octave.

In voice (P.A.) operation, the unclipped sine wave output shall be at least 55 watts RMS into a resistive load matching the nominal speaker system impedance at 1000Hz. The frequency response of the amplifier shall be from 500 to 3,000Hz +/- 3dB, when measured from 1000Hz reference. Total harmonic distortion shall not exceed 10 percent, at 20 watts RMS, over the specified frequency range when measured with the load shown above.

In addition, the electronic siren furnished with the exception of cancellation effects attributable to dual speakers shall comply to all the other requirements included in the State of California Vehicle Code Section 1020 through 1029, Title 13, Article 8, the latest issue for Class A sirens. The electronic siren shall be tested, approved, and listed with the Automotive Manufacturers Equipment Compliance Agency (AMECA).

3.15 ADDITIONAL SYSTEMS, EQUIPMENT, ACCESSORIES, AND SUPPLIES

3.15.1 ADDITIONAL AND OPTIONAL EQUIPMENT

When specified (see 6.2-ss), additional or optional system(s), equipment, accessories, and supplies shall be furnished, in addition to the standard ambulance component systems and devices specified herein. These items shall be selected by the purchasers. Optional items may be further described in authoritative procurement documents. In no event shall the specified or furnished optional item(s) reduce the quality and intent of the ambulance but shall enhance its design and purpose. The materials, devices, items, and fabrication, if not specifically described, shall be not less in quality, strength, performance, and service than those normally provided by the most reputable manufacturers.

3.15.2 STANDARD MANDATORY MISCELLANEOUS EQUIPMENT

Unless otherwise precluded in the contract, each ambulance shall be equipped with, but not limited to the following.

a. Fire extinguishers: Two, ABC dry chemical, carbon dioxide) minimum 2.3 kg (5 lb.) unit, in a quick-release bracket, one mounted in the driver/cab compartment or in the body reachable from outside the vehicle and one in the patient compartment. When located in either the driver or patient compartments, the mounting bracket shall be a stable design.
b. "No Smoking Oxygen Equipped" and "Fasten Seat Belts" signs: Conspicuously placed in the cab and patient compartment.
c. Overhead grab rail, minimum 152 cm (60 in.) long, maximum 10.2 cm (4 in.) depth, on the ceiling over the primary patient. Grab rail shall be stainless steel, aluminum, or other corrosion resistant material, and have padded or curved up ends, and rounded corners. Mounting brackets shall be chromed, stainless steel, polished east aluminum or other corrosion resistant materials. The grab rail(s) shall be installed in a manner that precludes loosening and shall meet a 136 kg (300 lb.) pull test in all three axis. The grab rail shall be tested in accordance with AMD Std. 009.
d. Backup alert alarm, (audible warning device) activated when the vehicle is shifted into reverse. Device must meet OSHA and SAE J994 requirements, and shall be rated (SAE) for Type C or B (97 or 107 dba at 122 cm (4 feet)) or purchaser may specify (see 6.2-ss) alarms that automatically adjust to ambient noise levels.

3.15.3 OPTIONAL EQUIPMENT

When specified (see 6.2-ss), the ambulance shall be equipped with, but not limited to the following. Additional weight, resulting from specified options, shall be deducted from the specified payload allowance (see 3.5.2).

1. Reserved
2. Tachograph; 145 km/ hour (90 mph), with 24-hour chart, warning light, and siren recording.
3. Intercom, Silent; The system shall consist of amber, red, and green lights in the driver's compartment controlled by the EMT. It shall include an alerting buzzer in the driver's compartment.
4. Intercom; voice intercommunication system between driver and technician compartments as per paragraph 3.14.5
5. Batteries, 2-12VDC, heavy-duty "high cycle life," maintenance-free automotive type per 3.7.7. Similar to Delco, Gould, or equal
6. 115 VAC onboard electric power supply, Inverter (specify brand or equal) per 3.7.8.3
7. 12 VDC battery charger or conditioner onboard, per 3.7.7.1 and 3.7.8

(*)8. High intensity cotlight; with flexible or adjustable shaft located near head of primary cot per 3.1.5.1

9. "Checkout light" for patient compartment (see 3.8.5.2)
10. Patient fan; located near primary cot, minimum 250 CFM rating, 12 VDC
11. Rear step storage compartment for Type I and III when available on specified chassis

12. Window; tinted, right side panel of patient compartment, minimum 2774 sq. cm (430 sq. in.) per 3.10.14
13. Drapes; fire retardant cloth drapes for each window of patient compartment except partition, installed on polished metal drapery hardware and rod(s) on both top and bottom

(*) 14. Drug kit compartment; locking, security compartment located near primary patient with piano hinge and non-exposed fasteners, door with key lock, minimum size 56 cm (22")W × 36 cm (14")D × 46 cm (18")H per 3.1.5.1

15. Stone/mud guard to protect lower front end of body (for Type I and III) (see 3.9.7.4)
16. Power plant heaters; minimum 34°C (-30°F) startability per 3.6.3.2.1 for severely cold areas only per 3.6.3.2.1
17. High altitude operations; engine adjustments and/or emission control(s), and test if available and applicable per 3.6.4.3
18. Extrication equipment, specified in 3.11.2.1. (This equipment may normally be carried on the ambulance unless it is routinely accompanied by a rescue vehicle. Select which items, (a through jj) and fully detail specific items wanted by brand, model, size, quantity, as required.
19. Body rubralls; for Types I and Ill, as specified in 3.9.7.5
20. Automatic transmission heavy-duty oil cooler, additional unit (for very hot areas only), per 3.6.5.2.1 when not furnished from the chassis manufacturer
21. Folding style step; at ambulance body rear loading doors, per 3.9.6
22. Front bumper override, grille, and radiator protection, per 3.9.6.1 (grille guard) and 3.9.6.2 (bumper guards)
23. Emergency safety kit; in sturdy metal/plastic carrying case containing: 3 bidirectional reflective triangles (approved FMVSS 125), 6 fusees, minimum 15 minutes each with holders, 2 sets of chassis circuit electrical system spare fuses
24. Towing devices (if permitted by chassis manufacturer), not less than 2 towing devices on the front of the vehicle
25. Spotlight, 360 degree horizontal and 90 degree vertical adjustment, remote controlled per 3.8.4
26. Radio-chassis manufactures, AM/FM, as specified. Additional speakers shall be furnished as specified.
27. Reserved

(*) 28. CPR seat on left, street side of vehicle (right side of patient) (see 3.1.5.1 and 3.11.4)

29. Cab Console between driver and passenger seat. Purchaser shall specify layout and dimensions (see 3.9.2 & 3.10.15.2).
30. Running boards (see 3.9.7.2)
31. Mud flaps (see 3.9.7.3)
32. Interior/exterior compartment, two shelves, at the front right comer of the body with exterior access per 3.11.1.2 (for Types I & III)
33. Map light, dash or ceiling mounted as required by the purchaser and shall be separately switched. Purchaser shall specify type or model required (see 3.8.5).
34. Siren system. Specify features e.g. foot switch or other requirements (see 3.14.6).
35. Audible warning indicator, in the cab, that sounds only when the patient or equipment compartment(s) are open and when the vehicle is placed in forward or reverse gear

(*)Items furnished as standard equipment when Configuration "A" (ALS) is specified (see 3.1.5.1).

3.15.4 MEDICAL, SURGICAL, AND BIOMEDICAL EQUIPMENT
When specified (see 6.2-ss), the ambulance shall be equipped with the following:

Ml. Rail system; a medical device(s) attaching aluminum rail, with quick release mounting mechanisms, fastened securely in the action area at the side of the primary cot. Rail length 152 cm to 229 cm (60 to 90 in.) long. Medical devices shall be rail mounted, and 3 spare mounting devices supplied per 3.11.1.1.

M2. I.V. telescoping pole for primary cot (see 3.11.9).

M3. Electric clock, 12 volt DC chassis powered, with sweep second hand, illuminated (when module is turned on), 10 cm (4 in.) minimum dial. Clock shall be located on the action wall and shall be visible to the EMT from both the EMT seat and squad bench. Other types of clocks as specified or approved by the purchaser may be furnished.

M4. Suction jar (clear plastic) for secondary patient with 20 disposable inserts or spare bottle (clear plastic) with holder (see 3.12.3.)

M5. Sphygmomanometer; 15 cm (6 in.) dial type, wall mounted or rail mounted, when Ml is specified, cuff; stethoscope, inflating bulb and mounting. Specify size, obese, adult, child, or pediatric.

M6. Squeeze bag-valve-mask system, portable, for artificial ventilation (see 3.12.2.1).

M7. Oxygen flowmeter (pressure compensated) and humidifier (see 3.12.1.1), with specified disconnect fittings. Specify quantity.

M8. Remote oxygen indicator for main bottle. Specify type (see 3.12.1).

M9. Oxygen outlet. A second, self sealing, duplex oxygen outlet for the secondary patient to be located at patients head area (see 3.12.1). Additional outlets may also be specified.

M10. Cot for primary and secondary patients (per 3.11.5). Specify desired models. Cot straps per 3.11.8.1

M11. A lower-extremity traction splint with all necessary straps, attachments, and case. Specify type and adult or pediatric size.

M12. Emergency medical jump kit or drug kit, specify model (case only).

M13. Scoop stretcher, specify model.

M14. Buckboards; wood, aluminum, or plastic, one long-minimum size 183 × 41 cm (72 × 16 in.), one short minimum size 81 cm × 41 cm (32 × 16 in.) with three straps. Boards shall be x-ray translucent. Specify type and quantity.

M15. Reserved

M16. Combination stretcher, in lieu of folding stretcher (see 3.11.5.1 and 3.11.4)

M17. Oxygen, portable unit (see 3.12.2)

M18. Suction aspirator, portable; battery or gas powered (see 3.12.4). Specify type.

M19. Mast, compartmentalized individually inflatable sections, pneumatic trouser kit, complete with carrying case. Specify adult, pediatric and model.

M20. Oxygen-powered, manually triggered, positive pressure valve with mask (see 3.12.2.2). Specify type of valve and quick disconnect fitting required.

M21. Dual position cot fastener mounting assembly (quick detachable type for center and side positions). Specify type (see 3.1.5 and 3.11.7).

M22. Vest type (wrap around) extrication device to provide spinal immobilization for seated patient, complete with all necessary belts, head straps, padding, etc., and case. Specify type.

M23. Extrication collar, specify type and size (pediatric, extra small, short, medium, tall, etc.).

M24. Reserved

M25. Cardiac monitor/defibrillator mounting bracket. Specify type of bracket.

M26. Pediatric, spinal immobilization/transport device with cot interface capability. Device shall be complete with all straps, attachment devices, and case. Specify desired model (see 3.11.8.2).

M27. Reserved

M28. Infant/child carrier per 3.11.8.2.

M29. Automatic Resuscitation/ventilator (see 3.12.2.3). Specify type desired and type oxygen system interface device (see 6.2-11).

(*)Items furnished as standard equipment when Configuration "A" (ALS) Is specified (see 3.1.5.1).

3.16 PREPARATION FOR PAINTING, COLOR, AND MARKINGS

3.16.1 PREPARATION FOR PAINTING

Ambulance body and all attached equipment exterior surfaces, except polished metal parts, shall be thoroughly cleaned, treated, and coated with a firm primer and preservative with rust inhibiting properties, and painted in the finish color as specified (see 6.2-tt). Ferrous metal interior surfaces shall be painted or, when not exposed for painting, shall be treated or coated to resist corrosion. Chassis and chassis frame components shall be preserved and finished in accordance to industry's standard practice.

3.16.2 COLOR, PAINT, AND FINISH

Unless otherwise specified, (see 6.2-tt), the exterior color of the ambulance shall be gloss white in combination with a solid uninterrupted orange stripe and blue lettering and emblems. The stripe should be as close to parallel as possible with the road but a stripe transition angle is acceptable to connect the module beltline stripe with the chassis stripe. The exterior finish on painted metal modular bodies and metal roofs on Type II ambulances shall be an acrylic composition urethane or polyurethane paint. The final stage manufacturer's painted components shall have a paint film not less than 1.25 mils. thick and a minimum total thickness of 4 mils., including primers. The orange stripe shall not be less than 15 cm (6 in.) wide, nor more than 36 cm (14 in.) wide and shall encircle the entire ambulance body at the belt line below the bottom edge of cab windows but may exclude the front of the hood panel. The orange stripe, as specified by the purchaser (see 6.2-tt) may be reflective tape conforming to Para. 3.16.4 or be painted and may be edged/pin striped in black or blue. This single, solid band (except when interrupted by windows, locks, etc.), when viewed horizontally, shall appear as a stripe near parallel to the road The interior finish shall be the manufacturer's standard light color harmonizing with the color of upholstery. The final film of painted surfaces shall be smooth and uniform, free of grit, streaks, blushing, runs, sagging, blisters, "fisheyes," "orange peel," pinholes, or other surface irregularities. Exterior finish paint shall not be required on the underbody and inside surface of the body skirting.

NOTE:USERS SHOULD AVOID SPECIFYING DARKER OR ALTERNATE COLORS BECAUSE OF THE ADVERSE EFFECT ON VEHICLE VISIBILITY, AND INCREASED HEAT LOAD THAT THE DARKER COLORS IMPOSE ON THE AIR CONDITIONING SYSTEM. SPECIFICALLY, THE WHITE/ORANGE COMBINATION PROVIDES THE MOST VISIBLE VEHICLE UNDER DIFFERENT LIGHT CONDITIONS, (E.G. MERCURY AND SODIUM VAPOR LIGHTS), AS WELL AS PROVIDING, ON AN INTERNATIONAL BASIS, THE MOST EASILY RECOGNIZED COLOR CONFIGURATION.

3.16.2.1 COLOR STANDARDS AND TOLERANCES

The exterior surface including the wheels shall be manufacturer's standard gloss white. The ambulance colors orange, stripe or band and blue markings shall be the same as specified Orange and Blue in American National Standard 753.1-1974, Safety Color Code for Marking Physical Hazards. They shall comply with the tolerances expressed in terms of Munsel hue, value (lightness), and chroma (saturation).

Color tolerance charts containing the color standards and tolerances for Ambulance Orange and Ambulance Blue are available from the Department of Transportation, Research and Special Programs Administration, Information Services Division (DMT-11). 400 7th Street. SW. Washington DC 20590.

3.16.3 SALT SPRAY RESISTANCE
Treated exterior sheet metal of the ambulance body (except OEM Type II van) shall be capable of withstanding 250 hours of salt spray tested in accordance with ASTM B 117-64. The specimen used for the salt spray test shall be run through all steps of the cleaning and treating process, including priming. The primed specimen shall be scored from corner to corner using a sharp knife. After the test, the specimen panels shall exhibit no failure and not more than 3 mm (1/8 in.) rust or blister creepage from the scored lines.

3.16.4 EMBLEMS AND MARKINGS (A 9.1)
The material for the emblems and markings shall be applied using reflectorized material conforming to Federal Specification L-S-300, Type I, Class 1 or 3 reflectivity 1. The reflective color used shall be blue (color a) and white (color I) when applicable. The emblems and markings shall be of the type, size, color, and location as follows:

A. Front markings
   1. The word "AMBULANCE," mirror imaged, shall be in block, blue, die cut style letters, not less than 10 cm (4 in.) high, centered above the grille, on the orange or white background. The placement of the word ambulance on the curved surface of the hood or on a flat plastic type bug screen is permitted.
   2. A "Star of Life" in 7.6 cm (3 in.), blue, die cut style, with a white border, conforming to Figure 4 (size A) [omitted] shall be located both to the right and left of the word "AMBULANCE."
B. Side and rear markings
   1. The word "AMBULANCE" shall be in block, blue, die cut style letters of not less than 15 cm (6 in.) in height, centered, with a white border, alongside or under the "Star of Life" on each side and rear of the vehicle body.
   2. A "Star of Life," not less than 40.6 cm (16 in.), in blue, die cut style, with a white border, conforming to Figure 4 (size C) [omitted], on the right and left side panels. A "Star of Life" emblem, size B, shall be provided on each rear door.
   3. All additional lettering and markings (required by the purchaser) should be below the word "AMBULANCE," except the unit's abbreviated identification lettering! number, which may be located on the uppermost white space of the front, rear, and sides.
C. Top markings
   1. A "Star of Life", of not less than 81 cm (32 in.) (size 0) in blue, die cut style, conforming to Figure 4 [omitted] (may be without the white Staff of Aesculapius), shall be provided on the ambulance rooftop.

The "Star of Life" is a six-barred cross upon which is superimposed the Staff of Aesculapius (es´´cu-la´pi-us) who, in both Greek and Roman Mythology, was the god of medicine and healing.

Dimensions (min.):

	Size A	Size B	Size C	Size D
Length of bar	76mm(3")	30.5 (12")	40.6cm(16")	81.3 cm(32")
Width of bar	25mm(.75")	76 (3")	10.2 cm(4")	20.3 cm(8")
Length of staff	63mm (2.5")	24.1 cm(9.5")	31.7 cm (12.5")	63.5 (25")
White border	(.25')	(.375')	(.5')	

All angles 60%. Deviations must be proportionate.

3.17 UNDERCOATING

Unless the ambulance is rustproofed, the vehicle shall be undercoated for sound deadening, corrosion, and stone damage protection. A commercial, sandless, petroleum base undercoating or other materials providing equivalent protection, shall be applied to the underbody and under chassis sheet metal surfaces to a thickness of 1.6 mm to 3.2 mm (1/ 16 to 1/8 in.), except to the drive shafts, drain holes, lubrication points, engine/transmission oil pans, fuel tanks, heavy castings, suspension components, heat shields, heat diffusing devices, catalytic converters, and areas 30.5 cm (12 in.) or less from the exhaust system(s) as well as other areas specifically excluded by the chassis manufacturer. These areas shall be kept free of coating material. Chassis frame, underside of engine compartment hood, and underbody surfaces in excess of 3.2 mm (1/8 in.) thickness, or that are inaccessible without removing vehicle fuel tank(s) or other major components shall not require undercoating.

3.18 RUSTPROOFING

When specified (see 6.2-uu), ambulance shall be rustproofed in accordance with Fed. Std. No.297.

3.19 MARKINGS, DATA PLATES, WARRANTY NOTICE, ETC.

Unless otherwise specified (see 6.2-vv), final stage manufacturer's caution plates and identification plates shall be conspicuously installed for all equipment, etc., furnished requiring such notices. Other than the manufacturer's trademark(s) names, no other identification than that authoritatively specified shall be shown on exterior of the vehicle. The ambulance/vehicle manufacturer's "Star of Life" certification shall be provided on a placard or label as shown below, permanently affixed and easily visible in the ambulance oxygen compartment (see 4.3).

All ambulances shall display a decal or sticker providing at least the following information: contract number; purchaser order number; date of delivery, month and year; and the warranty time, in months and miles (GSA Form 1398 or equal). Apply this data to the right or left front door lock face or doorjamb after final inspection and acceptance by the purchaser. As specified (see 6.2vv) by the procuring activity for the appropriate military service, identification markings and data plates, shall be provided conforming with MIL-STD-1223.

3.20 MANUALS, AND HANDBOOK OF INSTRUCTION

The supplier shall furnish with each ambulance, at the time of ambulance's acceptance, one copy of the reference handbook as specified in 6.8 and herein. This reference hand-

MFG BY		DATE OF MANUFACTURE MO, YR
ADDRESS		
CITY	STATE	ZIP
This Ambulance conforms to Federal Specification KKK-A-1 822 in effect on the date of manufacture shown above.		
AMBULANCE IDENTIFICATION NUMBER		
	TYPE CLASS - FLOOR PLAN - SERIAL NO.	
CURB WEIGHT	PAYLOAD	MAXGROSS WT. MAX
kg/lbs.	kg/lbs.	kg/lbs.
CERTIFIED "STAR OF LIFE" AMBULANCE		

Sample label.

book shall provide instructions for the operation, care, and repair for all ambulance related accessory, component equipment, and system(s) furnished as part of the emergency medical care vehicle. This handbook shall also contain all installation instructions, drawings, schematics wiring diagrams, illustrations, and safety precautions to insure proper management, operation and maintenance. The chassis manufacturer's repair manual and parts book shall be furnished when specified (see 6.2-ww).

3.21 PREDELIVERY INSPECTION AND SERVICING

The supplier prior to acceptance and inspection of the ambulance(s) shall service and inspect each vehicle in accordance with the chassis manufacturer's approved predelivery form, and the ambulance manufacturer's predelivery (test, inspection, and road test) form. A signed copy of these forms (check sheets) shall be furnished with the vehicle (see 6.8). Servicing shall comply to ambient temperatures and conditions applicable with the route of transport to the consignee's ultimate destination (see 5.1). Servicing shall include all tank(s) full of fuel; checking to determine satisfactory and complete operation of all mechanical and electrical features, equipment and system; elimination of rattles, noises, and squeaks; cleaning the interior and exterior. Thus the vehicle shall be delivered ready to use.

3.22 SPECIAL REQUIREMENTS

Purchaser shall specify (see 6.2-xx) additional requirements and state unusual operating conditions, miscellaneous items, permissible exceptions, marking, etc., not specified herein. If in conflict with the manufacturer's standards and options, those specified shall take precedence. Contractor shall comply to the requirements of paragraph 6.3.

3.22.1 OVERSEAS VEHICLE REQUIREMENTS.
When specified (see 6.2-yy), the following equipment changes, items, devices, and adjustments to systems shall be furnished:

a. Left dip headlights (as used on right-hand drive vehicles) adjusted for driving on the left side of the road (prior to shipment).
b. Speedometer-odometer, calibrated in kilometers.

3.22.2 EXPORT VEHICLE SAFETY, EMISSIONS, AND FUEL REQUIREMENTS
Unless otherwise specified (see 6.2-yy), export vehicle(s) shall conform to applicable safety standards of U.S.A. Vehicles exported to countries marketing only leaded gasoline fuel shall be capable of accepting (refueling nozzles), and operating on leaded gasoline. Vehicle modifications needed for leaded fuel operation shall be made prior to shipment. Vehicle(s) being offered/supplied with catalytic converters shall be identified in each bid.

3.22.3 TIEDOWN POINTS
When specified (see 6.2-yy), hard points (tow hooks, bumpers, axles, or other component points) suitable for use as tiedown points for the vehicle shall be identified. Identification of the points shall be provided by stenciling instructions and applying data plates to the vehicle or by supplementary instructions included with each copy of the owner's-operator's manual. Each identified tiedown or hard point shall be capable of withstanding its proportionate share of the GVWR.

3.22.3.1 LIFT POINTS
When specified (see 6.2-yy), a minimum of four hard points for use as lift-points shall be identified. Identification of the points shall be provided by stenciling instructions and applying data plates to the vehicle or by supplementary instructions included with each copy of the owner's-operator's manual. Each identified hard point shall withstand without permanent deformation a working load equal to its proportionate share of the GVW and shall have an ultimate strength of at least 1.5 times the GVW. In addition, slinging provisions shall be located so that

a. Attached sling legs shall converge over the center of gravity of the vehicle when at its GVW.
b. The attached sling apex does not exceed a height of 7.3 m (24 feet) above the lowest extremity of the equipment when suspended with each sling leg at a 45 degree maximum true angle.

3.23 WORKMANSHIP
Defective components shall not be furnished. Parts, equipment, and assemblies, which have been repaired or modified to overcome deficiencies shall not be furnished without the approval of the purchaser. Welded, bolted, and riveted construction utilized shall he in accordance with the highest standards of industry. Component parts and units shall be manufactured to definite standard dimensions with proper fits, clearances, and uniformity. General appearance of the vehicle shall not show any evidence of poor workmanship.

The following shall be reason for rejection.

1. Rough, sharp or unfinished edges, burrs, seams, sharp corners, joints, cracks, and dents.
2. Nonuniform panels. Edges that are not radiused, beveled, etc.
3. Paint runs, sags, orange peel, "fish eyes," etc., in addition to 3.16.2 and any other imperfection or lack of complete coverage.
4. Body panels that are uneven, unsealed, or have voids.
5. Misalignment of body fasteners, glass, viewing panels, light housings, other items with large or uneven gaps, spacing etc. such as door, body panels, and hinged panels.
6. Improper body design or interface with the chassis that could cause injury during normal use or maintenance.
7. Improperly fabricated and routed wiring or harnesses.
8. Improperly supported or secured hoses, wires, wiring harnesses, mechanical controls.
9. Loose, vibrating, abrading body parts, components, subassemblies, hoses, wiring harnesses or trim.
10. Interference of chassis components, body parts, doors, etc.
11. Leaks of any gas or fluid lines, (AC, coolant, oil, oxygen, etc.)
12. Noise, panel vibrations, etc.
13. Sagging, non-form-fitting upholstery or padding.
14. Incomplete or incorrect application of rustproofing.
15. Inappropriate or incorrect use of hardware, fasteners, components, or methods of construction.
16. Incomplete or improper welding, riveting.
17. Visual deformities.
18. Lack of uniformity and symmetry where applicable.
19. Unsealed appurtenances or other body components, gaskets, etc.

In addition, any deviation from specification requirements or any other item, whether or not stipulated herein, that affects form, fit, function, durability, reliability, safety, performance or appearance shall be cause for rejection.

4 QUALITY ASSURANCE PROVISIONS

4.1 RESPONSIBILITY FOR INSPECTION AND TESTS

The contractor is responsible for the performance of all inspections and test requirements specified herein. The supplier may use his own or any other facilities suitable for the predelivery and acceptance inspections unless disapproved by the purchaser. The purchaser reserves the right to perform any of the inspections and tests set forth in the specification where such inspections are deemed necessary to assure supplies and service conform to the specification and contract. The contractor shall provide the purchaser's inspection representatives with the manufacturers readily available instruments and all such assistance as they may find necessary.

4.1.1 PURCHASER VERIFICATION

Quality assurance operations performed by the supplier will be subject to purchaser verification at unscheduled intervals. Verification will consist of observation of the operations to determine that practices, methods, and procedures of the supplier's inspection are being properly applied. Failure of the supplier to promptly correct observed deficiencies shall be cause for suspension of acceptance of the ambulance(s) until conformance to specification criteria has been demonstrated.

4.2 CLASSIFICATION OF INSPECTION (FOR PURCHASER'S ACCEPTANCE)

Purchasers should select either or both of the following methods of inspections (to specify see 6.2-zz):

1. Source inspection (see 4.2.1).
2. Destination examination (see 4.2.2).

4.2.1 SOURCE INSPECTION

Ambulance(s\051 shall be inspected upon completion by the purchaser prior to shipment from manufacturer's factory or assembly plant and may consist of the following:

a. Workmanship inspection (see 3.23)
b. Quality conformance inspection (see 4.2.3)
c. First production ambulance inspection (see 4.2.5)

4.2.2 DESTINATION EXAMINATION

When specified (see 6.2-zz), the contracted ambulance shall be examined at the destination. The vehicle(s) shall be visually examined to determine compliance to the contract requirements and the quality conformance inspection of 4.2.3. Vehicle component chassis failures and other minor defects may be accepted subject to correction by the contractor manufacturer of those points.

4.2.3 QUALITY CONFORMANCE INSPECTION

Quality conformance inspection applies to all ambulance(s) offered for acceptance under the contract. Quality conformance inspection shall consist of (a) through (d) as follows.

a. Workmanship inspection (see 3.23).
b. Operational checks (see 4.2.4).
c. Water spray test (see 4.4.5).
d. Examine ambulance handbook (see 320).

4.2.4 OPERATION CHECKS

Operational checks of the ambulance shall cover all controls, electrical systems, and devices, doors, windows, cabinets, accessories, in and outside the ambulance. Ambulance shall be driven at highway speeds, turns made at minimum radii, brakes tested for dependability, checked for rattles and squeaks, and compliance to 3.21. All controls and mechanisms shall function and operate as intended.

4.2.5 FIRST PRODUCTION AMBULANCE INSPECTION

When specified (see 6.2-aaa), the first production ambulance of each Type and Class specified under contract shall be inspected. This first ambulance shall be completely representative of the balance of the vehicles to be delivered under the contract. The purpose of the inspection shall determine vehicle conformity to the contract. The ambulance shall be inspected by the contractor at his plant under the direction and in the presence of purchaser's representative(s). Inspections shall be conducted during normal weekday working hours. The cost of inspections, examination, all tests, and vehicle support, exclusive of personal expenses incurred by participating purchaser's representatives shall be borne by the contractor. Acceptance of the first production vehicle shall not constitute a waiver by the purchaser of any rights under the provisions of the contract.

4.2.6 INSPECTION FAILURE OF AMBULANCE(S)
Failure of a production ambulance to have the certifications required or successfully complete the examinations and tests shall be cause for nonacceptance of any of the contract quantity, until deficiencies and evidence of the corrective action preclude recurrence of similar deficiencies. Failure of the vehicle to successfully complete inspection shall not constitute an excusable delay in meeting scheduled deliveries.

4.2.1 PRODUCTION SAMPLE
Upon acceptance of the first production ambulance, it shall remain at the manufacturing facility as a production sample and shall be the last vehicle shipped on the contract. The contractor shall maintain the vehicle in a serviceable condition for the duration of the contract. Vehicle shall be then cleaned, serviced, and refurbished to the extent required for delivery of a new vehicle to the purchaser.

4.3 "STAR OF LIFE" CERTIFICATION REQUIREMENTS

4.3.1 QUALIFYING PROVISIONS
The manufacturer/supplier is obligated to certify to the Government/purchasers that the ambulance bearing the "Star of Life," its components, and equipment meet or exceed all the requirements and tests set forth in this specification. The certification and "Star of Life" label (3.19), verify that the ambulance conforms to this specification on the date of manufacture. Compliance for a "Star of Life" label is defined as certification backed by confirmed verifications of inspections and tests. The verifications shall be in possession of the issuer and presented if and when challenged. For the benefit of purchaser's procuring activity evaluation and review, prior to or with each proposed bid (solicitation), the bidder/supplier shall provide and forward representative material of their "Star of Life" ambulance(s). This material shall include: a letter certified by a company officer, stating that the delivered ambulance(s) shall comply with paragraphs 4.3.2 through 4.3.6. Failure to provide certification, at the time the vehicle is presented for inspection, will deem the vehicle unacceptable and shall constitute grounds for termination in accordance with the terms of the contract. Also included shall be: general specification data, exterior and interior pictures, dimensional drawings/data, etc., and other information as requested.

4.3.2 DOCUMENTATION OF "STAR OF LIFE" CERTIFICATION
The ambulance manufacturer shall compile complete certified documentation of verifications for all the tests required under 4.4 conforming to 4.3.4 and 4.3.6 in a booklet similar to that described in paragraph 6.8, for each Type and Class ambulance intended to be marketed to the Emergency Medical Care industry as a "Star of Life" ambulance.

4.3.4 CRITERIA OF CERTIFICATIONS

The testing and inspection(s) required for certification shall be performed by either an independent testing facility or the ambulance manufacturer's own testing facility. In-plant tests shall be supervised by a staff registered professional engineer, or consulting registered professional engineer. The individual certifications will remain valid so long as they are applicable to the vehicle model, component(s) and equipment offered as tested under this specification, and are furnished without alterations, for five years, then shall be recertified. Certifications that appear on the vehicle need not be resubmitted (i.e., Department of Transportation, (DOT); Environmental Protection Agency, (EPA), etc.). Certification(s) will be acceptable in lieu of actual verification test during inspections (4.2) providing supporting verifying data complying to 4.3.6 is on file for examination. Certification from chassis manufacturer and individual equipment manufacturers are acceptable providing they are not part of a system(s) or altered and in accordance with 4.3.5. Type certifications of individual components and equipment products are acceptable.

4.3.5 CERTIFICATION LETTER FORMAT

Certification letters submitted for the ambulance model, components, and equipment being certified shall contain the following information on supplier's letterhead stationery: to whom certifying, date, units or items, manufacturer and address, date product tested, model number, and specification data, applicable specification references and test requirement, summary of the test report, a certifying statement, and official signature.

4.3.6 CERTIFICATION VERIFICATION DATA REPORTS

The testing facility (see 4.3.4) for each certification shall supply supportive verification data and information on letterhead stationery, for whom tested, report date, name of sample product or device, manufacturer's address, serial and model number(s), specification referral and amendment number(s), and test requirement(s), test facilities used and location, test equipment used, test procedure, test results, verifying test data, photographs, test conclusion(s), witness(es), and authorized signature.

4.4 TESTS

4.4.1 TEST CRITERIA

The ambulance shall be prepared for operation in accordance with chassis manufacturer's recommendations and 3.21. The ambulance shall be manned and loaded when applicable, to simulate the maximum payload required for the Type, Class, and plan being tested. Road tests and test course(s) shall be documented with pictures and route map. Test room(s) designed and equipped to simulate ambient temperature environments shall be used to perform tests.

4.4.2 PERFORMANCE TESTS

The following ambulance items, Systems, and equipment shall be inspected, tested, and verified for conformance to the specification:

a. Ambulance physical dimensions (see 3.4.11 through 3.4.11.6 and 3.10 to 3.11.4)
b. Vehicle weight distribution (see 4.4.3 and 4.4.3.1)

c. Road test and performance tests (see 4.4.4)
d. Exterior mounted devices tested for temperature conditions (see 3.4.2)
e. Cooling system (see 3.6.4.5)
f. Electrical systems and components, inspect and verify compliance (see 3.7 to 3.8.5.2)
g. Electrical generating system (see 3.7.6 through 3.7.6.3)
h. Electromagnetic radiation and suppression (see 3.7.12), type approval acceptable
i. Warning light and siren system(s) (see 3.8.2 to 3.8.2.3, and 3.14.6)
j. Tests of ambulance body structure (see 3.10.5 and 3.10.9)
k. Patient compartment interior surfaces, FMVSS 302 (see 3.10.17)
l. Oxygen system and tests (see 3.11.3,3.12.1, and 3.12.1.1)
m. Litter fastener and anchorage test (see 3.11.7)
n. Suction aspiration system test (see 3.12.3 and 3.12.4)
o. Environmental systems tests (see 3.13.1 to 3.13.7, and 3.6.3.2)
p. Patient compartment sound level test (see 3.13.8)
q. Painting, color, and marking (see 3.16 to 3.16.4)
r. Manuals and handbook (see 3.20 and 6.8)

4.4.3 VEHICLE WEIGHTS

The vehicle shall be weighed to determine curb weight and distribution of curb weight on front and rear axle. The imposed loading on front and rear axle will be computed using the curb weight and the payload as specified in 3.5 through 3.5.6. Calculated imposed loads on front and rear axle will be utilized to ascertain that suspension, axles, and tires furnished are of adequate capacity to meet contract requirements.

4.4.3.1 WEIGHT DISTRIBUTION TOLERANCE

The mathematical method to calculate the weight distribution on a given axle shall be as follows.

a. Find the curb weight on each wheel of a given axle;
b. Divide the weight of each wheel by the total curb weight of both wheels (i.e., curb weight of the axle), times (×) 100 = percentage of weight on each side;
c. Subtract the smaller percentage from the larger percentage;
d. This is the percent difference between the left and right side of the axle. If this difference is 5 percent or less, that particular axle's weight distribution has met the provisions of 3.5.4.

4.4.4 ROAD TEST

The vehicle shall be subjected to a minimum 241 km (150 mile) road test of which 121 km (75 miles) shall be continuous miles on paved highways at highway speeds up to 113 km/hr (70 mph); 48 km (30 miles) on city streets; 24 km (15 miles) on gravel or dirt roads at speeds up to at least 56 km/hr (35 mph); and not less than 8 km (5 miles) in simulated or actual cross-country operation at speeds applicable to the terrain. Cross-country operation is defined as travel over open fields, rolling and side-sloping hills, rough and muddy terrain. Class 2 vehicles shall demonstrate cross-country operation in 4-wheel drive for an additional 32 km (20 miles). Ambulances shall meet performance requirements specified in 3.4.4 through 3.4.10 during road tests or by certification. Balance of the 241 km (150 miles) road test may be accumulated during other tests and checks requiring vehicle movements. After completion of the road test, vehicle shall be subjected to the water spray test (see 4.4.5). Road test may be performed at any ambient temperature.

4.4.4.1 TEST FAILURE OF FIRST AMBULANCE

Vehicle utilized for the road test shall successfully complete 241 km (150 miles) of test. Rejection of the test vehicle shall be for deficiencies, including but not limited to the following:

a. Damage caused by collision.
b. Failure of any major component.
c. Vibration due to misalignment of wheels, frame, drive shaft, or defective tires, etc.
d. Vibration due to type of body construction or mounting.
e. Evidence of abnormal tire wear due to misalignment or unbalanced wheels/tires.
f. Failure of any vehicular safety device such as brakes, steering assembly, windshield washers and wipers, or electrical circuits.
g. Evidence of structural weakness in any part of the vehicle, vehicle components, or accessories.
h. Loose mountings of parts or accessories due to workmanship or vehicular operation.
i. Failure of any vehicular performance requirements.

4.4.5 WATER SPRAY TEST.

The ambulance shall be subjected to a water spray test for approximately 15 minutes. The spray shall be delivered by nozzles operating at 172 kPa (25 psi) water pressure, sufficient in number and placed approximately 91 cm (3 ft) from the body to afford full coverage of sides, roof, front, rear, and undercarriage of the vehicle. Ambulance undergoing quality conformance inspection (see 4.2.3) shall be subjected to water spray test for not less than 2 minutes duration, Evidence of water leakage shall be cause for rejection until leaks are corrected.

4.4.6 OXYGEN SYSTEM TEST

The installed medical oxygen piping and outlet system (see 3.12.1) shall be leak tested at 1033 kPa (150 psi) pressure, for a time period of four hours. The system shall be tested with dry air or nitrogen gas or equal and kept decontaminated. After the successful completion of tests, the system shall be capped then tagged with date and signature of person and firm performing the tests.

5 PREPARATION FOR DELIVERY

5.1 PREPARATION

Unless otherwise specified (see 6.2-bbb), the ambulance(s) shall be preserved and packaged for mobile delivery in accordance with the supplier's standard commercial practice, insuring carrier acceptance and safe delivery to destination in compliance with regulations applicable to the mode of transportation.

6 NOTES

6.1 INTENDED USE OF SPECIFICATION

The intended use of this specification is to define and procure a certified "Star of Life" ambulance.

6.1.1 FEDERAL SPECIFICATION COVERAGE

This Federal specification does not include all the varieties of medical service vehicles commercially available as may be indicated by the title of the document. This specification covers only the ambulances approved to display the "Star of Life" symbols and purchased, to provide ambulance services, under contract, and funded by the Federal and State Governments of the United States.

6.1.2 PRECAUTIONS AND OBSERVATIONS

Purchasers should read the entire document before requisitioning an ambulance, in order to be knowledgeable of just what equipment is standard, and which options need to be exercised. Due to the variety of ambulance equipment or features, some options may be incompatible with the model desired (reference chassis and ambulance manufacturer's data books).

6.1.3 DEFINITION OF GOVERNMENT-PURCHASER

Government or purchaser as used in the context of this document means the Federal, State, or political subdivision Government, or any purchaser who cites this specification.

6.2 ORDERING DATA

Purchasers should prepare their procurement document(s) (specification, requisition, contract) in the following sequence. Select the preferred options and provide the necessary information requested:

a. Title, number, and date of this specification and amendment number, if any
b. Quantity, Type (1,11 or II I), and Class (1 or 2), of the ambulance required (see 1.2.1) and chassis brand if desired. See 3.1.2 through 3.1.4.
c. Patient compartment configuration A (ALS) or B (BLS). B (BLS is standard on all types). If ALS is required (on Types I & III), specify Configuration A (see 3.1.5).

NOTE: TYPE II VEHICLES MAY BE ORDERED WITH SOME ALS FEATURES DE PENDING ON MANUFACTURER.

d. Special performance requirements and/or other chassis that may be needed for special use applications such as carrying additional rescue/extrication equipment or heavily equipped patient/neonatal transport ambulances (see 3.4.1).

e. Fuel range. OEM manufacturers maximum fuel tanks shall be standard. Specify if reduced fuel capacity required (see 3.4.9 & 3.6.4.4).
f. Length. Standard length is 671 cm (22 ft.). Specify if additional length is required for specific applications (see 3.4.11.1).
g. Width. For modular units, wide bodies on wide track axles are standard. Specify if a narrower body on a dual or single rear wheel chassis is required (see 3.4.11.2).
h. Height. 279 cm (110") is the maximum standard height. Specify if other heights are required, eg for air transportability (see 3.4.11.3).
i. Payload. Purchaser may specify additional payload if available from manufacturer for a specific model or a medium-duty chassis when available (see 3.5.2).
j. Engine power unit (diesel is standard). Only if absolutely necessary should gasoline engine be specified. (see 3.6.3 to 3.6.3.4).
k. Air pollution controls. Cite where ambulance will operate, State and county, and/ or if for export, name country (see 3.6.4.3 and 3.22.2).
1. Drive train and transmission. Automatic transmission is standard for alt models. Specify if a manual transmission or other drive train options are required, other than that specified. Check manufacturer's data (see 3.6.5 through 3.6.5.9).
m. Wheels. Wide, dual rear wheels are standard equipment on all Type I and III units (3.6.7). Specify if narrow dual or single rear wheels are specifically desired.

NOTE: AMBULANCE BODY WIDTHS ARE ESTABLISHED IN 3.4.11.2. SPECIFY IF NARROW, MODULAR BODY TYPES WITH SINGLE OR DUAL WHEELS IN REQUIRED.

n. Tires. "All Season" steel belted radials are standard for all models. Mud and snow, or all purpose type tires may be desirable and specified on 4 × 4 vehicles. (see 3.6.8 to 3.6.12).
o. Spare. Purchaser may state specific location of spare/wheel, when furnished, or delete requirement for spare tire. NOTE: Purchaser should consult with ambulance manufacturer since moving spare can affect weight distribution of the vehicle (see 3.6.10).
p. Intermittent wipers. Specify if required. See 3.7,4.
q. Electrical generating system. Suppliers are required to provide adequate generating capacity in ambulances as manufactured; if reserve generating capacity is desired for owner installed electrical devices beyond that already provided, specify minimum, additional, generating capacity required (see 3.7.6).
r. Battery location. Specify other locations (see 3.7.7).
s. 12 Volt DC power. Specify if other than standard 12 volt DC outlets or current are required to power medical equipment, portable radio chargers, etc. (see 3.7.7.3).
t. Patient compartment controls. Specify if other than standard EMT panel/ controls, instruments, etc. are required (see 3.7.10).

u. Electromagnet Radiation Type Certification by equipment and ambulance manufacturers is standard. Specify if entire vehicle shall be tested. (see 3.7.12).

v. Warning lights. Red, clear, and amber lights are standard. Specify other color(s) only if required by State or local regulations. Specify strobe, or quartz halogen. Specify if other automatic switching arrangements are required (see 3.8 through 3.8.2.5).

NOTE: ADDITIONAL WARNING LIGHTS ARE NOT NECESSARY AND SHOULD NOT BE SPECIFIED IN ORDER TO AVOID OVERLOADING THE ELECTRICAL SYSTEM, AND REDUCING USABLE PAYLOAD.

w. Switching arrangements. Specify if automatic primary to secondary mode and automatic day/night switching arrangements are required. Standard system includes manual "Primary" and "Secondary" mode switch and, on strobe systems, a day/night dimming switch (see 3.8.2.3).

x. Seats. Specify if bench type seat is desired or permitted on Type I vehicles. Specify if high back bucket type seats are required and specify if other than vinyl type upholstery is desired or acceptable (see 3.9.3).

y. Mirrors. Specify if other than standard mirrors are required, eg. West Coast, integral convex, heated mirrors, etc. (see 3.9.5).

z. Bumpers and steps. On Class 2 vehicles, specify if other than standard step height is necessary for increased departure angles, eg. severe off road use, air transportability (see 3.9.6).

aa. Fenders. Required per specification. Specify type (see 3.9.7.1).

bb. EMT Seating. Specify other than standard requirements. Specify if integral EMT/child safety seat is required (see 3.10.3).

cc. Patient compartment length. 310 cm (122") is minimum required. Specify additional requirements (see 3.10.4).

NOTE: LENGTHS LESS THAN 310 CM (122") COULD RESULT IN REAR DOOR INTERFERENCE WITH TRACTION SPLINTS ON TALL PATIENTS.

dd. Patient compartment height. Minimum height of 152 cm (60") required, specify additional requirements (see 3.10.4).

ee. Body construction. Aluminum is standard for modular ambulances. Specify if other materials are required or permitted (see 3.10.5 and 3.10.6).

ff. Dual hinged doors are standard on Type II. Specify if sliding door is required (see 3.10.8). Specify if other door opening widths/configurations are required for specific applications.

gg. Windows. Specify if other than standard, tinted, opening windows are required or permitted (see 3.10.14).

hh. Walk-through on Type II & III is standard. Specify if walk-through deletion is required (see 3.10.15 through 3.10.15.2). When walk-through is deleted, a pass through window shall be furnished and additional compartments above and below the window can be furnished as specified by the purchaser.

ii. Interior and exterior storage accommodations. Specify any specific additional requirements (see 3.11.1 to 3.11.4).

jj. Stretchers, cots, and litters. Specify specific cot fastening device, item(s) required, and any upgrading features preferred (see 3.11.5 to 3.11.7, and 3.15.4 Mb, M16, M21). Cots are optional and not standard with vehicle. Cot fastener is standard. Specify specific model.

kk. Oxygen system (on board). Specify if other than "M" size bottle and other than manufacturer's standard location (see 3.12.1).

ll. Oxygen. Quick disconnect fitting for action wall and specified accessories. Specify type (see 3.12.1).

mm. Portable oxygen unit. Specify location of portable oxygen unit and spare bottle if other than location in 3.12.2.

nn. Suction aspirator system, electrically powered type, is standard. Gasoline powered vehicles may use engine vacuum systems (see 3.12.3 to 3.12.4).

oo. Air conditioning. Auxiliary Condenser. Specify if auxiliary condenser is required (see 3.13.5.1).

pp. Patient compartment sound level criteria. Specify if more stringent patient compartment sound level limits are required (see 3.13.8).

qq. Communication equipment. Two-way radio, telemetry modulator (local purchase of radio equipment and installations is recommended). Specify if contractor is to furnish and install the radio(s), and cite brand(s) acceptable, power output, frequencies needed, etc. Give complete details and provide a coordinator name, address, and telephone number (see 3.14 to 3.14.4). One antenna lead is standard. Specify additional antenna leads and locations.

rr. Siren. Specify if foot switch or other siren upgrading features are required (see 3.14.6).

ss. Additional systems, equipment, accessories, and supplies (see 3.15 through 3.15.4). Cite additional parameters, brand, model numbers, and/or equal, and details. User should be aware that additional equipment reduces usable payload capacity and if electrically operated reduces spare or reserve capacity that may be available and requires additional generating capacity (see 3.15.1).

    1. Specify if automatically adjustable backup alarm is required. (Fixed sound level is standard) (see 3.15.2 d.)
    2. Select optional vehicle equipment desired (see 3.15.3).
    3. Select optional medical equipment desired(see 3.15.4).

tt. Paint. White-Orange is standard. Only if absolutely necessary, specify other requirements. Reflective tape may be specified for the orange band (see 3.16.2).

uu. Rustproofing. If required. Undercoating is standard, (see 3.18).

vv. Markings, data plates, and warranty notices. Specify if additional or military markings are required (see 3.19).

ww. Manuals and handbooks. State if chassis manufacturer's complete repair manual and/or parts book(s) are required (see 3.20).

xx. Special requirements. State any other additions and permitted changes or deletions in the specification (see 3.22).

yy. Overseas vehicle requirements. Select requirements (see 3.6.3.4 and 3.22.1 through 3.22.3.1), include tiedown points, liftpoint requirements, additional safety requirements and fuel requirements for the consignees location.

zz. Quality assurance provisions. Specify classification of inspection desired (see 4 through 4.2.7).
 1. Source inspection (see 4.2.1) or
 2. Destination examination (see 4.2.2).

aaa. First production ambulance inspection. State if required. If quantity contract see 4.2.5.

bbb. Preparation for delivery. Specify if different (see 5.1). State mode of delivery preferred.

ccc. Procurement requirements (see 6.3). Specify other purchaser's contractual requirements.

ddd. Foreign use. State any additional or different warranty requirements for foreign use applications see 6.4.3).

eee. Statement of origin. Specify if other ownership documents are required and locations to which they shall be forwarded.

6.3 PROCUREMENT REQUIREMENTS

Unless otherwise specified (see 6.2-ccc) in the invitation for bids, contract, or orders, the following are contractual requirements (see 6.4 through 6.6), except in those instances where it is determined by the purchaser that inclusion thereof would not be to the best interests of the Government or purchaser.

6.4 WARRANTY

6.4.1 WARRANTY COVERAGE

The contractor shall warrant the vehicle and furnished equipment against parts failure or malfunction due to design, construction, or installation errors, defective workmanship, and missing or incorrect parts (exceptions, see 6.4.4) for a minimum period of 12 months, and 15 months for vehicles outside the contiguous (48) United States and District of Columbia from date of acceptance*, or 12,000 miles of operation, exclusive of any authorized accumulated driveway mileage, whichever occurs first. However, if the contractor received from any supplier or subcontractor additional warranty on the whole or any component of the vehicle, in the form of time and/or mileage, including any pro rata arrangements, or the contractor generally extends to his commercial customers a greater or extended warranty coverage, the Government/purchaser shall receive corresponding warranty benefits.

*The warranty begins when the Government/purchaser accepts the vehicle from the contractor FOB point of origin/destination (see 3.19).

6.4.2 DOMESTIC USE

When vehicles are used within the 50 States of the United States, the District of Columbia, Puerto Rico, and the Virgin Islands, the warranty shall include the furnishing, without cost to the Government/purchaser (FOB contractor's nearest dealer or branch to vehicle's location or station) new parts and assemblies to replace any that failed or malfunctioned within the warranty period. In addition, when the Government/purchaser elects to have the work performed at the contractor's plant, branch, dealer, or with the contractor's approval, (I) to correct the defect itself or (ii) to have it corrected by a commercial facility, the cost of the labor involved in the replacement of the failed or malfunctioned parts or assemblies shall be borne by the contractor.

6.4.3 FOREIGN USE

Unless otherwise specified (see 6.2-ddd), when vehicles are used outside the 50 States, the District of Columbia, Puerto Rico, and the Virgin Islands, the warranty shall include the furnishing of new parts or assemblies to replace any returned to the contractor which failed or malfunctioned within the warranty period. The replacement parts or assemblies shall be delivered by the contractor to the port of embarkation in the United States designated by the purchaser. The contractor shall not be required to bear the cost of the labor involved in correcting defects in vehicles operated in foreign countries.

6.4.4WARRANTY EXCLUSIONS

Unless within the additional coverage under 6.4.1, the following items are considered normal maintenance and repair for which the contractor need not assume liability for reimbursing the Government/purchaser regardless of the vehicle age or mileage:

a. Abuse, negligence, or unapproved alteration of original parts;
b. Damage from accidents;
c. Standard brake and clutch adjustments;
d. General tightening, headlamp adjustments;
e. Wheel alignment or tire balancing;
f. Tires, batteries, medical supplies and equipment, and radio(s) (if warranted by their manufacturers;
g. Miscellaneous expense such as fuel, towing, telephone, travel, lodging, or loss of personal property.

6.4.5 WARRANTY EXTENSION

In addition to the warranty provisions specified, the following shall also apply:

a. The ambulance chassis, chassis components (power train, etc.) and applicable body and cab shall be warranted and implemented in accordance with the original chassis manufacturer's standard warranty.
b. For vehicle(s) located within a country which has branch, dealer, or approved contractor service facilities, the cost of labor, parts, and materials involved in correcting warranted defects shall be borne by the contractor (see 6.4.3).
c. If action to effect repairs under warranty (except item a, chassis above) is not initiated within seven (7) working days by a dealer or supplier and completed within a reasonable length of time, or if the contractor does not have repair facilities in the city or county in which the vehicle operates, and in the event of emergency or dire need for the vehicle, the Government/purchaser reserves the right to make such repair, and be reimbursed by the contractor as follows: If work is performed at a Government/purchaser facility, at the rate of $25.00 per hour for labor based on the manufacturer's flat rate time schedule, and full cost of expenditures for parts and material; or actual labor time and materials, for repairs of nonscheduled operations, or when warranty services are performed at a commercial facility because no contractor facility is available, the new cost of the correction will be billed to the contractor.

6.5 REPAIR PARTS AND SERVICE

As a continuous operation of the vehicle described by this specification is of utmost importance for the successful bidder to be in a position to render prompt service and to furnish replacement parts. Accordingly, bidders shall indicate the extent of their ability to render prompt service by furnishing a list of branch offices or agencies where complete stocks of repair parts are maintained and can be secured within a reasonable time after ordering by part number from the manufacturer's part book and at such discount as may be quoted from year to year by the manufacturer of the vehicle purchased under this specification.

6.6 STATEMENT OF ORIGIN OR BILL OF SALE

A manufacturer's Statement of Origin or Bill of Sale showing the applicable purchase order number is required for each vehicle procured under this specification. Unless otherwise specified (see 6.2-eee), such documents shall be forwarded to the consignee.

6.7 OXYGEN TANKS

Following are the approximate dimensions, weights, and capacities of the various size oxygen cylinders referenced in this specification:

OXYGEN CYLINDERS	D	M	H
Outside diameter (cm/inch)	11/4.5	18/7	23/9
Overall heights, without valve (cm/inch)	51/20	119/47	142/56
Capacity (gallons)	95	800	1,825
Capacity (liters)	360	3,000	6,900
Capacity, flow @ 10 liters/minute (minutes)	30	325	700
Capacity (cubic feet)	13	106	244
Weight, empty (kg/pounds)	5.9/13	34/75	54/119
Weight, full (kg/pounds)	6.3/14	37.6/83	61.3/135

6.8 REFERENCE HANDBOOK

The handbook and all the material referenced in this specification (see 3.20) for the model ambulance furnished shall be provided in an 21.6 × 28 cm (8 1/2 × 11 in.), 3-ring, hard cover, looseleaf binder, inscribed with the complete address and telephone number of the manufacturer and shall contain the following.

1. Table of contents
2. Copy of supplier's invoice/date of delivery including chassis
3. Manufacturer's "Star of Life" certification of compliance statement
4. Ambulance manufacturer's illustrative, pictorial literature and a copy of predelivery inspection/test form signed by inspector
5. Complete specification KKK-A-1822D and all amendments including purchaser's specifications and modifications, in effect for date of manufacture
6. Conditions of sale and contract (unless otherwise specified)
7. Shipping papers
8. List of ambulance manufacturer's service points (see 6.5)
9. Copy of chassis manufacturer's warranty and owner manual
10. Final stage manufacturer's components and equipment information (hardware, fixture, etc.) including supplier's part numbers

11. Complete wiring diagrams and schematics (3.7.2)
12. Manufacturer's operating and servicing instructions for the entire ambulance, electrical system(s) of the ambulance, components, devices, and equipment aboard, etc., including each equipment manufacturer's data

All contents of the handbook that cannot be punched with three holes shall be secured in plastic leaves.

6.9 CHANGES AND AMENDMENTS

When a using agency or purchaser considers that this specification requires revision, a written request for change or additions to the document supported by adequate justification should be sent to the General Services Administration, Automotive Commodity Center, Engineering Branch (FCAE-E), Washington, DC 20406, for appropriate action. The agency will be informed of action taken. New and revised information regarding this specification may be issued from time to time under an amendment to the Federal Specification. These amendments are identified by the same number and title as the document and are on green paper. Amendments should be retained until such time as the entire document is revised.

Military Coordination Activity:

Army - AT

Activities:

Air Force - 84-99
Army - AT-EL-CR-ER-MD
Navy - YD-MC

Custodian and Preparing Activity:

GSA-FSS-FCAE

Civil Agency Activities:

USDA - APHIS
Interior - BIA
State Department - AID
DOT - NHTSA-EMS
DOE
HHS-EMS-IHS
VA-OSS
DC Government

Non-Federal Activities:

Ambulance Manufacturers Division, National Truck Equipment Association (AMD/ NTEA);

Index to Appendix II

A

B

C

D

E

F

G

H

I

L

M

N

O

P

Q

R

S

T

U

V

W

About the Authors

Stephen S. Solomon, B.S., O.D. Dr Stephen S. Solomon has been a practicing optometrist for thirty-seven years and a researcher in visibility and perceptual issues for about as long. He has written numerous articles on how visibility affects perception. Among his studies are juried or referred research papers published by the National Safety Council, the American Optometric Association, and the Human Factors Engineering Society.

Dr. Solomon has acted as a consultant for attorneys and industry on issues of visual perception and its relation to accidents and accident prevention. His recent presentations include one which addresses the relationship of vision and emergency vehicle accidents for the National Academy of Science's Transportation Research Board and another for the NYS Department of Municipal Police for advanced accident reconstructionists.

Dr. Solomon has been awarded New York State Optometrist of the Year in 1976 and received their Distinguished Achievement Award in 1996.

Paul F. Hill is a retired law school librarian from the Creighton University School of Law, Omaha, Nebraska. He has coauthored books published by Lawyers & Judges covering litigation on bicycle, motorcycle, pedestrian, bus and recreational vehicle, and electrical accidents. He is a graduate of Kenyon College, State University of New York at Albany, and has his J.D. from Capital University.

George L. Ellis, M.D., FACEP. After completing his emergency medicine residency training at the Medical College of Pennsylvania in Philadelphia, Pennsylvania, Dr. Ellis spent eleven years as faculty at the University of Pittsburgh Affiliated Residency in emergency medicine. He is currently chairman of Emergency Medicine at the Robert Packer Hospital/Guthrie Healthcare System in Sayre, Pennsylvania. He serves as medical director for Guthrie Emergency Medical Services which includes the Guthrie One helicopter as well as ground units. In addition, he is medical director for several volunteer services in New York State and is a member of the Susquehanna Regional Medical Advisory Committee in New York. He is the medical director for Bradford and Susquehanna counties in Pennsylvania and chairs the regional Medical Advisory Committee in that region. He is also a member of the Pennsylvania State Medical Advisory Committee.

James G. King has BS and AB degrees in electrical engineering and linguistics from the University of Rochester. He is employed at Loral Federal Systems Company as a senior engineer/manager and currently manages systems engineering (including human engineering and systems safety engineering) for a military helicopter program. He has published several technical disclosures and holds patents dealing with electronic circuits implementing numeric algorithms. He is a twenty-five year veteran of the volunteer fire service and emergency medical service.

Robert J. Simpson, Esq. was born Ithaca, New York on January 18, 1944. He received his Bachelor of Arts degree from University of Colorado in 1965 and was awarded his law degree from Albany Law School in 1969. He served as Tioga assistant district attorney from 1970 to 1978, and has been the Tioga County district attorney from 1978 to the present.

Stephen D. Smyk, Esq. Educated at Syracuse University, Stephen Smyk has practiced personal injury law in New York state since 1971. He frequently lectures on various topics relating to litigation, including trial practice skills, expert witnesses, products liability, evidence, medical malpractice, voir dire, and summation. He has served as an officer and in other capacities for the Broome County (New York), and New York State Bar Associations. He is also an active member of the ABA, ATLA and the New York Trial Lawyers Association.

Index

D

E

K

L

M

N

S

T

U

V

W